Reviews of Plasma Physics

VOPROSY TEORII PLAZMY

ВОПРОСЫ ТЕОРИИ ПЛАЗМЫ

Translated from Russian by **Dave Parsons**

Translation Editor **Herbert Lashinsky**
University of Maryland

Reviews of
Plasma Physics

Edited by Acad. M. A. Leontovich

Volume **8**

 SPRINGER SCIENCE+BUSINESS MEDIA, LLC

The Library of Congress cataloged the first volume of this title as follows:

Reviews of plasma physics. v. 1—
　New York, Consultants Bureau, 1965—
　/v. illus. 24 cm.
　Translation of Voprosy teorii plasmy.
　Editor: v. 1 —　M. A. Leontovich.

　1. Plasma (Ionized gases) — Collected works. I. Leontovich, M. A., ed. II. Consultants Bureau Enterprises, inc., New York. III. Title: Voprosy teorii plasmy. Eng.
QC718.V63　　　　　　　　　　　　　　　　　　　　　　64—23244

The original text, published by Atomizdat in Moscow in 1974,
has been corrected and updated by the authors.

Library of Congress Catalog Card Number 64-23244
ISBN 978-1-4615-7816-1　　　　ISBN 978-1-4615-7814-7 (eBook)
DOI 10.1007/978-1-4615-7814-7

©1980 Springer Science+Business Media New York
Originally published by Plenum Publishing Corporation, New York in 1980
Softcover reprint of the hardcover 1st edition 1980

TRANSLATOR'S PREFACE

In the interest of speed and economy the notation of the original text has been retained so that the cross product of two vectors **A** and **B** is denoted by [AB], the dot product by (AB), the Laplacian operator by Δ, etc. It might also be worth pointing out that the temperature is frequently expressed in energy units in the Soviet literature so that the Boltzmann constant will be missing in various familiar expressions. In matters of terminology, whenever possible several forms are used when a term is first introduced, e.g., magnetoacoustic and magnetosonic waves, "probkotron" and mirror machine, etc. It is hoped in this way to help the reader to relate the terms used here with those in existing translations and with the conventional nomenclature. In general the system of literature citation used in the bibliographies follows that of the American Institute of Physics "Soviet Physics" series. Except for the correction of some obvious misprints the text is that of the original.

We wish to express our gratitude to Academician Leontovich for kindly providing the latest corrections and additions to the Russian text.

CONTENTS

Steady-State Plasma Flow in a Magnetic
Field
A. I. Morozov and L. S. Solov'ev

Calculation of Two-Dimensional Plasma
Flows in Channels
K. V. Brushlinskii and A. I. Morozov

Two-Dimensional Magnetohydrodynamic
Model for the Dense Plasma Focus
of a Z Pinch
V . F . D'yachenko and V. S. Imshennik

Plasma Optics
A. I. Morozov and S. V. Lebedev

STEADY-STATE PLASMA FLOW IN A MAGNETIC FIELD

A. I. Morozov and L. S. Solov'ev

INTRODUCTION

Since the beginning of research on the magnetic fusion reactor, attention has focused on the development of what might be called the "statics" of high-temperature plasmas, that is to say, the analysis of equilibrium and stability of a hot plasma confined by a magnetic field. Another area which is now attracting increasing attention is the field of "plasma dynamics," which is the theory of high-amplitude nonlinear waves and plasma flow. Many authors have surveyed work on nonlinear waves and on low-temperature plasma dynamics, primarily in connection with the problem of MHD generators; therefore, in this review we will only consider the flow of a fully ionized, conducting plasma. Since we cannot hope to cover all possible varieties of the flow of such plasmas, we will only consider certain simple cases which are of interest in connection with the acceleration of comparatively dense plasma $(n \sim 10^{14}\text{-}10^{17} \text{ cm}^{-3})$ to an energy of 0.1-10 keV.

Plasma accelerators [1-5] have much in common with charged-particle accelerators, gasdynamic accelerators, and Laval nozzles. The similarity to charged-particle accelerators lies in the use of electric and magnetic fields, which can accelerate the ions to energies substantially above the thermal level. The medium being accelerated remains neutral, as in the gasdynamic case; however, the interaction between the particles of the medium is of fundamental importance,† in contrast with the situation in a conventional accelerator, in which this interaction can be neglected.

The interaction between particles means that a variety of waves can propagate in the plasma stream; it also rules out direct control of the particles in the stream by means of external fields [6]. The problems involved in the acceleration of a plasma turn out to be far more complicated than those of gasdynamics or conventional accelerators.

A plasma can be accelerated in the steady state or as the result of a transient process.‡ In the present review we will be dealing with steady-state accelerators only; that is, it is assumed that $\partial/\partial t = 0$ at any point in the stream. Thus we ignore systems in which standing waves, traveling waves, or instabilities are used to achieve acceleration.

A variety of operating principles are available for the design of plasma accelerators; the method used for a particular accelerator depends on the velocity with which the plasma leaves the accelerator, v (and, correspondingly, the kinetic energy of the particles), and the mass flow rate m.

Chapter 1

ACCELERATION MECHANISMS

§1. Microscopic Picture of Plasma Acceleration [2, 3, 6]

The ions and electrons in the plasma stream leaving and accelerator have the same directed velocity, but the ion energy is three or four orders of magnitude higher than the electron energy. Thus, in dealing with velocities of 10^8 cm/sec, we can ignore the electron acceleration completely. The problem of plasma acceleration is thus really the problem of ion acceleration under the condition of neutrality. Assuming that all the plasma ions experience essen-

---

†As follows simply from neutrality itself, it for no other reason.

‡This distinction is similar to that between flow from a nozzle and an explosion in ordinary gasdynamics.

tially the same acceleration,[†] we can write the following equation for the motion of a "typical" ion:

$$M \frac{dv_i}{dt} = e \left(E + \frac{1}{c} [v_i\, H] \right) + F_{ii} + F_{ie}. \tag{1.1}$$

Here, M and e are the ion mass and charge, F_{ii} and F_{ie} are the forces exerted on a given ion by collisions with ions and electrons, respectively, and E and H are the average microscopic fields that act on the ion in question.

Of the four forces which appear in Eq. (1.1), only three can accelerate ions directly, i.e., increase the ion energy; the magnetic field does not affect the ion energy.

Collisions of ions with other ions (the force F_{ii}) can only result in a redistribution of the energy over the various degrees of freedom, in particular, the conversion of thermal energy into directed energy. From the hydrodynamic model we have [7]

$$(F_{ii})_\alpha = - \left(\frac{1}{n_i}\, \nabla p_i \right)_\alpha - \frac{1}{n_i} \cdot \frac{\partial \pi_{i\alpha\beta}}{\partial x_\beta}, \tag{1.2}$$

where $\pi_{i\alpha\beta}$ is the viscous stress tensor, and p_i is the ion pressure. If the force F_{ii} is to play an important role in the acceleration, the ions entering the system must be at a high temperature or else some mechanism must be used to heat ions during the acceleration.

The first case is not interesting since the system operates as an ordinary gasdynamic nozzle. The second case requires that the following chain of processes occur rapidly:

Ohmic heating of electrons → Heating of ions by electrons
→ Conversion of ion thermal energy into directed   (1.3)
energy.

These processes operate in the plasmatron.

Let us estimate the ranges of parameters in which the chain in (1.3) is effective, assuming that the calculations can be carried out with two-body collisions. In this case the slowest step is energy

---

[†] As noted above, we do not consider systems in which distinct groups of particles are accelerated as a result of "turbulence," as in fast pinches.

transfer from electrons to ions, since the characteristic time for this process is [7]

$$\tau_{ie} \sim \tau_e M/m, \qquad (1.4)$$

where

$$\tau_e = \frac{3\sqrt{m}\,(kT_e)^{3/2}}{4\sqrt{2\pi}\lambda e^4 Z^2 n_i} = \frac{3.5 \cdot 10^4}{(\lambda/10)}\frac{(T_e)^{3/2}}{Zn} \qquad (1.5)$$

is the characteristic time of electron−electron collisions, m is the electron mass, Z is the ion charge ($Zn_i = n_e = n$), and λ is the Coulomb logarithm. We must compare the time τ_{ie} with the transit time

$$\tau_0 \sim L/v_{max} \qquad (1.6)$$

required for the ion to traverse the accelerating system, of length L. If $\tau_{ie} \ll \tau_0$, the chain in (1.3) can be important for acceleration. As an example, consider hydrogen, with L = 10 cm, Z = 1, $v_{max} = 10^7$ cm/sec,† and n = 10^{16} cm^{-3}. For T_e = 1 eV we have

$$\tau_0/\tau_{ie} \approx 100. \qquad (1.7)$$

Under these conditions the force F_{ii} can be effective, since the minimum required value is $\tau_0/\tau_{ie} \approx \mathscr{E}_i/kT_e$, where $\mathscr{E}_i$ is the ion energy.

The force F_{ie} arises from collisions between electrons and ions (the "electron wind"). In the hydrodynamic model of a plasma, this force is [7]

$$F_{ie} = -enj/\sigma \equiv -enE^*, \quad E^* \equiv j/\sigma, \qquad (1.8)$$

where j is the current density and σ is the electrical conductivity. If the relative electron velocity satisfies $|v_i - v_e| \gtrsim v_{max}$, this force is important even when $\tau_0 \sim \tau_{ie}$, since the characteristic time is again τ_{ie}. If, on the other hand, the electron velocity is much higher than v_{max}, the electron wind can become important at lower densities. At higher relative velocity various "turbulent" processes can develop and these processes greatly intensify the entrainment of ions by electrons.

---

†The corresponding ion energy is about 50 eV.

Substituting numerical values, we can estimate the efficiency of the electron wind under the conditions that correspond to Coulomb conductivity, rather than turbulent conductivity. If j = 300 A/cm², T_e = 1 eV, and $\sigma = 10^{13}$ abs. units, we find that the effective field is $E^* = j/\sigma \sim 30$ V/cm. If, on the other hand, T_e increases, the field E^*_{Cou} weakens rapidly and E^*_{turb} can become important.

In addition to the two collisional mechanisms for acceleration, another mechanism exists by which the ions are accelerated by the electric field in a plasma (the eE force). This mechanism can be used to accelerate a plasma of any density. In this case the ac-acceleration problem reduces to the problem of producing an electric field of the required intensity and configuration in the appropriate volume.

§ 2. Conditions for the Existence of an Electric

Field in a Plasma

The difficulties which arise in trying to generate a strong electric field in a plasma result primarily from the high electron mobility; thus, to determine the necessary conditions for maintaining an electric field in a plasma we must examine the equation that describes the electron dynamics [6].

The high mobility of electrons over broad ranges of the parameters means that the electrons will usually have a Maxwellian distribution. Thus, neglecting the electron viscosity, we can write the following simplified hydrodynamic equation [7]:

$$m\frac{dv_e}{dt} = -\frac{1}{n}\,\nabla p_e - e\left(E + \frac{1}{c}\,[v_e\,H]\right) + e\,\frac{j}{\sigma}\,. \tag{1.9}$$

If the electron inertia is neglected in (1.9) (electron inertia is only important if there are discontinuities, in particular, near the walls), we find the electric field can be maintained in the plasma by virtue of the electron temperature, the electron—ion friction, and the Lorentz force:

$$E = -\frac{1}{en}\,\nabla p_e + \frac{j}{\sigma} - \frac{1}{c}\,[v_e\,H]. \tag{1.10}$$

Let us examine these possibilities separately.

1. When $j/\sigma \approx 0$ and $\mathbf{H} \approx 0$, we have

$$\mathbf{E} = -\frac{1}{en}\nabla p_e, \tag{1.11}$$

where $p_e = nkT_e$. In this case, hot ions can only be produced if there are hot electrons. Substituting (1.11) into the equation of motion of the cold ions,

$$M\frac{d\mathbf{v}}{dt} = e\mathbf{E}, \tag{1.12}$$

we find the usual hydrodynamic equation

$$\rho\frac{d\mathbf{v}}{dt} = -\nabla p, \tag{1.13}$$

where $\rho = Mn$, $\mathbf{v} = \mathbf{v}_i$, and $p = p_e$.

This case is equivalent to classical gasdynamic flow [8] with electron pressure and ion inertia.

The use of a plasma with a high electron temperature is apparently only feasible for hydrogen, since the radiation loss is substantial for other gases. To analyze this case we must know T_e, so that dissipative processes can be taken into account.

2. If the electric field is maintained in the plasma by the ohmic friction in (1.10), and if $T_e \rightarrow 0$ and $[\mathbf{v}_e\,\mathbf{H}] \rightarrow 0$, the ions will not be accelerated. Equations (1.1), (1.8), and (1.10) show that the combination of terms

$$\mathbf{E}_{\text{eff}} = \mathbf{E} + j/\sigma \tag{1.14}$$

appears in both the ion and electron equations. If this combination is zero in the electron equation, (1.10), it is then zero in the ion equation. The reason is that the acceleration of ions by the electric field is offset by the acceleration in the opposite direction caused by the electron wind.[†]

It follows from (1.10) that the ions will be accelerated by $\mathbf{E}$ or j/σ only if ∇p_e or $\mathbf{H}$ is nonvanishing.

[†]Some ion acceleration occurs, but this of order m/M.

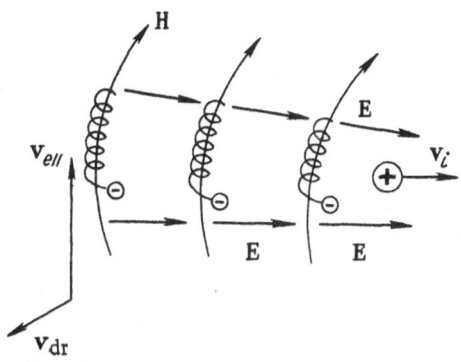

Fig. 1

3. Finally, an electric field can exist in the plasma if $T_e \to 0$ and $j/\sigma \to 0$; in this case we can write

$$E = -\frac{1}{c} [v_e \, H]. \qquad (1.15)$$

We will refer to plasma accelerators which operate on the basis of the electric field in (1.15) as "Lorentz accelerators."

Equation (1.15) shows that a fundamental requirement for a Lorentz accelerator is a magnetic field whose lines of force act as a sort of grid that interferes with the motion of electrons along the electric field (Fig. 1). According to (1.15), the electron drift velocity across the fields **E** and **H** is

$$v_{e\perp} = c/[EH]/H^2. \qquad (1.16)$$

We can draw two important conclusions from Eq. (1.15).

1. The magnetic lines of force are equipotentials since

$$EH = 0. \qquad (1.17)$$

This equation is correct to within the neglected terms in Eq. (1.15). In particular [9], a magnetic line of force is an equipotential to quantities of the order of the electron temperature.†

---

†For a more detailed discussion of this point, see the review by A. L. Morozov and S. V. Lebedev in this volume (p. 301).

2. The electrons drift along the equipotential surfaces since

$$\mathbf{E}\mathbf{v}_e = 0 \qquad (1.18)$$

or

$$(\mathbf{v}_e \, \nabla)\varphi = 0, \qquad (1.18a)$$

where φ is the electric potential.

Depending on the nature of the electron drift, Lorentz accelerators can be classified as those in which the electron drift circuit is closed or open. Systems of the first kind are analyzed in detail in [9]. In the present review we will be concerned primarily with axisymmetric systems in which the electron drift circuit is open (Fig. 2). In systems of this kind there are longitudinal electric and azimuthal magnetic fields which produce a radial electron drift which is terminated at electrodes 1 and 2.

§3. Basic Equations; the Replacement Factor ξ

We only consider dissipationless processes, ignoring the ohmic resistance as well as the viscosity and the thermal conductivity. To describe the processes we use the simplified two-fluid hydrodynamic equations:

$$
\begin{aligned}
&\frac{\partial n_i}{\partial t} + \operatorname{div} n_i \mathbf{v}_i = 0 \ \text{(a)}; \quad \frac{\partial n_e}{\partial t} + \operatorname{div} n_e \mathbf{v}_e = 0 \ \text{(b)}; \\
&\frac{M}{e} \cdot \frac{d_i \mathbf{v}_i}{dt} = -\frac{\nabla p_i}{e n_i} + \left(\mathbf{E} + \frac{1}{c} [\mathbf{v}_i \, \mathbf{H}] \right) \ \text{(c)}; \\
&\frac{m}{e} \cdot \frac{d_e \mathbf{v}_e}{dt} = -\frac{\nabla p_e}{e n_e} - \left(\mathbf{E} + \frac{1}{c} [\mathbf{v}_e \, \mathbf{H}] \right) \ \text{(d)}; \\
&d_e \, S_e / dt = 0 \ \text{(e)}; \quad d_i \, S_i / dt = 0 \ \text{(f)}.
\end{aligned}
\qquad (1.19)
$$

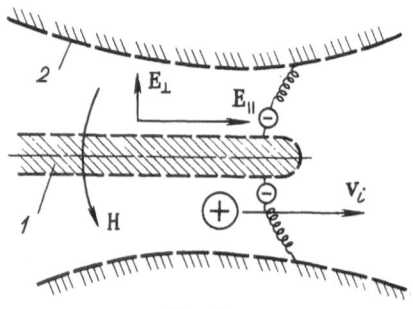

Fig. 2

This system is supplemented by Maxwell's equations:

$$\left.\begin{aligned} &\text{curl } \mathbf{H} = \frac{4\pi}{c} e \left(n_i \mathbf{v}_i - n_e \mathbf{v}_e\right) \text{ (a);} \\ &\text{div } \mathbf{H} = 0 \text{ (b);} \quad \text{curl } \mathbf{E} = -\frac{1}{c} \cdot \frac{\partial \mathbf{H}}{\partial t} \text{ (c);} \\ &\text{div } \mathbf{E} = 4\pi e \left(n_i - n_e\right) \text{ (d).} \end{aligned}\right\} \tag{1.20}$$

Here, S_e and S_i are the electron entropy and the ion entropy, respectively.

In using hydrodynamic equations for the calculations we are actually assuming that the plasma is dense so that the electron and ion mean free paths are both short. This assumption is reasonable at densities $n > 10^{15}$ cm^{-3} for the temperatures T_i and T_e that are observed experimentally (see p. 105 in this volume).

In a low-density plasma it is obviously necessary to resort to the kinetic equations. If the ions are cold, however, and if the ion velocity spread is neglected, (1.19b), (1.19c) are equivalent to a collisionless kinetic equation for the ions.

In addition to using the two-fluid model we will also make use of the single-fluid model. The transformation from the two-fluid model to the single-fluid model is accomplished by assuming

$$\left.\begin{aligned} &m \to 0; \quad n_i \to n_e; \\ &\mathbf{v} = \mathbf{v}_i \to \mathbf{v}_e, \end{aligned}\right\} \tag{1.21}$$

so that only a single continuity equation remains. Combining Eqs. (1.19c) and (1.19d), we find the Euler equation

$$Mn \frac{d\mathbf{v}}{dt} = -\nabla p + \frac{1}{c} [\mathbf{jH}], \quad p = p_i + p_e. \tag{1.22a}$$

Equation (1.19d) becomes Ohm's law:

$$\frac{\nabla p}{en} + \mathbf{E} + \frac{1}{c} [\mathbf{vH}] = 0. \tag{1.22b}$$

We also find an equation for the total entropy:

$$dS/dt = 0, \quad S = S_i + S_e. \tag{1.22c}$$

A model which is frequently used for fully ionized plasma is the model which is found from (1.19) by taking a single limit rather than the two limits in (1.21):

$$m \to 0, \; n_i \to n_e. \tag{1.23}$$

This model is frequently called the single-fluid Hall-effect model

$$
\left.
\begin{aligned}
&\frac{\partial n}{\partial t} + \operatorname{div} nv = 0 && \text{(a)}; \\[2mm]
&Mn \frac{dv}{dt} = -\nabla p_i + en \left(E + \frac{1}{c} [v_i \, H] \right) && \text{(b)}; \\[2mm]
&0 = \frac{\nabla p_e}{en} + \left(E + \frac{1}{c} [v_e \, H] \right) && \text{(c)}; \\[2mm]
&\frac{d_e S_e}{dt} = 0 \;\; \text{(d)}; \qquad \frac{d_i S_i}{dt} = 0 \;\; \text{(e)}.
\end{aligned}
\right\}
\tag{1.24}
$$

We can draw an important conclusion from (1.24). The magnetic field is frozen in the electrons rather than the ions [10–13]. To show this we take the curl of Ohm's law in (1.24c) and use Eq. (1.24d) and the condition

$$\partial n/\partial t + \operatorname{div} nv_e = 0; \tag{1.25}$$

to find the desired "freezing condition"†

$$\frac{d_e}{dt} \left(\frac{H}{\rho} \right) = (v_e \, \nabla) \frac{H}{\rho}. \tag{1.26}$$

The single-fluid gasdynamics equations in (1.22) do not incorporate this feature, nor do they incorporate the entropy transport along the electron and ion trajectories.

Equation (1.21) shows that the actual condition that ensures that the Hall effect is weak is the inequality

$$\frac{|v_i - v_e|}{|v_i|} \ll 1. \tag{1.27}$$

This local condition can be replaced by the integral condition [14]

$$\xi = \frac{I_d}{I_{\dot{m}}} \ll 1. \tag{1.28}$$

Here I_d is the current, and $I_{\dot{m}} = e\dot{m}/M$ is the flow rate of a material with atomic mass M, expressed in units of the current. This definition of ξ is justified in detail below (Chap. 4, §3) [15].

---

†Subtracting (1.25) from (1.24a), we find the equation j = 0 which holds for processes which are stationary (in the electrodynamic sense).

We call ξ the "replacement factor," since it tells us how many times the electrons that neutralize the space charge of the ions that traverse the accelerator channel are replaced. The reason for introducing the replacement factor is illustrated in Fig. 3, which shows that two mass fluxes, I_d and $I_m^{\cdot}$ are incident on a plasma accelerator. It is thus reasonable to introduce a dimensionless parameter which is the ratio of the magnitudes of these fluxes. In conclusion we note that a useful equation relates the voltage V applied to the accelerator, and the energy of the particles leaving the accelerator ($Mv_{max}^2 / 2$). This equation follows from the energy conservation relation $\dot{m}v_{max}^2 / 2 = IV\eta$ and can be written

$$\frac{Mv_{max}^2}{2} = eV\xi\eta. \qquad (1.29)$$

Here η is the efficiency of the accelerator; the product $\xi\eta$ is the "transformation coefficient."

§4. Perturbations of the Plasma Stream

In conventional particle accelerators the fields that act on a particular particle are governed by external currents and external charges, so that these fields can be assumed to be specified. The distortion of these fields due to the interaction of the particles is usually treated as a small perturbation which "overloads" the accelerator. In the theory of plasma accelerators the situation is qualitatively different: In this case the interaction between particles is so important that it must be incorporated in the initial equations as a self-consistent system consisting of Maxwell's equations and the equations of motion. Correspondingly, a particle in a plasma stream only can be influenced by perturbing the entire flow. The problem of designing a plasma accelerator is thus primarily the problem of arranging conditions at the surface of the

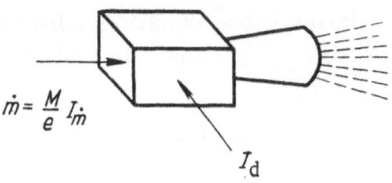

$$\dot{m} = \frac{M}{e} I_m^{\cdot}$$

$$I_d$$

Fig. 3

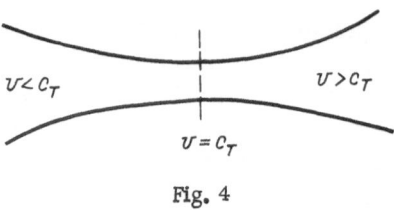

Fig. 4

plasma stream which, by perturbing the entire flow, produce fields which provide the desired ion acceleration within the plasma.

The general study of these questions is still at a very early stage, so that we will only discuss the situation qualitatively.

In the gasdynamic case, a supersonic gas flow with a particular velocity distribution is produced by choosing a nozzle profile for which the effect of the walls on the flow results in the desired flow pattern. The clearest picture of gasdynamic flow is obtained by using the "method of characteristics" to analyze steady-state supersonic flow [6]. These "characteristics" are the lines along which perturbations are transported. At the present time we are far from this stage in plasma accelerators, but it would would be useful to exploit this approach.

1. Inversion Law. It was noted back in the last century [17] that changing the geometry of a channel has different effects on sonic and supersonic flows. For example, this feature is responsible for the fact that a gradual transition from subsonic to supersonic flow requires a nozzle-shaped channel (Fig. 4). We can explain this result by a simple calculation. The equations that describe the stationary flow of an ideal gas in a narrow channel can be written

$$\left.\begin{array}{l} \rho v f = \text{const}; \quad \dfrac{v^2}{2} + W\,(\rho) = \text{const}; \\[2mm] S = \text{const}; \quad W\,(\rho) = \displaystyle\int \dfrac{dp}{\rho} \end{array}\right\} \qquad (1.30a)$$

where f is the channel cross section, S is the entropy, and W is the enthalpy. Differentiating these equations, introducing the sound velocity $c_T^2 = (\partial p / \partial \rho)_{S=\text{const}}$, and eliminating $d\rho$, we find a relation[†] between dv and df:

$$(1 - v^2/c_T^2)(dv/v) = - df/f. \qquad (1.30b)$$

_____

[†]Equation (1.30b) is called the Hugoniot equation.

We see immediately that acceleration of a subsonic flow required that the channel cross section be reduced, while acceleration of a supersonic flow requires that it be increased. This law was generalized to the case of arbitrary agents acting on the flow in a narrow channel in [18].

We can now formulate the inversion law for a flow of arbitrary nature and show that the inversion is of a resonant nature, being analogous to the change in the oscillation phase of a pendulum driven by a periodic external force at resonance.

We begin with the simple example in which a homogeneous gas is subjected to a weak periodic force

$$F = A \sin \omega \, (t - x/v_0), \tag{1.31}$$

due to a wave traveling with velocity v_0.

The linear system of equations for the corresponding one-dimensional process,

$$\frac{\partial \rho}{\partial t} + \rho_0 \frac{\partial v}{\partial x} = 0; \quad \rho_0 \frac{\partial v}{\partial t} = -c_T^2 \frac{\partial \rho}{\partial x} + F, \tag{1.32}$$

has the solution

$$\left. \begin{aligned} v &= -\frac{A}{\omega \rho_0} \cdot \frac{\cos \omega \, (t - x/v_0)}{1 - c_T^2/v_0^2}; \\ \rho &= -\frac{\rho_0}{v_0} \cdot \frac{A}{\omega} \cdot \frac{\cos \omega \, (t - x/v_0)}{1 - c_T^2/v_0^2}. \end{aligned} \right\} \tag{1.33}$$

We note that as the phase velocity of the perturbation, v_0, passes through the velocity of sound the sign of the perturbation reverses. Forces F of a given magnitude cause greater perturbations, the closer the phase velocity v_0 is to the velocity of sound. The effect thus has a clearly defined resonant nature.

To determine how this simple example is related to the inversion law, we transform to a coordinate system fixed with the force F, i.e., the "fixed" coordinate sysmte†: $x \to x - v_0 t$, $t \to t$. In this

---

† The force F is naturally assumed to be fixed with respect to the walls of the channel through which the gas flows.

case, Eq. (1.32) is rewritten as

$$v_0 \frac{\partial \rho}{\partial x} + \rho_0 \frac{\partial v}{\partial x} = 0; \quad \rho_0 v_0 \frac{\partial v}{\partial x} = -c_T^2 \frac{\partial \rho}{\partial x} + F. \qquad (1.34a)$$

We then find

$$(v_0^2 - c_T^2) \, \partial v / \partial x = v_0 F / \rho_0, \qquad (1.34b)$$

which is nothing more than the Hugoniot equation for this case.

To now find a general formulation of the inversion law, we consider a flow with slow variation in time and space which is subjected to a high-frequency perturbation

$$F \sim \exp (i\omega t - i\mathbf{k}\mathbf{r}). \qquad (1.35a)$$

Here F represents an n-component perturbation.

We write the system of linearized equations of motion of the medium, which describe the influence of the perturbations, in the form

$$L \, (\partial/\partial t, \, \partial/\partial x)u = F, \, u = (\rho, \, v_x, \, \ldots). \qquad (1.35b)$$

The coefficients in (1.35b) can be assumed to be constant within a comparatively small part of the flow.

If F = 0, the equation

$$L \, (\partial/\partial t, \, \partial/\partial x)u = 0 \qquad (1.36)$$

describes the natural oscillations of the medium. Assuming

$$u \sim \exp (i\omega t - i\mathbf{k}\mathbf{r}),$$

we find the dispersion relation

$$D \, (\omega, k) \equiv \mathrm{Det} \, L = 0, \qquad (1.37a)$$

which has the n roots

$$\omega_i = \omega_i \, (\mathbf{k}), \, i = 1, \, \ldots, \, n. \qquad (1.37b)$$

When a given $\mathbf{k}$ corresponds to a real value of ω, the wave can be treated as a steady-state flow of the fluid. To see this feature, we transform to the coordinate system moving with velocity $\mathbf{v}_0$ with respect to the medium; in this coordinate system the frequency ω

becomes

$$\omega' = \omega - kv_0, \tag{1.37c}$$

and if ω is real we can choose† a velocity $v_0 \| k$ such that the per-turbation becomes a stationary perturbation, i.e., such that $\omega' = 0$. If ω is complex, then we can use transformation (1.37c) to elim-inate the real part alone; in this coordinate system the wave will be damped monotonically or will grow monotonically.

We now assume $F \neq 0$. Assuming all quantities to be propor-tional to $\exp(i\omega t - ikr)$, we find

$$LU = \mathcal{F}, \tag{1.38}$$

where U and $\mathcal{F}$ are the amplitudes and where $L = L(\omega, k)$. The solution of Eq. (1.38) for some u_α is

$$u_\alpha = \mathrm{Det}\,(L, \mathcal{F})_\alpha / \mathrm{Det}\, L. \tag{1.39}$$

Here $\mathrm{Det}\,(L, \mathcal{F})_\alpha$ is the determinant obtained from Det L by re-placing column α by the right side of Eq. (1.38). Obviously

$$\mathrm{Det}\, L = \prod_{i=1}^{n} (\omega_1 - \omega_i(k)), \tag{1.40}$$

where $\omega_i(k)$ are the roots of the dispersion relation. In the fixed coordinate system $\omega = 0$, and $\omega_i = \omega_i(k) - kv_0$.

If, on the given k, all the ω_i are real and different, then by changing v_0 we can cross n critical velocities in succession, so that Det L changes sign n times.

When a critical velocity is crossed, the determinant $\mathrm{Det}\,(L\mathcal{F})_\alpha$, does not change sign in general, so that the u_α have different signs, depending on whether the flow is subcritical or supercritical with respect to the given harmonic (k).

If the medium shows pronounced dispersion, for each $|k|$ an inversion occurs at a distinct flow velocity v_0, even if the direc-tion of k remains the same. If, on the other hand, the dispersion is negligible (as in the case of acoustic waves), for all $|k|$ with fixed

---

†We assume that the phase velocity v_ϕ is lower than the speed of light.

$\mathbf{k}^0 = \mathbf{k}/|\mathbf{k}|$ the inversion occurs at the same velocity $\mathbf{v}_0$. We note in this connection that for the flow of an ideal gas in a channel the inversion need not occur at $v_0 = c_T$, as would follow from the Hugoniot equation. To see this feature we note that if $\mathbf{k}^0$ is directed at some angle with respect to the flow, rather than parallel to the flow, the velocities required for inversion are $v_0 > c_T$, where

$$v_0 = \frac{c_T}{\cos(\widehat{\mathbf{k}, \mathbf{v}_0})}. \tag{1.41}$$

The Hugoniot equation, (1.30b), is derived for a narrow channel, in which the vector $\mathbf{k}$ is parallel to the flow.

2. Acceleration at Different Modes. Various plasma waves or modes can propagate in a plasma; one can then set up boundary conditions such that only a single mode is generated in the plasma to provide acceleration. Such an accelerator is called "single-mode" accelerator. For example, within the framework of single-fluid magnetohydrodynamics, one can conceive of the following single-mode accelerators: Alfvén accelerator (A accelerator), fast magnetosonic (magnetoacoustic) accelerator (R accelerator), and slow magnetosonic accelerator (S accelerator) (see Chaps. 2 and 3 below). "Hybrid" accelerators are also possible in which the acceleration process is governed by perturbations that correspond to more than one mode.

Let us put these arguments in a more concrete form and describe a method, possible in principle, for designing an accelerator for operation with specified modes. Assume we are given a channel in which a homogeneous plasma stream moves with a constant velocity $\mathbf{v}_0$. Assume that beginning at $x = 0$, we seek an increase in the velocity of the stream at the x axis in accordance with the relation

$$\mathbf{v}\big|_{y=z=0} = \mathbf{v}_0 + \mathbf{V}(x). \tag{1.42}$$

Assume that the velocity increases very slowly, so that $(V/v_0) \ll 1$. Under this condition, we can find the changes in all properties in the flow by means of perturbation theory:

$$L((\mathbf{v}_0 \nabla), \nabla)u = 0, \quad u = (\rho, v, \ldots). \tag{1.43}$$

The dispersion relation of system (1.43) is[†]

$$D(\mathbf{k}\mathbf{v}_0, \mathbf{k}) = 0. \tag{1.44}$$

If the flow is three-dimensional, in general we can specify two components k; the third will be governed by Eq. (1.44). In two-dimensional flow, only a single component k is arbitrary. Accordingly, in the three-dimensional case we can specify the velocity variation in (1.42) in some plane (e.g., the x, y plane); in the two-dimensional case we can only specify this variation at the x axis.

We choose the solution of the dispersion relation corresponding to the mode in which we are interested and find the variation $\mathbf{v}_1$ throughout the channel. Furthermore, using system (1.43), we can express the perturbations of all other properties of the stream in terms of the variation $\mathbf{v}_1$.

Writing Eq. (1.43) in the form

$$(L(\mathbf{k})u)_i = q_i V_i, \quad i = 1, \ldots, n-3, \tag{1.45}$$

we find a system of $n-3$ linear equations which can be solved if there is no degeneracy. In (1.45), L(k) is the matrix $L((\mathbf{v}\nabla), \nabla)$ after its arguments have been replaced by $i(\mathbf{v}\mathbf{k})$ and ik.

Knowing the changes in all quantities within the stream and, thus, at its boundary, we can invert the problem; i.e., after specifying the boundary conditions, we can find the acceleration of the stream in one mode.

This discussion shows the possibility of only a small increase in the velocity of a certain part of the stream by means of the given mode. With some general assumptions regarding the properties of the mode, we can arrange a sequence of such small perturbations in order to produce the desired change in velocity, at least over the major part of the stream.

---

†Written in this manner, this equation corresponds to the replacement of ω by $\mathbf{k}\mathbf{v}_0$.

Chapter 2

AXISYMMETRIC FLOW OF AN IDEAL PLASMA IN A MAGNETIC FIELD WITH $\xi = 0$

§1. Conservation Laws and Equations for the Stream Functions

In this and the following chapters we will be dealing with the axisymmetric flow of an ideal plasma (i.e., a plasma which is inviscid, has a vanishing thermal conductivity, and infinite electrical conductivity), in which the Hall effect is ignored, i.e., the factor ξ is equal to zero. Although this model is a crude approximation of the actual system (for a more detailed discussion of this point, see [15], Chap. 1, §3), it is useful to begin a theoretical analysis of steady flow with this simple model.

There are three reasons for taking this approach: First, the physics and mathematics of this idealized model are highly pertinent. Second, many of the methodological questions which are so important in the more realistic models arise in this simple model. Third, some of the conclusions reached with this model apply in more complicated models and are confirmed experimentally. One of these is the existence of current eddies.

Within the framework of ideal single-fluid magnetohydrodynamics, the plasma motion is described by the system of equations

$$
\left.
\begin{aligned}
&\frac{\partial \rho}{\partial t} + \operatorname{div} \rho \mathbf{v} = 0; \quad \operatorname{div} \mathbf{H} = 0; \\
&\rho \frac{d\mathbf{v}}{dt} = \nabla p + [\mathbf{j}\mathbf{H}]; \quad \frac{\partial \mathbf{H}}{\partial t} = \operatorname{curl}[\mathbf{v}\mathbf{H}]; \\
&\frac{dS}{dt} = 0; \quad \mathbf{j} = \operatorname{curl} \mathbf{H}.
\end{aligned}
\right\}
\qquad (2.1)
$$

Here ρ is the density, p is the pressure, $\mathbf{v}$ is the velocity, $\mathbf{H}\sqrt{4\pi}$ is the magnetic field, $(c/\sqrt{4\pi})\mathbf{j}$ is the current density, and S is the entropy. The first two of these equations are the continuity equations for the fluid and for the magnetic field; the third is the equation of motion; the fourth is an equation for the magnetic field; and the fifth is the entropy equation. Equation (2.1) must be supplemented with an equation of state. The total time derivative is the substantive derivative.

$$d/dt = \partial/\partial t + (\mathbf{v}\nabla). \tag{2.2}$$

The "freezing" of the magnetic field is conveniently described along with the condition for continuous flow in the following equation:

$$\frac{d}{dt} \cdot \frac{\mathbf{H}}{\rho} = \left(\frac{\mathbf{H}}{\rho} \nabla \right) \mathbf{v}. \tag{2.3}$$

Introducing the magnetic vector potential **A**, which is defined by the equation **H** = curl **A**, we can write the freezing condition as curl $\{\partial\mathbf{A}/\partial t - [\mathbf{vH}]\} = 0$. Then we find

$$\partial\mathbf{A}/\partial t - [\mathbf{vH}] = \nabla\Phi. \tag{2.4}$$

The quantity Φ is proportional to the potential of electric field **E**.

Axial symmetry is assumed: in the coordinate system r, φ, z, the derivatives of all quantities with respect to φ vanish. The magnetic stream function in the axisymmetric case can be written

$$\psi(r, z, t) = rA_\varphi. \tag{2.5}$$

The components of the field **H** which lie in the meridional planes can be written in terms of ψ:

$$H_r = -\frac{1}{r} \cdot \frac{\partial\psi}{\partial z}; \quad H_z = \frac{1}{r} \cdot \frac{\partial\psi}{\partial r}, \tag{2.6}$$

while the azimuthal component of **H** is determined from the r and z components of **A**:

$$H_\varphi = \frac{\partial A_r}{\partial z} - \frac{\partial A_z}{\partial r}. \tag{2.7}$$

It is easy to show that the magnetic lines of force lie on the magnetic surfaces $\psi(r, z, t)$ = const. Since $\partial\Phi/\partial\varphi = 0$, Eq. (2.4) tells us that

$$d\psi/dt = 0. \tag{2.8}$$

The conservation of ψ means that the axisymmetric magnetic surfaces ψ in an ideally conducting medium move along with the material.

The vanishing of the total derivative of the entropy means that the entropy is also frozen in the material, so that

$$S = S(\psi). \qquad (2.9)$$

The azimuthal components of (2.3) and of the equation of motion are written in the following forms, respectively:

$$\frac{d}{dt} \cdot \frac{I}{\rho r^2} = \frac{\mathbf{H}}{\rho} \, \nabla \frac{I_0}{r^2} ; \qquad (2.10)$$

$$\frac{d}{dt} I_0 = \frac{\mathbf{H}}{\rho} \, \nabla I, \qquad (2.11)$$

where I and I_0 are the stream functions for $\mathbf{j} = \mathrm{curl}\,\mathbf{H}$ and $\mathbf{j}_0 = \mathrm{curl}\,\mathbf{v}$, equal to $I = H_\varphi$ and $I_0 = rv_\varphi$.

The equation in (2.3) for the r and z components of the magnetic field and the velocity is written in a form analogous to (2.10), (2.11):

$$\left.\begin{aligned}
\frac{d}{dt} \cdot \frac{H_r}{\rho} &= \frac{\mathbf{H}}{\rho} \, \nabla v_r; \\
\frac{d}{dt} \cdot \frac{H_z}{\rho} &= \frac{\mathbf{H}}{\rho} \, \nabla v_z.
\end{aligned}\right\} \qquad (2.12)$$

We write the remaining two equations for the r and z components of the equation of motion as

$$\left.\begin{aligned}
\rho \frac{dv_r}{dt} - \rho \, \frac{I_0^2}{r^3} + \frac{1}{2r^2} \cdot \frac{\partial I^2}{\partial r} - \frac{1}{r} \, j_\varphi \frac{\partial \psi}{\partial r} + \frac{\partial p}{\partial r} &= 0; \\
\rho \frac{dv_z}{dt} + \frac{1}{2r^2} \cdot \frac{\partial I^2}{\partial z} - \frac{1}{r} \, j_\varphi \frac{\partial \psi}{\partial z} + \frac{\partial p}{\partial z} &= 0,
\end{aligned}\right\} \qquad (2.13)$$

where

$$j_\varphi = \frac{\partial H_r}{\partial z} - \frac{\partial H_z}{\partial r} = -r\Delta^* \psi \equiv -\left(\frac{\partial}{\partial r} \cdot \frac{1}{r} \cdot \frac{\partial \psi}{\partial r} + \frac{\partial}{\partial z} \cdot \frac{1}{r} \cdot \frac{\partial \psi}{\partial z} \right). \qquad (2.14)$$

It is a complicated task to analyze this system of equations for nonstationary axisymmetric motion, so we restrict the analysis below to the steady-state flow [26].

In the steady state ($\partial/\partial t = 0$) the continuity equation for the flow, div $\rho\mathbf{v} = 0$, permits us to introduce the stream function ψ_0, in terms of which we can express the longitudinal components of the velocity,

$$v_z = \frac{1}{\rho r} \cdot \frac{\partial \psi_0}{\partial r}, \quad v_r = -\frac{1}{\rho r} \cdot \frac{\partial \psi_0}{\partial z}, \tag{2.15}$$

so that the streamlines of the fluid lie on the surfaces ψ_0 = const. It follows from Eq. (2.8) that $\psi_0 = \psi_0(\xi)$; the same result is found from Eq. (2.12). For symmetry purposes we introduce the new stream function ξ, which is ψ or ψ_0; any other integral of motion which is a function of ψ alone will also serve the purpose. It is assumed that

$$\psi = \psi(\xi); \quad \psi_0 = \psi_0(\xi). \tag{2.16}$$

The longitudinal components of the velocity and the magnetic field,

$$\mathbf{v}_{\parallel} = \mathbf{v} - \mathbf{v}_{\varphi}; \quad \mathbf{H}_{\parallel} = \mathbf{H} - \mathbf{H}_{\varphi},$$

are parallel and their ratio is obviously

$$\frac{v_{\parallel}}{H_{\parallel}} = \frac{v_r}{H_r} = \frac{v_z}{H_z} = \frac{1}{\rho} \cdot \frac{d\psi_0}{d\psi} = \frac{1}{\rho} \cdot \frac{\psi_0'(\xi)}{\psi'(\xi)}. \tag{2.17}$$

Using the steady-state condition ($\partial/\partial t = 0$) and the condition for axial symmetry ($\partial / \partial \varphi = 0$), we can write Eqs. (2.10) and (2.11) as

$$\mathbf{v}_{\parallel} \, \nabla \frac{I}{\rho r^2} - \frac{\mathbf{H}_{\parallel}}{\rho} \, \nabla \frac{I_0}{r^2} = 0; \tag{2.18}$$

$$\mathbf{v}_{\parallel} \, \nabla I_0 - \frac{\mathbf{H}_{\parallel}}{\rho} \, \nabla I = 0. \tag{2.19}$$

Using the condition $\mathbf{v}_{\parallel} \parallel \mathbf{H}_{\parallel}$ and (2.17), we can now find the momentum and "freezing" integrals:

$$I_0 \psi_0' - I\psi' = A(\xi); \tag{2.20}$$

$$\frac{I}{\rho r^2} \psi_0' - \frac{I_0}{r^2} \psi' = B(\xi). \tag{2.21}$$

For steady-state flow we can write the fourth equation in (2.1) as

$$[\mathbf{v}\mathbf{H}] = -\nabla \Phi. \tag{2.22}$$

Taking the scalar product of this equation with $\mathbf{v}$, we see that $\Phi = \Phi(\xi)$; substituting in the expressions for $\mathbf{v}$ and $\mathbf{H}$ in terms of I_0, I, and ξ, we find

$$\Phi'(\xi) = B(\xi). \tag{2.23}$$

The integral in (2.21) can thus be thought of as an equation that describes the equipotential nature of the magnetic surfaces $\psi = $ const.

To find the energy integral we use the thermodynamic relation

$$\nabla p/\rho = \nabla W - T\nabla S, \tag{2.24}$$

where W is the enthalpy. Using this equation, the equation $\mathbf{v}\nabla S = 0$, and (2.22) and multiplying the equations of motion (2.1) by $\mathbf{v}$ we find

$$\rho \mathbf{v} \, \nabla \left(\frac{v^2}{2} + W \right) = \mathbf{j} \, \nabla \Phi. \tag{2.25}$$

Now substituting in the expressions for $\mathbf{v}$ and $\mathbf{j}$ in terms of ψ_0 and I, we can reduce Eq. (2.25) to the Jacobian equation $D(U, \xi)/D(r, z) = 0$, where $U = U(\xi)$ is the desired energy integral:

$$\frac{v^2}{2} + W + \frac{I\Phi'(\xi)}{\psi_0'(\xi)} = U(\xi). \tag{2.26}$$

System (2.20), (2.21) can be solved for I_0 and I:

$$\left. \begin{aligned} I_0 &= \frac{1}{s} \left(A \frac{\psi_0'}{\rho} + r^2 B\psi' \right); \\ I &= \frac{1}{s} (A\psi' + r^2 B\psi_0'), \end{aligned} \right\} \tag{2.27}$$

where

$$s = \psi_0'^2/\rho - \psi'^2. \tag{2.28}$$

After we substitute (2.27) into (2.13), we can transform the latter to a single equation [26] for the function ξ, which contains ρ:

$$\frac{s}{\rho} \Delta^* \xi + \frac{1}{2\rho r^2} \cdot \frac{\partial s}{\partial \xi} (\nabla \xi)^2 - \frac{\psi_0'^2}{\rho^3 r^2} (\nabla \rho \, \nabla \xi) + \frac{1}{2\rho r^2} \cdot \frac{\partial}{\partial \xi} \times$$

$$\times \frac{A^2}{s} + \frac{r^2}{2} \cdot \frac{\partial}{\partial \xi} \cdot \frac{B^2}{s} + \frac{\partial}{\partial \xi} \cdot \frac{AB\psi'}{s\psi_0'} + TS' - U' = 0. \tag{2.29}$$

Here the operator Δ^* is given by (2.14), and the partial derivative with respect to ξ is evaluated at a fixed density ρ. Equation (2.29) contains the integrals of motion ψ, ψ_0, A, B, U, and S, which can all be written as arbitrary functions of ξ. Equation (2.29) must be solved together with Eq. (2.26) and the equation of state.

These equations can be generalized to the case in which the system shows helical symmetry. Omitting the calculations, we write the result [26] [see Eq. (A.41)]:

$$\frac{1}{\rho r}\left(\frac{\partial}{\partial r}\cdot\frac{rs}{\beta}\cdot\frac{\partial\xi}{\partial r}+\frac{\partial}{\partial\theta}\cdot\frac{s}{r}\cdot\frac{\partial\xi}{\partial\theta}\right)-\frac{1}{2\rho\beta}\cdot\frac{\partial s}{\partial\xi}(\nabla\xi)^2-\frac{2\alpha A}{\beta^2\rho}+$$
$$+\frac{1}{2\rho\beta}\cdot\frac{\partial}{\partial\xi}\cdot\frac{A^2}{s}+\frac{\beta}{2}\cdot\frac{\partial}{\partial\xi}\cdot\frac{B^2}{s}+\frac{\partial}{\partial\xi}\cdot\frac{AB\psi_0'}{\rho s\psi'}+TS'-U'=0; \qquad (2.30)$$

$$W+\frac{v^2}{2}+\Phi+\frac{\beta B^2}{s}+\frac{AB\psi_0'}{\rho s\psi'}=U; \qquad (2.31)$$

$$\begin{pmatrix}\rho v_r\\H_r\end{pmatrix}=\begin{pmatrix}\psi_0'\\\psi'\end{pmatrix}\frac{1}{r}\cdot\frac{\partial\xi}{\partial\theta}; \quad \alpha r\begin{pmatrix}\rho v_z\\H_z\end{pmatrix}-\begin{pmatrix}\rho v_\varphi\\H_\varphi\end{pmatrix}=(\psi_0')\frac{\partial\xi}{\partial r}; \qquad (2.32)$$

$$\begin{pmatrix}\rho v_z\\H_z\end{pmatrix}+\alpha r\begin{pmatrix}\rho v_\varphi\\H_\varphi\end{pmatrix}=\frac{A}{s}\begin{pmatrix}\psi_0'\\\psi'\end{pmatrix}+\frac{\beta B}{s}\begin{pmatrix}\rho\psi_0'\\\psi'\end{pmatrix}. \qquad (2.33)$$

Here ξ is a function of the two coordinates r and θ, $\theta=\varphi-\alpha z$, $\alpha=2\pi/L$, $\beta=1+\alpha^2 r^2$, and L is the pitch of the helix.

§2. Axisymmetric Flow across an Azimuthal

Magnetic Field

An important case of MHD flow is the flow across an azimuthal magnetic field with velocity $\mathbf{v}=\mathbf{v}_{\parallel}$ and field $\mathbf{H}=\mathbf{H}_\varphi$. The equations describing flow of this type can be found from Eqs. (2.20)-(2.29) by setting $\psi=0$ and A = 0. By choosing $\xi=\psi_0$, we find the following "freezing" condition for the field†:

$$\frac{H_\varphi}{\rho r}=\sqrt{4\pi}\,B(\xi). \qquad (2.34)$$

---

†Here and below, H and j are respectively the true magnetic field and the current density (without the factors of $\sqrt{4\pi}$ and $\sqrt{4\pi}/c$, respectively).

In this case the entropy $S = S(\xi)$ is conserved, as is the generalized Bernoulli integral

$$v^2/2 + W(S, \rho) + \rho r^2 B^2\,(\xi) = U\,(\xi). \tag{2.35}$$

It is evident that the Euler equation (2.1) can be written

$$[v\,\mathrm{curl}\,v] = -\rho r^2\,\nabla\frac{B^2}{2} + \nabla U - T\,\nabla S. \tag{2.36}$$

It follows that if the flow is isentropic and isomagnetic (i.e., if $S =$ const and $B =$ const), then this flow can be a potential flow, provided that the Bernoulli integral U is constant over the entire flow. This conclusion is a generalization of the Crocco theorem [21] for flow across a magnetic field. In axisymmetric, isomagnetic flow, the ratio of the magnetic flux within an annular force tube to the mass of the tube remains constant over the entire flow. In the analysis below we assume the flow to be isentropic everywhere; thus, $W = W(\rho)$.

In this case the function ξ satisfies [19, 20]

$$\frac{1}{\rho r}\cdot\frac{\partial}{\partial r}\cdot\frac{1}{\rho r}\cdot\frac{\partial\xi}{\partial r} + \frac{1}{\rho r}\cdot\frac{\partial}{\partial z}\cdot\frac{1}{\rho r}\cdot\frac{\partial\xi}{\partial z} + \frac{\rho z^2}{2}\cdot\frac{dB^2}{d\xi} - \frac{dU}{d\xi} = 0. \tag{2.37}$$

Equations (2.35) and (2.37) must be solved together for ρ and ξ; the functions $U(\xi)$ and $B(\xi)$ and the equations of the boundary surfaces must be given. If the field H_φ is known we can write the current density j in terms of H_φ by means of

$$rj_z = \frac{c}{4\pi}\cdot\frac{\partial}{\partial r}rH_\varphi;$$

$$rj_r = -\frac{c}{4\pi}\cdot\frac{\partial}{\partial z}(rH_\varphi). \tag{2.38}$$

We consider plasma flow in an axisymmetric device consisting of two coaxial electrodes (Fig. 5). If the plasma density ρ tends to—

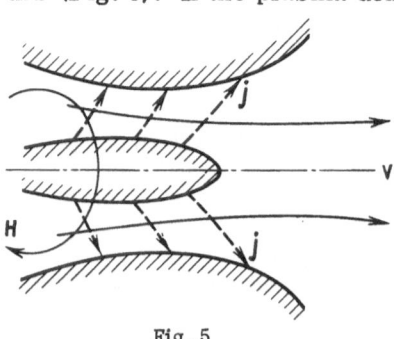

Fig. 5

wards zero at the exit from this system, we conclude from the Bernoulli equation (2.35) that the maximum exit velocity is

$$v_{max} = \sqrt{2U(\xi)}.$$ (2.39)

Accordingly, if U = const at the input the plasma velocity at the exit (i.e., in the limit $\rho \to 0$) is the same, regardless of the distance from the given streamline to the axis. In addition to the exit velocity v_{max} we must know such parameters as the mass flow rate $\dot{m}$, the thrust F, the energy flux N, the voltage U, and the current through the system. All these quantities can be found from ξ, $U(\xi)$, $B(\xi)$, and H_φ. Specifically, using the equations written above, we find

$$\dot{m} = \int_{r_1}^{r_2} \rho v_z \, 2\pi r \, dr = 2\pi (\xi_2 - \xi_1);$$ (2.40)

$$N = \int_{r_1}^{r_2} \rho \frac{v^2}{2} v_z \, 2\pi r \, dr \bigg|_{z=L} = 2\pi \int_{\xi_1}^{\xi_2} U(\xi) \, d\xi;$$ (2.41)

$$F = \int_{r_1}^{r_2} \rho v_z^2 \, 2\pi r \, dr \bigg|_{z=L} = 2\pi \int_{\xi_1}^{\xi_2} v_z \, d\xi.$$ (2.42)

Here ξ_1 and ξ_2 are the values of $r = r_1(z)$ and $r = r_2(z)$. If the flow divergence at the exit is small, i.e., if $v_z \approx v$, then we can use (2.35) to write

$$F = 2\pi \int_{\xi_1}^{\xi_2} \sqrt{2U(\xi)} \, d\xi.$$ (2.43)

The voltage between the electrodes, ξ_1 = const and ξ_2 = const, is

$$V = \frac{1}{c} \int_{r_1}^{r_2} v_z H_\varphi \, dr = \frac{1}{c} \int_{\xi_1}^{\xi_2} B(\xi) \, d\xi,$$ (2.44)

and the total current through the system is

$$\mathscr{I} = \int_0^{r_2} j_z \, 2\pi r \, dr = \frac{c}{2} r_2 H_\varphi(r_2, 0),$$ (2.45)

where H_φ is taken at the input cross section (z = 0).

§3. Plasma Flow in a Narrow Flux Tube ($H_{\parallel} = 0$)

1. Two Types of Flow. Many of the properties of the plasma flow in an axisymmetric channel can be understood by partitioning the flow into narrow coaxial tubes bounded by the streamlines ξ = const (Fig. 6). These tubes have width $f(z)$ and average radius r(z). Within a tube all properties are assumed to be independent of the coordinate n, transverse to the tube walls. It follows from Eqs. (2.9), (2.16), (2.34), and (2.35) that the equations that describe the flow in such a tube reduce to the algebraic equations

$$\rho v r f = \delta \xi = \text{const}; \tag{2.46}$$

$$W(\rho) + \frac{v^2}{2} + \rho r^2 B^2 = U = \text{const}; \tag{2.47}$$

$$\frac{H}{\rho r} = \sqrt{4\pi}\, B = \text{const}; \tag{2.48}$$

$$S = \text{const}. \tag{2.49}$$

The first of these equations describes conservation of matter, the second describes conservation of energy, and the third and fourth describe conservation of magnetic flux and entropy, respectively. The constancy of the entropy is incorporated in (2.47) through the introduction of the enthalpy $W(\rho)$.

Since we are neglecting the change in the plasma properties in the direction transverse to the tube, we can ignore Eq. (2.37).

The system of algebraic equations in (2.46)–(2.48) contains three equations for the five unknown functions r(z), $f(z)$, v(z), $\rho(z)$, and H(z). Accordingly, two of these functions, e.g., $\rho(z)$ and H(z), can be specified arbitrarily.

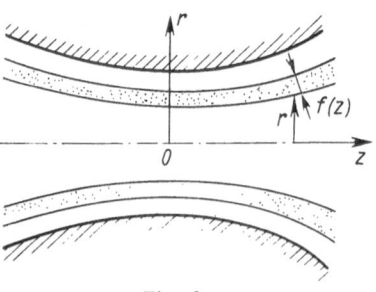

Fig. 6

Let us examine the Bernoulli equation, (2.47), in more detail. This equation shows that there can be two extreme limits in the flow: Either the process occurs in such a way that the density of the medium decreases without bound, and the velocity tends toward a certain limit v_{max},

$$\rho \to 0, \quad v \to \sqrt{2U} = v_{max}. \tag{2.50}$$

or the velocity tends to zero, while the density tends toward some maximum ρ_{max}:

$$v \to 0, \ \rho \to \rho_{max}. \tag{2.51}$$

where

$$W(\rho_{max}) = U. \tag{2.52}$$

We call the first flow regime the "pure acceleration" mode and the second the "pure compressional" mode [15, 21].

Let us examine in detail the conditions under which each of these regimes occurs. Obviously, the acceleration mode arises when the flow radius r remains bounded at the exit from the channel and the density goes to zero, $\rho \to 0$. In this case the magnetic field also vanishes, $\mathbf{H} \to 0$, and by virtue of the continuity equation the tube cross section f increases without bound.

In purely compressional flow (Fig. 7), the tube radius r must tend toward zero. In this case, as follows from the freezing condition, (2.48), the magnetic field also vanishes since ρ remains finite.

We see from the continuity equation, (2.46), that in the compression zone the flow width f should increase without bound. Noting

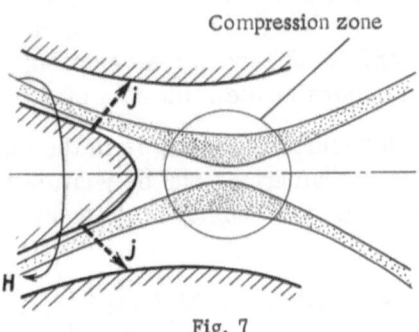

Fig. 7

that $f \lessgtr r$, and using $r \to 0$, we conclude that we must consider infinitesimally narrow tubes for a rigorous discussion. This circumstance requires analysis of the compression zone as a whole (Subsection 3, § 4 of this chapter).

2. Hugoniot Equations. To derive the Hugoniot equation for flow in a narrow tube, we differentiate Eqs. (2.46)–(2.48):

$$\frac{d\rho}{\rho} + \frac{df}{f} + \frac{dv}{v} + \frac{dr}{r} = 0; \tag{2.53}$$

$$c_T^2 \frac{d\rho}{\rho} + v\,dv + c_A^2 \frac{d\rho}{\rho} + 2c_A^2 \frac{dr}{r} = 0; \tag{2.54}$$

$$\frac{dH}{H} = \frac{d\rho}{\rho} + \frac{dr}{r}. \tag{2.55}$$

Here

$$c_T^2 = \left(\frac{\partial p}{\partial \rho}\right)_S \quad \text{and} \quad c_A^2 = \frac{H^2}{4\pi\rho} \tag{2.56}$$

are the squares of the sound velocity and the Alfvén velocity.

Eliminating dH and dρ from (2.53)–(2.55), and carrying out some simple manipulations, we find [20]

$$(v^2 - c_s^2)\frac{dv}{v} = c_T^2 \frac{d(fr)}{fr} + c_A^2 \frac{d(f/r)}{f/r}. \tag{2.57}$$

Equation (2.57) shows that the effect vanishes at a velocity v equal to the signal-propagation velocity $c_s = \sqrt{c_T^2 + c_A^2}$ across the magnetic field. In the limit $c_A \to 0$, the signal velocity is crossed at the minimum of fr, the channel cross section; while in the limit $c_T \to 0$ the crossing occurs at the minimum of f/r.

Equation (2.57) describes the changes in the plasma velocity in the flux tube. Analogous equations can be written for the other flow properties. In particular, the density is given by

$$(v^2 - c_s^2)\frac{d\rho}{\rho} = -v^2 \frac{d(rf)}{rf} + 2c_A^2 \frac{dr}{r}. \tag{2.58}$$

Comparing (2.47) and (2.58), we see that in gasdynamic flow ($c_A = 0$) the quantities $d\rho$ and dv always have opposite signs; i.e., the density decreases during acceleration and increases during deceleration. If $c_A \neq 0$, however, the signs of dv and $d\rho$ are the same under certain conditions. An anomalous density change of this type was first detected in numerical computer calculations for two-dimensional flows by Brushlinskii et al. [23]. Let us consider this effect in more detail. Assume that the flow velocity is lower than the signal velocity ($v^2 < c_s^2$); then with $d\rho > 0$ and $dv > 0$ the following inequalities hold:

$$\left.\begin{array}{l} c_T^2 \dfrac{d\,(fr)}{fr} + c_A^2 \dfrac{d\,(f/r)}{f/r} < 0; \\[2mm] -v^2 \dfrac{d\,(fr)}{fr} + 2c_A^2 \dfrac{dr}{r} < 0. \end{array}\right\} \qquad (2.59)$$

If the flow velocity is higher than the signal velocity, and if $dv > 0$ and $d\rho > 0$, then the inequalities in (2.59) are reversed. We rewrite (2.59) as

$$\left.\begin{array}{l} c_s^2 \dfrac{df}{f} < (c_A^2 - c_T^2) \dfrac{dr}{r} ; \\[2mm] (2c_A^2 - v^2) \dfrac{dr}{r} < v^2 \dfrac{df}{f} . \end{array}\right\} \qquad (2.60)$$

Multiplying these inequalities we find the condition under which they both hold to be $(c_s^2 - v^2)dr\,df < 0$ or, since the flow velocity is assumed below the signal velocity, $dr\,df < 0$.

If we assume accelerated flow at a velocity above the signal velocity, we find the same condition. If the flow is decelerated, on the other hand, then the requirement $d\rho\,dv > 0$ leads to the condition $dr\,df > 0$.

Using (2.60), we find the conditions that correspond to the anomalous behavior of the density. For velocities below the signal velocity we find

$$\left.\begin{array}{l} \text{a) } dr < 0, \quad df > 0; \qquad \text{b) } c_A^2 < c_T^2; \\[3mm] \text{c) } \dfrac{|dr|}{r} > \dfrac{df}{f} \cdot \dfrac{c_A^2 + c_T^2}{c_T^2 - c_A^2}, \qquad \text{for} \quad 2c_A^2 > v^2; \\[3mm] \dfrac{df}{f} \cdot \dfrac{v^2}{|2c_A^2 - v^2|} > \dfrac{|dr|}{r} > \dfrac{c_A^2 + c_T^2}{c_T^2 - c_A^2} \cdot \dfrac{df}{f}, \qquad \text{for} \quad 2c_A^2 < v^2. \end{array}\right\} \qquad (2.61)$$

Accordingly, in this case the magnetic pressure must be small in comparison with the gas pressure.

For flow at a velocity above the signal velocity the anomalous density behavior occurs if

$$
\left.\begin{array}{l}
\text{a) } dr > 0, \quad df < 0; \qquad \text{b) } 2c_A^2 < v^2; \\[2mm]
\text{c) } \dfrac{|2c_A^2 - v^2|}{v^2} \cdot \dfrac{dr}{r} > \dfrac{|df|}{f}, \qquad \text{for} \quad c_A^2 > c_T^2; \\[3mm]
\dfrac{|2c_A^2 - v^2|}{v^2} \cdot \dfrac{dr}{r} > \dfrac{|df|}{f} > \dfrac{|c_T^2 - c_A^2|}{c_s^2} \cdot \dfrac{dr}{r}, \qquad \text{for} \quad c_A^2 < c_T^2.
\end{array}\right\} \qquad (2.62)
$$

3. **Acceleration Regimes.** We assume that a plasma enters the channel at z = 0, more precisely, that it enters a tube selected for consideration. The plasma has a low velocity, $v_0 \to 0$, a density ρ_0, and a temperature T_0. The magnetic field in this cross section of the tube is denoted by H_0, while the tube parameters r and f are denoted by r_0 and f_0.

The boundary surfaces of the channel (not the tube) must be electrodes since the azimuthal magnetic field must decay along the channel and this is possible only if a current flows between the inner and outer walls.

As noted above, we do not consider the electrode processes proper; to do this explicitly it would be necessary to take the finite conductivity and the electron mass into account.

We first consider the simple case in which the radius of the tube remains constant, r = const (Fig. 8). In this case the sys-

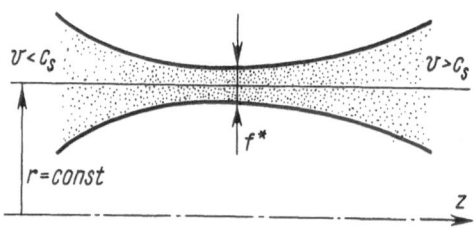

Fig. 8

tem of equations (2.46)-(2.48) simplifies, becoming

$$\rho v f = \text{const (a)};$$
$$W(\rho) + v^2/2 + \rho k^2 = \text{const (b)};$$
$$H/\rho = k = \text{const (c)}.$$

(2.63)

Similarly, the dependence on the ratio of the Alfvén and sound velocities disappears from the Hugoniot equation:

$$(v^2 - c_s^2)\frac{dv}{v} = c_s^2 \frac{df}{f}.$$

(2.64)

Denoting the ratio of specific heats of the plasma by γ and assuming

$$p = p_0 (\rho/\rho_0)^\gamma,$$

(2.65)

we find

$$W(\rho) = \frac{c_p T}{\gamma - 1} = \frac{c_T^2}{\gamma - 1}.$$

(2.66)

Here, c_p is the specific heat at constant pressure. In this case the maximum exit velocity is found from (2.47) and (2.50):

$$v_{\max} = \sqrt{2\left(\frac{c_{T0}^2}{\gamma - 1} + c_{A0}^2\right)}.$$

(2.67)

Accordingly, if $c_{A0}^2 \gg c_{T0}^2$, i.e., if the parameter $\beta = 8\pi p/H^2$ is much smaller than unity ($\beta \ll 1$), we have $v_{\max} \approx c_{A0}\sqrt{2}$. If, on the other hand, $\beta \gg 1$, then the exit velocity is given by the usual gasdynamic equation $v_{\max} = c_{T0}\sqrt{\frac{2}{\gamma - 1}}$.

Let us examine the continuity equation, (2.63a). Assuming $v \to 0$, $\rho \to \rho_0$ at the input to the channel and $\rho \to 0$, $v \to v_{\max}$ at the exit, we conclude that the channel width must increase without bound near both the entrance and exit. Accordingly, for a constant tube radius r, regular flow is possible only if the tube is nozzle-shaped (Fig. 8).

It follows from the Hugoniot equation, (2.64), that the minimum cross section $f*$ has the remarkable property that the local plasma velocity v* in it is equal to the local signal velocity: $v* = c_s^*$.

Accordingly, in this case (r = const) we find a property of an ordinary gasdynamic nozzle: The sound velocity and the flow velocity are equal at the critical cross section. If, on the other hand, r ≠ const, then we see from (2.58) that the signal velocity cannot be crossed at the minimum of f.

For high-velocity flows we are primarily interested in the case $c_A^2 \gg c_T^2$. In this case, the Bernoulli equation (2.63b) takes the simple form

$$v^2/2 + \rho k^2 = \rho_0 k^2. \tag{2.68}$$

Hence, using (2.63c), we can relate v and H to ρ:

$$v = v_{max} \sqrt{1 - \rho/\rho_0}; \qquad H = H_0 \rho/\rho_0. \tag{2.69}$$

Substituting these equations into continuity equation (2.63a), we find

$$f = \text{const}/\rho \sqrt{1 - \rho/\rho_0}. \tag{2.70}$$

These equations show that if the velocity increases monotonically in the channel, then ρ and H decrease monotonically. If a particular function v(z) is specified, we find a single-valued dependence on the coordinate z and on the other parameters. In particular, if $\rho = az$, then the current density (j ~ $\partial H/\partial z$) is constant within the plasma volume.

We turn now to the general case in which f and r are allowed to vary. If $\beta \gg 1$, the changes in ρ and v are described by the ordinary hydrodynamic equations as follows from Eqs. (2.46)–(2.48) with the term β^{-2} neglected. These variations are governed solely by the channel area, which is proportional to the product rf.

If $\beta \ll 1$, we can neglect the gas pressure, i.e., the term W(ρ) in Eq. (2.47); we then find that the plasma acceleration is governed by the magnetic field. In this case the geometric factor governing the flow is the ratio f/r [see Eq. (2.57)].

Accordingly, it is possible to accelerate a plasma in a channel with a constant value of f. When $\beta \gg 1$, the channel radius r should increase; when $\beta \ll 1$ the radius r should decrease monotonically. The reason is that the magnetic field H not only performs work during its expansion (in accordance with the term $\nabla H^2/8\pi$ in the Euler equation), but also in the contraction of the lines of force, as described by the term $(1/4\pi)(H\nabla)H$.

4. Compressional Regimes. Let us assume, as in the preceding case, that the plasma velocity at z = 0 is negligibly small. Then the constant of the Bernoulli integral is

$$U = \frac{H_0^2}{4\pi\rho_0}(1+\mu), \quad \mu = \frac{W_0}{c_{A0}^2}. \tag{2.71}$$

For compression to occur, the radius r must tend toward zero (Fig. 7). How the velocity varies is unimportant; the only important consideration is that this velocity not exceed the signal velocity (otherwise, shock waves can be generated). To obtain maximum compression at r = 0 the velocity in the compression zone must be negligibly small. In this case the density is governed by (2.52). Substituting (2.71) into (2.52), and assuming the process to be polytropic [see Eq.(2.66)], we find the following equation for the maximum compression [22]:

$$v_{max} \equiv \frac{\rho_{max}}{\rho_0} = \left[(\gamma-1)\frac{c_{A0}^2}{c_{T0}^2}(1+\mu)\right]^{1/(\gamma-1)}. \tag{2.72}$$

In adiabatic ($\gamma = 5/3$) compression of hydrogen, with $c_{A0} = 10^8$ cm/sec and $c_{T0} = 10^6$ cm/sec we have

$$v_{max} \equiv \frac{\rho_{max}}{\rho_0} \approx 5.5 \cdot 10^5. \tag{2.72a}$$

Equation (2.72) for the compression coefficient shows that, for given values of H_0 and ρ_0, this coefficient increases without bound as $\gamma \to 0$, $c_{T0}^2/c_{A0}^2 \to 0$.

Accordingly, in an isothermal process with $\gamma = 1$, the compression is pronounced:

$$v_{max} = \rho_{max}/\rho_0 = \rho_0 \exp\frac{c_{A0}^2}{c_{T0}^2}(1+\mu), \tag{2.73}$$

because the enthalpy is

$$W(\rho) = c_{T0}^2 \ln(\rho/\rho_0). \tag{2.74}$$

Equation (2.72) has another important consequence: if H_0 and c_{T0} are constant, then ρ_{max} is described as a function of ρ_0 by

$$\rho_{max} \sim \rho_0^{(\gamma-2)/(\gamma-1)}; \tag{2.75}$$

that is, when $1 < \gamma < 2$, decreasing the initial density increases the absolute value of the maximum density. If $\gamma > 2$, on the other hand, then ρ_{max} decreases with decreasing ρ_0.

Under actual conditions the value of γ is not the exponent that corresponds to an adiabat since plasma radiation reduces γ while Joule heating increases it. These factors act in opposite directions so that it is difficult to determine the actual value of γ.

During the compression process the plasma is heated as well as compressed. If this process is polytropic,

$$W(\rho) = \frac{k}{M} \cdot \frac{\gamma}{\gamma - 1} T. \qquad (2.76)$$

Here, k is the Boltzmann constant and M is the ion mass. In a fully ionized plasma with z = 1 we have $T = T_i + T_e$. Substituting (2.76) into (2.52), we find the maximum temperature for γ = const:

$$T_{max} = \frac{\gamma - 1}{\gamma} \cdot \frac{H_0^2}{4\pi\rho_0} \cdot \frac{M}{k} (1 + \mu). \qquad (2.77)$$

With the numerical values in (2.72a) we find $T_{max} \approx 5$ keV for compression of hydrogen.

To conclude this section we consider the magnetic fields in compression under the assumption of an infinite conductivity, in accordance with the present model. We emphasize that the assumption of an ideal conductivity in the compression zone is restrictive: With the small diameter of the compressed plasma configuration an extremely high conductivity is required to prevent the field from leaving the plasma. Nevertheless at thermonuclear temperatures the assumption $\sigma = \infty$ is still reasonable. Let us estimate the magnetic fields. If the process occurs in such a way that $v \to 0$, then

$$\frac{c_{T0}^2}{\gamma - 1} \left(\frac{\rho}{\rho_0}\right)^{\gamma - 1} + \frac{H^2}{4\pi\rho} = \frac{H_0^2}{4\pi\rho_0}(1 + \mu). \qquad (2.78)$$

Thus, with $\rho \sim \rho_{max}$ and

$$\frac{H^2}{4\pi\rho} \sim \frac{c_{T0}^2}{\gamma - 1}\left(\frac{\rho}{\rho_0}\right)^{\gamma - 1} \qquad (2.79)$$

we have

$$H_{max} \sim H_0 \sqrt{\frac{\rho_{max}}{\rho_0}}. \tag{2.80}$$

An exact expression for the maximum magnetic field is given in [22].

§4. Integrable Flows Which Vary Slowly along the z Axis

1. **General Equations.** Let us consider the class of approximate solutions of system (2.35), (2.37) for the case in which the dependence on the coordinate z is weak, i.e., $\xi = \xi(r, \varepsilon z)$, $\rho = \rho(r, \varepsilon z)$, where ε is a small parameter. Neglecting terms in Eqs. (2.35) and (2.37) which are quadratic in ε, we find a simplified system of differential equations [20]:

$$\frac{1}{\rho r} \cdot \frac{\partial}{\partial r} \cdot \frac{1}{\rho r} \cdot \frac{\partial \xi}{\partial r} + \frac{\rho r^2}{2} \cdot \frac{dB^2}{d\xi} - \frac{dU}{d\xi} = 0; \tag{2.81}$$

$$W(\rho) + \frac{1}{2} \left(\frac{1}{\rho r} \cdot \frac{\partial \xi}{\partial r} \right)^2 + \rho r^2 B^2 = U. \tag{2.82}$$

As noted above, B and U are arbitrary functions of ξ alone. Equations (2.81)–(2.82) only contain partial derivatives with respect to r. In general, solution of this system of equations depends on the two arbitrary functions $c_1(z)$ and $c_2(z)$, so that $\xi = \xi(r, c_1, c_2)$, $\rho = \rho(r, c_1, c_2)$. The integration of Eqs. (2.81)–(2.82) can correspond to various problems. If the electrode geometry $r = r_1(z)$ and $r = r_2(z)$ is specified, the conditions $\xi = \xi_1 = $ const and $\xi = \xi_2 = $ const at the electrodes imply a problem which is a nonlinear analog of the Sturm–Liouville problem. If one of the electrodes is specified, $\xi = \xi_1 = $ const at $r = r_1(z)$ and the flow is specified near this electrode, i.e., $\partial \xi / \partial n$ at $r = r_1(z)$, we have an analog of the Cauchy problem.

In practice, the values of $B(\xi)$ and $U(\xi)$ can be determined from the conditions at the entrance and exit of the accelerator. If the values of $\rho(r)$, $v(r)$, and $H_\varphi(r)$ are specified at the entrance (z = 0), we can find ξ from Eq. (2.15) ($\xi = \psi_0$): $\xi = \int_{r_1}^{r} \rho v r dr \big|_{z=0} = \xi(r, 0)$.

We can thus find $B(\xi)$ and $U(\xi)$. However, even if we specify all these properties at the entrance as well as the electrode geometry, we generally do not find solutions.[†] Accordingly, in the specific examples below we pose the Cauchy problem, i.e., we specify one of the electrodes and the value of $\partial \xi / \partial r$ at this electrode and specify v, ρ, p, and H_φ at the entrance to the accelerator; we seek then ξ and the geometry of the second electrode in the form $\xi(r, z) = \xi_2 = \text{const}$.

Equations (2.81)–(2.82) are called the equations of the smooth-flow approximation, since they describe flows which vary slowly along the z axis. Multiplying Eq. (2.8) by $\partial \xi / \partial r$ and subtracting Eq. (2.82) from it, after (2.82) has been differentiated with respect to r, we find

$$\frac{\partial W}{\partial r} + \frac{1}{2\rho r^2} \cdot \frac{\partial}{\partial r} (\rho r^2 B)^2 = 0. \qquad (2.83)$$

This equation can be written

$$\frac{\partial}{\partial r} \left(p + \frac{H^2}{8\pi} \right) + \frac{H^2}{4\pi r} = 0 \qquad (2.83a)$$

or

$$\frac{\partial p}{\partial r} = \frac{1}{c} [\mathbf{j} \mathbf{H}]_r. \qquad (2.83b)$$

Accordingly, the equilibrium condition does hold in the radial direction in the adiabatic approximation.

Equation (2.83) or (2.83a) can be integrated; we can treat three cases: 1) W = 0; 2) B = const; 3) $r \to \infty$ (the two-dimensional case). If the plasma is cold and the gas pressure is negligible the first integral of Eq. (2.83) is

$$\rho r^2 B\, (\xi) = c_1\, (z), \qquad (2.84)$$

where $c_1(z)$ is an arbitrary function. Determining ρ from this equation, and substituting it into Eq. (2.82), we find

$$\int_0^\xi \frac{B\, (\xi)\, d\xi}{\sqrt{U\, (\xi) - c_1\, (z)\, B\, (\xi)}} = \sqrt{2}\, c_1\, (z) \ln \frac{r}{c_2\, (z)}, \qquad (2.85)$$

[†] The reason is that the signal velocity is reached in the flow.

where $c_2(z)$ is a secondary arbitrary function. The functions $c_1(z)$ and $c_2(z)$ must vary slowly.

When B = const, i.e., isomagnetic flow, the integral of Eq. (2.83) is

$$W(\rho) + \rho r^2 B^2 = c_3(z), \qquad (2.86)$$

which gives implicitly ρ as a function of r and $c_3(z)$. Substituting $\rho(r, c_3)$ into Eq. (2.82) we find

$$\int_0^\xi \frac{d\xi}{\sqrt{U(\xi) - c_3(z)}} = \sqrt{2} \int_{c_4(z)}^r \rho r \, dr. \qquad (2.87)$$

It is not difficult to evaluate the integral on the right side of this equation for special values of the polytrope exponent, e.g., $\gamma = 2$.

Finally, in the two-dimensional problem, with $(H/\rho) = \sqrt{4\pi} B(\xi)$, $\rho v_z = \partial\xi/\partial x$, $\rho v_x = (-\partial\xi/\partial z)$, the equations of a slowly varying flow,

$$\left. \begin{array}{c} \dfrac{1}{\rho} \cdot \dfrac{\partial}{\partial x} \cdot \dfrac{1}{\rho} \dfrac{\partial\xi}{\partial x} + \dfrac{\rho}{2} \cdot \dfrac{dB^2}{d\xi} = \dfrac{dU}{d\xi}; \\[2mm] W(\rho) + v_z^2/2 + \rho B^2 = U, \end{array} \right\} \qquad (2.88)$$

admit the first integral

$$p + H^2/8\pi = P(z), \qquad (2.89)$$

where $P(z)$ is a slowly varying function. Assuming $\gamma = 2$, $p = p_0(\rho/\rho_0)^2$ for simplicity, we find an explicit expression for the square of the density:

$$\rho^2(\xi, z) = \frac{P(z)}{p_0/\rho_0^2 + B^2/2}. \qquad (2.89a)$$

Substituting this expression into the Bernoulli integral in (2.88) and proceeding as before, we find

$$\int_0^\xi \frac{d\xi}{\sqrt{\rho} \sqrt{U(\xi) - W(\rho) - \rho B^2(\xi)}} = \sqrt{2}\, x. \qquad (2.89b)$$

We can now determine the stream function $\xi(x, z)$ for two-dimensional flow when $H = H_y$, $(\partial/\partial y) = 0$.

<u>2. Flow of Cold Plasma in a Channel of Slow-</u>
<u>ly Varying Cross Section.</u> Let us examine in more de-
tail the flow of a cold plasma, $W = 0$, described by integrals (2.84)
and (2.85) [20]. Different functions $B(\xi)$ and $U(\xi)$ correspond to
different flows. The appropriate functions are determined by specifying the distribution $\rho(r)$ and $v(r)$ at the entrace ($z = 0$) or in
some other cross section ($z = $ const). It is easiest to give the initial values of $\rho(r)$ and $v(r)$ as the power functions

$$\rho = \rho_0 \, (r/R_0)^{\nu}; \quad v = v_0 \, (r/R_0)^{\mu}, \tag{2.90}$$

where ρ_0 and v_0 are the density and velocity at the point $r = R_0$,
$z = 0$, which lies on some average streamline $\xi = 0$, whose equation $r = R(z)$ is assumed to be given (Fig. 1). The quantities ν
and μ are arbitrary. Using $(\partial \xi / \partial r) = \rho v r$ and Eqs. (2.90), at
$z = 0$ we have

$$\xi = \xi_0 \left[\left(\frac{r}{R_0} \right)^{\mu + \nu + 2} - 1 \right], \xi_0 = \frac{\rho_0 \, v_0 \, R_0^2}{\mu + \nu + 2}. \tag{2.91}$$

Substituting these equations into (2.82) and (2.84) we find the
following equations for the unknown functions $B(\xi)$ and $U(\xi)$ when
$W = 0$:

$$B(\xi) = \frac{c_1(0)}{\rho_0 R_0^2} \left(\frac{\xi + \xi_0}{\xi_0} \right)^{-\frac{\nu + 2}{\mu + \nu + 2}}; \tag{2.92}$$

$$U(\xi) = \frac{v_0^2}{2} \left(\frac{\xi + \xi_0}{\xi_0} \right)^{\frac{2\mu}{\mu + \nu + 2}} + \frac{c_1^2(0)}{\rho_0 R_0^2} \left(\frac{\xi + \xi_0}{\xi_0} \right)^{-\frac{\nu + 2}{\mu + \nu + 2}} \tag{2.93}$$

A general result for this case of cold plasma flow is the function

$$H_\varphi(r, z) = c_1(z)/r, \tag{2.94}$$

which follows from the definition of $B(\xi)$ and from Eq. (2.84). This
function shows that the longitudinal current only flows along the
electrodes. On the other hand, the values of ρ and v in the flow
depend on the choice of the functions $B(\xi)$ and $U(\xi)$.

We assume that the velocity $v_R(z)$ is given at $r = R(z)$. Noting
that $r = R(z)$ is a streamline, so that $B(\xi)$ and $U(\xi)$ are constant
at $r = R(z)$, (2.82) and (2.84) yield the function $c_1(z)$ introduced

above:

$$c_1(z) = c_1(0) \frac{v_{max}^2 - v_R^2(z)}{v_{max}^2 - v_0^2} ; \quad c_1^2(0) = \frac{\rho_0 R_0^2}{2}(v_{max}^2 - v_0^2). \qquad (2.95)$$

Using these equations, we write the following equations for ρ and v^2:

$$\rho = \rho_0 \frac{R_0^2}{r^2} \cdot \frac{v_{max}^2 - v_R^2(z)}{v_{max}^2 - v_0^2} \left(\frac{\xi + \xi_0}{\xi_0}\right)^{\frac{v+2}{\mu + v + 2}} ; \qquad (2.96)$$

$$v^2 = v_0^2 \left(\frac{\xi + \xi_0}{\xi_0}\right)^{\frac{2\mu}{\mu + v + 2}} + [v_R^2(z) - v_0^2] \left(\frac{\xi + \xi_0}{\xi_0}\right)^{-\frac{v+2}{\mu + v + 2}}. \qquad (2.97)$$

Let us examine in greater detail the case of isomagnetic and iso-Bernoulli flow, with B = const and U = const. This case is degenerate† and leads to a logarithmic, rather than power-law, dependence of ξ on r. From Eq. (2.87) we find $\xi(r, z)$ to be

$$\xi(r, z) = \rho_0 R_0^2 v_R(z) \frac{v_{max}^2 - v_R^2(z)}{q_{max}^2 - v_0^2} \ln \frac{r}{R(z)}. \qquad (2.98)$$

By equating ξ to a constant we can find the electrode shape. When $v_0 \to 0$ (2.98) yields the equations for the electrodes $r_1(z)$ and $r_2(z)$:

$$\frac{r_{1,2}(z)}{R} = \left(\frac{r_{1,2}^*}{R}\right)^{\frac{2}{3\sqrt{3}\,x\,(1-x^2)}} ; \quad x = \frac{v_R(z)}{v_{max}}. \qquad (2.99)$$

If we take r = R = const as one of the electrodes, the channel has the shape shown in Fig. 9a when $r_2^* > R$ or that shown in Fig. 9b when $r_1^* > R$. The critical cross section which, in this case, coincides with the minimum distance between the electrodes, is governed by the equation $v_R^2(z*) = v_{max}^2/3$.

The density ρ is proportional to r^{-2}:

$$\rho = \rho_0 \frac{R_0^2}{r^2} \cdot \frac{v_{max}^2 - v_R^2(z)}{v_{max}^2 - v_0^2}, \qquad (2.100)$$

while the velocity $v_R(z)$ is a function of z only.

---

†Formally this case corresponds to $\mu = 0$, $v = -2$.

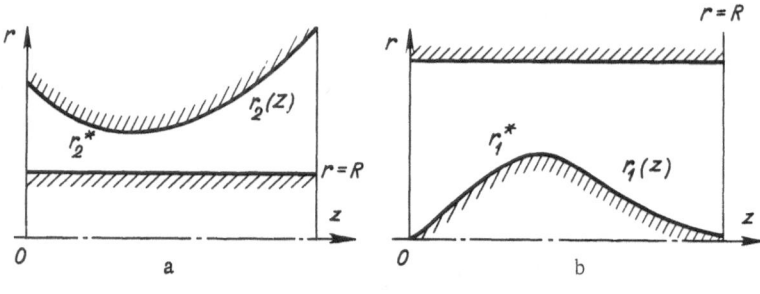

Fig. 9

If the profiles of both electrodes are specified, rather than $v_R(z)$, it is necessary to solve cubic equation (2.99) for v_R, and the roots of this equation can be complex. This result explains our claim that, in general, there is no solution of the Sturm–Liouville problem for nonlinear equations.

Isomagnetic flows (B = const) are distinguished from the many other possible types of flow of a cold plasma by the fact that the radial density gradient remains small during the flow. The situation is quite different in highly nonisomagnetic flows (i.e., $|\nu + 2| > 1$). In the limit $v_0 \to 0$ in these cases the radial density gradients increase without bound. As an example we consider a flow with $\mu = -1$, $\nu = 0$, with a constant density at the entrance cross section ($z = 0$), and velocity proportional to r^{-1}. In this case U is proportional to B, and ξ is given by

$$\frac{\xi + \xi_0}{\xi_0} = \left(\frac{r}{R}\right)^{\frac{v_R}{v_0} \cdot \frac{v_{max}^2 - v_R^2}{v_{max}^2 - v_0^2}} \; ; \quad \xi_0 = \rho_0 v_0 R_0^2. \qquad (2.101)$$

In this case the electrode shape is the same as in the case B = const, U = const (see above). The coordinate dependences of ρ and v are

$$\rho = \rho_0 \frac{R^2}{r^2} \cdot \frac{v_{max}^2 - v_R^2}{v_{max}^2 - v_0^2} \left(\frac{r}{R}\right)^{\frac{2v_R}{v_0} \cdot \frac{v_{max}^2 - v_R^2}{v_{max}^2 - v_0^2}} \; ;$$

$$v = v_R \left(\frac{r}{R}\right)^{-\frac{v_R}{v_0} \cdot \frac{v_{max}^2 - v_R^2}{v_{max}^2 - v_0^2}} . \qquad (2.102)$$

Here a strong dependence of ρ and v on r is exhibited if v_0 is sufficiently small and if r is appreciably different from R; this result
implies that the initial density distribution strongly affects the characteristics of the flow in channels of large cross section.

We note in conclusion that in nonisomagnetic flows such characteristics as the currents $\mathcal{J}$ and the thrust F do not exhibit
anomalous features in the limit $v_0 \to 0$ whereas the mass flow rate
$(\dot{m} \to \infty)$ and the voltage $(V \to 0)$ do. For example, for the flow described by Eqs. (2.101)-(2.102) these characteristics are, respectively,

$$\mathcal{J} = \frac{c}{2} \sqrt{2\pi\rho_0} \, R v_{\max}; \quad F = \frac{4\pi}{3\sqrt{3}} \, \rho_0 \, R^2 \, v_{\max}^2 \ln \frac{r_2^*}{R} ; \left.\begin{array}{c}\\[2em]\\\end{array}\right\} \quad (2.103)$$
$$\dot{m} = 2\pi\rho_0 \, R^2 \, v_0 \left(\frac{r_2^*}{R} \right)^{\frac{2v_{\max}}{3\sqrt{3}\,v_0}} ; \quad V = \frac{\sqrt{2\pi\rho_0}}{c} \, R v_{\max} \, v_0.$$

where R = const is the radius of the inner electrode, r_2 is the radius of the outer electrode, and we have assumed $v_0 \ll v_{\max}$. These
features obviously derive from the assumption W = 0.

3. Isomagnetic Flows Which Vary Slowly along
the z Axis. Let us examine the case in which B = const but
in which the plasma is not cold, i.e., W $\neq$ 0. It follows from Eqs.
(2.82) and (2.86) that

$$(v^2/2) = U(\xi) - c_3(z). \tag{2.104}$$

If the velocity vanishes $(v \to 0)$ in the entrance cross section $(z = 0)$,
we find U = const. Accordingly, in the limit $v_0 \to 0$ isomagnetic
flows which vary slowly along the z axis are also iso-Bernoulli
flows (U = const). The dependence of U on ξ in this case is governed solely by the radial profile of the initial velocity.

We restrict the discussion of flows with U = const. In the limit
of a cold plasma, with W $\to$ 0, we obviously find the flow (see the
discussion above) for which $\rho \sim r^2$, rH_φ = const, i.e., this is the
case in which the current flows through the central electrode. If,
on the other hand, the plasma is hot, so that W $\gg$ $B^2 r^2 \mu$ Eqs. (2.34)
and (2.86) show that $\rho \approx \rho(z)$, $H_\varphi \sim r$. Accordingly, the j_z component of the current density is constant over the entire entrance
cross section and can be realized without a central electrode.

We rewrite Eq. (2.86) in the form

$$w_0 \rho^{\nu-1} + \rho r^2 B^2 = c_3(z), \tag{2.105}$$

where $w_0 = $ const. To calculate ξ explicitly with the help of Eq. (2.87), we must solve Eq. (2.105) for ρ. Since this cannot be done easily[†] for real values of γ, we restrict the analysis to the model with $\gamma = 2$. We introduce the notation

$$c_3 = \frac{1+\mu}{4\pi} \cdot \frac{H_0^2}{\rho_0} \eta(z); \quad \mu = \frac{c_{T0}^2}{c_{A0}^2}; \tag{2.106}$$

obviously $\eta(0) = 1$, $\eta_{max} = 0$, and

$$\eta_{max} = \frac{4\pi\rho_0 U}{(1+\mu) H_0^2} = \frac{1+\mu+n}{1+\mu}; \quad n = \frac{2\pi\rho_0 v_0^2}{H_0^2}. \tag{2.107}$$

Using this notation we can rewrite (2.105) in a form which can be solved for ρ:

$$\frac{\rho}{\rho_0} = \frac{1+\mu}{(r/r_0)^2+\mu} \eta(z). \tag{2.108}$$

Substituting this equation into (2.87), we find

$$\xi = r_0^2 \rho_0 (1+\mu) \eta(z) v(z) \ln \frac{\mu+(r/r_0)^2}{\mu+[a(z)/r_0]^2}. \tag{2.109}$$

Here r_0 and ρ_0 are the radius and density at the entrance to the system taken on the surface of the inner electrode, and $v(z)$ is the plasma velocity in this cross section. Under these assumptions, this velocity is independent[‡] of ξ. The actual flow pattern can be determined by specifying the explicit function $v(z)$ or $\eta(z)$ and the geometry of the inner electrode, $a(z)$. The equation of the outer electrode, $b(z)$, is determined from the condition $\xi(b(z), z) = \xi_0 = $ const.

This model is of primary interest in analyzing the compressional regime [22]. Let us assume that the inner electrode exhibits a profile $a = r_0(1 - z/l)$ at $z < l$ and $a = 0$ at $z > l$.

[†]For an arbitrary value of γ, Eq. (2.88) can only be solved in the cases $\beta \gg 1$ and $\beta \ll 1$ [20].

[‡]The functions $\eta(z)$ and $v(z)$ are obviously related by $v^2(z) = \frac{H_0}{2\pi\rho_0} [(1 + \mu$

$+ n) - (1+\mu) \eta(z)]$.

The maximum density is reached at the z axis at $z \geq l$:

$$\frac{\rho_{max}}{\rho_0} = \left(\frac{1+\mu}{\mu}\right) \eta(z).$$ (2.110)

This equation agrees with (2.72) if $\gamma = 2$ is substituted into the latter equation and we note that $\eta(z) = \eta_{max}$. The maximum magnetic field is reached at $r^* = r_0\sqrt{\mu}$

$$H\mid_{r=r^*} = H_0\left(\frac{1+\mu}{2\sqrt{\mu}}\right)\eta(z).$$ (2.111)

The geometry of the outer electrode will depend on whether the compression occurs with a velocity $v \to 0$, a nonvanishing constant velocity, or an increasing velocity. The lines of the electric current are governed by the equation

$$Hr = \text{const.}$$ (2.112)

§5. Flow Velocity Equal to Signal Velocity at Some Point

We now treat the case in which the flow velocity reaches the signal velocity at some point. It turns out that the critical surface, at which $v = v_s$, has certain distinctive features. In particular, if there exists a planar cross section of the channel $z = z^* = \text{const}$ in which the velocity has only the single component v_z, then the critical velocity, which is equal to the signal-propagation velocity, is reached in this cross section [20]. Taking the partial derivative with respect to z of the equation $\rho r v_z = \partial\xi/\partial r$ and of the Bernoulli integral in (2.35), and setting $v = v_z$, $\partial\xi/\partial z = 0$, we find

$$v\frac{\partial\rho}{\partial z} + \rho\frac{\partial v}{\partial z} = 0; \quad (c_T^2 + c_A^2)\frac{1}{\rho} \cdot \frac{\partial\rho}{\partial z} - v\frac{dv}{dz} = 0,$$ (2.113)

where c_T is the sound velocity and c_A is the Alfvén velocity. If the velocity continues to increase at $z = z^*$, i.e., $(\partial v/\partial z) = 0$ in this cross section, then Eq. (2.113) yields the following equation for v^*:

$$v^* = \sqrt{c_T^2 + c_A^2}.$$ (2.114)

This is the signal-propagation velocity, c_s.

Ordinarily, however, the critical velocity is reached on some surface which is not planar. This conclusion can also be reached on the basis of the "smooth-flow approximation" used here.

A. We assume $U = \text{const}$ and $B = \text{const}$; then, as noted above, in "smooth" flow we have $v = v(z)$. Using the familiar equation $c_T^2 = (\gamma - 1)W$ and the Bernoulli equation (2.35), we can write $c_s^2 = c_T^2 + c_A^2$ as

$$c_s^2 = (\gamma - 1)\left(U - \frac{v^2(z)}{2}\right) - (\gamma - 2)\, r^2 B^2 \rho. \qquad (2.115)$$

For a hot plasma, with $\beta \gg 1$, in the zeroth approximation in the parameter β^{-1}, the density $\rho = \rho_0(z)$ is governed by the equation $W(\rho_0) = U - v^2/2$. In the first approximation, we can replace ρ in Eq. (2.115) by $\rho_0(z)$. Equating the flow velocity $v(z)$ to the critical velocity c_s, we find the equation of the surface at which the critical velocity is reached:

$$(1 + \gamma)\,\frac{v^2(z)}{2} - (2 - \gamma)\,B^2 r^2 \rho_0(z) = (\gamma - 1)\,U. \qquad (2.116)$$

Since $v(z)$ is an increasing function, while $\rho_0(z)$ is a decreasing function, the critical surface is located at progressively larger values of z as the radius increases (Fig. 10a).

B. In the other limiting case, $\beta \ll 1$, a cold plasma, in the zeroth approximation in the parameter β, the density is $\rho_0(r, z) = (U - v^2/2)/r^2 B^2$. We now write c_s^2 as

$$c_s^2 = U - v^2/2 + (\gamma - 2)W\,(\rho). \qquad (2.117)$$

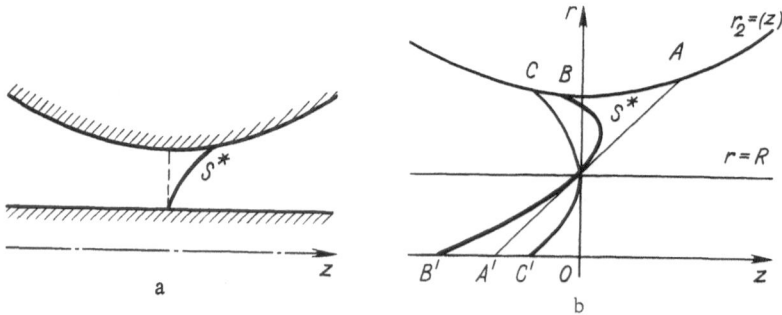

Fig. 10

The last term in (2.117) is small; hence, in the first approximation in β we can set $\rho = \rho_0(r, z)$. Writing $c_s^2 = v^2$, we find the equation of the critical surface:

$$(3/2)v^2 (z) + (2 - \gamma) W (\rho_0) = U. \qquad (2.118)$$

We see again that the critical surface is inclined in the direction of increasing velocity (see Fig. 10a).[†]

In the limiting cases of small and large values of β treated here, the critical surface makes small angles with the surface r = const. The slope of the critical surface is obviously maximized when $\beta \sim 1$.

§6. Current Eddies and Critical Surfaces

Brushlinskii et al. [23] have detected current eddies (or loops) in numerical analyses of two-dimensional flow in coaxial accelerators. The current drawn from the first electrode returns to this electrode without reaching the second electrode. Below we analyze the conditions for the formation of such current eddies and give an illustrative analytic calculation of the overall picture [24], using the mathematical apparatus of slowly varying flows. In addition, we use the same mathematical apparatus in this section for a calculation (more detailed than that above) of the shape of the critical surfaces at which the flow velocity reaches the local value of the signal-propagation velocity.

1. Lines of Electric Current. If the plasma velocity has two components, v_r and v_z, and if the magnetic field has only the single component $H = H_\varphi$, then for the axisymmetric problem the equations of ideal single-fluid hydrodynamics for an isotropic (S = const) flow reduce to the following system of equations, as follows from (2.34)–(2.37):

$$\frac{1}{\rho r} \cdot \frac{\partial}{\partial r} \cdot \frac{1}{\rho r} \cdot \frac{\partial \xi}{\partial r} + \frac{1}{\rho r} \cdot \frac{\partial}{\partial z} \cdot \frac{1}{\rho r} \cdot \frac{\partial \xi}{\partial z} + \frac{\rho r^2}{2} \cdot \frac{dB^2}{d\xi} = \frac{dU}{d\xi}; \qquad (2.119)$$

$$W (\rho) + v^2/2 + \rho r^2 B^2 = U (\xi); \qquad (2.120)$$

$$\frac{1}{\rho r} \, H_\varphi = 4\pi I/c\rho r^2 = \sqrt{4\pi} B (\xi). \qquad (2.121)$$

[†] The inclinations of the critical surface S^* in the opposite direction (Fig. 10b), which has been established in conventional hydrodynamics [17], is of the order of ε^2/β with respect to that described here (p. 54).

The plasma streamlines are governed by the equation $\xi\,(r,\,z) = $ const, while the lines of electric current are governed by the equation $I(r,\,z) = $ const. The components of the velocity and the current density are expressed in terms of ξ and I by Eqs. (2.15) and (2.38) with $\xi = \psi_0$.

Near a singularity of the electric-current lines (this singularity can be elliptical or hyperbolic) both components of the current density $\mathbf{j}$ must vanish. We note that the vanishing of the $\mathbf{j}$ component normal to the streamlines can occur only at a point at which the following condition holds:

$$r\,\frac{dv^2}{ds} - 4\rho W'\,(\rho)\,\frac{dr}{ds} = 0. \tag{2.122}$$

This condition is found from the condition $j_\perp \sim (I/r)(\partial I/\partial s) = 0$ if (2.120) and (2.121) are differentiated along the streamline $r = r(s)$. Equation (2.122) shows that under the condition of accelerated flow, $(dv^2/ds) > 0$, the current can only form an eddy if $(dr/ds) > 0$, i.e., only at an expanding part of the electrode.

To calculate the overall flow pattern we restrict the discussion to the case $U = $ const and assume that the magnetic pressure is small in comparison with the gas pressure ($\beta = 8\pi p/H^2 \gg 1$). Assuming a flow which varies slowly along the z axis, and neglecting terms of order B^2, $(\partial\xi/\partial z)^2$, and $\partial^2\xi/\partial z^2$ in (2.119) and (2.120), we find

$$\frac{\partial}{\partial r}\cdot\frac{1}{\rho r}\cdot\frac{\partial\xi}{\partial r} = 0; \qquad W\,(\rho) + \frac{1}{2}\left(\frac{1}{\rho r}\cdot\frac{\partial\xi}{\partial r}\right)^2 = U. \tag{2.123}$$

Integration of these equations yields

$$\xi = \frac{\rho V}{2}\,(r^2 - R^2); \qquad W\,(\rho) + \frac{1}{2}\,V^2 = U. \tag{2.124}$$

Here $V(z)$ and $R(z)$ are slowly varying but otherwise arbitrary functions of the variable z. By specifying these functions we find various configurations of the surfaces, $\xi\,(r,\,z) = $ const. It follows from the second equation in (2.124) that in this approximation $\rho = \rho(z)$.

Let us set $R = $ const; i.e., we assume that among the streamlines $\xi\,(r,\,z) = $ const there is a straight line $r = R$. The correspond-

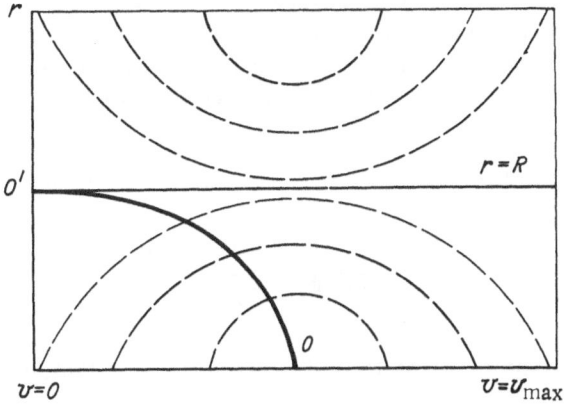

Fig. 11

ing pattern of streamlines is shown in Fig. 11, where the velocity $V(z)$ is plotted along the z axis. The streamlines are equipotentials, so that any pair can be taken as the electrodes. Near the planes $z = z_0$ and $z = z_{max}$, where the velocity $V(z) = 0$ and where the maximum value $V_{max} = \sqrt{2U}$, the flow does not satisfy the slowness condition; thus these results only apply in the central part of the nozzle, where the streamlines are smooth.

To find the singularity of the family of electric current lines $I(r, z) = const$, we set the derivatives I_r and I_z equal to zero:

$$\frac{\partial I}{\partial r} \sim \rho r \, [2B + \rho r^2 \, VB'] = 0; \quad \frac{\partial I}{\partial z} \sim r^2 \left[\frac{\partial \rho}{\partial z} \, B + \right.$$
$$\left. + \rho \frac{r^2 - R^2}{2} \cdot \frac{d(\rho V)}{dz} \, B' \right] = 0. \quad (2.125)$$

Evidently a necessary condition for the existence of a singularity is $BB' < 0$; in the isomagnetic case, with $B = const$, there are no singularities. Eliminating $B'(\xi)$ from (2.125), using the second equation in (2.124) after it has been differentiated with respect to z, and using $\rho W'(\rho) = c_T^2$, where $c_T = \sqrt{\gamma p / \rho}$ is the sound velocity, we find

$$r^2 = R^2 \, (1 - V^2/c_T^2). \quad (2.126)$$

This equation defines the curve OO' (Fig. 11), on which the component of the current density normal to the streamlines, $j_\perp$, vanishes.

According to (2.126), this curve connects the point $(r = 0, V = c_T)$, in the critical cross section of the nozzle, to the point $(r = R, V = 0)$. The singularities of the family of current lines $I = $ const can only lie on the curve OO'; hence they lie below the line $r = R$ on streamlines with $(dr/ds) > 0$.

To determine the nature of the singularity, we now calculate the second derivatives of $I(r, z)$. Using

$$W'' = \frac{\gamma - 2}{\rho} \quad W' = \frac{\gamma - 2}{\rho^2} c_T^2,$$

where γ is the exponent for the adiabat, and using Eqs. (2.124)-(2.126), we find

$$
\left.
\begin{aligned}
\frac{\partial^2 I}{\partial r^2} &\sim -8\rho B + \rho^3 r^4 V^2 B''; \\
\frac{\partial^2 I}{\partial r\, \partial z} &\sim -\frac{2\rho r V' B}{V}\left(1 - \frac{2V^2}{c_T^2}\right) - \frac{\rho^3 r^5 V^3}{2c_T^2} V' B''; \\
\frac{\partial^2 I}{\partial z^2} &\sim -\frac{\rho R^2 V'^2 B}{c_T^2}\left[1 + (\gamma + 2)\frac{V^2}{c_T^2} - \frac{2V^4}{c_T^4}\right] + \\
&\quad + \frac{\rho^3 r^6 V^4}{4 c_T^4} V'^2 B''.
\end{aligned}
\right\}
\qquad (2.127)
$$

The sign of the invariant $\Sigma = I_{rr} I_{zz} - I_{rz}^2$ determines the nature of the singularity (if the analysis is restricted to a linear function $B(\xi)$, the sign of the invariant is the same as the sign of the quantity $(-1 + 7V^2/c_T^2 + 2(\gamma - 2)V^4/c_T^4)$. Accordingly, if the singularity lies on a part of curve OO' below the point $r_c = 0.92R$, $V_c = 0.22 \times V_{max}$, this singularity is elliptic; if it lies above r_c, it is hyperbolic (Fig. 11). In this case,

$$B = b(\xi + c); \quad I = \rho r^2 b(\xi + c), \qquad (2.128)$$

and one of the streamlines $\xi = -c$ is also a current line. The magnetic field $H = H_\varphi$ vanishes on this line. Substituting the expressions for ρ and ξ into (2.128), we find the equation of the family of electric current lines:

$$V(V_{max}^2 - V^2)^{\frac{2}{\gamma-1}} r^2(r^2 - R^2) + c(V_{max}^2 - V^2)^{\frac{1}{\gamma-1}} r^2 = \text{const}; \qquad (2.129)$$

this equation can be solved for r^2.

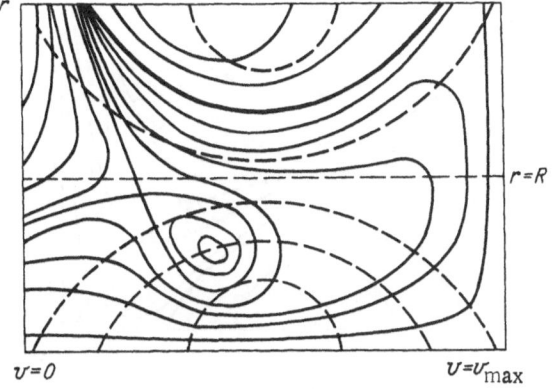

Fig. 12

For negative values of the constant c the family of current lines exhibits two singularities: one is elliptic and the other is hyperbolic (Fig. 12). If c = 0 the hyperbolic point moves toward the entrance to the channel, at which V = 0, and the elliptic point is located at $r = R/\sqrt{2}$ (Fig. 13). If c > 0 only the elliptic singularity remains (Fig. 14).

Figures 12-14 show the electric current lines for the linear function B = b(ξ + c). The dashed curves show the streamlines of the plasma; any one of these streamlines can be taken as an elec-

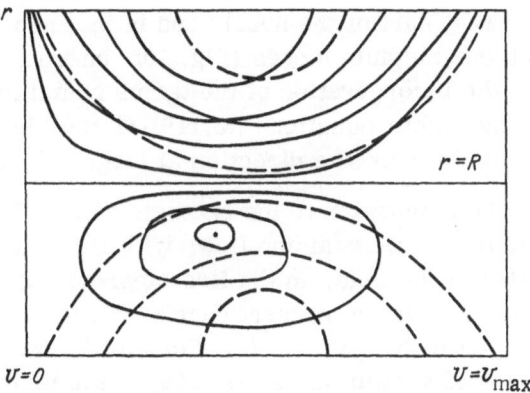

Fig. 13

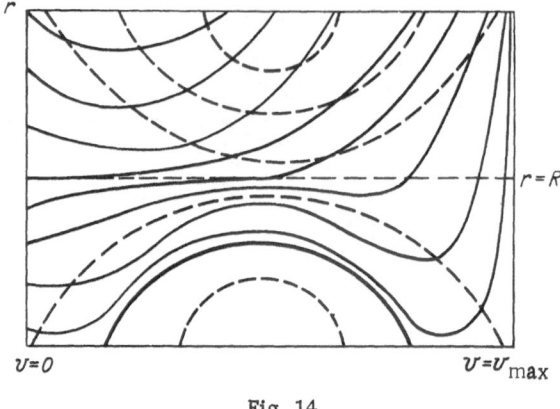

Fig. 14

trode. The heavy solid curves show the streamlines $\xi = -c$ at which the magnetic field changes sign.

It should be noted that the current eddies have been observed experimentally under conditions similar to those predicted by the present theory.

2. Critical Surfaces. The theory of slowly varying flows can be used effectively to study the shape of the critical surfaces, i.e., the surface at which the velocity reaches the signal velocity. The simplest case are found in weak ($\beta \gg 1$) or strong ($\beta \ll 1$) magnetic fields. The calculation (with only the first derivatives with respect to z being taken into account) has been carried out above (see also [24]). We note that the critical surface lies in the narrowest part of the nozzle and is inclined in the direction in which the medium moves (Fig. 10b, curve CA). We show below that the incorporation of the terms containing $(\partial \xi / \partial z)^2$ and $\partial^2 \xi / \partial z^2$ in the basic equations [(2.119)-(2.120)] bends the "sonic surface" in the opposite direction at large values of r.

Restricting the discussion to the extreme cases of weak and strong magnetic fields, we examine flows with U = const, taking the variation $B(\xi)$ into account in the first approximation. Equation (2.119) shows that in the present case (U = const, B = const) the flows are irrotational (curl $\mathbf{v}$ = 0). For slowly varying flows we seek a solution of system (2.119)-(2.120) in the form

$$\xi = \xi_0 (r,\ \varepsilon z) + \xi_1 (r,\ \varepsilon z) + \ldots (\xi_1 \sim \varepsilon^2 \ll 1). \qquad (2.130)$$

Correspondingly, we have $v_r \sim \varepsilon$ and $v_z = V + v_{1z}$, where $v_{1z} \sim \varepsilon^2$. We also assume $B = B_0 + B'\xi$, where $B'\xi \ll B_0$, and only take account of first-order quantities in B'. Terms containing products of B' and other small parameters are neglected. In a first approximation, neglecting terms of order ε^2 in Eq. (2.119) we find $V = V(z)$.

It is assumed that $V(z)$ is a specified velocity at the electrode, $r = R = $ const. By solving Eqs. (2.119)–(2.120) we can find the function $\xi(r, z)$ within the specified error; the condition $\xi = $ const is then used to determine the system of electrodes for the specified velocity $V(z)$. The analysis is restricted to the determination of the sonic surfaces; for this purpose it is sufficient to know only the velocity $\mathbf{v}$, i.e., we only need to find the first integral of Eqs. (2.119) and (2.120). In the first approximation we find

$$\xi_0 = \int_R^r \rho V(z)\, r\, dr. \tag{2.131}$$

Substituting this equation into (2.119) we find the correction† to the longitudinal velocity which is proportional to ε^2:

$$v_{1z} = -\int_R^r dr' \frac{\partial}{\partial z}\left[\frac{1}{\rho r'}\int_R^{r'} \frac{\partial}{\partial z}(\rho V)\, r\, dr\right]. \tag{2.132}$$

The square of the radial velocity v_r can be determined with the necessary accuracy by differentiating (2.131). We now determine the function $\rho(r, z)$, which is different for weak $(\beta \gg 1)$ and strong $(\beta \ll 1)$ magnetic fields. In the zeroth approximation we have

$$W(\rho) + \frac{V^2}{2} = U; \quad \rho = \rho(z) \ (\text{as } B^2 \to 0); \tag{2.133}$$

$$\frac{V^2}{2} + \rho r^2 B^2 = U; \quad \rho r^2 = \left(U - \frac{V^2}{2}\right)\frac{1}{B^2} \equiv f(z) \ (\text{as } W \to 0). \tag{2.134}$$

---

† The corrections to ξ and ρ due to B' are taken into account later [see Eq. (2.138) and the discussion which follows].

Substituting these equations into (2.131) and (2.132), respectively, we find

$$v_r = - \frac{(\rho V)'}{\rho} \cdot \frac{r^2 - R^2}{2r}; \quad v_{1z} = - \frac{1}{4} \left(\frac{(\rho V)'}{\rho} \right) \left(r^2 - R^2 - R^2 \ln \frac{r^2}{R^2} \right);$$

(2.133a)

$$v_r = - \frac{(fV)'}{f} r \ln \frac{r}{R}; \quad v_{1z} = \frac{1}{4} \left(\frac{(fV)'}{f} \right)' \left(r^2 - R^2 - r^2 \ln \frac{r^2}{R^2} \right). \quad (2.134a)$$

The surface at which the velocity v reaches the local signal velocity $c_s = \sqrt{c_T^2 + c_A^2}$ ($c_A^2 = H_\varphi^2 / 4\pi\rho$) is found from the equation $v^2 = c_s^2$. Using $c_s^2 = (\gamma - 1)W + \rho r^2 B^2$ and Eq. (2.120), we find the equations of the critical surfaces $v^2 = c_s^2$ for weak and strong magnetic fields:

$$v^2 = \frac{\gamma - 1}{\gamma + 1} V_{max}^2 + 2 \frac{2 - \gamma}{1 + \gamma} \rho r^2 B^2;$$

$$v^2 = \frac{V_{max}^2}{3} - 2 \frac{2 - \gamma}{3} W(\rho).$$

(2.135)

The second term in each of these equations is small and we can substitute ρ from (2.133) and (2.134).

In the first approximation in ε in the limiting cases $B^2 \to 0$ and $W \to 0$, the critical surface is a plane which is governed by the following equations for the two cases:

$$V^2(z) = \frac{\gamma - 1}{\gamma + 1} V_{max}^2; \quad V^2(z) = \frac{V_{max}^2}{3}. \quad (2.136)$$

Terms of order B^2 in the first equation in (2.135) and those of order W in the second equation cause the inclination of this surface (at r > R) toward increasing W(z).

To determine the effect of the quantities of order ε^2, we must substitute v^2 from (2.134) into the left sides of (2.135). It follows from (2.133)–(2.134) that can write the following expressions for $\beta \gg 1$ and $\beta \ll 1$:

$$\frac{(\rho V)'}{\rho} = \left(1 - \frac{V^2}{c_T^2} \right) V'; \quad \frac{(fV)'}{f} = \frac{V_{max}^2 - 3V^2}{V_{max}^2 - V^2} V'. \quad (2.137)$$

These quantities (and thus the radial components of the velocity) vanish at the critical surface found in the zeroth approximation. This result is obvious since the critical surface of the zeroth approximation coincides with the plane of the minimum nozzle cross section.

We can now find the complete expression for the velocity, incorporating the correction to V due to $B(\xi)$. According to (2.120), this correction is $\delta V = -BB' \int_R^r \rho^2 r^3 dr$, where ρ must be replaced by the expression from (2.133) and (2.134). Accordingly, the equations of the critical surfaces for $\beta \gg 1$ and $\beta \ll 1$ become

$$V^2 + \frac{\gamma+1}{2} V'^2 \left(r^2 - R^2 - R^2 \ln \frac{r^2}{R^2} \right) =$$
$$= c_T^{*2} + 2 \frac{2-\gamma}{1+\gamma} \rho r^2 B^2 - \frac{BB' \rho^2 V}{2} (r^4 - R^4); \qquad (2.138)$$

$$V^2 - \frac{3}{2} V'^2 \left(r^2 + R^2 - R^2 \ln \frac{r^2}{R^2} \right) =$$
$$= c_A^{*2} - 2 \frac{2-\gamma}{1+\gamma} W - \frac{B' c_A^3}{2B^3} \ln \frac{r^2}{R^2}. \qquad (2.139)$$

Restricting the discussion to the expansion of V(z) accurate to the linear term in the neighborhood of the sonic surface of the zeroth approximation ($z = 0$), and expanding the functions of r in powers of ($r - R$), we find

$$V'z = \frac{2-\gamma}{1+\gamma} \cdot \frac{2\rho B^2 R}{c_T} (r-R) -$$
$$- \frac{1+\gamma}{2c_T} V'^2 (r-R)^2 + BB' \rho^2 R^3 (r-R); \qquad (2.140)$$

$$V'z = \frac{2-\gamma}{3} \cdot \frac{\gamma-1}{c_A} \cdot \frac{2W_R}{R} (r-R) -$$
$$- \frac{3V'^2}{2c_A} (r-R)^2 + \frac{B' c_A^2}{2B^3R} (r-R). \qquad (2.141)$$

Terms of order V'^2 for both (2.140) and (2.141) lead to a curvature of the sonic surface in the direction opposite the motion of the medium ($z < 0$). The influence of these terms increases with distance from the streamline $r = R$. Terms of order B' partially cancel the basic terms which are linear in ($r - R$) if $BB' < 0$ or strengthen them if $BB' > 0$.

In both cases, weak and strong magnetic fields, the sonic surfaces have the shape shown schematically in Fig. 10b, where line AA' corresponds to the case in which terms of order V'^1 are neglected, while lines BB' and CC' show the shape of the sonic surface for the cases of weak and extremely strong magnetic fields. When B = 0 (2.140) yields the familiar expression for the sonic surface found in conventional gasdynamics [16].

Chapter 3

FLOW IN AXISYMMETRIC CHANNEL WITH A LONGITUDINAL MAGNETIC FIELD

§1. Integrated Characteristics

In the discussion above we have treated acceleration of a nonrotating plasma by an azimuthal field ($\psi' = 0$, A = 0). When a longitudinal magnetic field exists, the plasma will generally rotate about the z axis; hence, in this case we must treat flows in which $v_\varphi \neq 0$. Furthermore, if U(ξ) is replaced by the function $U_1(\xi) = U + AB/\psi_0'\psi'$, the Bernoulli equation (2.26) can be written

$$W(\rho, S) + \frac{v^2}{2} + \frac{r v_\varphi B}{\psi'} = U_1(\xi), \tag{3.1}$$

It is evident that with a nonvanishing longitudinal field ($\psi' \neq 0$), acceleration along a streamline with $v_\varphi = 0$ can only result from the conversion of thermal energy.

The quantities ψ_0 and ψ are proportional to the fluxes of the fluid and the magnetic field between the axisymmetric surfaces $\xi(r, z) = $ const. Using (2.22) and the equations

$$d\psi_0/d\psi = \rho v_z/H_z = \rho v_z/H_r, \tag{3.2}$$

which follows from it, we can write the Bernoulli integral in the two forms that correspond to (2.26) and (3.1):

$$W + \frac{v^2}{2} + \frac{[H[vH]]_q}{4\pi \rho v_q} = U(\xi);$$

$$W + \frac{v^2}{2} + \frac{[v[vH]]_q}{H_q} = U_1(\xi) \tag{3.3}$$

$$(q = r, z).$$

Choosing $\xi = \psi_0$, we can write integrals A and B as

$$A = r\left(v_\varphi - \frac{H_\varphi H_q}{4\pi \rho v_q}\right); \quad B = \frac{1}{\rho r}\left(H_\varphi - \frac{v_\varphi H_q}{v_q}\right)\frac{1}{4\pi}. \qquad (3.4)$$

These quantities can also be written in other forms, if (3.2) is used.

Equations (2.26) and (3.1) show that when $H_\varphi \to 0$ at the accelerator exit the maximum velocity that correspond to $\rho \to 0$ is $v_{max} = \sqrt{2U}$; on the other hand, if $v_\varphi \to 0$, then $v_{max} = \sqrt{2U_1}$.

As with acceleration by an azimuthal field, the properties at the accelerator exit can be expressed in a simple way with the help of the integrals ψ_0, U, and B. For example, the mass flow rate is

$$\dot{m} = \int_{r_1}^{r_2} \rho v_z\, 2\pi r dr = 2\pi\{\psi_0(\xi_2) - \psi_0(\xi_1)\}. \qquad (3.5)$$

To calculate the energy flux N and the thrust F, we note that Eq. (2.26) shows that $v \to v_{max}$ in the case $\rho \to 0$, $H_\varphi \to 0$; it follows from the Bernoulli equation (2.26) that $v^2 \to 2U(\xi)$ at the accelerator exit. Integrating over the exit cross section we find

$$N = \int_{r_1}^{r_2} \rho\frac{v^2}{2}v_z\, 2\pi r dr = 2\pi \int_{\psi_0(\xi_1)}^{\psi_0(\xi_2)} U d\psi_0; \qquad (3.6a)$$

$$F = \int_{r_1}^{r_2} \rho v_z^2\, 2\pi r dr = 2\pi \int_{\psi_0(\xi_1)}^{\psi_0(\xi_2)} v_z\, d\psi_0. \qquad (3.6b)$$

When $v_z \approx v$, this equation can be rewritten as

$$F = 2\pi \int_{\psi_0(\xi_1)}^{\psi_0(\xi_2)} \sqrt{2U}\, d\psi_0. \qquad (3.6c)$$

Furthermore, the voltage between the electrodes is

$$U = \frac{1}{c}\int_{\xi_1}^{\xi_2}(v_z H_\varphi - v_\varphi H_z)\, dr = \frac{\sqrt{4\pi}}{c}\int_{\xi_1}^{\xi_2} B(\xi)\, d\xi, \qquad (3.7a)$$

and the current through the system can be expressed as an integral of j_z over the entrance cross section:

$$\mathcal{I} = \int_0^{r_2} j_z \, 2\pi r \, dr = c r_2 H_\varphi (r_2, \, 0). \tag{3.7b}$$

§2. Flow in Narrow Axisymmetric Channel

1. Hugoniot Equation. If the flow is partitioned into narrow coaxial tubes, the flow in a tube can be described by a system of algebraic equations [25] as in the case $H_{\parallel} = 0$. We denote the velocity and field components in the meridional cross section by $\mathbf{v}_{\parallel} = \mathbf{v} - \mathbf{v}_\varphi$ and $\mathbf{H}_{\parallel} = \mathbf{H} - \mathbf{H}_\varphi$, while

$$c_{\parallel} = H_{\parallel}/\sqrt{4\pi\rho}, \quad c_\varphi = H_\varphi/\sqrt{4\pi\rho} \tag{3.8}$$

are the components of the Alfvén velocity $c_A = \mathbf{H}/\sqrt{4\pi\rho}$. Obviously,

$$\rho v_{\parallel} = \psi_0' \, \partial\xi/\partial n, \quad r H_{\parallel} = \psi' \partial\xi/\partial n, \tag{3.9}$$

where the derivative is taken along the normal n to the streamline, as shown in Fig. 15. Multiplying Eqs. (3.9) by the distance f between the two nearest streamlines, replacing $\partial \xi f/\partial n$ by the increment $\delta \xi$ = const, and assuming that the integrals $U(\xi)$, $A(\xi)$, and $B(\xi)$ in Eqs. (2.20), (2.21) and (2.26) are constant [equal to their values at some average streamline ξ = const, whose equation is $r = r(z)$], we find

$$\rho v_{\parallel} f = \alpha = \text{const}; \tag{3.10a}$$

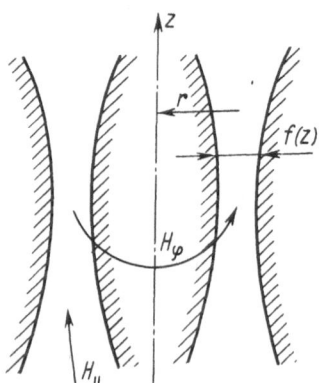

Fig. 15

$$\sqrt{\rho}\, rc_{\parallel}f = \beta = \text{const}; \tag{3.10b}$$

$$r\left(\alpha v_\varphi - \beta \sqrt{\rho}\, c_\varphi\right) = a = \text{const}; \tag{3.10c}$$

$$\frac{1}{r}\left(\alpha \frac{c_\varphi}{\sqrt{\rho}} - \beta v_\varphi\right) = b = \text{const}; \tag{3.10d}$$

$$W(\rho) + \frac{v^2}{2} + \frac{b}{\alpha}\sqrt{\rho}\, rc_\varphi = U = \text{const}. \tag{3.10e}$$

The first two of these equations describe conservation of the longitudinal fluxes of the plasma and the magnetic field, while the other equations follow from momentum conservation, the "freezing in" of the magnetic field, and energy conservation. Equations (3.10) are obviously exact if the distance f between adjacent streamlines is infinitesimally small. By introducing the quantity f in an analysis of flows by means of system (3.10), we can avoid the integration of Eq. (2.29), which essentially serves for determining possible configurations of the streamlines between two electrodes which are not close together. Since system (3.10) contains five equations with seven unknowns (f, r, ρ, $v_\parallel$, v_φ, $c_\parallel$, and c_φ), two of the quantities, for example, r and f, can be specified arbitrarily.

Differentiating Eqs. (3.10), and eliminating $d\rho$, we find the Hugoniot equation

$$\left(v_\parallel^2 - c_T^2 - \frac{c_\varphi^2}{1 - c_\parallel^2/v_\parallel^2}\right)\frac{dv_\parallel}{v_\parallel} = \left(c_T^2 + \frac{c_\varphi^2}{1 - c_\parallel^2/v_\parallel^2}\right)\frac{d(rf)}{rf} +$$

$$+ \left(v_\varphi^2 - \frac{2c_\varphi}{v_\parallel}\cdot\frac{[c_A\, v]_n}{1 - c_\parallel^2/v_\parallel^2}\right)\frac{dr}{r}. \tag{3.11}$$

The coefficient of $dv_\parallel/v_\parallel$ vanishes at the velocity $v_\parallel$ found from the equation

$$v_\parallel^4 - \left(c_T^2 + \frac{H^2}{4\pi\rho}\right)v_\parallel^2 + c_T^2\frac{H_\parallel^2}{4\pi\rho} = 0, \tag{3.12}$$

i.e., at $v_\parallel = c_s$; the signal-velocity is

$$c_s^2 = \frac{1}{2}\left(c_T^2 + \frac{H^2}{4\pi\rho}\right) \pm \sqrt{\frac{1}{4}\left(c_T^2 + \frac{H^2}{4\pi\rho}\right)^2 - \frac{c_T^2 H_\parallel^2}{4\pi\rho}}. \tag{3.13}$$

Here c_s^+ and c_s^- are the velocities of the fast and slow magnetosonic waves, respectively. Accordingly, in a channel of constant average radius $r = $ const the critical velocity $v_\parallel = c_s$ is reached at the minimum cross-sectional area of the channel, $2\pi r f$; in a channel of constant cross section this critical velocity is reached at the minimum of $r(z)$. In a cylindrical coaxial system, with r and f constant, we have $dv_\parallel = 0$, and the plasma is not accelerated.

Equations (2.27) show that the azimuthal components of the velocity and magnetic field become infinite if $s = (\psi_0'^2/\rho) - \psi'^2 = 0$. This situation occurs when the density reaches the critical value

$$\rho_{cr} = (d\psi_0/d\psi)^2 = \rho \cdot v_\parallel^2/c_\parallel^2. \tag{3.14}$$

Evidently the plasma density is equal to ρ_{cr} if the local longitudinal plasma velocity is equal to the local Alfvén velocity for the field $H_\parallel$. Continuous flow can only exist if ρ does not pass through ρ_{cr}, in other words, only if $v_\parallel$ remains either larger than $c_\parallel$, smaller than $c_\parallel$, or equal to $c_\parallel$ everywhere. In the first case ($\rho < \rho_{cr}$, $v_\parallel > c_\parallel$), the flow is "subcritical"; in the second case ($\rho > \rho_{cr}$, $v_\parallel < c_\parallel$) the flow is "transcritical"; and in the third case ($\rho = \rho_{cr}$, $v_\parallel = c_\parallel$) it is "critical." In acceleration by an azimuthal field ($c_\parallel = 0$) the acceleration is obviously always subcritical.

2. **Flow with Infinitesimally Small Initial Velocity at $r = $ const.** We first consider flow in a channel with a constant average radius r under the assumption that the velocity at the entrance to the system is negligible. Assuming that the initial values of v_φ and $v_\parallel$ are zero and that $r = $ const, we can write (3.10) in the form

$$\rho v_\parallel f = \alpha_1; \qquad \sqrt{\rho}\, c_\parallel f = \beta_1; \tag{3.15a}$$

$$\alpha_1 v_\varphi - \beta_1 \sqrt{\rho}\, c_\varphi = -\beta_1 \sqrt{\rho_0}\, c_{\varphi_0}; \quad \alpha_1 \frac{c_\varphi}{\sqrt{\rho}} - \beta_1 v_\varphi = \alpha_1 \frac{c_{\varphi_0}}{\sqrt{\rho_0}}; \tag{3.15b}$$

$$W(\rho) + \frac{v_\varphi^2 + v_\parallel^2}{2} + c_{\varphi_0} c_\varphi \sqrt{\frac{\rho}{\rho_0}} = U, \tag{3.15c}$$

where the subscript "0" represents the value of the corresponding quantity at the entrance.

It follows from Eqs. (3.15b) that

$$c_\varphi = c_{\varphi_0} \sqrt{\frac{\rho}{\rho_0}} \cdot \frac{1 - \rho_0/\rho_{cr}}{1 - \rho/\rho_{cr}}; \quad v_\varphi = -c_{\varphi_0} \sqrt{\frac{\rho_0}{\rho_{cr}}} \cdot \frac{1 - \rho/\rho_0}{1 - \rho/\rho_{cr}}. \tag{3.16}$$

Substituting these expressions into Bernoulli integral (3.15c), we find the dependence of $v_{\parallel}$ on ρ:

$$v_{\parallel}^2 = 2 \{W(\rho_0) - W(\rho)\} + c_{\varphi_0}^2 G(\rho),\tag{3.17}$$

where

$$G(\rho) = \frac{\rho_{cr}}{\rho_0}\left[1 - \left(\frac{1-\rho_0/\rho_{cr}}{1-\rho/\rho_{cr}}\right)^2\right] \qquad (\rho_{cr} = \alpha_1^2/\beta_1^2).\tag{3.18}$$

Equations (3.16) and (3.17) show that singularities appear when the density ρ approaches the critical value ρ_{cr}.

a. We first consider critical flow. In this case $\rho = \rho_{cr} = \text{const}$ and $v_{\parallel} = c_{\parallel}$. Equations (3.16b) reduce to the single equation

$$c_{\varphi} = c_{\parallel} - c_{\varphi_0},\tag{3.19}$$

and the Bernoulli equation (3.15c) leads to

$$v_{\parallel}^2 = c_{\varphi_0}^2 - c_{\varphi}^2.\tag{3.20}$$

Hence, using (3.15a), we find the channel profile to be

$$f = f_1 c_{\varphi_0}/\sqrt{c_{\varphi_0}^2 - c_{\varphi}^2},\tag{3.21}$$

where f_1 is the cross section of the channel at the entrance, and $c_{\varphi} \to 0$.

In the critical flow regime the density of the medium remains constant, so that the medium can only be accelerated by Alfvén waves; this critical regime might be called the "Alfvén" regime.

b. We now consider subcritical flow ($\rho < \rho_{cr}$). If $r = \text{const}$, then the Hugoniot equation (3.11) becomes

$$\left(v_{\parallel}^2 - c_T^2 - \frac{c_{\varphi}^2}{1-c_{\parallel}^2/v_{\parallel}^2}\right)\frac{dv_{\parallel}}{v_{\parallel}} = \left(c_T^2 + \frac{c_{\varphi}^2}{1-c_{\parallel}^2/v_{\parallel}^2}\right)\frac{df}{f}.\tag{3.22}$$

In the subcritical regime $d_{\parallel}^2 > c_{\parallel}^2$, and the coefficient of df/f is always positive and nonvanishing. The quantity in parentheses on the left side of (3.22) can be written

$$(v_{\parallel}^2 - c_{s+}^2)(v_{\parallel}^2 - c_{s-}^2)/(v_{\parallel}^2 - c_{\parallel}^2).\tag{3.23}$$

Since c_{s-}^2 is always smaller than $c_{\parallel}^2$, as follows from (3.13), the quantity in (3.23) vanishes only at $v_{\parallel} = c_{s+}$ if $v_{\parallel}^2 > c_{\parallel}^2$. Accordingly, in subcritical flow the channel has only a single constriction, where

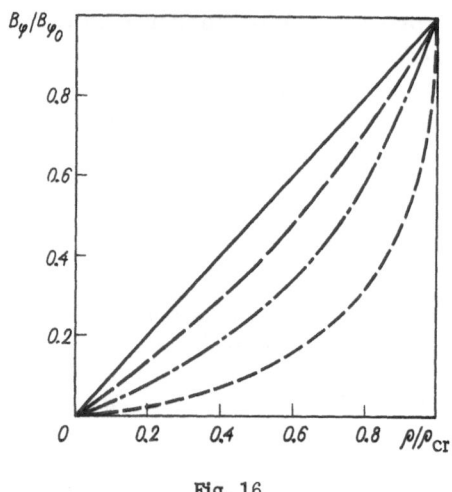

Fig. 16

the velocity reaches the velocity of the fast magnetosonic wave; thus this channel has the shape of an ordinary nozzle.

The ratio $B_\varphi/B_{\varphi 0}$ is plotted as a function of ρ/ρ_{cr} on the basis of the first equation in (3.16) in Fig. 16.† The figure shows that in the limit $\rho_{cr} \to \infty$, i.e., $H_\parallel \to 0$, the usual freezing condition $B_\varphi/\rho = B_{\varphi 0}/\rho_0$ holds. As ρ_{cr} decreases, the "sag" of the curves becomes progressively more pronounced, so that at $\rho_{cr} = \rho_0$ the entire change of B_φ occurs at a constant density (acceleration by Alfvén perturbations).

Assuming the plasma to be cold, we now examine the change in the longitudinal velocity of the accelerated plasma. When $W \to 0$, Eq. (3.17) yields

$$v_\parallel^2 = c_{\varphi_0}^2 G = c_{\varphi_0}^2 \frac{\rho_{cr}}{\rho_0} \left[1 - \left(\frac{1 - \rho_0/\rho_{cr}}{1 - \rho/\rho_{cr}} \right)^2 \right]. \qquad (3.24)$$

Here, G is a monotonically increasing function of ρ/ρ_{cr} which assumes the following particular values:

$$G = \begin{cases} 0 & \text{for} \quad \rho = \rho_0; \\ 2 - \rho_0/\rho_{cr} & \text{for} \quad \rho = 0; \\ 1 & \text{for} \quad \rho_0 = \rho_{cr}; \\ 2(1 - \rho/\rho_0) & \text{for} \quad \rho_{cr} \to \infty. \end{cases} \qquad (3.25)$$

---

† In Figs. 16 and 17, the parameter of the curves is the quantity $\varkappa = \rho_0/\rho_{ci}$. Specifically, these curves correspond (from top to bottom) to the values $\varkappa = 0$; 0.3, 0.6, and 0.9.

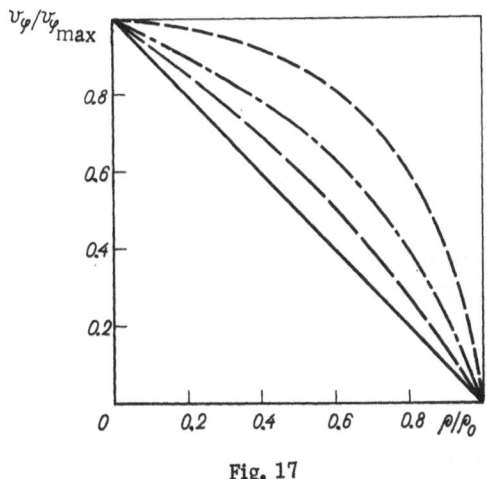

Fig. 17

Evidently the maximum velocity is reached in the limit $\rho \to 0$:

$$(v_\parallel)_{max} = c_{\varphi_0} \sqrt{2 - \rho_0/\rho_{cr}}. \tag{3.26}$$

The velocity at the exit from the accelerating system decreases because part of the energy supplied to the system is expended in imparting an azimuthal rotation to the stream [see (3.16)].

The azimuthal velocity increases monotonically along the channel; at the exit it tends toward the following value (with $\rho = 0$):

$$(v_\varphi)_{max} = -c_{\varphi_0} \sqrt{\rho_0/\rho_{cr}}. \tag{3.27}$$

The ratio $v_\varphi /(v_\varphi)_{max}$ is shown as a function of ρ/ρ_0 in Fig. 17.

Knowing $v_\parallel = v_\parallel(\rho)$ we can construct the function $f(\rho)$ defined by

$$f = \frac{\alpha_1}{\rho v_\parallel} = \frac{\alpha_1}{\rho c_{\varphi_0} \sqrt{G}}. \tag{3.28}$$

The quantity f reaches a minimum at $\rho^*/\rho_{cr} = 1 - (1 - \rho_0/\rho_{cr})^{2/3}$; the velocity at this critical cross section is

$$v_\parallel^* = c_{\varphi_0} \sqrt{G^*} = c_{\varphi_0} \sqrt{\frac{\rho_{cr}}{\rho_0}} \sqrt{1 - \left(1 - \frac{\rho_0}{\rho_{cr}}\right)^{2/3}} \tag{3.29}$$

Using these results we can find a convenient equation for the mass flow rate (per second) $\dot{m} = 2\pi r \alpha_i$:

$$\dot{m} = 2\pi r f^* \rho_0 \, c_{\varphi_0} \left[\frac{1 - (1 - \rho_0/\rho_{cr})^{2/3}}{\rho_0/\rho_{cr}} \right]^{3/2}. \qquad (3.30)$$

If $\rho_0/\rho_{cr} \to 0$, (3.29) and (3.30) become the corresponding equations derived in the preceding chapter.

Finally, we can write an equation for the tangent of the angle θ between the velocity and the axis of the system:

$$\tan \theta = \frac{v_{\varphi}}{v_{\parallel}} = - \sqrt{\frac{\rho_0}{\rho_{cr}} \frac{1 - \rho/\rho_0}{(1 - \rho/\rho_{cr})\sqrt{G}}}. \qquad (3.31)$$

At the accelerator exit

$$\tan \theta_{max} = - \sqrt{\frac{\rho_0/\rho_{cr}}{2 - \rho_0/\rho_{cr}}}.$$

c. Let us now consider the transcritical regime, with $\rho/\rho_{cr} > 1$, $v_{\parallel} < c_{\parallel}$. We note at the outset that arguments analogous to those in the preceding subsection (b), based on Eqs. (3.22) and (3.23), now lead to the conclusion that the accelerator again is shaped like an ordinary nozzle; however, at the critical cross section the slow magnetosonic velocity is reached rather than the fast magnetosonic velocity.

We first analyze the flow of a cold plasma. It follows from Eq. (3.24) that ρ must increase if the plasma is to be accelerated. Then the velocity $v_{\parallel}$ also increases monotonically; in the limit $(\rho/\rho_{cr}) \to \infty$ this velocity approaches the limit

$$(v_{\parallel})_{max} = c_{\varphi_0} \sqrt{\frac{\rho_{cr}}{\rho_0}}. \qquad (3.32)$$

This result shows that as the ratio ρ_0/ρ_{cr} decreases the velocity $v_{\parallel}$ increases for a given value of c_{φ_0} at the exit. From Eqs. (3.16) we can find B_{φ} and v_{φ} in the limit $\rho \to \infty$:

$$B_{\varphi} \to B_{\varphi_0} \frac{\rho_0/\rho_{cr} - 1}{\rho_0/\rho_{cr}}; \quad v_{\varphi} \to - c_{\varphi_0} \sqrt{\frac{\rho_{cr}}{\rho_0}}. \qquad (3.33)$$

In the flow under consideration here the value of B_φ at the exit does not tend toward zero, so that the velocities are comparatively low. At the same time, the channel cross section f obviously tends toward zero monotonically because of the increase in the density and velocity of the stream.

The situation becomes more complicated when the plasma temperature is taken into account. It is of fundamental importance to take this temperature into account in the transcritical regime since the plasma density increases during the acceleration and the enthalpy becomes important. Taking account of the enthalpy on the right side of the Bernoulli equation, (3.17), we now introduce two terms, $2\{W(\rho_0) - W(\rho)\}$ and $c_{\varphi_0}^2 G(\rho)$. The first falls off with increasing density ρ, while the second increases. Taking the derivative of the velocity with respect to the density, we find

$$v_\parallel \frac{dv_\parallel}{d\rho} = -\frac{c_T^2}{\rho} - \frac{c_\varphi^2}{\rho_0} \cdot \frac{(1-\rho_0/\rho_{cr})^2}{(1-\rho/\rho_{cr})^3}. \tag{3.34}$$

In the transcritical regime ($\rho > \rho_{cr}$) the energy which accelerates the plasma comes from thermal or magnetic energy, depending on which of the terms on the right side of this equation is larger. If

$$c_T^2 > \frac{c_\varphi^2}{\rho_0/\rho_{cr} - 1}, \tag{3.35}$$

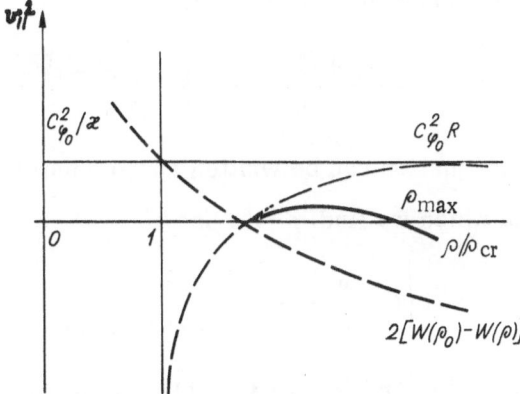

Fig. 18

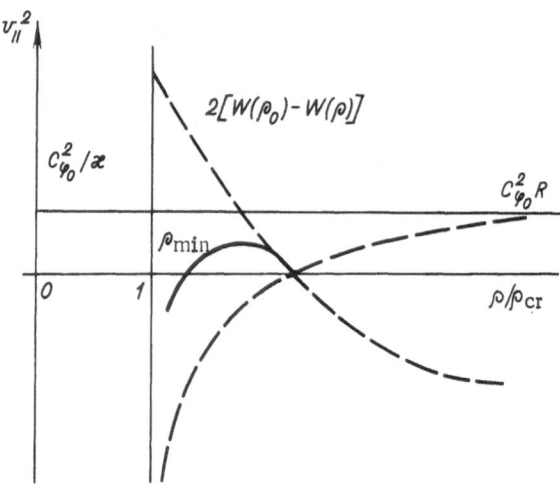

Fig. 19

the acceleration results from the expenditure of thermal energy with decreasing ρ (Fig. 18); if this inequality is reversed, the acceleration occurs with increasing ρ (Fig. 19) and is due to the magnetic energy. Obviously, the density can neither reach ρ_{cr}, nor become infinite, since the velocity becomes fixed at a certain density ρ, for which the right side of Eq. (3.35) vanishes.

As noted above, in the transcritical flow the channel can only have a single constriction which corresponds to the slow magnetosonic wave. The signal velocity is reached at the density $\rho *$, which satisfies the condition $v_{\parallel}^2 = c_{s-}^2$ or

$$W(\rho_0) - W(\rho) + c_{\varphi_0}^2 G(\rho) = \frac{c_T^2 + c_{\parallel}^2 + c_{\varphi}^2}{2} - \sqrt{\left(\frac{c_T^2 + c_{\parallel}^2 + c_{\varphi}^2}{2}\right)^2 - c_{\parallel}^2 c_T^2}.$$

If $v_{\parallel}^2 = c_{s-}^2$, this equation can be written approximately as $c_{\varphi_0}^2 G(\rho) = \left(\frac{c_{\parallel}^2 c_T^2}{c_{\parallel}^2 + c_{\varphi}^2}\right)_0$, from which we find $\rho *$ to be

$$\rho* = \rho_0 \left\{ 1 + \frac{\rho_0/\rho_{cr} - 1}{3} \left[\frac{c_{\parallel}^2 c_T^2}{c_{\varphi}^2 (c_{\parallel}^2 + c_{\varphi}^2)} \right]_0 \right\}. \tag{3.36}$$

3. Flow with Infinitesimally Low Initial Velocity with r $\neq$ const. We now treat the case in which the average channel radius is not constant; however, it is also as-

sumed that $v_\varphi \to 0$, $v_\parallel \to 0$ at the entrance to the system. If ρ_0 is replaced by ρ_{eff} in accordance with

$$\rho_{eff} = \rho_0 \, (r_0/r)^2, \tag{3.37}$$

the equations for flow in a narrow channel (3.10) can be written in a form analogous to (3.15) for flow in a channel of constant average radius $r = r_0$:

$$\rho v_\parallel \, rf = \alpha; \quad \sqrt{\rho} \, c_\parallel \, rf = \beta; \tag{3.38a}$$

$$c_\varphi = c_{\varphi_0} \sqrt{\frac{\rho}{\rho_{eff}} \cdot \frac{1 - \rho_{eff} / \rho_{cr}}{1 - \rho / \rho_{cr}}};$$

$$v_\varphi = - c_{\varphi_0} \sqrt{\frac{\rho_{eff}}{\rho_{cr}} \cdot \frac{1 - \rho / \rho_{eff}}{1 - \rho / \rho_{cr}}}; \tag{3.38b}$$

$$v_\parallel^2 = 2 \{ W (\rho_0) - W (\rho) \} + c_{\varphi_0}^2 \, R, \tag{3.38c}$$

where

$$R = \frac{\rho_{cr}}{\rho_{eff}} \left[1 - \left(\frac{1 - \rho_{eff} / \rho_{cr}}{1 - \rho / \rho_{cr}} \right)^2 \right],$$

and $\rho_{cr} = \rho v_\parallel^2 / c_\parallel^2 = \alpha^2 / \beta^2$. The equations for c_φ, v_φ, and $v_\parallel$ differ from the corresponding equations in the preceding case in that the initial density ρ_0 is replaced by the effective density ρ_{eff}. The classification of flows as subcritical, critical, and transcritical obviously remains valid, as does the conclusion that there is only a single crossing of the signal velocity in the subcritical and transcritical regimes. In this case, however, a new variable parameter exists, ρ_{eff}, so that flows with new properties can be obtained.

Figure 20 shows the regions in which $R(\xi, \varkappa)$ is positive and negative; here $\xi = \rho / \rho_{cr}$ and $\varkappa = \rho_{eff} / \rho_{cr}$. Also shown in this figure are lines of $R = const$.

a. If the flow is subcritical, the corresponding point on this diagram, emerging from the bisector $\varkappa - \xi = 0$, can move in an arbitrary manner in region I (Fig. 20). As this point moves toward the ordinate the flow velocity increases, and the flow acquires its maximum velocity $(v_\parallel)_{max} = \sqrt{2} \sqrt{W (\rho_0) + c_{\varphi_0}^2}$ when the operating point arrives at the origin, or ordinate.

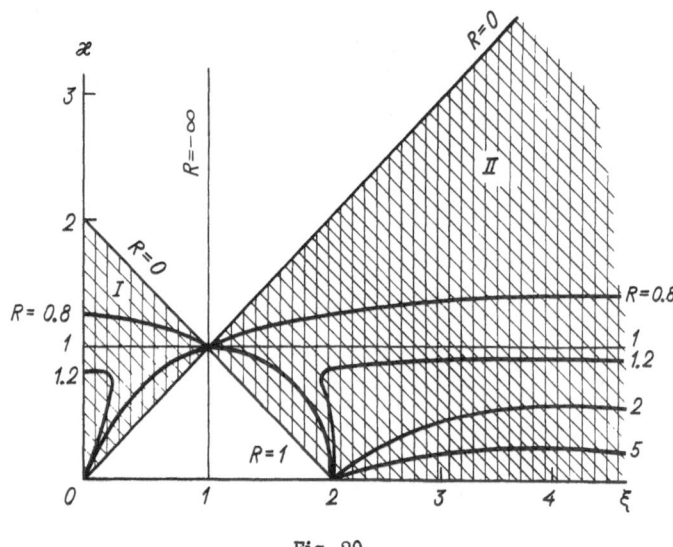

Fig. 20

b. In critical flow the density ρ should remain constant. However, the system (3.10c), (3.10d) is consistent only if $r = r_0 =$ const, which corresponds to the flow considered above.

c. We now consider the features of transcritical flow with variable $r(z)$, assuming that the thermal processes are relatively unimportant. In this case the operating point can lie in region II (Fig. 20). After emerging from the points on the bisector $\varkappa - \xi = 0$ with $\varkappa > 1$, the operating point should move toward the ξ axis if the velocity is to increase, while the density remains fixed. When the line $\varkappa = 1$ is crossed the field component B_φ vanishes, and the velocity components become equal: $v_\parallel = v_\varphi = c_{\varphi 0}$. In moving to smaller values of $\varkappa$, we find $B_\varphi \to -\infty$, and the velocity components v_φ and $v_\parallel$ increase without bound. This result is opposite to that found in the subcritical case.

Physically, increasing $v_\parallel$ at a negative value of B_φ decreases $\varkappa$ with increasing r; because of the centrifugal force the longitudinal acceleration of the plasma is stronger than the electromagnetic deceleration. The current distribution is shown schematically in Fig. 21.

4. General Flows in Narrow Channels. Finally, we consider briefly general MHD flows in narrow channels. If the

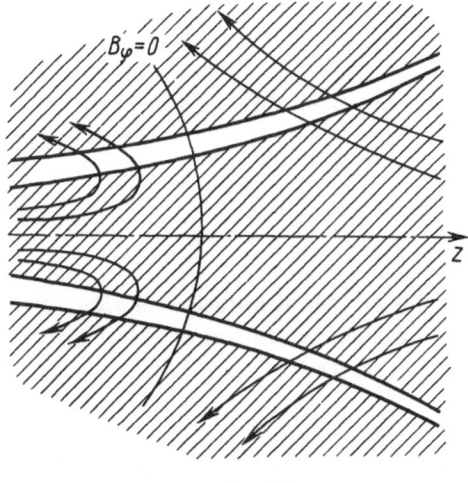

Fig. 21

average channel radius, $r(z)$, and the distance between the electrodes $f(z)$ are given, all quantities in Eq. (3.10) can be expressed in terms of ρ:

$$c_\varphi = \frac{\sqrt{\rho}}{r} \cdot \frac{\beta a + \alpha b r^2}{\alpha^2 - \beta^2 \rho}; \quad v_\varphi = \frac{1}{r} \cdot \frac{\alpha a + \beta b r^2 \rho}{\alpha^2 - \beta^2 \rho}; \left. \begin{array}{c} \\ \\ \\ \end{array} \right\} \quad (3.39)$$

$$c_\| = \frac{\beta}{\sqrt{\rho}\, rf}; \quad v_\| = \frac{\alpha}{\rho rf}. \cdot$$

The density ρ is determined from

$$W(\rho) + \frac{\alpha^2}{2\rho^2 f^2 r^2} + \frac{\alpha^3 a^2 + \rho b r^2 (2\alpha^2 - \beta_\rho^2)(2\beta a + \alpha b r^2)}{2(\alpha^2 - \beta^2 \rho)^2 \alpha r^2} = U. \quad (3.40)$$

In flow of a c 1 plasma with $W \to 0$ Eq. (3.40) is an algebraic equation of food degree in ρ. Specifying the velocity $v_\|(z)$ and the radius $r(z)$, we find a quadratic equation for f, whose solution, for $a = 0$, is

$$f_\pm = \frac{1}{\alpha r v_\|} \left\{ \beta^2 + \frac{b^2 r^2}{2U - v_\|^2} \pm \sqrt{ \left(\beta^2 + \frac{b^2 r^2}{2U - v_\|^2} \right) \frac{b^2 r^2}{2U - v_\|^2} } \right\}. \quad (3.41)$$

If the longitudinal field is weak, in the limit $\beta^2 \to 0$ the quantities

f_+ and f_- are

$$f_+ \approx \frac{2b^2 r}{\alpha v_{\parallel} (2U - v_{\parallel}^2)}; \quad f_- \approx \frac{\beta^2}{2\alpha v_{\parallel} r}. \tag{3.42}$$

§3. Flow of Cold Plasma in Channel of Slowly Varying Cross Section

Let us examine the simplest solution of Eqs. (2.26) and (2.29), which corresponds to a slow variation of the properties along the axis of the system. We write Eq. (2.26) as

$$\frac{1}{\rho r} \left(\frac{\partial}{\partial r} \cdot \frac{s}{r} \cdot \frac{\partial \xi}{\partial r} + \frac{\partial}{\partial z} \cdot \frac{s}{r} \cdot \frac{\partial \xi}{\partial z} \right) - \frac{\partial s}{\partial \xi} \cdot \frac{(\nabla \xi)^2}{2\rho r^2} + \frac{1}{2\rho r^2} \cdot \frac{\partial}{\partial \xi} \cdot \frac{A^2}{s} +$$

$$+ \frac{r^2}{2} \cdot \frac{\partial}{\partial \xi} \cdot \frac{B^2}{s} + \frac{\partial}{\partial \xi} \cdot \frac{AB\psi'}{s\psi_0} + TS' - U' = 0. \tag{3.43}$$

When all the integrals ψ_0', ψ', A, B, S, and U are constant, this equation becomes

$$\frac{\partial}{\partial r} \cdot \frac{s}{r} \cdot \frac{\partial \xi}{\partial r} + \frac{\partial}{\partial z} \cdot \frac{s}{r} \cdot \frac{\partial \xi}{\partial z} = 0. \tag{3.44}$$

If the adiabatic approximation is used and terms of second order of $\partial^2 \xi / \partial z^2$ and $(\partial \xi / \partial z)^2$ are dropped, we find the integral

$$\frac{s}{r} \cdot \frac{\partial \xi}{\partial r} = F(z), \tag{3.45}$$

where $F(z)$ is a slowly varying but otherwise arbitrary function of z.

Restricting the discussion to the case $W = 0$, we find the following result from the Bernoulli equation (2.26):

$$(\rho s)^2 = \psi_0'^2 \frac{F^2 \psi'^2 + \left(rB\psi_0' + \frac{1}{r} A\psi' \right)^2}{2U\psi'^2 + r^2 B^2}. \tag{3.46}$$

Substituting this equation into (3.45), we find the stream function

$$\xi = \frac{F}{\psi'^2} \left\{ \psi_0' \int_R^r r^2 \, dr \sqrt{\frac{2U\psi'^2 + r^2 B^2}{r^2 F^2 \psi'^2 + (B\psi_0' r^3 + A\psi')^2} - \frac{r^2 - R^2}{2}} \right\}. \tag{3.47}$$

Here $r = R(z)$ is the equation of the streamline of the fluid, which is assumed to be given. The integral in (3.47) is expressed in terms of elliptic functions, but in the case of interest here, $A = 0$, the stream function ξ can be expressed in terms of elementary functions. Setting $\xi = \psi_0$, we find

$$\xi = \frac{F}{2\psi'^2} \left\{ \sqrt{(r^2 + \gamma_+)^2 - \gamma_-^2} - \sqrt{(R^2 + \gamma_+)^2 - \gamma_-^2} - \right.$$

$$\left. - (r^2 - R^2) + \gamma_- \ln \frac{r^2 + \gamma_+ + \sqrt{(r^2 + \gamma_+)^2 - \gamma_-^2}}{R^2 + \gamma_+ + \sqrt{(R^2 + \gamma_+)^2 - \gamma_-^2}} \right\}, \qquad (3.48)$$

where

$$\gamma_\pm = \left(U \pm \frac{1}{2} F^2 \right) \frac{\psi'^2}{B^2}. \qquad (3.49)$$

The longitudinal velocity is

$$v_z = F \sqrt{\frac{r^2 B^2 + 2U\psi'^2}{r^2 B^2 + F^2 \psi'^2}}. \qquad (3.50)$$

If it is assumed that the velocity $v_z(z)$ is given at $r = R(z)$, then

$$F^2 = \frac{v_z^2}{1 + \dfrac{1}{r^2 B^2} (2U - v_z^2) \psi'^2}. \qquad (3.51)$$

Using $F(z)$ and $v_z(z)$, we can also write equations for all the other characteristics of the flow:

$$\rho = \frac{1}{\psi'^2} \left(1 - \frac{F}{v_z} \right); \quad H_z = \frac{F}{\psi'} \left(\frac{v_z}{F} - 1 \right);$$

$$v_\varphi = \frac{rB}{\psi'} \left(\frac{v_z}{F} - 1 \right); \quad H_\varphi = \frac{rB}{\psi'^2} \left(\frac{v_z}{F} - 1 \right). \qquad (3.52)$$

Here $\psi' = d\psi/d\psi_0$ is the ratio of the longitudinal fluxes of the magnetic field and the fluid. If ψ' is small we can neglect quadratic terms in this quantity, so that

$$F \approx v_z; \quad \xi \approx \frac{F}{B^2} \left(U - \frac{F^2}{2} \right) \ln \frac{r}{R};$$

$$\rho \approx \frac{U - \dfrac{1}{2} F^2}{r^2 B^2}; \quad H_\varphi \approx \frac{U - \dfrac{1}{2} F^2}{rB};$$

$$v_\varphi \approx \frac{U - \dfrac{1}{2} F^2}{rB} \, \psi'; \quad H_z \approx \frac{U - \dfrac{1}{2} F^2}{r^2 B^2} \, F\psi'. \qquad (3.53)$$

It follows that to first order in ψ', the electrode geometry is the same as in the case $\psi' = 0$. The quantities ρ and H_φ are also the same, while v_φ and H_z are proportional to ψ'; for the flow considered here they vanish in the limit $v \to v_{max} = \sqrt{2U}$.

Chapter 4

AXISYMMETRIC HALL FLOW OF AN IDEAL PLASMA

§1. Conservation Laws and Integral Parameters

Under laboratory conditions, the replacement factor is not always negligibly small.[†] Indeed, a high replacement factor is needed to achieve high velocities or high compression. In these cases the single-fluid model developed above is generally not applicable, and a theory for the flow of a two-component plasma is needed.

1. Conservation Laws. If the flow is stationary the most general system of equations of ideal two-fluid hydrodynamics, (1.19), can be written

$$\operatorname{div} n_i \mathbf{v}_i = 0 \ (a); \quad \operatorname{div} n_e \mathbf{v}_e = 0 \ (b); \tag{4.1}$$

$$M \left(\mathbf{v}_i \, \nabla \right) \mathbf{v}_i = -\frac{\nabla p_i}{n_i} + e \left(\mathbf{E} + \frac{1}{c} \, [\mathbf{v}_i \, \mathbf{H}] \right); \tag{4.2}$$

$$m \left(\mathbf{v}_e \, \nabla \right) \mathbf{v}_e = -\frac{\nabla p_e}{n_e} - e \left(\mathbf{E} + \frac{1}{c} \, [\mathbf{v}_e \, \mathbf{H}] \right); \tag{4.3}$$

$$p_i = p_i \left(n_i, \, S_i \right) \ (a); \quad \left(\mathbf{v}_i \, \nabla \right) S_i = 0 \ (b); \tag{4.4}$$

$$p_e = p_e \left(n_e, \, S_e \right) \ (a); \quad \left(\mathbf{v}_e \, \nabla \right) S_e = 0 \ (b); \tag{4.5}$$

$$\operatorname{curl} \mathbf{H} = \frac{4\pi}{c} \, e \left(n_i \, \mathbf{v}_i - n_e \, \mathbf{v}_e \right) \ (a); \quad \operatorname{div} \mathbf{H} = 0 \ (b); \tag{4.6}$$

$$\mathbf{E} = - \nabla \Phi \ (a); \quad \operatorname{div} \mathbf{E} = 4\pi e \left(n_i - n_e \right) \ (b). \tag{4.7}$$

If the flow is axisymmetric system (4.1)-(4.7) can be simplified considerably.

[†]Appropriate experiments are analyzed in detail in [15].

For this purpose we introduce the three stream functions ψ_i, ψ_e, ψ, which are defined by

$$r\,(nv_r)_{i,\,e} = -\frac{\partial\psi_{i,\,e}}{\partial z}; \quad r\,(nv_z)_{i,\,e} = \frac{\partial\psi_{i,\,e}}{\partial r}\,;$$
$$rH_r = -\frac{\partial\psi}{\partial z}; \quad rH_z = \frac{\partial\psi}{\partial r}, \tag{4.8}$$

so that the continuity equations (4.1) and (4.6b) are satisfied identically.

Introducing the Bernoulli functions

$$\frac{v_i^2}{2}+W_i+\frac{e\Phi}{M} = U_i; \quad \frac{v_e^2}{2}+W_e-\frac{e\Phi}{m} = U_e;$$
$$W_{i,\,e} = \int \frac{dp_{i,\,e}}{m_{i,\,e}\,n_{i,\,e}} \tag{4.9}$$

and the effective magnetic fields

$$\mathbf{H}_i^* = \mathbf{H}+\frac{Mc}{e}\,\mathrm{rot}\,\mathbf{v}_i; \quad \mathbf{H}_e^* = \mathbf{H}-\frac{mc}{e}\,\mathrm{rot}\,\mathbf{v}_e, \tag{4.10}$$

and assuming an isentropic system, we can rewrite equations of motion (4.2) and (4.3) as

$$\nabla U_i = \frac{e}{Mc}\,[\mathbf{v}_i\,\mathbf{H}_i^*]; \quad \nabla U_e = -\frac{e}{mc}\,[\mathbf{v}_e\,\mathbf{H}_e^*]. \tag{4.11}$$

Multiplying the equations in (4.11) by $\mathbf{v}_i$ and $\mathbf{v}_e$, respectively, and using (4.8), we find

$$U_i = U_i\,(\psi_i); \quad U_e = U_e\,(\psi_e). \tag{4.12}$$

This result means that the total energy of the electrons and the ions U is conserved along the trajectories of these particles.

Noting that

$$rH_z^{*\,i,\,e} = \frac{\partial D_{i,e}}{\partial r}; \quad rH_r^{*\,i,\,e} = -\frac{\partial D_{i,\,e}}{\partial z}, \tag{4.13}$$

where

$$D_i = \psi+\frac{Mc}{e}\,rv_\varphi^i; \quad D_e = \psi-\frac{mc}{e}\,rv_\varphi^e, \tag{4.14}$$

and substituting these expressions into the φ components of (4.11), we find conservation laws for the angular momenta:

$$D_i = D\,(\psi_i);\quad D_e = D_e\,(\psi_e). \tag{4.15}$$

From the r components[†] of Eqs. (4.11) and conservation laws (4.12) we find equations for the functions ψ_i and ψ_e:

$$\left.\begin{aligned}
\frac{M_i c}{e} U_i' &= \frac{v_\varphi^i D_i'}{r} - \frac{1}{rn_i} H_\varphi^{*i}; \\
-\frac{mc}{e} U_e' &= \frac{v_\varphi^e D_e'}{r} - \frac{1}{rn_e} H_\varphi^{*e};
\end{aligned}\right\} \tag{4.16}$$

$$U_{i,\,e}' \equiv \frac{dU}{d\psi_{i,e}};\quad D_{i,\,e}' \equiv \frac{dD}{d\psi_{i,\,e}}. \tag{4.17}$$

In particular, when $D = 0$, $m = 0$, $n_i = n_e$ the flow of the ion fluid is irrotational if $M_i U_i' = -U_e' m_e$. Finally, using Maxwell's equations (4.6) and (4.7), we find

$$I = 2\pi e\,(\psi_i - \psi_e);\quad I = \frac{cr}{2}\,H_\varphi; \tag{4.18a}$$

$$\Delta^* \psi = -\frac{4\pi e}{c}\,r\,(n_i v_\varphi^i - n_e v_\varphi^e); \tag{4.18b}$$

$$\Delta\Phi = -4\pi e\,(n_i - n_e); \tag{4.18c}$$

$$\Delta^* \equiv r\,\frac{\partial}{\partial r}\cdot\frac{1}{r}\cdot\frac{\partial}{\partial r} + \frac{\partial^2}{\partial z^2}.$$

For greater clarity, we rewrite the mechanical equations, eliminating the quantities $\text{curl}_\varphi\,\mathbf{v}_{i,e}$ from (4.16) and (4.17) with the help of (4.8). As a result, we find

$$\left.\begin{aligned}
\frac{v_i^2}{2} + W_i + \frac{e}{M}\Phi &= U_i\,(\psi_i);\quad \frac{v_e^2}{2} + W_e - \frac{e}{m}\Phi = U_e\,(\psi_e); \\
\psi + \frac{Mc}{e} rv_\varphi^i &= D_i\,(\psi_i);\quad \psi - \frac{mc}{e} rv_\varphi^e = D_e\,(\psi_e); \\
\frac{c}{e}\left(\frac{1}{rn_i}\cdot\frac{\partial}{\partial r}\cdot\frac{1}{rn_i}\cdot\frac{\partial\psi_i}{\partial r}\right.&\left.+\frac{1}{rn_i}\cdot\frac{\partial}{\partial z}\cdot\frac{1}{rn_i}\cdot\frac{\partial\psi_i}{\partial z}\right) - \\
-\frac{H_\varphi}{Mrn_i} + \frac{v_\varphi^i D_i'}{Mr} &= \frac{c}{e} U_i'\,(\varphi_i); \\
-\frac{c}{e}\left(\frac{1}{rn_e}\cdot\frac{\partial}{\partial r}\cdot\frac{1}{rn_e}\cdot\frac{\partial\psi_e}{\partial r}\right.&\left.+\frac{1}{rn_e}\cdot\frac{\partial}{\partial z}\cdot\frac{1}{rn_e}\cdot\frac{\partial\psi_e}{\partial z}\right) - \\
-\frac{H_\varphi}{mrn_e} + \frac{v_\varphi^e D_e'}{mr} &= -\frac{c}{e} U_e'\,(\psi_e).
\end{aligned}\right\} \tag{4.19}$$

[†] Or from the z components of these equations.

These equations for the six quantities ψ_i, ψ_e, n_i, n_e, v_φ^i, v_φ^e, contain the four functions $U_i(\psi_i)$, $U_e(\psi_e)$, $D_i(\psi_i)$, $D_e(\psi_e)$, which can be specified arbitrarily.

2. **Integrated Accelerator Parameters.** Knowing the conservation laws, we can find parameters which describe the operation of the accelerator. Let us examine the most important of these parameters.

1. The ion velocity at the exit [i.e., $I = 2\pi e(\psi_i - \psi_e) \to 0$] is governed by the Bernoulli integral in (4.9), and when $W_{i,e} \to 0$ we have

$$v_{\max} = \sqrt{2\left[U_i(\psi_i) + U_e(\psi_i)\frac{m}{M}\right]}.$$ (4.20)

If $m \to 0$, $I \to 0$, and $W_{i,e} \to 0$, the potential at the exit point is $\Phi_{out}(\psi_i) = -m/eU(\psi_i)$.

2. The mass flow rate is

$$\dot{m} = 2\pi M \psi_{\max}.$$ (4.21)

We assume that ψ_i lies in the range $0 < \psi_i < \psi_{i\,\max}$.

3. The momentum carried away per unit time by the stream leaving the accelerator is obviously

$$P = \dot{m}(v_z)_{\max} = \int_{r_{\min}}^{r_{\max}} v_{\max z}(r)\,2\pi r\,M n v_{\max z}\,dr\,|_{z\to\infty} =$$
$$= 2\pi \int_{\psi_{\min}}^{\psi_{\max}} M v_{\max z}(\psi_i)\,d\psi_i.$$ (4.22a)

If $v_{\max z} \approx v_{\max} = \sqrt{2\left(U_i + U_e\frac{m}{M}\right)}$, at the exit, then

$$F = 2\sqrt{2}\,\pi M \int_{\psi_{\min}}^{\psi_{\max}} \sqrt{U_i(\psi_i) + U_e(\psi_i)\frac{m}{M}}\,d\psi_i.$$ (4.22b)

4. The power in the accelerated stream is

$$N = \int \frac{v_{\max}^2}{2}\,d\dot{m} = 2\pi M \int_{\psi_{i\,\min}}^{\psi_{i\,\max}} \left[U_i(\psi_i) + U_e(\psi_i)\frac{m}{M}\right]d\psi_i.$$ (4.23)

5. The azimuthal angular momentum carried by the stream is

$$P_\varphi = \int r v_\varphi \, d\dot{m} = 2\pi \frac{e}{c} \int_{\psi_i \, min}^{\psi_i \, max} [D_i(\psi_i) - D_e(\psi_i)] \, d\psi_i. \qquad (4.24)$$

6. The voltage between the electrodes, $\delta\Phi$, is generally not determined since the potential of a point on an electrode is generally a variable (see § 2 below).

In a cold plasma, with $m \to 0$, the voltage between the electrodes is

$$\delta\Phi = -\frac{M}{e} \, \delta\mathring{U}_e(\psi_e), \qquad\qquad \mathring{U}_e \equiv \lim_{m \to 0} \frac{m}{M} \, U_e \qquad (4.25)$$

where the difference is evaluated for $\psi_{e \, min}$ and $\psi_{e \, max}$ corresponding to the given value of z.

7. The current through the system is

$$\mathcal{I} = 2\pi e \, | \, \psi_{i \, max} - \psi_{e \, min}|_{z \, = \, 0}. \qquad (4.26)$$

We assume that all the current enters the system (the system consisting of the plasma and the electrodes) at z = 0 and that there are no additional current inlets, i.e., that $I_{max}(z)$ is a monotonic function of z.

§ 2. Qualitative Analysis of Equations (4.18) and (4.19)

1. Cold Plasma. We assume that the plasma is cold $(W_i = W_e = 0)$, that the electron inertia is negligible (m = 0), and that the plasma is neutral. Then the equations of motion (4.19) become [10]

$$
\left.
\begin{aligned}
&\frac{v_i^2}{2} + \frac{e}{M} \Phi = U_i(\psi_i) \text{ (a)}; \qquad -\frac{e}{M} \Phi = \mathring{U}_e(\psi_e) \text{ (b)}; \\[4pt]
&\psi + \frac{Mc}{e} r v_\varphi^i = D_i(\psi_i) \text{ (c)}; \qquad \psi = D_e(\psi_e) \text{ (d)}; \\[4pt]
&\frac{Mc}{enr^2}\left(r \frac{\partial}{\partial r} \cdot \frac{1}{rn} \cdot \frac{\partial \psi_i}{\partial r} + \frac{\partial}{\partial z} \cdot \frac{1}{n} \cdot \frac{\partial \psi_i}{\partial z}\right) = \frac{cM}{e} U_i' + \frac{H_\varphi}{rn} - \frac{v_\varphi^i D_i}{r} \text{ (e)}; \\[4pt]
&0 = -\frac{cM}{e} \mathring{U}_e'(\psi_e) + \frac{H_\varphi}{rn} - \frac{v_\varphi^e D_e'}{r} \text{ (f)}.
\end{aligned}
\right\} \qquad (4.27)
$$

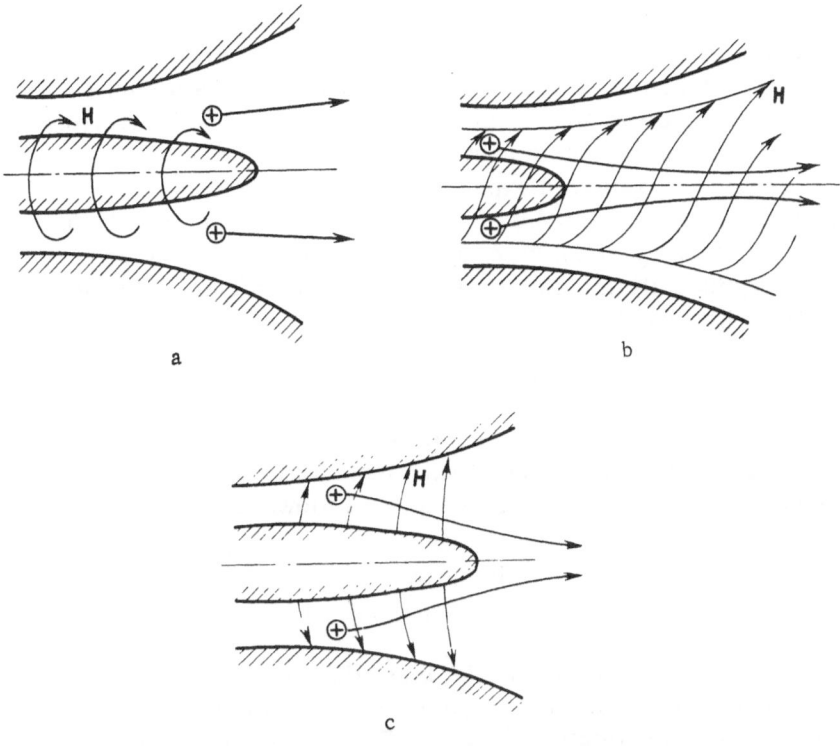

Fig. 22

Equations (4.27b) and (4.27d) show that the magnetic lines of force are equipotentials and that the electrons are "tied" to the lines of force.

We should distinguish among axisymmetric systems in which a) the magnetic field only has an azimuthal (φ) component and the plasma does not rotate ($D = 0$, $\psi = 0$) (Fig. 22a); b) the field has all components, but the lines of force do not intersect the channel walls (Fig. 22b); and c) the lines of force intersect the channel walls (Fig. 22c).

Let us consider the features of each case.

a. Equation (4.27b) shows that the electrons move along equipotentials. This simple fact has an extremely important consequence: Plasma acceleration is very difficult in a system with solid, coaxial, conducting electrodes in the absence of instabilities and dissipative processes.

Since the electrode surfaces are equipotentials, no regular motion of electrons from the cathode into the plasma or from the plasma to the anode occurs. In this situation, certain singularities must always appear in the current distribution at both electrodes, one of which represents a source of electrons and the other a sink. This feature is observed experimentally [15]. Accordingly, if electrons are to be supplied from the entire cathode, and if they are to be drawn uniformly to the entire anode, the electrodes cannot be equipotentials. For example, segmented electrodes or low-conductivity electrodes can be used.

In the case under consideration here, with $H_r = H_z = 0$, we cannot immediately specify the potential distribution in the system, since the potential of each annular line of force is not directly related to the boundary conditions. Instead, it is governed by the discharge as a whole. In this case, as follows from (4.27):

$$\left.\begin{aligned} \Phi &= \Phi\,(\psi_e); \\ H_\varphi/rn &= -c\,d\Phi/d\psi_e. \end{aligned}\right\} \tag{4.28}$$

This equation shows that the ratio H_φ/rn is constant along an electron trajectory in the axisymmetric case without a longitudinal field. This is the familiar "freezing" integral. However, it is very important to note that the magnetic field is frozen in the electron fluid.

This feature (see below) makes it possible to extract the ions from the magnetic field without the electrons if the replacement factor ξ is sufficiently large.

b. The case shown by Fig. 22b is described by the complete system of equations in (4.27). In this case there are magnetic surfaces ψ = const which are parallel to the "lateral" walls of the accelerator channel. These surfaces intersect the entrance surface of the accelerator; hence in principle the potential distribution is governed by the distribution at the entrance.[†] Since the magnetic surfaces are equipotentials ion acceleration can only occur in this case if the ion trajectories intersect magnetic surfaces. These configurations can be of interest in an "end-fire" system

---

[†]The shape of the surfaces ψ = const is governed by the currents in the plasma.

with a strongly diverging longitudinal field in which most of the acceleration occurs beyond the electrodes. In this case the magnetic axis of the system must be at the lowest potential in the system, and the geometric dimensions must be such that the ion gyroradius calculated from the total particle energy is larger than the radius of the system:

$$R_{ig} > a. \tag{4.29}$$

In this case the accelerated ions can leave the accelerator freely, while the electrons (in the absence of dissipation) are rigidly tied to the associated magnetic surfaces. If the space charge of the ions leaving the accelerator is to be neutralized, there must be an electron source near the exit (a neutralizer). Otherwise, the excitation development of intense oscillatory and dissipative processes can be expected.

c. The third type of accelerator (Fig. 22c) is also described by the complete system in (4.28). In this case the potential distribution in the plasma volume is fixed by the potential distribution at the tangential walls and, to some extent, by the distribution at the entrance.

All the discussion for the preceding case can be extended to this case, and condition (4.29) can again be derived. Systems of this type have been studied in detail in [9].

2. Characteristic Diagrams for an Accelerator with an Azimuthal Magnetic Field. The characteristics of a coaxial accelerator can be shown with the help of certain diagrams [10]. We represent the accelerating channel as a rectangle[†] and draw lines of $\psi_i = $ const, $\psi_e = $ const, I = const. This diagram is the "characteristic diagram" of the accelerator (Fig. 23).

First assume that the electrodes are segmented. In this case there is no difficulty with the distribution of the current over the electrodes, and the lines I = const can be drawn as a system of equidistant vertical lines. The line I_0 corresponds to the total current through the accelerator, while I = 0 corresponds to the exit from the accelerator. If we assume that the ions are not lost

---

†For definiteness we treat the flow between electrodes.

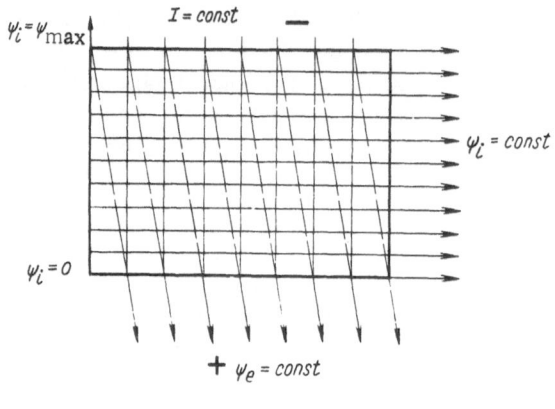

Fig. 23

at the electrodes, then the lines ψ_i can be drawn as horizontal lines; the lines $\psi_i = 0$ and $\psi_i = \psi_{i\,max}$ coincide with the anode and cathode, respectively.

In terms of the coordinates we have chosen, the electron trajectories $\psi_e = \psi_i - 1/2\pi e$ are sloping lines. If electron inertia is neglected the "thermalized" potential $\Phi^* = \Phi - mW_e/e$ and the "freezing" parameter $H_\varphi/rn = -c\Phi^{*\prime}(\psi_e)$ are conserved along lines $\psi_e = \text{const}$, while U_i is conserved along a line $\psi_i = \text{const}$.

A distinction should obviously be made between two types of characteristic diagrams, which will be called "hydrodynamic" and "Hall" characteristic diagrams (Figs. 24a and 24b, respectively). In the former case (a), most of the electrons emerging from the accelerator have entered it along with the ions. In the latter case (b), the electrons entering the accelerator are lost at the anode near the channel entrance, and the electrons which neutralize the space charge beyond this point come from the cathode. In the former case the freezing parameter does not vary along the accelerator; in the second it can vary.

The transition from one regime to the other is evidently governed by the quantity $J_0/2\pi e\psi_{i\,max}$ (Fig. 23), i.e., the replacement factor $\xi = (I_0/2\pi e\psi)_{max} = (I_d/I_m^*)$.

What is the potential distribution in the system? The function $\Phi^*(\psi_e)$ cannot be specified at the outset since the function $\Phi^*(\psi_e)$ must be monotonic and the presence of maxima and minima leads

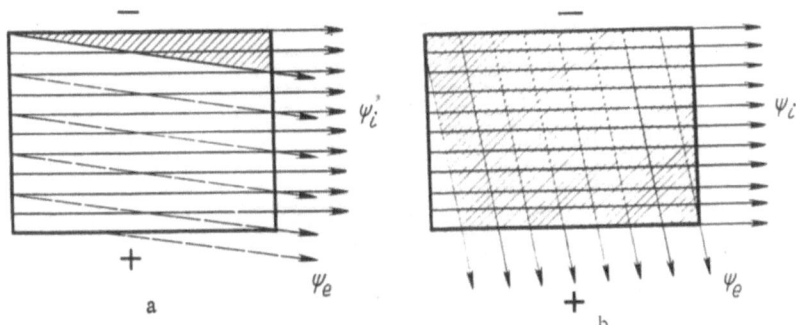

Fig. 24

to irregularities in the flow. Let us assume that the function $\Phi^*(\psi_e)$ is as shown in Fig. 25. Then at ψ_{e1} and ψ_{e2} the freezing parameter Φ^* vanishes and it becomes negative between these points. Since H_φ does not vary significantly over this interval (I is the independent variable), the implication is that the density becomes infinite and even changes sign. Again, this behavior is evidence of the appearance of certain discontinuities, which must change the electron enthalpy in such a way that the potential Φ^* becomes a monotonic function of ψ_e.

The characteristic diagrams described above correspond to systems in which the current is carried by the electrons only. If the electrodes are solid (i.e., equipotentials), however, cases can occur in which current is carried by ions (Fig. 26). In this case the electron trajectories ψ_e are horizontal lines, while the ion trajectories ψ_i are sloping lines. In order to achieve this case

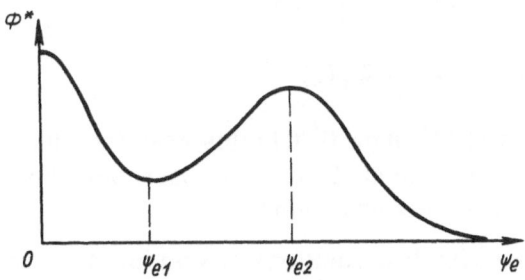

Fig. 25

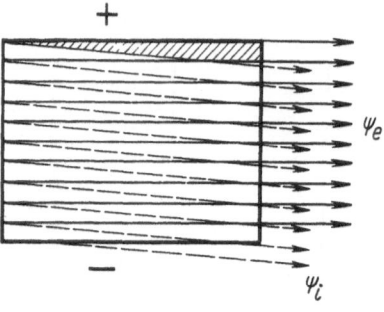

Fig. 26

it is necessary to supply the medium through the anode. Taking account of the loss of ions at the cathode, however, we conclude that current transport by ions is only useful if $\xi \ll 1$.

3. Thermal Acceleration. We turn now to the other extreme cases: Assume a plasma with a high temperature T_e and assume that the acceleration results from a change in the electron enthalpy W_e [8]. As before, the ion enthalpy is assumed to be zero. We also neglect the electron inertia. In this case system (4.29) becomes

$$
\left.
\begin{aligned}
& \frac{Mv_i^2}{2} + e\frac{\Phi}{M} = U_i(\psi_i) \ \text{(a)}; \quad \mathring{W}_e - e\frac{\Phi}{M} = \mathring{U}_e(\psi_e) \ \text{(b)}; \\[4pt]
& \psi + \frac{Mc}{e}\,rv_\varphi^i = D_i(\psi) \ \text{(c)}; \quad \psi = D_e(\psi_e) \ \text{(d)}; \\[4pt]
& \frac{Mc}{e}\left(\frac{1}{rn}\cdot\frac{\partial}{\partial r}\cdot\frac{1}{rn}\cdot\frac{\partial \psi_i}{\partial r} + \frac{1}{rn}\cdot\frac{\partial \psi_i}{\partial z}\right) = \frac{cM}{e}\,U_i'(\psi_i) + \\[4pt]
& \qquad\qquad + \frac{H_\varphi}{rn} - \frac{v_\varphi^i}{r}\,D_i' \ \text{(e)}; \\[4pt]
& \frac{H_\varphi}{rn} - \frac{v_\varphi^e}{r}\,D_e' = \frac{cM}{e}\,\mathring{U}_e'(\psi_e) \ \text{(f)}; \\[4pt]
& \mathring{W}_e = \lim_{m\to 0}\frac{m}{M}\,W_e \ \text{(g)}.
\end{aligned}
\right\}
\qquad (4.30)
$$

Equations (4.30) show that in this case the electrons are again "tied" to the magnetic surfaces $\psi = $ const; now, however, the magnetic surfaces are not equipotentials.

One of the simplest solutions of system (4.30) is that of a flow with $\psi_i = \psi_e$, i.e., one in which the ion and electron fluxes along the system are equal and the flow occurs along magnetic surfaces

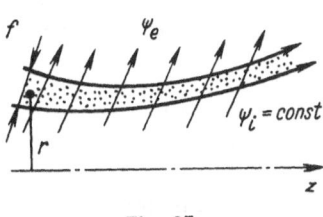

Fig. 27

ψ = const. If $D_i = D_e$, then

$$v_\varphi^i = 0; \quad H_\varphi = 0. \tag{4.31}$$

This is the regime of "pure thermal" acceleration. Substituting (4.31) into (4.30f), we find the electron revolution velocity to be

$$v_\varphi^e = -\frac{c}{e} \, r \, \frac{dU_e}{dD_e} \, m. \tag{4.32}$$

In general, the ions do not move along magnetic surfaces; they revolve in the azimuthal direction at a velocity governed by (4.30c).

Because of the high electron thermal conductivity along the magnetic lines of force, it is assumed that the electron temperature is constant over most of the acceleration region: T_e = const.

The analysis of plasma loss from the magnetic field in such a system is the same as in the cases considered above with $\psi \neq 0$.

§3. Flow of a Two-Component Plasma in Narrow Tube with $H_\parallel = 0$

We assume an azimuthal magnetic field in the system and that the plasma does not rotate. The flow between the electrodes is partitioned into a system of narrow concentric tubes whose surfaces are formed by the ion trajectories, i.e., the lines ψ_i = const (Fig. 27). Equations (4.18), (4.19) show that the flow in one of these tubes is described by the following set of algebraic equations:

$$nvrf = \text{const}; \tag{4.33a}$$

$$H/rn = -c \, d\Phi^*/d\psi_e; \tag{4.33b}$$

$$\Phi^* = \Phi^*(\psi_e); \quad \Phi^* = \Phi - W_e \frac{m}{\rho}; \tag{4.33c, d}$$

$$\frac{v^2}{2} + W_i(\rho) + \frac{e}{M}\,\Phi = U = \text{const;} \qquad (4.33e)$$

$$rH = \frac{4\pi e}{c}(\psi_i - \psi_e) + (rH)^0 \approx -\frac{4\pi e}{c}\,\psi_e + (rH)^0. \qquad (4.33f)$$

In this last equation we assume $\psi_i = 0$ since the tube is narrow; $(rH)^0$ is a possible constant component.

If the plasma flow in a narrow channel is always isomagnetic in the single-fluid model, the flow in a narrow tube generally becomes nonisomagnetic when the Hall effect is taken into account, i.e., when the nonvanishing value of the replacement factor is taken into account. In other words, the freezing parameter can vary substantially along z.

Isomagnetic flows are also possible in the two-fluid case; the properties turn out to be identical to those of isomagnetic flows in the single-fluid model.† The complete formal equivalence of the isomagnetic flows in the two cases is illustrated clearly as follows: From the isomagnetic conditions $(d\Phi*/d\psi_e) = k = \text{const}$, and using (4.33b) and (4.33e), we find

$$\Phi* = k\psi_e = -\frac{kc}{4\pi e}\,Hr = \frac{H^2}{4\pi ne}, \qquad (4.34)$$

but since $\Phi = \Phi* + W_e m/e$ the Bernoulli equation, (4.33d), takes the usual "single-fluid" form [see Eq. (2.120)]:

$$\frac{v^2}{2} + W + \frac{H^2}{4\pi nM} = U = \text{const;} \quad W = W_i + W_e\frac{m}{M}.$$

Accordingly, the results found in Chap. 2 regarding isomagnetic flow remain valid. In particular, the conclusion that accelerated and compressional flows occur still holds.

For an ion to move along the average surface of the tube $r = r(z)$, the sum of the forces acting on it in the direction perpendicular to its velocity must vanish. Accordingly, the following equilibrium condition must hold in a cold plasma ($W_i = W_e = 0$) (Fig. 28):

$$\frac{Mv_i^2}{\rho} - \frac{e}{c}\,v_i H_\varphi + eE_\perp = 0. \qquad (4.35)$$

---

†We do not treat the problem of matching the flow to the electrodes; this problem is different in the single-fluid and two-fluid models.

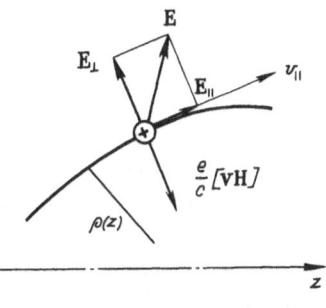

Fig. 28

Here ρ is the radius of curvature of the line $r(z)$, and v_i is the total ion velocity. On the other hand, with $m = 0$, the electron equation of motion, (4.3), leads to

$$-\frac{1}{c} v_e H_\varphi + E_\perp = 0. \tag{4.36}$$

Evidently the main trajectory is a straight line ($R = \infty$), the longitudinal velocities of the electrons and ions are the same, and the longitudinal current in the tube vanishes. Otherwise, there is a longitudinal current which balances the centrifugal force.

In conclusion we derive the Hugoniot equation for system (4.33). We write the variations of Eqs. (4.33), assuming $W_e = 0$:

$$\left. \begin{array}{l} \dfrac{dn}{n} + \dfrac{dv}{v} + \dfrac{dr}{r} + \dfrac{df}{f} = 0; \quad \dfrac{dH}{H} - \dfrac{dr}{r} - \dfrac{dn}{n} - \dfrac{\Phi''}{\Phi'} d\psi_e = 0; \\[3mm] v\,dv + c_{Ti}^2 \dfrac{dn}{n} + \dfrac{e}{M} \Phi'\, d\psi_e = 0; \quad r\,dH + H\,dr + \dfrac{4\pi e}{c} d\psi_e = 0. \end{array} \right\} \tag{4.37}$$

Eliminating dn, dH, and $d\psi_e$, we find the desired equation [10]:

$$(v^2 - c_s^2) \frac{dv}{v} = c_s^2 \frac{df}{f} + (c_{Ti}^2 - c_s^2) \frac{dr}{r}, \quad c_s^2 = c_{Ti}^2 + \frac{c_A^2}{1 - (\psi_e \Phi'')/\Phi'} . \tag{4.38}$$

When the Hall effect is taken into account in flow in a narrow tube the critical velocity is a function of the second derivative of the potential Φ''. In isomagnetic flow ($\Phi'' = 0$), Eq. (4.38) is the same as the corresponding equation for the single-fluid model (2.57).

§ 4. Flow in Smooth Channel with $H_{\parallel} = 0$

If the flow properties vary quite rapidly along r but slowly along z, we can neglect terms in Eqs. (4.18) and (4.19) which are proportional to $(\partial/\partial z)^2$. Then we find a system of ordinary differential equations, which can be solved numerically. In the present section, restricting the discussion to the flow of a nonrotating, quasineutral plasma with $H_{\parallel} = 0$ and $m_e = 0$, we treat a certain class of flows for which the calculations can be reduced to quadratures. Neglecting terms in (4.18), (4.19) which are proportional to $(\partial/\partial z)^2$, and taking account of the restrictions listed here, we find the basic system of equations to be

$$\frac{1}{2n^2 r^2}\left(\frac{\partial \psi_i}{\partial r}\right)^2 + W_i(n) + \frac{e\Phi}{M} = U_i(\psi_i); \qquad (4.39a)$$

$$W_e(n) - \frac{e\Phi}{m} = U_e(\psi_e); \qquad (4.39b)$$

$$\frac{1}{rn}\cdot\frac{\partial}{\partial r}\left(\frac{1}{rn}\cdot\frac{\partial \psi_i}{\partial r}\right) = U_i' + \frac{eH}{Mcnr}; \qquad (4.39c)$$

$$0 = -\frac{c}{e}U_e' + \frac{H}{mnr}; \qquad (4.39d)$$

$$rH = \frac{4\pi e}{c}(\psi_i - \psi_e) + (rH)_0. \qquad (4.39e)$$

Combining the first four equations in (4.39) in pairs,

$$\frac{1}{2n^2 r^2}\left(\frac{\partial \psi_i}{\partial r}\right)^2 + W_i(n) = U_i(\psi_i) + U_e(\psi_e)\frac{m}{M}; \qquad (4.40a)$$

$$\frac{1}{nr}\cdot\frac{\partial}{\partial r}\left(\frac{1}{nr}\cdot\frac{\partial \psi_i}{\partial r}\right) = U_i'(\psi_i) + U_e'(\psi_e)\frac{m}{M}; \qquad (4.40b)$$

$$W = W_i(n) + W_e(n)\frac{m}{M}, \qquad (4.40c)$$

differentiating the first equation with respect to r, multiplying the second by $\partial \psi_i/\partial r$, and then subtracting the second from the first, we find

$$\frac{\partial W}{\partial r} = U_e'(\psi_e)\frac{\partial}{\partial r}(\psi_e - \psi_i)\frac{m}{M}. \qquad (4.41)$$

Equation (4.41) replaces the second equation in (4.40), and the system obtained from (4.40a) and (4.41) in this way turns out to be more convenient, since it only contains first-order equations. Equation (4.41) is the equation for radial equilibrium

$$\frac{\partial}{\partial r}\left(p + \frac{H^2}{8\pi}\right) = -\frac{H^2}{4\pi r},$$ (4.42)

as follows directly from (4.39d) and (4.39e) and the definition of the enthalpy. In the two-dimensional problem Eq. (4.41) can be integrated in the general form

$$p + (H^2/8\pi) = P(z).$$ (4.43)

In the axisymmetric case this equation can be integrated in two cases:

$$W = 0;$$ (4.44)

$$U'_e(\psi_e) = \text{const.}$$ (4.45)

Accordingly, by analogy with single-fluid flow, system (4.39) can be reduced to quadratures if a) the flow is two-dimensional, b) the plasma is cold, and c) the flow is isomagnetic. In Appendices 1 and 2, the corresponding problems are treated in an arbitrary curvilinear coordinate system.

Appendix 1

TWO-PARAMETER STEADY-STATE HYDROMAGNETIC FLOW OF IDEALLY CONDUCTING MEDIUM IN ARBITRARY CURVILINEAR COORDINATE SYSTEM

The study of MHD flow can be simplified substantially if the flow is independent of one of the spatial coordinates. Such flows are called "symmetric flows." The simplest symmetry is translational symmetry, in which the physical properties are constant along the z axis; in axisymmetric symmetry the system is invariant with respect to rotation of the coordinate system around the z axis. Finally, we have helical symmetry, in which all physical properties are invariant with respect to displacement along the z axis and simultaneous rotation around this axis. In the first case the flow is independent of z, in the second it is independent of the

azimuthal angle φ, and in the third it is independent of the curvilinear coordinate reckoned along the helices $\theta = \varphi - \alpha z = $ const
with a constant pitch $L = 2\pi/\alpha$. The equations for symmetric
steady-state flows in a cylindrical coordinate system are derived in [26]. In the derivation of the corresponding equations for
an arbitrary curvilinear coordinate system, certain geometric
properties of the potentials are introduced and the integrals of
motion obtained are found. It is then possible to write the equations for symmetric flows in a general form which holds regardless of the coordinate system.†

Using the thermodynamic relations

$$dW = (dp/\rho) + TdS, \qquad (A1.1)$$

where ρ is the density, p is the pressure, T is the temperature, S
is the entropy, and W is the enthalpy, we can write the system of
MHD equations for steady-state, dissipationless flows in the ra-
tionalized system of units in the following form:

$$\mathrm{div}\,\mathbf{H} = 0; \qquad (A1.2)$$

$$\mathrm{div}\,\rho\mathbf{v} = 0; \qquad (A1.3)$$

$$[\mathbf{vH}] = -\nabla\Phi; \qquad (A1.4)$$

$$\mathbf{v}\,\nabla\,S = 0; \qquad (A1.5)$$

$$[\mathbf{j_0\,v}] - \frac{1}{\rho}\,[\mathbf{jH}] = -\nabla\,w + T\,\nabla\,S. \qquad (A1.6)$$

Here

$$\mathbf{j} = \mathrm{curl}\,\mathbf{H}; \qquad (A1.7)$$

$$\mathbf{j_0} = \mathrm{curl}\,\mathbf{v}; \qquad (A1.8)$$

$$w = W + (v^2/2) + G; \qquad (A1.9)$$

$\mathbf{v}$ is the velocity, $\mathbf{H}$ is the magnetic field, Φ is a function proportional to the electric potential, and G is the potential representing
forces of nonelectromagnetic origin.

We carry out the calculations in the curvilinear coordinate
system x^1, x^2, x^3, in which the square of the element of length is‡

$$dl^2 = g_{ik}\,dx^i\,dx^k, \qquad (A1.10)$$

---

† The mathematical tools developed below can evidently also be used for stationary
waves.
‡ A repeated index implies summation from 1 to 3.

where g_{ik} is the metric tensor. In the calculations we use the following equations from vector analysis:

$$
\left.
\begin{aligned}
(\nabla\varphi)_i &= \frac{\partial\varphi_i}{\partial x^i}; \quad \mathbf{ab} = a^i\, b_i; \\[2mm]
\operatorname{div}\mathbf{a} &= \frac{1}{\sqrt{g}}\cdot\frac{\partial}{\partial x^i}\,(\sqrt{g}a^i); \quad [\mathbf{ab}]_j = \varepsilon_{ijk}\sqrt{g}\, a^i\, b^k; \\[2mm]
(\operatorname{curl}\mathbf{a})^j &= \frac{\varepsilon^{ijk}}{\sqrt{g}}\cdot\frac{\partial a_k}{\partial x^i}; \quad a_i = g_{ik}\, a^k; \quad a^i = g^{ik}a_k.
\end{aligned}
\right\}
\tag{A1.11}
$$

Here the subscript represents the covariant component of the vector, the superscript represents the contravariant component, ε^{ijk} is the Levi–Civita density, and $g^{ik} = G^{ik}/g$, where G^{ik} are the minors of g_{ik} and g is the determinant of g_{ik}.

We assume that all physical properties as well as the components of the metric tensor g_{ik} are independent of the coordinates x^3. Continuity equations (A1.2) and (A1.3) make it possible to introduce vector potentials $\mathbf{H}$ and $\rho\mathbf{v}$:

$$
\mathbf{H} = \operatorname{curl}\mathbf{A}; \quad \rho\mathbf{v} = \operatorname{curl}\mathbf{A_0}.
\tag{A1.12}
$$

We choose $\mathbf{A}$ and $\mathbf{A_0}$ to be independent of x^3; then the contravariant components of $\mathbf{H}$ and $\rho\mathbf{v}$ with indices 1 and 2 can be written in terms of the covariant components A_3 and A_{03}. We introduce the notation

$$
\psi = A_3; \quad \psi_0 = A_{03}.
\tag{A1.13}
$$

Here

$$
\left.
\begin{aligned}
\sqrt{g}\,H^1 &= \frac{\partial\psi}{\partial x^2}; \quad \sqrt{g}\,H^2 = -\frac{\partial\psi}{\partial x^1}; \\[2mm]
\rho\sqrt{g}\,v^1 &= \frac{\partial\psi_0}{\partial x^2}; \quad \rho\sqrt{g}\,v^2 = -\frac{\partial\psi_0}{\partial x^1}.
\end{aligned}
\right\}
\tag{A1.14}
$$

From Eqs. (A1.7) and (A1.8) we find analogous equations for the contravariant components of the vectors $\mathbf{j}$ and $\mathbf{j_0}$:

$$
\left.
\begin{aligned}
\sqrt{g}\,j^1 &= \frac{\partial H_3}{\partial x^2}; \quad \sqrt{g}\,j^2 = -\frac{\partial H_3}{\partial x^1}; \\[2mm]
\sqrt{g}\,j_0^1 &= \frac{\partial v_3}{\partial x^2}; \quad \sqrt{g}\,j_0^2 = -\frac{\partial v_3}{\partial x^1}.
\end{aligned}
\right\}
\tag{A1.15}
$$

Below we will use the vector components a^1, a^2, a^3. It is evident that the other components of an arbitrary vector $\mathbf{a}$ can be expressed in terms of a^1, a^2, and a^3 by means of the equations

$$\left.\begin{array}{l} g_{33}\, a_1 = gg^{22}\, a^1 - gg^{12}\, a^2 + g_{13}\, a_3; \\ g_{33}\, a_2 = -gg^{12}\, a^1 + gg^{11}\, a^2 + g_{23}\, a_3; \\ g_{33}\, a^3 = -g_{31}\, a^1 - g_{32}\, a^2 + a_3. \end{array}\right\} \qquad (A1.16)$$

Equations (A1.2) and (A1.3) are satisfied when the functions ψ and ψ_0 are introduced. We now write Eq. (A1.4) in component form:

$$\left.\begin{array}{l} \dfrac{\partial \Phi}{\partial x^1} = \sqrt{g}\,(v^2\, H^3 - v^3\, H^2); \\[2mm] \dfrac{\partial \Phi}{\partial x^2} = \sqrt{g}\,(v^3\, H^1 - v^1\, H^3); \\[2mm] 0 = v^1\, H^2 - v^2\, H^1. \end{array}\right\} \qquad (A1.17)$$

Substituting (A1.14) into these equations, we find that the third equation in (A1.17) yields the Jacobian equality

$$\partial\,(\psi,\ \psi_0)/\partial\,(x^1,\ x^2) = 0. \qquad (A1.18)$$

It follows that $\psi = \psi(\psi_0)$. In the solution of different problems it becomes convenient to choose both ψ and ψ_0 as the basic function; hence, for the sake of symmetry we introduce a new unknown function ξ, which is functionally related to ψ and ψ_0, and we use the following assumption everywhere:

$$\psi = \psi_0\,(\xi); \quad \psi_0 = \psi_0\,(\xi), \qquad (A1.19)$$

where the function ξ can be chosen arbitrarily. In this case the first two equations in (A1.17) can be written

$$\frac{\partial \Phi}{\partial x^i} = \left(\frac{H^3}{\rho}\,\psi_0' - v^3\,\psi'\right)\frac{\partial \xi}{\partial x^i} \equiv B\frac{\partial \xi}{\partial x^i}. \qquad (A1.20)$$

Here the prime denotes the total derivative with respect to ξ. From Eq. (A1.20) we find the Jacobian equality $\partial\,(B,\ \xi)/\partial\,(x^1, x^2) = 0$ and thus $B = B(\xi)$, $\Phi = \Phi(\xi)$, with $B(\xi) = \Phi'(\xi)$. Writing H^3 and v^3 with the help of the covariant components, we find

$$B = \frac{1}{\rho}\, H^3\,\psi_0' - v^3\,\psi' = \left(\frac{1}{\rho}\, H_3\,\psi_0' - v_3\,\psi'\right)\frac{1}{g_{33}}. \qquad (A1.21)$$

It follows from (A1.5) that the entropy S is a function of ψ_0, so that

$$S = S (\xi). \tag{A1.22}$$

Writing Eq. (A1.6) in component form, we find

$$\left.\begin{array}{l} T\dfrac{\partial S}{\partial x^1} - \dfrac{\partial w}{\partial x^1} = \sqrt{g}\,(j_0^2\,v^3 - j_0^3\,v^2) - \dfrac{\sqrt{g}}{\rho}\,(j^2\,H^3 - j^3\,H^2); \\[2mm] T\dfrac{\partial S}{\partial x^2} - \dfrac{dw}{dx^2} = \sqrt{g}\,(-j_0^1\,v^3 + j_0^3\,v^1) - \dfrac{\sqrt{g}}{\rho}\,(-j^1\,H^3 + j^3\,H^1); \\[2mm] 0 = (j_0^1\,v^2 - j_0^2\,v^1) - \dfrac{1}{\rho}\,(j^1\,H^2 - j^2\,H^1). \end{array}\right\} \tag{A1.23}$$

The last of these equations leads to the equation $\partial(A,\ \xi)/\partial(x^1, x^2) = 0$, where

$$A = v_3\,\psi_0' - H_3\psi' \tag{A1.24}$$

and thus $A = A(\xi)$.

Accordingly, working from the equation stating that the magnetic lines of force are frozen-in, (A1.4), we find integrals (A1.19), which show that the fluid flows along the magnetic surfaces $\Phi =$ const; we also find the integrals $B = B(\xi)$ and $\Phi = \Phi(\xi)$, which mean that the magnetic surfaces are electric equipotentials. Finally, the integral $A = A(\xi)$ is an expression of the conservation of the momentum corresponding to the cyclic coordinate x^3.

We need another integral — the analog of the Bernoulli integral in ordinary hydrodynamics. It is also necessary to find the equation which the function ξ satisfies. To find this integral and this equation we rewrite the first two equations in (A1.23), introducing (A1.15) and (A1.22):

$$-\frac{\partial w}{\partial x^i} = -v^3\frac{\partial v_3}{\partial x^i} + \frac{H^3}{\rho}\cdot\frac{\partial H_3}{\partial x^i} + \left(\frac{j_0^3\,\psi_0' - j^3\,\psi'}{\rho} - TS'\right)\frac{\partial\xi}{\partial x^i}. \tag{A1.25}$$

Using (A1.21) and (A1.24), we can rewrite Eq. (A1.25) as

$$\frac{\partial}{\partial x^i}\left(\frac{Bv_3}{\psi'} + w\right) = \left\{v_3\left(\frac{B}{\psi'}\right)' + \frac{H^3}{\rho}\left(\frac{A}{\psi'}\right)' - \frac{v_3 H^3}{\rho}\left(\frac{\psi_0'}{\psi'}\right)' - \right.$$
$$\left. - \frac{j_0^3\,\psi_0' - j^3\,\psi'}{\rho} + TS'\right\}\frac{\partial\xi}{\partial x^i}. \tag{A1.26}$$

Proceeding in the same way, we see that the quantity in parentheses on the left is a function of ξ. We denote this function by
$U(\xi)$. The result is the two equations

$$w + \frac{Bv_3}{\psi'} = U(\xi); \tag{A1.27}$$

$$\frac{1}{\rho}(j_0^3 \psi_0' - j^3 \psi') - v_3 \left(\frac{B}{\psi'}\right)' - \frac{H^3}{\rho}\left(\frac{A}{\psi'}\right)' + \frac{v_3 H^3}{\rho}\left(\frac{\psi_0'}{\psi'}\right)' - TS' + U' = 0. \tag{A1.28}$$

The first of these equations is the required Bernoulli integral;
the second is the equation for ξ, but this latter equation must
still be converted to a form explicitly containing partial derivatives with respect to ξ.

Equation (A1.27) and (A1.28) contain the quantity ψ' in the
denominator. This feature represents an inconvenience in taking
the limit $\psi = 0$. However, using Eq. (A1.26) we can find equivalent
equations which contain ψ_0' rather than ψ' in the denominator:

$$w + \frac{BH_3}{\psi_0'} = U_1(\xi); \tag{A1.27'}$$

$$\frac{1}{\rho}(j_0^3 \psi_0' - j^3 \psi') - H_3\left(\frac{B}{\psi_0'}\right)' - v^3\left(\frac{A}{\psi_0'}\right) - v^3 H_3\left(\frac{\psi'}{\psi_0'}\right)' - TS' + U_1' = 0. \tag{A1.28'}$$

Let us rewrite Eqs. (A1.28) and (A1.28'). Equations (A1.4)–
(A1.6) lead to

$$j_0^3 \psi_0' - j^3 \psi' = -\frac{g_{31}}{\sqrt{g}\, g_{33}}\left(\frac{\partial v_3}{\partial x^2}\psi_0' - \frac{\partial H_3}{\partial x^2}\psi'\right) + \frac{g_{32}}{\sqrt{g}\, g_{33}} \times$$

$$\times \left(\frac{\partial v^3}{\partial x^1}\psi_0' - \frac{\partial H_3}{\partial x^1}\psi'\right) + \frac{j_{03}\psi_0' - j_3\psi'}{g_{33}}.$$

On the other hand, using (A1.24), we find

$$\left[-\left(\frac{A}{\psi'}\right)' + v_3\left(\frac{\psi_0'}{\psi'}\right)'\right]\frac{\partial \xi}{\partial x^i} = \frac{\partial H_3}{\partial x^i} - \frac{\psi_0'}{\psi'} \cdot \frac{\partial v_3}{\partial x^i}$$

or the following equivalent equation, which contains ψ_0' in the denominator:

$$\left[\left(\frac{A}{\psi_0'}\right)' + H_3\left(\frac{\psi'}{\psi_0'}\right)'\right]\frac{\partial \xi}{\partial x^i} = \frac{\partial v^3}{\partial x^i} - \frac{\psi'}{\psi_0'} \cdot \frac{\partial H_3}{\partial x^i}.$$

The intermediate equations are

$$v_3\left(\frac{B}{\psi'}\right)' + \frac{H_3}{\rho g_{33}}\left[-\left(\frac{A}{\psi'}\right)' + v_3\left(\frac{\psi_0'}{\psi'}\right)\right] + \frac{j_{03}\psi_0' - j_3\psi'}{\rho g_{33}} - TS' + U' = 0;$$

$$H_3\left(\frac{B}{\psi_0'}\right)' - \frac{v_3}{g_{33}}\left[\left(\frac{A}{\psi_0'}\right)' + H_3\left(\frac{\psi'}{\psi_0'}\right)\right] + \frac{j_{03}\psi_0' - j_3\psi'}{\rho g_{33}} - TS' + U_1' = 0.$$

To transform these equations we introduce the equation

$$j_{03} = -\frac{g_{33}}{\sqrt{g}}\cdot\frac{\partial}{\partial x^i}\left(\frac{\sqrt{g}\,g^{ik}}{\rho g_{33}}\cdot\frac{\partial\psi_0}{\partial x^k}\right) + \frac{g_{33}\,v_3}{\sqrt{g}}\left(\frac{\partial}{\partial x^1}\cdot\frac{g_{23}}{g_{33}} - \frac{\partial}{\partial x^2}\cdot\frac{g_{13}}{g_{33}}\right) \quad \text{(A1.29a)}$$

and the analogous equation for j_3, obtained from this equation through the substitutions $\psi_0 \to \psi'$, $\rho \to 1$, $v_3 \to H_3$:

$$\frac{j_{03}\psi_0' - j_3\psi'}{\rho g_{33}} = -\frac{1}{\rho\sqrt{g}}\left\{s\frac{\partial}{\partial x^i}\cdot\frac{\sqrt{g}\,g^{ik}}{g_{33}}\cdot\frac{\partial\xi}{\partial x^k} + \frac{\sqrt{g}\,g^{ik}}{g_{33}}\cdot\frac{\partial\xi}{\partial x^k}\times\right.$$

$$\left.\times\left(\psi_0'\frac{\partial}{\partial x^i}\cdot\frac{\psi_0'}{\rho} - \psi'\frac{\partial\psi'}{\partial x^i}\right)\right\} + \frac{A}{\rho\sqrt{g}}\left(\frac{\partial}{\partial x^1}\cdot\frac{g_{23}}{g_{33}} - \frac{\partial}{\partial x^2}\cdot\frac{g_{13}}{g_{33}}\right). \quad \text{(A1.29b)}$$

Here we have introduced the quantity

$$s = \frac{\psi_0'^2}{\rho} - \psi'^2, \quad \text{(A1.30)}$$

which appears in the solutions of Eqs. (A1.21) and (A1.24) for v_3 and H_3:

$$v_3 = \frac{1}{s}\left(A\frac{\psi_0'}{\rho} + g_{33}B\psi'\right); \quad H_3 = \frac{1}{s}(A\psi' + g_{33}B\psi_0'). \quad \text{(A1.31)}$$

Referring to (A1.31), where v_3 and H_3 depend on the three variables g_{33}, ρ, and ξ, and denoting the partial derivative with respect to ξ for fixed g_{33} and ρ by a prime, we find

$$v_3\left(\frac{v_3}{g_{33}} - \frac{H_3\psi_0'}{\rho g_{33}\psi'}\right)' + \frac{H_3}{\rho g_{33}}\left[H_3' - \left(\frac{v_3\psi_0'}{\psi'}\right)' + v_3\left(\frac{\psi_0'}{\psi'}\right)'\right] =$$

$$= \frac{\rho v_3\,v_3' + H_3 H_3'}{\rho g_{33}} - \frac{1}{\rho g_{33}}\left(v_3 H_3\frac{\psi_0'}{\psi'}\right)' = -\frac{1}{2\rho g_{33}}\cdot\frac{\partial}{\partial\xi}\cdot\frac{A^2}{s} -$$

$$- \frac{g_{33}}{2}\cdot\frac{\partial}{\partial\xi}\cdot\frac{B^2}{s} - \frac{\partial}{\partial\xi}\cdot\frac{AB\psi_0'}{\rho s\psi'}.$$

Similarly,

$$H_3\left(-\frac{H_3}{\rho g_{33}} + \frac{v_3\psi'}{g_{33}\psi_0'}\right)' + \frac{v_3}{g_{33}}\left[-v_3' + \left(\frac{H_3\psi'}{\psi_0'}\right)' - H_3\left(\frac{\psi'}{\psi_0'}\right)'\right] =$$

$$= -\frac{1}{2g_{33}} \cdot \frac{\partial}{\partial \xi}\left(v_3^2 + \frac{H_3^2}{\rho} - \frac{2v_3 H_3 \psi'}{\psi_0'}\right) = -\frac{1}{2\rho g_{33}} \cdot \frac{\partial}{\partial \xi} \cdot \frac{A^2}{s} -$$

$$- \frac{g_{33}}{2} \cdot \frac{\partial}{\partial \xi} \cdot \frac{B^2}{s} - \frac{\partial}{\partial \xi} \cdot \frac{AB\psi'}{s\psi_0'} ,$$

where $\partial/\partial\xi$ denotes the derivative with respect to ξ for fixed ρ.

Finally the complete system of equations that describe the two-parameter flow of an ideally conducting medium in a magnetic field is

$$\frac{1}{\rho\sqrt{g}} \cdot \frac{\partial}{\partial x^i} \cdot \frac{\sqrt{g}g^{ik}s}{g_{33}} \cdot \frac{\partial \xi}{\partial x^k} - \frac{|\nabla \xi|^2}{2\rho g_{33}} \cdot \frac{\partial s}{\partial \xi} - \left(\frac{\partial}{\partial x^1} \cdot \frac{g_{23}}{g_{33}} - \frac{\partial}{\partial x^2} \cdot \frac{g_{13}}{g_{33}}\right)\frac{A}{\rho\sqrt{g}} +$$

$$+ \frac{1}{2\rho g_{33}} \cdot \frac{\partial}{\partial \xi} \cdot \frac{A^2}{s} + \frac{g_{33}}{2} \cdot \frac{\partial}{\partial \xi} \cdot \frac{B^2}{s} + \frac{\partial}{\partial \xi} \cdot \frac{AB\psi_0'}{\rho s\psi'} + TS' - U' = 0;$$

$$\text{(A1.32)}$$

$$W + \frac{v^2}{2} + G + \frac{g_{33}}{s}\frac{B^2}{s} + \frac{AB\psi_0'}{\rho s\psi} = U; \qquad \text{(A1.33)}$$

$$\binom{\rho v^1}{H^1} = \frac{1}{\sqrt{g}}\binom{\psi_0'}{\psi'} \quad \binom{\rho v^2}{H^2} = -\frac{1}{\sqrt{g}}\binom{\psi_0'}{\psi'}\frac{\partial \xi}{\partial x^1}; \qquad \text{(A1.34)}$$

$$\binom{\rho v_3}{H_3} = \frac{A}{s}\binom{\psi_0'}{\psi'} + \frac{g_{33}B}{s}\binom{\rho\psi'}{\psi_0'}; \quad s = \frac{\psi'^2}{\rho} - \psi'^2. \qquad \text{(A1.35)}$$

Here, the primes denote the total derivative with respect to ξ; the partial derivative with respect to ξ is taken for a fixed density ρ; and the functions ψ_0, ψ, A, B, S, and U are arbitrary functions of the variables ξ. Equation (A1.32) and the Bernoulli equation (A1.33) are used to determine the functions ξ and ρ. Equations (A1.34) and (A1.35) can be used to find the components of the velocity and the magnetic field from the known values of ξ and ρ. For completeness (A1.32)–(A1.35) are supplemented by the equation of state.

To convert to an equivalent form containing ψ_0' instead of ψ' in the denominator, we use the following substitutions in Eqs. (A1.32) and (A1.33):

$$\frac{AB\psi_0'}{\rho s\psi'} \to \frac{AB\psi'}{s\psi_0'}; \quad U \to U_1 = U + \frac{AB}{\psi'\psi_0'}. \qquad \text{(A1.36)}$$

In orthogonal coordinates (A1.32)-(A1.35) become

$$\frac{1}{\rho h_1 h_2 h_3}\left(\frac{\partial}{\partial x_1}\cdot\frac{h_2\,s}{h_1 h_3}\cdot\frac{\partial\xi}{\partial x_1}+\frac{\partial}{\partial x_2}\cdot\frac{h_1\,s}{h_2 h_3}\cdot\frac{\partial\xi}{\partial x_2}\right)-\frac{|\nabla\xi|^2}{2\rho h_3^2}\cdot\frac{\partial s}{\partial\xi}+$$

$$+\frac{1}{2\rho h_3^2}\cdot\frac{\partial}{\partial\xi}\cdot\frac{A^2}{s}+\frac{h_3^2}{2}\cdot\frac{\partial}{\partial\xi}\cdot\frac{B^2}{s}+\frac{\partial}{\partial\xi}\cdot\frac{AB\psi_0'}{\rho s\psi'}+TS'-U'=0;$$

$$(A1.37)$$

$$W+\frac{v^2}{2}+G+\frac{h_3^2 B^2}{s}+\frac{AB\psi_0'}{\rho s\psi'}=U; \qquad (A1.38)$$

$$\left(\begin{array}{c}\rho v_{x_1}\\ H_{x_1}\end{array}\right)=\frac{1}{h_2 h_3}\left(\frac{\psi_0'}{\psi'}\right)\frac{\partial\xi}{\partial x_2}; \quad \left(\begin{array}{c}\rho v_{x_2}\\ H_{x_2}\end{array}\right)=-\frac{1}{h_1 h_3}\left(\frac{\psi_0'}{\psi'}\right)\frac{\partial\xi}{\partial x_1}; \quad (A1.39)$$

$$\left(\begin{array}{c}\rho v_{x_3}\\ H_{x_3}\end{array}\right)=\frac{A}{h_3 s}\left(\frac{\psi_0'}{\psi'}\right)+\frac{h_3 B}{s}\left(\begin{array}{c}\rho\psi'\\ \psi_0'\end{array}\right); \quad s=\frac{\psi_0'^2}{\rho}-\psi'^2. \quad (A1.40)$$

For the case of helical symmetry, in which all quantities are functions of $x^1 = r$, $x^2 = \theta = \varphi - \alpha z$ only, and the third coordinate $x^3 = z$ is cyclic, system (A1.32)-(A1.35) becomes

$$\frac{1}{\rho r}\left(\frac{\partial}{\partial r}\cdot\frac{rs}{\beta}\cdot\frac{\partial\xi}{\partial r}+\frac{\partial}{\partial\theta}\cdot\frac{s}{r}\cdot\frac{\partial\xi}{\partial\theta}\right)-\frac{|\nabla\xi|^2}{2\rho\beta}\cdot\frac{\partial s}{\partial\xi}-\frac{2\alpha A}{\beta^2\rho}+\frac{1}{2\rho\beta}\cdot\frac{\partial}{\partial\xi}\cdot\frac{A^2}{s}+$$

$$+\frac{\beta}{2}\cdot\frac{\partial}{\partial\xi}\cdot\frac{B^2}{s}+\frac{\partial}{\partial\xi}\cdot\frac{AB\psi_0'}{\rho s\psi'}+TS'-U'=0; \qquad (A1.41)$$

$$W+\frac{v^2}{2}+G+\frac{\beta B^2}{s}+\frac{AB\psi_0'}{\rho s\psi'}=U; \qquad (A1.42)$$

$$\left(\begin{array}{c}\rho v_r\\ H_r\end{array}\right)=\left(\frac{\psi_0'}{\psi'}\right)\frac{1}{r}\cdot\frac{\partial\xi}{\partial\theta}; \quad \alpha r\left(\begin{array}{c}\rho v_z\\ H_z\end{array}\right)-\left(\begin{array}{c}\rho v_\varphi\\ H_\varphi\end{array}\right)=\left(\frac{\psi_0'}{\psi'}\right)\frac{\partial\xi}{\partial r}; \quad (A1.43)$$

$$\left(\begin{array}{c}\rho v_z\\ H_z\end{array}\right)+\alpha r\left(\begin{array}{c}\rho v_\varphi\\ H_\varphi\end{array}\right)=\frac{A}{s}\left(\frac{\psi_0'}{\psi'}\right)+\frac{\beta B}{s}\left(\begin{array}{c}\rho\psi_0'\\ \psi'\end{array}\right); \quad \beta=1+\alpha^2 r^2. \quad (A1.44)$$

According to (A1.21) and (A1.24), the momentum and freezing integrals $A(\xi)$ and $B(\xi)$ are the following combinations of the components of the magnetic field, velocity, and density:

$$A\,(\xi) = I_0\,\psi_0' - I\psi'; \quad B\,(\xi) = I\,\frac{\psi_0'}{\rho} - I_0\,\psi', \qquad (A1.45)$$

where $I = H_z + \alpha r H_\varphi$ and $I_0 = v_z + \alpha r v_\varphi$. Equations (A1.41)-(A1.45)
are the same as the corresponding equations given in [26] in terms
of cylindrical coordinates. The cases of axial and translation symmetry are found from (A1.41)-(A1.45) by taking the limits $\alpha \to \infty$
and $\alpha \to 0$.

Appendix 2

STEADY-STATE SYMMETRIC FLOW IN TWO-FLUID MAGNETOHYDRODYNAMICS

§1.   Flow of the Electron and Ion Fluids

By analogy with the single-fluid case, we can exploit the steady-
state nature of the system and the symmetry to find several integrals of motion; these integrals are then used to simplify the system of equations for two-fluid magnetohydrodynamics.

Assuming that there are no dissipative processes, we write
the equations for the electron and ion fluids as

$$m_\alpha\,\frac{d_\alpha\,v_\alpha}{dt} = e_\alpha\,\mathbf{E} + \frac{e_\alpha}{c}\,[\mathbf{v}_\alpha\mathbf{B}] - \frac{\nabla p_\alpha}{n_\alpha}\,; \qquad (A2.1)$$

$$\frac{\partial n_\alpha}{\partial t} + \operatorname{div}\,n_\alpha\,\mathbf{v}_\alpha = 0; \qquad (A2.2)$$

$$\frac{d_\alpha\,S_\alpha}{dt} = 0; \qquad (A2.3)$$

$$\operatorname{curl}\,\mathbf{B} = \frac{4\pi}{c}\,(e_i\,n_i\,\mathbf{v}_i + e_e\,n_e\,\mathbf{v}_e); \quad \operatorname{curl}\,\mathbf{E} = -\frac{1}{c}\cdot\frac{\partial\mathbf{B}}{\partial t}\,; \qquad (A2.4)$$

$$\operatorname{div}\,\mathbf{B} = 0; \quad \operatorname{div}\,\mathbf{E} = 4\pi\,(e_i\,n_i + e_e\,n_e). \qquad (A2.5)$$

Here m_α, e_α, and n_α are the mass, charge, and density of the electrons or ions, respectively; $d_\alpha/dt = \partial/\partial t + \mathbf{v}_\alpha \nabla$ $(\alpha = i, e)$; $\mathbf{B}$ is the magnetic field; and the remaining notation is the same as above.

Replacing $\nabla p_\alpha/n_\alpha$ by its expression in terms of the enthalpy W_α and the entropy S_α

$$(\nabla p_\alpha/n_\alpha) = \nabla W_\alpha - T_\alpha \nabla S_\alpha, \tag{A2.6}$$

and writing $\mathbf{E}$ and $\mathbf{B}$ in terms of the scalar and vector potentials Φ and A,

$$\mathbf{E} = -\nabla \Phi - \frac{1}{c} \cdot \frac{\partial \mathbf{A}}{\partial t} \; ; \quad \mathbf{B} = \text{rot } \mathbf{A}, \tag{A2.7}$$

we can write the equations of motion (A2.1) for electrons and ions in the following form:

$$\frac{dP_\alpha}{dt} = \nabla \left(-e_\alpha \Phi + \frac{e_\alpha}{c} v_\alpha A - W_\alpha \right) + T_\alpha \nabla S_\alpha, \tag{A2.8}$$

where

$$\mathbf{P}_\alpha = m_\alpha \mathbf{v}_\alpha + \frac{e}{c} \mathbf{A} \tag{A2.9}$$

is the canonical momentum of the ion or electron, and the operator ∇ is used for fixed values of $\mathbf{v}_\alpha$. For calculations in an arbitrary curvilinear coordinate system x^i we can rewrite Eqs. (A2.8) in Lagrangian form:

$$\frac{d}{dt} \cdot \frac{\partial L_\alpha}{\partial v^k} - \frac{\partial L_\alpha}{\partial x^k} = T_\alpha \frac{\partial S_\alpha}{\partial x^k} \tag{A2.10}$$

with the Lagrangian† $L_\alpha(x^k, v^k)$ given by

$$L_\alpha = \frac{m_\alpha}{2} v_\alpha^k v_{\alpha k} - e_\alpha \Phi + \frac{e_\alpha}{c} v_\alpha^k A_k - W_\alpha. \tag{A2.11}$$

If the physical properties and the components of the metric tensor g_{ik} are independent of one of the coordinates x^v, the corresponding generalized momentum $P_{\alpha k} = \partial L_\alpha / \partial v_\alpha^k = m_\alpha v_{\alpha k} + e_\alpha/c A_k$ is considered. Accordingly, when there is no dependence

---

† A repeated Latin index implies summation from 1 to 3.

on x^3 we find the two integrals

$$m_\alpha v_{\alpha 3} + \frac{e}{c} A_3 = P_{\alpha 3}, \tag{A2.12}$$

which describe the conservation of the generalized momenta of the ions and electrons. This result indicates that these integrals hold for both steady-state and transient conditions, being a result of the spatial symmetry of the problem.

To find the energy integrals we assume that the process is steady state. Taking the scalar product of Eq. (A2.1) and $\mathbf{v}_\alpha$ and using Eqs. (A2.3), (A2.6), and (A2.7), we find two more integrals,

$$(1/2)\, m_\alpha v_\alpha^2 + e_\alpha \Phi + W_\alpha = U_\alpha, \tag{A2.13}$$

which express the conservation of the ion and electron energies. The existence of the integrals in (A2.13), in contrast with (A2.12), is not a consequence of the spatial symmetry, but is a consequence of the steady-state nature of the system, i.e., the temporal symmetry.

We note that the energy-momentum conservation law of a system of charged particles and an electromagnetic field also holds in the general case of relativistic particles and time-varying fields.

We restrict the discussion to symmetric, steady-state flows. The integrals $P_{\alpha 3}$ and U_α are conserved along the trajectories of the fluid particles of the ion or electron fluid so that these integrals are functions of the corresponding stream functions ψ_α, which are introduced through the equations

$$v_\alpha^1 = \frac{1}{\sqrt{g}\, n_\alpha} \cdot \frac{\partial \psi_\alpha}{\partial x^2}; \quad v_\alpha^2 = -\frac{1}{\sqrt{g}\, n_\alpha} \cdot \frac{\partial \psi_\alpha}{\partial x^1}. \tag{A2.14}$$

Writing v_α^1 and v_α^2 in terms of ψ_α, we satisfy the equations div $n_\alpha \mathbf{v}_\alpha = 0$. It is not difficult to see that $\psi_\alpha = $ const are the integrals of the equations $dx^1/v_\alpha^1 = dx^2/v_\alpha^2 = dx^3/v_\alpha^3$ for the streamlines of the corresponding fluids. Introducing the notation

$$\text{curl } v_\alpha = j_\alpha; \quad \text{curl } \mathbf{B} = j \tag{A2.15}$$

and using (A2.13) for U_α, we can convert equations of motion (A2.1) to

$$\left[\mathbf{v}_\alpha, \ m_\alpha \mathbf{j}_\alpha + \frac{e_\alpha}{c} \mathbf{B} \right] = \nabla U_\alpha - T_\alpha \nabla S_\alpha. \tag{A2.16}$$

From (A2.7), (A2.12), and (A2.15), we find

$$\begin{pmatrix} B^1 \\ j_\alpha^1 \end{pmatrix} = \frac{1}{\sqrt{g}} \cdot \frac{\partial}{\partial x^2} \begin{pmatrix} A_3 \\ v_3 \end{pmatrix}; \quad \begin{pmatrix} B^2 \\ j_\alpha^2 \end{pmatrix} = -\frac{1}{\sqrt{g}} \cdot \frac{\partial}{\partial x^1} \begin{pmatrix} A_3 \\ v_3 \end{pmatrix};$$

$$m_\alpha \begin{pmatrix} j_\alpha^1 \\ j_\alpha^2 \end{pmatrix} + \frac{e_\alpha}{c} \begin{pmatrix} B^1 \\ B^2 \end{pmatrix} = \frac{1}{\sqrt{g}} \begin{pmatrix} \dfrac{\partial P_{\alpha 3}}{\partial x^1} \\[2mm] -\dfrac{\partial P_{\alpha 3}}{\partial x^2} \end{pmatrix}. \tag{A2.17}$$

Substituting these equations into (A2.16), and using $P_{\alpha 3} = P_{\alpha 3}(\psi_\alpha)$, $U_\alpha = U_\alpha(\psi_\alpha)$, and $S_\alpha = S_\alpha(\psi_\alpha)$, we find

$$\frac{m_\alpha}{n_\alpha} j_\alpha^3 + \frac{e_\alpha}{n_\alpha c} B^3 - v_\alpha^3 P'_{\alpha 3} + U'_\alpha - T_\alpha S'_\alpha = 0, \tag{A2.18}$$

where the prime denotes the derivative with respect to the argument.

These equations are equations for ψ_α. We can convert them by using Eqs. (A1.16):

$$\frac{m_\alpha}{n_\alpha g_{33}} j_{\alpha 3} + \frac{e_\alpha}{n_\alpha g_{33} c} B_3 - \frac{v_{\alpha 3}}{g_{33}} P'_{\alpha 3} + U'_\alpha - T_\alpha S'_\alpha = 0. \tag{A2.19}$$

Here, $j_{\alpha 3}$ is expressed in terms of ψ_α and $v_{\alpha 3}$ by the operator [see Eq. (A1.29a)]:

$$j_{\alpha 3} = -\frac{g_{33}}{\sqrt{g}} \cdot \frac{\partial}{\partial x^i} \left(\frac{\sqrt{g} g^{ik}}{n_\alpha g_{33}} \cdot \frac{\partial \psi_\alpha}{\partial x^k} \right) + \frac{g_{33} v_{\alpha 3}}{\sqrt{g}} \left(\frac{\partial}{\partial x^1} \cdot \frac{g_{23}}{g_{33}} - \frac{\partial}{\partial x^i} \cdot \frac{g_{13}}{g_{33}} \right). \tag{A2.20}$$

Writing $A_3 = \psi$ we find an analogous expression for j_3 in terms of ψ and B_3:

$$j_3 = -\frac{g_{33}}{\sqrt{g}} \cdot \frac{\partial}{\partial x^i} \left(\frac{\sqrt{g} g^{ik}}{g_{33}} \cdot \frac{\partial \psi}{\partial x^k} \right) + \frac{g_{33} B_3}{\sqrt{g}} \left(\frac{\partial}{\partial x^1} \cdot \frac{g_{23}}{g_{33}} - \frac{\partial}{\partial x^i} \cdot \frac{g_{13}}{g_{33}} \right). \tag{A2.21}$$

From Eq. (A2.4) we can find two other equations, which relate B_3 and j_3 to ψ_α and $v_{\alpha 3}$:

$$B_3 = \frac{4\pi}{c} (e_i \psi_i + e_e \psi_e);$$

(A2.22)

$$j_3 = \frac{4\pi}{c} (e_i n_i v_{i3} + e_e n_e v_{e3}).$$

(A2.23)

Also assuming $e_i = -e_e = e$, where e is the electron charge, and using the neutrality condition $n_i e_i = n_e e$, we can write this system of equations as

$$
\left.
\begin{aligned}
&\frac{m_\alpha}{n g_{33}} j_{3\alpha} + \frac{e_\alpha B_3}{g_{33} n c} - \frac{v_{3\alpha} P'_{3\alpha}}{g_{33}} + U'_\alpha - T_\alpha S'_\alpha = 0; \\[4pt]
&\frac{1}{n} j_3 - \frac{4\pi e}{c} (v_{i3} - v_{e3}) = 0; \\[4pt]
&\frac{4\pi e}{c} (\psi_i - \psi_e) = B_3; \\[4pt]
&m_\alpha v_{\alpha 3} + \frac{e_\alpha}{c} \psi = P_{\alpha 3} (\psi_\alpha); \\[4pt]
&\frac{1}{2} m_\alpha v_\alpha^2 + W_\alpha + e_\alpha \Phi = U_\alpha (\psi_\alpha).
\end{aligned}
\right\}
$$

(A2.24)

When $n_i \neq n_e$, these equations are supplemented by the Poisson equation

$$\Delta \varphi = -4\pi e (n_i - n_e).$$

We also use the equation of state and the equation for the adiabat of an ideal gas:

$$p_\alpha = n_\alpha T_\alpha; \quad p_\alpha n^{-\gamma} = \exp [(\gamma - 1) S].$$

(A2.25)

Then system (A2.24) constitutes the complete system of equations for the stream function for the magnetic field, ψ, the stream functions for the ion and electron fluids, ψ_i and ψ_e, and the particle density n. This system of equations contains the six arbitrary functions $P_{\alpha 3}(\psi_\alpha)$, $U_\alpha(\psi_\alpha)$, $S_\alpha(\psi_\alpha)$.

In orthogonal curvilinear coordinates with an element of length defined by $dl^2 = h_1^2 dx_1^2 + h_2^2 dx_2^2 + h_3^2 dx_3^2$, Eq. (A2.24) becomes

$$\left.\begin{aligned}
&\frac{m_\alpha}{nh_3} j_{\alpha x_3} + \frac{e_\alpha}{nch_3} B_{x_3} - \frac{v_{\alpha x_3}}{h_3} P'_{\alpha 3} + U'_\alpha - T_\alpha S'_\alpha = 0; \\
&\frac{1}{n} j_{x_3} - \frac{4\pi e}{c} (v_{ix_3} - v_{ex_3}) = 0; \\
&\frac{4\pi e}{c} (\psi_i - \psi_e) = h_3 B_{x_3}; \\
&m_\alpha h_3 v_{\alpha x_3} + \frac{e_\alpha}{c} \psi = P_{\alpha 3} (\psi_\alpha); \\
&\frac{1}{2} m_\alpha v_\alpha^2 + W_\alpha + e_\alpha \varphi = U_\alpha (\psi_\alpha),
\end{aligned}\right\} \quad \text{(A2.26)}$$

where $j_{\alpha x_3}$ and j_{x_3} are given by

$$\left.\begin{aligned}
j_{\alpha x_3} &= -\frac{1}{h_1 h_2} \left(\frac{\partial}{\partial x_1} \cdot \frac{h_2}{h_1 h_3 n_\alpha} \cdot \frac{\partial \psi_\alpha}{\partial x_1} + \frac{\partial}{\partial x_2} \cdot \frac{h_1}{h_2 h_3 n_\alpha} \cdot \frac{\partial \psi_\alpha}{\partial x_2} \right); \\
j_{x_3} &= -\frac{1}{h_1 h_2} \left(\frac{\partial}{\partial x_1} \cdot \frac{h_2}{h_1 h_3} \cdot \frac{\partial \psi}{\partial x_1} + \frac{\partial}{\partial x_2} \cdot \frac{h_1}{h_2 h_3} \cdot \frac{\partial \psi}{\partial x^2} \right).
\end{aligned}\right\}$$

$$\text{(A2.27)}$$

An important particular case is that of flow across a magnetic field. If we set $\psi = 0$ and $P_{\alpha 3} = 0$ in Eqs. (2.25), then $v_{\alpha x_3} = 0$ and we can rewrite Eq. (A2.26) as

$$\left.\begin{aligned}
&\frac{m_\alpha}{nh_3} j_{\alpha x_3} + \frac{e_\alpha}{nch_3} B_{x_3} + U'_\alpha (\psi_\alpha) - T_\alpha S'_\alpha (\psi_\alpha) = 0; \\
&B_{x_3} = \frac{4\pi e}{ch_3} (\psi_i - \psi_e); \\
&(1/2) m_\alpha v_\alpha^2 + W_\alpha + e_\alpha \varphi = U_\alpha (\psi_\alpha).
\end{aligned}\right\} \quad \text{(A2.26')}$$

Here the magnetic field $\mathbf{B}$ has only the single component B_{x_3} and the velocities of the ion and electron fluids, $\mathbf{v}_\alpha (\alpha = i, e)$, are perpendicular to $\mathbf{B}$.

§2. Symmetric Flows of Two-Fluid Magneto-hydrodynamics with Slow Variation along One of the Coordinates

We have been dealing with flows which depend on the two co-ordinates x^1 and x^2. If the dependence on one of these coordinates, say x^2, is weak, we can neglect terms containing the second derivatives and the squares of the first derivatives with respect to x^2, thus simplifying the equations for two-parameter flows. The simplified system of equations obtained in this way contains another "adiabatic" integral in certain cases, so that this system can be integrated completely.

We write the system of equations for two-parameter flows in (A2.24) (here we do not assume neutrality) in the following form:

$$
\left.
\begin{aligned}
&\frac{1}{g_{33}}\left(\frac{m_\alpha}{n_\alpha}\,j_{\alpha 3}+\frac{e_\alpha B_3}{n_\alpha c}-v_{\alpha 3}P'_{\alpha 3}\right)+U'_\alpha-T_\alpha S'_\alpha=0;\\[4pt]
&j_3=\frac{4\pi}{c}\,(e_i\,n_i\,v_{i3}+e_e\,n_e\,v_{e3});\\[4pt]
&B_3=(4\pi/c)\,(e_i\,\psi_i+e_e\,\psi_e);\\[4pt]
&m_\alpha v_{\alpha 3}+\frac{e_\alpha}{c}\,\psi=P_{\alpha 3};\\[4pt]
&\frac{1}{2}\,m_\alpha v_\alpha^2+e_\alpha\Phi+W_\alpha=U_\alpha;\\[4pt]
&\Delta\Phi=-4\pi\,(e_i\,n_i+e_e\,n_e).
\end{aligned}
\right\}\qquad\text{A2.27'})
$$

The "adiabatic" approximation used below reduces to the neglect of the terms of the order of the second derivative or the square of the first derivative with respect to x^2 of the quantities n_α, ψ_α, ψ, and Φ; hence, in a first approximation, in orthogonal curvilinear coordinates the terms $j_{\alpha 3}$, j_3, v_α, and $\Delta\Phi$ in system (A2.27') can be replaced by

$$
\left.
\begin{aligned}
&j_{\alpha 3}\approx-\frac{h_3}{h_1 h_2}\cdot\frac{\partial}{\partial x_1}\left(\frac{h_2}{h_1 h_2 n_\alpha}\cdot\frac{\partial \psi_\alpha}{\partial x_1}\right);\quad j_3\approx-\frac{h_3}{h_1 h_2}\cdot\frac{\partial}{\partial x_1}\times\\[4pt]
&\qquad\qquad\times\left(\frac{h_2}{h_1 h_3}\cdot\frac{\partial \psi}{\partial x_1}\right);
\end{aligned}
\right.
$$

$$v_a^2 \approx v_{ax_s}^2 + \left(\frac{1}{h_1 h_2 n_a} \cdot \frac{\partial \psi_a}{\partial x_1} \right)^2 ; \quad \Delta\Phi \approx \frac{1}{h_1 h_2 h_3} \times$$

$$\times \frac{\partial}{\partial x_1} \left(\frac{h_2 h_3}{h_1} \cdot \frac{\partial \Phi}{\partial x_1} \right),$$

(A2.28)

where h_1, h_2, and h_3 are the Lamé coefficients.

Multiplying the first equation in (A2.27') by $\partial \psi_a / \partial x_1$, and combining it with the fifth equation in (A2.27') after the latter has been differentiated with respect to x_1 we find

$$m_a \frac{\partial}{\partial x_1} \cdot \frac{v_{ax_s}^2}{2} + \frac{e_a B_3}{n_a c g_{33}} \frac{\partial \psi_a}{\partial x_1} - \frac{v_{ax_s}}{g_{33}} \cdot \frac{\partial P_{a3}}{\partial x_1} + e_a \frac{\partial \Phi}{\partial x_1} +$$

$$+ \frac{\partial W_a}{\partial x_1} - T_a \frac{\partial S_a}{\partial x_1} = 0.$$

(A2.29)

Multiplying the ion form of Eq. (A2.29) by n_i, multiplying the electron form by n_e, and then combining the results, using the other equations in (A2.27'), we find

$$\frac{1}{8\pi g_{33}} \cdot \frac{\partial B_3^2}{\partial x_1} - \frac{1}{2g_{33}} \cdot \frac{\partial g_{33}}{\partial x_1} \left(n_i m_i v_{ix_s}^2 + m_e n_e v_{ex_s}^2 \right) -$$

$$- \frac{j_3}{4\pi g_{33}} \cdot \frac{\partial \psi}{\partial x_1} - \frac{\Delta\Phi}{4\pi} \cdot \frac{\partial \Phi}{\partial x_1} + \frac{\partial}{\partial x_1} (p_i + p_e) = 0.$$

(A2.30)

Substituting Eqs. (A2.28) we find the final expression in terms of curvilinear orthogonal coordinates:

$$\frac{\partial}{\partial x_1} \left(\frac{h_3^2 B_{x_s}^2}{8\pi} \right) + \frac{h_3^2}{h_2^2} \cdot \frac{\partial}{\partial x_1} \left(\frac{h_2^2 B_{x_2}^2}{8\pi} \right) - \frac{1}{h_2^2} \cdot \frac{\partial}{\partial x_1} \left(\frac{h_2^2 h_3^2 E_{x_1}^2}{8\pi} \right) -$$

$$- \frac{1}{2} \cdot \frac{\partial h_3^2}{\partial x_1} \left(n_i m_i v_{ix_s}^2 + n_e m_e v_{ex_s}^2 \right) + h_3^2 \frac{\partial}{\partial x_1} (p_i + p_e) = 0.$$

(A2.31)

This "adiabatic" equation in (A2.31) is obviously exact for the one-dimensional problem, in which $\partial / \partial x_2 = \partial / \partial x_3 = 0$. It describes the local pressure balance along the coordinate x_1. In the two-dimensional problem in terms of the Cartesian coordinates $x_1 = -x$, $x_2 = z$, $x_3 = y$, Eq. (A2.31) takes the following form for a flow which varies slowly along z and is independent of y:

$$\frac{\partial}{\partial x} \left[\frac{1}{8\pi} (B_y^2 + B_x^2 - E_x^2) + p_i + p_e \right] = 0.$$

(A2.32)

The integral of this equation is defined by

$$\frac{1}{8\pi}\left(B_y^2 + B_x^2 - E_x^2\right) + p_i + p_e = P(z), \qquad (A2.33)$$

where $P(z)$ is a slowly varying, arbitrary function of z.

The corresponding equation for axisymmetric flow in terms of the cylindrical coordinates $x_1 = r$, $x_2 = z$, $x_3 = -\varphi$ can be written

$$\frac{1}{r^2} \cdot \frac{\partial}{\partial r}\left(r^2 \frac{B_\varphi^2 - E_r^2}{8\pi}\right) + \frac{\partial}{\partial r}\left(\frac{B_z^2}{8\pi} + p_i + p_e\right) - \frac{1}{r}\left(n_i m_i v_{i\varphi}^2 + n_e m_e v_{e\varphi}^2\right) = 0.$$
$$(A2.34)$$

This equation holds if the flow is independent of the azimuthal angle φ and varies slowly along the z axis.

This equation cannot be integrated in its general form; however, in certain cases it is possible to find a first integral by analogy with the two-dimensional problem. For example, with $B_z = 0$ and $v_{i\varphi} = v_{e\varphi} = 0$, neglecting the pressures p_i and p_e, we find

$$r^2 \left(B_\varphi^2 - E_r^2\right) = Q(z), \qquad (A2.35)$$

where $Q(z)$ is a slowly varying arbitrary function of z.

In the corresponding one-dimensional problems the values of P and Q are obviously constant.

REFERENCES

1. L. A. Artsimovich, S. Yu. Luk'yanov, I. M. Podgornyi, and S. A. Chuvatin, Zh. Eksp. Teor. Fiz., 33:3 (1957) [Sov. Phys. JETP 6:1 (1958)].

2. A. I. Morozov, Zh. Eksp. Teor. Fiz., 32:305 (1957) [Sov. Phys. JETP 5:215 (1957)].

3. G. P. Wood and A. F. Carter, in: Ion, Plasma, and Arc Rockets [Russian translation], Gosatomizdat, Moscow (1961).

4. A. I. Morozov, Zh. Prikl. Mekhan. i Tekh. Fiz., No. 2:30 (1966).

5. A. V. Zharinov and Yu. S. Popov, Zh. Tekhn. Fiz. 37:294 (1967) [Sov. Phys. Tech. Phys. 12:208 (1967)].

6. A. I. Morozov, E. V. Artyushkov, L. S. Solov'ev, and A. P. Shuvin, in: Low-Temperature Plasmas [in Russian], Mir, Moscow (1967).

7. S. I. Braginskii, Reviews of Plasma Physics, Vol. 1 (ed. M. A. Leontovich), Consultants Bureau, New York (1965), p. 205.

8. L. A. Artsimovich (editor), Plasma Accelerators [in Russian], Mashinostroenie, Moscow (1972).

9. A. I. Morozov and S. V. Lebedev, "Plasma optics," this volume p. 301.

10. A. I. Morozov and L. S. Solov'ev, Zh. Tekhn. Fiz., 34:1141-1153 (1964) [Sov. Phys. Tech. Phys. 9:889 (1965)].

11. A. I. Morozov and A. P. Shubin, Zh. Eksp. Teor. Fiz., 46:710 (1964) [Sov. Phys. JETP 19:484 (1964)].

12. V. I. Bryzgalov and A. I. Morozov, Zh. Eksp. Teor. Fiz., 49:1789 (1965) [Sov. Phys. JETP 22:1223 (1966)].

13. A. I. Morozov and A. P. Shubin, Teplofiz. Vys. Temp., 3:827-837 (1965).

14. A. I. Morozov and L. S. Solov'ev, Dokl. Akad. Nauk SSSR, 164:80-83 (1965) [Sov. Phys. Dokl. 10:834 (1966)].

15. K. V. Brushlinskii and A. I. Morozov, "Calculation of two-dimensional plasma flows in channels," this volume, p. 105.

16. L. D. Landau and E. M. Lifshitz, Fluid Mechanics, Addison-Wesley, Reading Mass. (1959).

17. R. Sauer, Flow of Compressible Fluids [Russian translation], Izd. Inostr. Lit., Moscow (1954).

18. L. A. Vulis, Thermodynamics of Gas Flows [in Russian], Gosenergoizdat, Moscow (1950).

19. A. I. Morozov, Dokl. Akad. Nauk SSSR, 154:306-309 (1964) [Sov. Phys. Dokl. 8:1086 (1964)].

20. A. I. Morozov and L. S. Solov'ev, Zh. Tekhn. Fiz. 34:429-443 (1964) [Sov. Phys. Tech. Phys. 9:337 (1964)].

21. L. Crocco, Z. Angew. Math. Mech., 17:1-7 (1937); see also [17].

22. A. I. Morozov, Zh. Tekhn. Fiz., 37:2147 (1967) [Sov. Phys. Tech. Phys. 12:1580 (1968)].

23. K. V. Brushlinskii, N. I. Gerlakh, and A. I. Morozov, Izv. Akad. Nauk SSSR Mekhan. Zhidk. i Gaza, No. 2:189-192 (1966).

24. L. M. Alekseeva and L. S. Solov'ev, Zh. Tekhn. Fiz., Mekh., 28:987-995 (1964).

25. A. I. Morozov and L. S. Solov'ev, Zh. Tekh. Fiz., 34:1154-1169 (1964) [Sov. Phys. Tech. Phys. 9:898 (1965)].

26. L. S. Solov'ev, Reviews of Plasma Physics, Vol. 3, (ed. M. A. Leontovich), Consultants Bureau, New York (1967), p. 277.

27. A. I. Morozov and L. S. Solov'ev, Zh. Tekhn. Fiz. 30:1104-1108 (1960) [Sov. Phys. Tech. Phys. 9:898 (1965)].

CALCULATION OF TWO-DIMENSIONAL PLASMA FLOWS IN CHANNELS

K. V. Brushlinskii and A. I. Morozov

INTRODUCTION

Computer calculations are assuming a stature equivalent to that of theory and experiment in modern physics research. Numerical integration of the complicated nonlinear equations of mathematical physics can yield detailed information about the solutions of these equations in situations in which the traditional theoretical methods are capable of only rough, qualitative results at best. A thorough series of computer calculations can take the place of experiments which are laborious, expensive, and, in certain cases, impossible.

This situation is illustrated by the many calculations in gasdynamics which have been made possible by progress in calculation techniques. These calculations have led to the development of interesting new approaches in mechanics, nonlinear differential equations, and difference methods in the solution of these equations.

Numerical methods are finding increasing application in plasma physics. The calculations have been based on both the kinetic equations (for low-density plasmas) and the magnetogasdynamic equations (for relatively dense plasmas). As an example of the work in low-density plasmas we note the interesting paper by Gel'fand et al. [28], who studied the nonlinear one-dimensional oscillations of a low-density plasma. For an introduction to the non-Soviet work on the numerical solution of the kinetic equation we refer the reader to the collection edited by Alder et al. [69] and to the Proceedings of the APS Topical Conference on Numerical Simulation of Plasma, held at Los Alamos [73]. Since the kinetic description of a plasma is complicated, the work so far has been limited to one-dimensional problems.

In the magnetohydrodynamic approach, on the other hand, it is possible to treat both one-dimensional and two-dimensional processes. Calculations are being carried out for the plasma flow in plasma accelerators and MHD generators and for problems related to thermonuclear fusion, astrophysics, etc. Most of the calculations have been based on one-dimensional models. One of the first such papers was a calculation of the one-dimensional pinch carried out in 1958 by Braginskii et al. [12].

Generally speaking, calculations on the basis of the two-dimensional magnetogasdynamic equations represent the next step in the research. Although these equations are more complicated, they are of important because of the qualitatively new results which they yield. Although the first two-dimensional calculations for plasma flows were carried out in the early 1960s (Brushlinskii et al. [18]), comparatively little has been published in this field. Certain problems in this field constitute the subject matter of the present paper, but we would also like to cite several other series of studies by Soviet investigators. Calculations carried out for the pinch effect on the basis of the two-dimensional model by D'yachenko and Imshennik [32] have yielded a satisfactory explanation for several experimental results reported by Filippov [66]. Gubarev et al. [29] have studied the properties of a two-dimensional supersonic plasma flow in the channel of an MHD generator. Velikhov et al. [25] have analyzed numerically the nonlinear stages of the ionizational instability. Two-dimensional calculations for steady-state plasma flows in end-fire accelerators have been carried out by Ostretsov et al. [54].

The non-Soviet work has focused on pinch and accelerator experiments. The two-dimensional calculation carried out by Potter et al. [70, 72, 74] for the Z pinch, including the plasma-focus stage, deserves special mention. Freeman and Lane [65] have carried out calculations on the basis of an axisymmetric model of a θ pinch, and Duchs [64] has carried out calculations for the dynamics of the azimuthal rotation of this pinch. A numerical analysis of the plasma flow in a coaxial Marshall gun has been carried out [63,

71]. Roberts and Potter [74] have published a comprehensive review of non-Soviet calculations on magnetogasdynamics.

In this paper we review calculations on the plasma flow in coaxial channels in the azimuthal magnetic field produced by the plasma. This work has been carried out over a period of several years by N. M. Zueva, N. I. Gerlakh, V. V. Paleichik, and the present authors, and the results have been published in several papers [15-23].

Interest in this type of flow is related to the development of high-current, steady-state plasma accelerators. The purpose of these calculations is to reveal qualitatively new aspects of these processes, rather than to obtain quantitative characteristics. This approach is useful because existing theoretical models are relatively ineffective for such complicated conditions, so that the inferences drawn from numerical studies are very important in the development of an appropriate theory. On the other hand, the models for which calculations were carried out have been too simple to justify far-reaching quantitative conclusions.

Plasma flow in channels is now attracting the interest of many investigators and is the subject of a large number of analytic studies. This analytic work is reviewed thoroughly by Vatazhin et al. in [24], where the number of citations in the bibliography runs to four figures. However, the difficulties involved in analytic methods for solving the nonlinear equations are such that many questions remain unresolved. It is this circumstance which has stimulated the calculations described in the present paper. These calculations were based on a numerical solution of the nonstationary two-dimensional magnetogasdynamic equations for the classical single-fluid model and for the two-fluid model (incorporating the Hall effect). The steady-state flow regimes are found by an asymptotic method.

In the first chapter we review preliminary ideas inferred from the experimental data, report a detailed physical formulation of the problem, and outline the mathematical description of the problem on the basis of the transport equations for a two-component plasma. Then we use certain natural simplifying assumptions to reduce the problem to a series of simpler problems, in particular, the problem of solving the magnetogasdynamic equations. We choose units for the unknown properties and specify all the dimen-

sionless parameters on which the flow properties depend. Some of
the material is of a referential nature and is included for a complete discussion of all the questions bearing on the reasoning behind the calculations. The chapter ends with a description of the
numerical methods used to solve the problem, including the various
methods used to choose the system of curvilinear spatial coordinates in the channel and the difference scheme for solving the
equations.

In the second chapter we review the physical results obtained
in calculations for the flow of a fully ionized plasma. Several series
of calculations have been carried out under various assumptions
regarding the properties of the plasma, for various channel shapes,
and for various values of the dimensionless parameters. These
calculations show how the flow properties vary with these factors
and reveal several new properties: a nonmonotonic variation of
the density along the nozzle, current eddies, and an instability of
the flow near the anode in the case in which the Hall effect is important. An interesting effect is a steady-state plasma compression near the channel; this was first discovered through the numerical calculations.

The third chapter takes account of the ionization of the gas in
the channel. The basic result of these calculations is the demon-
stration of nonstationary fluctuating flow regimes at the high plasma
conductivity that prevails after ionization.

Chapter 1

FORMULATION OF THE PROBLEM AND
METHOD OF ANALYSIS

§ 1.1.   Physical Formulation of the Problem

These calculations for plasma flow in coaxial systems are
carried out in connection with experiments on stationary (quasi-
stationary) coaxial plasma accelerators in which the magnetic
field produced by the plasma is important and which were carried
out at the Kurchatov Institute of Atomic Energy, Moscow [1, 36-38,
44, 46].

The accelerator (Fig. 1) consists of two coaxial electrodes,
an inner electrode 3 and a specially shaped outer electrode 2, sep-

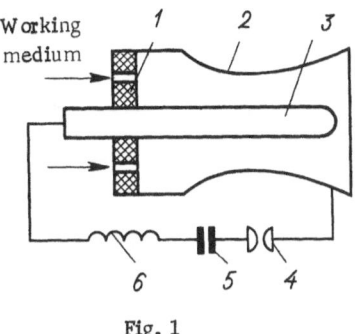

Fig. 1

arated by dielectric insert 1, through apertures in which the working medium flows into the interelectrode gap. The accelerator is supplied from a capacitor bank 5, connected to a "stretching" inductance 6 and switch 4. The current pulse length is $t_p \approx 1$ msec, which is at least two orders of magnitude longer than the transit time

$$t_0 = L/v. \tag{1.1}$$

Here, L is the length of the system and v is the typical flow velocity. Since the ratio t_p/t_0 is large, we can treat the process as a steady-state process.

The discharge current is varied from 20 to 60 kA. The working medium enters the system through a pulsed injection system; the amount of medium which is injected is 10 cm³ of nitrogen or hydrogen at standard conditions. The discharge voltage is several hundred volts. The flow rate at the exit from the accelerator is in the range 10^6-10^7 cm/sec.

In planning the experiments, the experimentalists have been guided by the quasi-one-dimensional theory for plasma flow in a narrow plane channel in a transverse magnetic field (Fig. 2) developed by A. I. Morozov in 1959. In this case the equations of ideal magnetogasdynamics reduce to the three conservation laws [52]

$$\rho v f = \text{const}; \tag{1.2a}$$

$$H/\rho = \text{const}; \tag{1.2b}$$

$$v^2/2 + i + H^2/4\pi\rho = U = \text{const}; \tag{1.2c}$$

$$p = p(\rho).$$

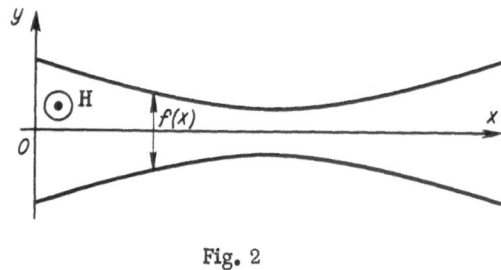

Fig. 2

Here ρ, p, v, and $i = \int dp/\rho$ are the plasma density, pressure, velocity, and enthalpy, respectively; H is the magnetic field, and $f(x)$ is the cross-sectional area of the channel. Equation (1.2a) describes conservation of matter along the channel, (1.2b) describes conservation of magnetic flux, and (1.2c) describes energy conservation.[†] The number of integrals of Eqs. (1.2) is one less than the number of variables (ρ, v, H, and f), so that three of these variables can be expressed in terms of the fourth, e.g., the channel cross section f. It follows, in particular, that continuous acceleration is not possible in a narrow channel of constant cross section. It can be shown that for continuous plasma acceleration the channel profile must be nozzle-shaped (Fig. 2). At the minimum cross section of the nozzle the plasma velocity reaches the local value of the magnetosonic velocity [52].

It follows from the energy-conservation equation, (1.2c), that the largest possible plasma velocity at the exit from the channel is

$$v_{\max} = \sqrt{2U} = \sqrt{\frac{H_0^2}{2\pi\rho_0} + 2i_0}. \tag{1.3}$$

Here, the subscript "0" denotes the value of a quantity at the "entrance" to the accelerator, where the "entrance" is understood to mean the cross section of the channel at which v_0^2 can be neglected in comparison with, for example the square of the Alfvén velocity, $C_{A0}^2 = H_0^2/4\pi\rho_0$. If we are interested in producing plasma streams with high velocities we are particularly interested in the case in which the magnetic pressure is substantially stronger than the gas pressure and in which the quantity i_0 in Eq. (1.3) can be neglected. In

---

[†]Or conservation of the generalized Bernoulli integral.

this case the maximum exit velocity is

$$v_{max} \approx \sqrt{2}\, C_{A0}. \qquad (1.4)$$

For example, the maximum velocity corresponding to a discharge current of 100 kA in a channel with an average radius of 5 cm at an ionized-hydrogen density of about 10^{16} cm^{-3} near the entrance is v $_{max} \approx 10^7$ cm/sec. This attractive possibility for producing intense, high-velocity, steady-state plasma streams stimulated the experimental studies. We note that the estimate in (1.4) is based on energy conservation; experiments have basically confirmed this estimate [1, 37].

The choice of a model for the medium and the formulation of the problem are guided by certain considerations which follow from the experimental data. Let us examine these considerations.

Equation (1.2) does not incorporate a dependence on the transverse coordinate; it describes a simple flow model in which the anode and the cathode that form the channel are treated symmetrically. The invariance of the processes in the channel under reversal of electrode polarity is a general property of the magnetogasdynamic equations:

$$\left.\begin{array}{l} \partial\rho/\partial t + \operatorname{div} \rho\mathbf{v} = 0; \\ \rho dv/dt = -\nabla p + [\mathbf{j},\, \mathbf{H}]/c; \\ \partial\mathbf{H}/\partial t = \operatorname{curl}[\mathbf{v},\, \mathbf{H}] + (c^2/4\pi\sigma)\,\Delta\mathbf{H}; \\ \operatorname{curl} \mathbf{H} = (4\pi/c)\,\mathbf{j}. \end{array}\right\} \qquad (1.5)$$

These equations are obviously not affected by the substitutions $\mathbf{H} \rightarrow -\mathbf{H}$, $\mathbf{j} \rightarrow -\mathbf{j}$, which are equivalent to a reversal of the electrode polarity. It follows that the flow can be described by Eqs. (1.5) only if the discharge configuration is not affected† by such a reversal.

The very first experimental results demonstrated that the criterion above is not satisfied. Polarity reversal not only changes the electrode processes proper (as is evident from the erosion tracks) but also causes a substantial change in the flow configuration throughout the channel. An external manifestation of these changes is the fact that the plasma stream always presses against the cathode, regardless of whether this cathode is the central

---

†At least within an error corresponding to the thin electrode sheaths, which clearly do not obey Eqs. (1.5).

Potential

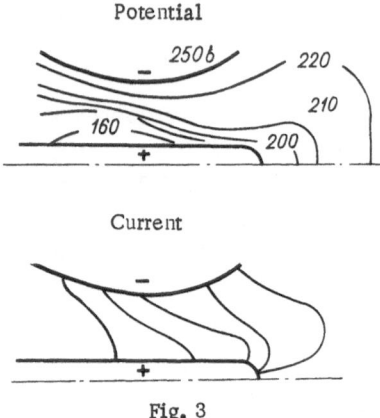

Fig. 3

electrode or the outer electrode. The asymmetry of the discharge
with respect to the anode and cathode is particularly clear from
the experimental distributions of the electric potential [36] and the
lines of electric current [38] in the channel (Figs. 3 and 4; the
potential is given in volts). We note the large potential drop near
the anode and the clearly defined "grazing" motion of the current
along the anode. The longitudinal component of the electric current
tends to squeeze the plasma toward the cathode and leads to the
appearance of large "ejection" currents, i.e., currents which flow
beyond the end of the accelerator. Since the current lines in the
channel are not parallel, the flow is fundamentally two-dimensional.

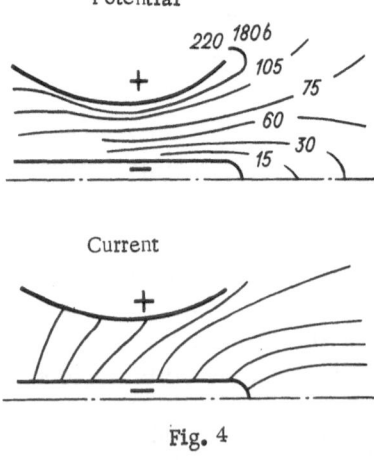

Fig. 4

Two extreme explanations for this misalignment of the current lines are possible. It may be a volume effect resulting from the Hall effect, or it may be a consequence of the appearance of a sheath near the anode which conducts very poorly.

Let us examine this first possibility. When the Hall effect is taken into account Ohm's law can be written

$$\frac{j}{\sigma} = E + \frac{1}{c}[v, H] - \frac{1}{enc}[j, H] + \frac{\nabla p_e}{en}. \qquad (1.6)$$

When the electrode polarity is reversed, i.e., when we use the substitutions $H \rightarrow -H$, $E \rightarrow -E$, $j \rightarrow -j$, Eq. (1.6) changes substantially because of the Hall term and because of the gradient of the electron pressure on the right side. Under actual experimental conditions, the quantity ∇p_e in Ohm's law is unimportant and the misalignment of the current line is due to the Hall effect. The direction of this misalignment predicted by Eq. (1.6) is the same as that observed experimentally. A more detailed analysis of flows incorporating the Hall effect is reported in [52].

We turn now to the other possibility — the formation of an anode sheath. If, for some reason, the current component normal to the anode encounters a high resistance, while that at the cathode does not, a misalignment of the current is unavoidable. What is the nature of this "blocking" anode sheath? We will show that the Hall effect again yields a likely explanation. These experiments use solid copper electrodes, so that the boundary condition of a vanishing tangential component of the electric field is well satisfied:

$$E_\tau = 0. \qquad (1.7)$$

If the magnetic field in the accelerator is purely azimuthal, then condition (1.7) means that in the crossed E and H fields the electrons drift along the anode, rather than toward it. The incidence of electrons on the anode is the result of the finite (Coulomb or anomalous) conductivity of the plasma. Let us first assume that this conductivity is the usual Coulomb conductivity; then the current density drawn by the anode is

$$j_\perp = \frac{\sigma E_\perp^*}{1 + (\omega\tau)_e^2}; \quad (\omega\tau)_e = \frac{\sigma H}{enc}; \quad E^* = E + \frac{1}{c}[v, H]. \qquad (1.8)$$

As a result of the volume Hall effect noted above (the presence of a longitudinal current), the plasma is squeezed away from the anode. The reduction in the plasma density increases $(\omega\tau)_e$ and, thus, increases the resistance in the anode sheath. This higher resistance, in turn, intensifies the grazing of the current along the anode, contributing to a further squeezing of plasma away from the anode, etc.[†]

Accordingly, in the immediate vicinity of the anode a sheath is formed with thickness of the order of the electron gyroradius [53]; most of the potential drop, δU, is concentrated in this sheath. In the sheath, electrons with energy of the order of $l\delta U$ drift along the anode, undergoing essentially no collisions. This drift continues up to a certain singularity, where the electrons do strike the anode. If the anode is the central electrode, this singularity is the end of the electrode. It is thus not surprising to find that a round spot is left at the end of the central electrode (when this electrode is the anode) after each discharge [38]; this spot is not observed when the electrode polarity is reversed.

Study of the anode and cathode surfaces after a discharge reveals that when the anode has an appreciable roughness there are deep fused grooves at certain places. If the anode is carefully polished before an experiment these grooves do not appear and an inspection of the anode surface under a microscope reveals a pattern typical of chemical etching [9]; i.e., grain boundaries and slip lines are clearly visible at the surface. The evident explanation for these results is that the stream of hot electrons drifting along the cathode (at a temperature corresponding to the anode potential drop) causes the evaporation of small irregularities on the anode. If this electron stream strikes large surface defects it causes intense evaporation of the material. As a result, the plasma density increases rapidly at such points, the resistance decreases, and the high local current density produces a deep groove. On the cathode surface, on the other hand, there is usually a large number of erosion tracks, which are not deep [9, 38]; these tracks are typical of most high-current, cold-cathode arcs. These processes have been studied in detail by Kesaev [35]. These results imply that the flow model should incorporate the Hall effect.

[†]It can be shown that this process also increases the electron temperature.

A remarkable feature of the discharges at these discharge currents is the constancy of the electron temperature throughout the channel and in the current-ejection region. Specifically, this temperature remains within a few percent of 2 eV everywhere [1, 37]. This fact has not been fully explained at the present time. The constancy seems to be due to an abrupt increase in the radiation power with increasing electron temperature. Unfortunately, no systematic study of the ion temperature as a function of the discharge parameter is available.

Another guiding consideration is based on the ionization of the gas in the channel. The working medium enters the discharge gap as a neutral gas (nitrogen, argon, or hydrogen). Experiments show that the ionization occurs in a thin sheath (less than a few millimeters wide) and that for the specified values of the current and amount of working medium, the neutral gas is almost completely ionized. Specifically, a spectrogram of the region ahead of the ionization front reveals no ion lines, while a spectrogram of the region behind the front reveals no lines of neutrals [1, 37]. The dependence of the position of the ionization front on the discharge current and the amount of working medium is curious. As the current is raised or the amount of medium supplied is reduced, the front moves further into the interior of the accelerator; when the current is reduced or the amount of working medium increased the front moves toward the entrance of the accelerator [1, 37].

The neutral component is detected in the discharge gap only in a thin sheath near the anode (3-5 mm wide); the neutral component consists of copper atoms and nitrogen.

It follows from this discussion that a thorough calculation for plasma flow in the accelerator must incorporate ionization of the neutral gas. The flow model should describe the dynamics of the three plasma components — the neutrals, ions, and electrons. However, we do not deal with the three-component problem in the present paper. The calculations are carried out in two simpler cases. In the first, the plasma is fully ionized; this case covers the problems formulated in this chapter and the results of the calculations described in Chap. 2. In the second problem the ionization can vary over the channel. The ionization is simulated by an abrupt increase in the conductivity when the temperature crosses a specified critical value. The plasma is treated as a single-com-

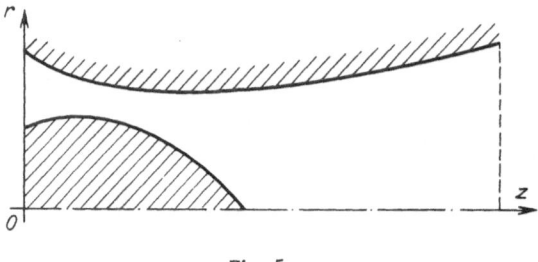

Fig. 5

ponent plasma and the problem is solved in the quasi-one-dimensional approximation. The corresponding refinements in the formulation of the problem and the results found in the calculations are described in Chap. 3.

We now consider the model for the plasma flow used for all the two-dimensional calculations (Chap. 2) and the general formulation of the problem.

Let us examine the plasma flow in the channel formed by two coaxial solid metal† electrodes of arbitrary shape, as shown schematically in Fig. 5. The plasma is assumed to be fully ionized and to consist of two components: electrons and singly charged ions. In accordance with the experiments discussed above, in which the particle density lies between 10^{14} and 10^{16} cm^{-3}, we assume that the plasma is neutral:

$$n^i = n^e. \tag{1.9}$$

To describe the dynamics of both components we begin with the Braginskii equations [13]. In the equation of motion for the electron component we neglect the electron inertia, $m^e dv^e/dt$. This simplification is equivalent to ignoring processes whose characteristic frequency is of the order of the plasma frequency or the electron gyrofrequency‡; this simplification is clearly justified for a numerical study of the overall flow pattern. The electron inertia, however, governs the spatial scales as well as the temporal scales; specifically, it governs the Debye length and the elec-

---

†More precisely, we assume that the ohmic resistance of the electrodes can be neglected.

‡For an average density of $n \approx 10^{15}$ cm^{-3} and a magnetic field of $H \approx 2$ kOe, the plasma frequency is $\omega_0 \approx 1.5 \cdot 10^{12}$ sec^{-1}, and the gyrofrequency is $\omega_{eH} \approx 3.5 \cdot 10^{10}$ sec^{-1}.

tron gyroradius, both of which are important in the structure of the anode and cathode sheaths [53]. Accordingly, neglect of the electron inertia can have important consequences since, as noted above, the resistance of the anode sheath affects the flow pattern in the volume. We also neglect the displacement current $(1/c)(\partial E/\partial t)$ in Maxwell's equations. Then the flow is described by the magnetogasdynamic equations with the Hall effect. These equations are discussed specifically below. At this point we simply note that the gasdynamic equations are strictly applicable only when the mean free path of the particles is much shorter than the scale lengths of the flow inhomogeneities, although when there is a magnetic field this condition is apparently replaced by a less stringent condition.

It is clear that the exact condition holds for the electrons. Specifically, we note that at $T^e \approx 2$ eV the Coulomb collision cross section is $\sigma_{Cou} \approx 3 \cdot 10^{-14}$, so that with $n^e \approx 10^{15}$ cm^{-3} the mean free path is $\lambda^e \approx 0.5$ mm, in comparison with a scale length for the flow inhomogeneities which is of the order of a centimeter.

The situation with regard to the ions is more complicated. In the first place, since the ion temperature is higher than the electron temperature the ion mean free path can be an order of magnitude larger than the electron mean free path; in some cases, it can approach the characteristic length of the inhomogeneity. In the second place, the ions which strike an electrode and are then reflected back into the channel have a velocity with respect to the flow particles which is comparable to the flow velocity; their relative energy is of the order of a few tens of electron volts. At this energy the Coulomb cross section is 10^{-16}–10^{-17} cm^2, being comparable to the gaskinetic cross section for the heavy particles. Accordingly, the interaction of the ions which "drop out" of the acceleration with the ions which continue in normal acceleration (the "bulk" ions) turns out to be very weak.

These estimates show that the accuracy of the gasdynamic description of the ions is limited. Nevertheless, we are forced to use this model since a kinetic calculation of the ion dynamics is highly complicated.

In choosing a mathematical model for the process we must take account of the oscillations observed in the accelerator. Special measurements show that the oscillations of the flow properties are comparatively slight (ranging from 10% to 20%) over broad ranges of the experimental parameters. The discharge as a whole can be assumed to be axisymmetric. When the amount of working medium injected is small, especially in experiments with hydrogen, axisymmetric longitudinal oscillations of the ionization front† are observed.

These experimental results show that an axisymmetric flow model is appropriate. Thus, using the cylindrical coordinates (z, r, φ), we can assume

$$\mathbf{v} = (v_z, v_r, 0); \quad \mathbf{H} = (0, 0, H_\varphi); \quad \partial/\partial\varphi \equiv 0. \qquad (1.10)$$

The second of these equations is also related to the fact that there is no external magnetic field, while the magnetic field produced by the plasma is directed across the flow. Furthermore, it follows from (1.10) that the electric field $\mathbf{E}$ and the electric current $\mathbf{j}$ lie in the (z, r) plane.

To arrange plasma flow in the desired direction (from left to right in Fig. 5), the initial distributions of mass, temperature, and magnetic field in the channel are specified so that the resultant pressure (the gas pressure plus the magnetic pressure) falls off from left to right. Then the medium can be supplied to the channel by specifying constant values of the number density n_0 (or mass density ρ_0) and the temperature T_0 at the left boundary of the channel. Consequently, the gas pressure is also specified at the entrance to the channel. At $z = 0$ we have

$$n = n_0; \; \rho = \rho_0; \; T = T_0; \; p = p_0 = kn_0T_0 = (c_p - c_v)\rho_0T_0. \qquad (1.11)$$

Boundary conditions of this type can be interpreted by imagining an infinite plasma reservoir with specified values of the density and pressure connected to the entrance of the channel.

We assume further that the entire current I which flows through the system is carried by the electrons alone and is held constant. This requirement gives the following boundary condition on the azimuthal magnetic field $H = H_\varphi$ produced by the current at the

---

†In certain intermediate regimes there is an axially asymmetric instability of the ionization front, called the "spin instability." In this case a discontinuity appears at the ionization front and rotates azimuthally at a frequency of 20-30 kHz. An instability of this kind causes a strong modulation of the density of the medium directly behind the front and a comparatively weak modulation in the exit part of the channel.

channel entrance:

$$H_\varphi (0, r) = 2I/cr.$$

We use the notation $H_0 = 2I/cr_0$, where r_0 is the scale value of the radius in the entrance cross section. Then H_0 is the scale value of the magnetic field at the entrance, and the boundary condition above becomes

$$H (0, r) = H_0 r_0/r. \tag{1.12}$$

In other words, at $z = 0$

$$j_z = (c/4\pi r)(\partial Hr/\partial r) = 0; \tag{1.13}$$

and the current at the entrance is strictly radial. Condition (1.13) is largely formal in nature, since, as we see from Fig. 6, the ionization front is included and the electric current has a longitudinal component throughout the accelerator volume.

It follows that the plasma flow in the channel is maintained by the specified gas pressure and the magnetic field at the entrance. Under the actual experimental conditions, the right end of the accelerator faces a large evacuated volume into which the flow expands (ideally this expansion would be unbounded) so that the magnetic field carried out of the accelerator is lost.

Let us examine the boundary conditions at the channel walls. There is apparently no perfect way to specify these boundary conditions within the hydrodynamic approach for the density range in which we are interested. In order to close the problem it is necessary to impose a condition on the velocity of each plasma component at the electrodes or to impose an equivalent condition.

However, in view of the comparatively long mean free path of the ions in the flow we evidently cannot adopt the usual gasdynamic boundary condition, i.e., the vanishing of the normal velocity:

$$v_n = 0. \tag{1.14}$$

Ionization front

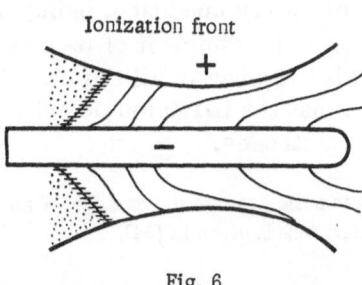

Fig. 6

Clearly, the ions can strike the cathode and thereby contribute to the current transport between the electrodes. It is unlikely that the ions will strike the anode, however, because of the potential drop near the anode, so that condition (1.14) is apparently valid for the ion velocity at the anode.

In hydrodynamics, condition (1.14) is related to the fact that the flux of particles incident on the wall is cancelled by the flux leaving the wall. In the case under consideration here, the ions that strike the cathode are generally reflected back as neutrals, which are subsequently ionized in the flow. The mean free path of the neutrals against ionization is evidently

$$\lambda_{\text{ion}} = v^N / n^e \langle \sigma v^e \rangle_{\text{ion}},$$

where v^N is the velocity of the neutral component. Substituting the typical numbers from above we find the mean free path of the neutrals against ionization to be a few millimeters, in agreement with the thickness of the cathode sheath that contains the neutrals which is observed experimentally.

Once ionized, however, a particle reflected from the cathode interacts weakly with the bulk of the flow since the cross section for a Coulomb interaction between this particle and the "bulk" ions is very small (see the discussion above). Those particles which do collide with the cathode have a tendency to remain[†] near the cathode, forming a distinctive plasma sheath near the cathode which flows out of the accelerator slowly. Although this scheme has not been directly tested experimentally, it is consistent with the observed facts and seems to be plausible. In this case the ion flow near the cathode exhibits at least two velocities; this state is difficult to describe in the hydrodynamic approach.

Nevertheless, wishing to remain within the framework of the hydrodynamic description of the ions, we are forced to impose a somewhat arbitrary boundary condition, being guided by purely formal considerations. The simplest of these considerations is condition (1.14). It also figures in all the calculations; in effect it only reflects the fact that the particles do not enter or leave the channel through the electrodes.

[†]The reason is that an ion is pulled toward the cathode. A more detailed analysis of the behavior of the retarded ions is given in [44].

The same difficulties arise in the formulation of the second boundary condition for the ion velocity, which must incorporate the viscosity. When the viscosity is incorporated in the equation of motion the slip condition in (1.14) is replaced by the more stringent condition that the plasma is attracted toward the electrode:

$$v = 0. \tag{1.15}$$

We hope that some future accurate kinetic analysis of the cathode sheath and detailed experimental studies will make it possible to formulate a boundary condition for the ion velocity on the basis of less formal considerations.

With regard to the electron component of the plasma, we note that when we neglect the ion inertia and the violation of quasi-neutrality it is sufficient to use a single boundary condition that contains the tangential component of the electric field E_τ.

For formal simplicity in the calculations it is required that E_τ vanish at the electrodes, that is to say, boundary condition (1.7) is imposed. Support for the use of this condition comes from the high longitudinal conductivity of copper electrodes, so that the potential drop along the surfaces of these electrodes can be neglected. However, this argument only works at the cathode, near which there is no important experimental potential drop [36]. Near the anode, on the other hand, because of the large potential drop, condition (1.7) is clearly questionable. Here, again, we need refinement of the existing theory and the formulation of appropriate experiments to get a better boundary condition (1.7).

Finally, in calculations which incorporate the thermal conductivity of the plasma the electrodes are assumed to be thermal insulators or to absorb a heat flux proportional to the boundary temperature.

In addition to carrying out calculations for a coaxial accelerator, we have carried out a numerical analysis of the flow in a magneto-plasma compressor. The concept of this compressor arose during the first two-dimensional calculations. The basic features of the model in the compressor are the same as in the accelerator, so that a special discussion is not required.

These are the general features of the physical formulation of the problem for which these numerical solutions are carried out.

§1.2. Mathematical Models

__Basic Equations.__ The mathematical description of the problem formulated above is based on the two-fluid model of a plasma. The discussion below is based on the ion and electron transport equations taken from Braginskii's work [13]. For the case under consideration here, flow across a magnetic field, these equations are

$$\frac{\partial n^i}{\partial t} + \operatorname{div} n^i \mathbf{v}^i = 0;$$

$$
\left.
\begin{aligned}
& m^i n^i \frac{d^i \mathbf{v}^i}{dt} = -\nabla p^i + \operatorname{Div} \mathbf{\Pi}^i + e n^i \left(\mathbf{E} + \frac{1}{c} [\mathbf{v}^i, \mathbf{H}] \right) - \mathbf{R}; \\
& \frac{k n^i}{\gamma - 1} \cdot \frac{d^i T^i}{dt} + p^i \operatorname{div} \mathbf{v}^i = \operatorname{div}(\varkappa^i \nabla T^i) + \sum_{\alpha\beta} \Pi^i_{\alpha\beta} \frac{\partial v^i_\alpha}{\partial x_\beta} + Q; \\
& \frac{\partial n^e}{\partial t} + \operatorname{div} n^e \mathbf{v}^e = 0; \\
& m^e n^e \frac{d^e \mathbf{v}^e}{dt} = -\nabla p^e + \operatorname{Div} \mathbf{\Pi}^e - e n^e \left(\mathbf{E} + \frac{1}{c} [\mathbf{v}^e, \mathbf{H}] \right) + \mathbf{R}; \\
& \frac{k n^e}{\gamma - 1} \cdot \frac{d^e T^e}{dt} + p^e \operatorname{div} \mathbf{v}^e = \operatorname{div}(\varkappa^e \nabla T^e) + \\
& \qquad\qquad + \sum_{\alpha\beta} \Pi^e_{\alpha\beta} \frac{\partial v^e_\alpha}{\partial x_\beta} - Q + \frac{j^2}{\sigma},
\end{aligned}
\right\} \tag{1.16}
$$

where

$$p^{i,\,e} = k n^{i,\,e} T^{i,\,e}; \quad \mathbf{j} = e(n^i \mathbf{v}^i - n^e \mathbf{v}^e);$$

$$\mathbf{R} = \frac{e^2 n^i n^e}{\sigma}(\mathbf{v}^i - \mathbf{v}^e); \quad Q = B(T^e - T^i);$$

$$\Pi^{i,\,e}_{\alpha\beta} = \eta^{i,\,e}\left(\frac{\partial v_\alpha}{\partial x_\beta} + \frac{\partial v_\beta}{\partial x_\alpha}\right)^{i,\,e} + \left(\zeta^{i,\,e} - \frac{2}{3}\eta^{i,\,e}\right)\delta_{\alpha\beta}\operatorname{div}\mathbf{v}^{i,\,e};$$

$$\frac{d^{i,\,e}}{dt} = \frac{\partial}{\partial t} + (\mathbf{v}^{i,\,e}, \nabla).$$

Here n is the density of particles of the given species, m is the particle mass, p is the pressure, T is the temperature, **v** is the velocity, **H** and **E** are the magnetic and electric fields, **j** is the current density, σ is the conductivity, $\varkappa$ is the thermal conductivity, k is the Boltzmann constant, e is the electron charge, and c is the velocity of light. Also, **R** is the force of the mutual friction between electrons and ions, Q is the heat transferred from electrons to ions in collisions, and the viscous-stress tensor Π has the form used in ordinary hydrodynamics [41] which is frequently used in magnetohydrodynamic problems [24, 74]. In (1.16), this tensor is written in Cartesian coordinates (x_1, x_2, x_3). The coefficients η and ζ are measures of the first and second viscosities. The subscripts "i" and "e" refer to ions and electrons, respectively. The first three equations in (1.16) are the continuity equation, the equation of motion for the ions, and the energy equation for the ions; the last three equations are the equivalent equations for the electrons.

The electromagnetic field vectors $\mathbf{H}$, $\mathbf{E}$, and $\mathbf{j}$ obey the Maxwell equations

$$\left. \begin{array}{c} \dfrac{1}{c} \cdot \dfrac{\partial \mathbf{H}}{\partial t} + \operatorname{curl} \mathbf{E} = 0; \\[2mm] \dfrac{1}{c} \cdot \dfrac{\partial \mathbf{E}}{\partial t} + \dfrac{4\pi}{c}\, \mathbf{j} = \operatorname{curl} \mathbf{H}. \end{array} \right\} \tag{1.17}$$

Equations (1.16) and (1.17) form a closed system; i.e., the number of equations is equal to the number of unknowns. Here we are making use of the fact that the transport coefficients η, ζ, $\varkappa$, σ, and B can be expressed in terms of the unknown functions [13].

<u>Units, Dimensionless Equations, and Parameters</u>. The units for all the quantities in Eqs. (1.16)-(1.17) are the corresponding scale values. Specifically, we use the dimensional physical parameters involved in the formulation of the problem: the ion and electron densities n_0 specified at the entrance to the channel, the temperatures T_0^i and T_0^e (and thus the pressures p_0^i and p_0^e) specified at the entrance of the channel, the scale value of the magnetic field, H_0, the length of the channel, L, and the known constants m^i, m^e, e, and c.

We adopt m^i as the unit of mass, L as the unit of length, n_0 as the unit of density, H_0 as the unit of magnetic field, and $T_0 = T_0^i + T_0^e$ as the unit of temperature. From these units we form the unit of velocity, $v_0 = H_0/(4\pi m^i n_0)^{1/2}$ (the characteristic value of the Alfvén velocity), the unit of time, $t_0 = L/v_0$, the unit of pressure,†

---

†Here the unit of pressure is the scale magnetic pressure at the entrance to the channel. Another possibility is to take the quantity $\rho_0 = \rho_0^i + \rho_0^e$ [19, 20]. Still another possibility is to take $\sqrt{\rho_0}$ as the unit of magnetic field [17, 18].

$H_0^2/4\pi$, the unit of electric field, $E_0 = H_0 v_0/c$, and the unit of current density, $j_0 = cH_0/4\pi L$. Furthermore, in certain (simplified) versions of the equations we take account of the entropy; the unit of entropy is the specific heat at constant volume, c_v.

In terms of these units, Eqs. (1.16)-(1.17) can be written in the following dimensionless form† :

$$\frac{\partial n^i}{\partial t} + \operatorname{div} n^i \mathbf{v}^i = 0; \tag{1.18a}$$

$$n^i \frac{d^i \mathbf{v}^i}{dt} = -\nabla p^i + \operatorname{Div} \mathbf{\Pi}^i + \frac{n^i}{\xi}\left(\mathbf{E} + [\mathbf{v}^i, \mathbf{H}] - \frac{\nu}{\xi} n^e (\mathbf{v}^i - \mathbf{v}^e)\right); \tag{1.18b}$$

$$\frac{\beta n^i}{2(\gamma-1)} \cdot \frac{d\, T^i}{dt} + p^i \operatorname{div} \mathbf{v}^i =$$
$$= \frac{\beta}{2(\gamma-1)}\, [B(T^e - T^i) + \operatorname{div}(\varkappa^i \nabla T^i)] + \sum_{\alpha,\,\beta} \Pi^i_{\alpha\beta} \frac{\partial v^i_\alpha}{\partial x_\beta}; \tag{1.18c}$$

$$\frac{\partial n^e}{\partial t} + \operatorname{div} n^e \mathbf{v}^e = 0; \tag{1.18d}$$

$$\mu n^e \frac{d^e \mathbf{v}^e}{dt} = -\nabla p^e + \operatorname{Div} \mathbf{\Pi}^e -$$
$$-\frac{n^e}{\xi}\left(\mathbf{E} + [\mathbf{v}^e, \mathbf{H}] - \frac{\nu}{\xi} n^i (\mathbf{v}^i - \mathbf{v}^e)\right); \tag{1.18e}$$

$$\frac{\beta n^e}{2(\gamma-1)} \cdot \frac{d^e T^e}{dt} + p^e \operatorname{div} \mathbf{v}^e = \frac{\beta}{2(\gamma-1)}\, [-B(T^e - T^i) + \operatorname{div}(\varkappa^e \nabla T^e)]$$
$$+ \sum_{\alpha\beta} \Pi^e_{\alpha\beta} \frac{\partial v^e_\alpha}{\partial x_\beta} + \nu j^2; \tag{1.18f}$$

$$\frac{\partial \mathbf{H}}{\partial t} + \operatorname{curl} \mathbf{E} = 0; \tag{1.18g}$$

$$\varepsilon \cdot \frac{\partial \mathbf{E}}{\partial t} + \mathbf{j} = \operatorname{curl} \mathbf{H}; \tag{1.18h}$$

$$p^{i,\,e} = \frac{\beta}{2} n^{i,\,e} T^{i,\,e}; \quad \mathbf{j} = \frac{1}{\xi}(n^i \mathbf{v}^i - n^e \mathbf{v}^e);$$

$$\Pi^{i,\,e}_{\alpha\beta} = \eta^{i,\,e}\left(\frac{\partial v_\alpha}{\partial x_\beta} + \frac{\partial v_\beta}{\partial x_\alpha}\right)^{i,\,e} + \left(\zeta^{i,\,e} - \frac{2}{3}\eta^{i,\,e}\right)\delta_{\alpha,\,\beta} \operatorname{div} \mathbf{v}^{i,\,e}.$$

---

† Below we only see these equations in their dimensionless form, so that the same notation can be retained for all quantitites. In the rare cases in which a dimensional quantity is used in the text, this feature is pointed out specifically or an appropriate index is supplied.

The original dimensional constants appear in (1.18), so that they only enter the problem in the following dimensionless combinations:

$$\beta = 8\pi p_0/H_0^2,\qquad (1.19)$$

which is the ratio of the gas pressure and the magnetic pressure specified at the entrance ($p_0 = p_0^i + p_0^e$); and

$$\xi = \frac{c}{eL}\sqrt{\frac{m^i}{4\pi n_0}},\qquad (1.20)$$

which is a dimensionless constant whose value corresponds to the "replacement factor" [51], which is a measure of the influence of the Hall effect (see § 2.5 for details); $\mu = m^e/m^i$, the ratio of the electron and ion masses; and $\varepsilon = v_0^2/c^2$, the square of the ratio of the scale flow velocity to the velocity of light. Equations (1.18) also contain dimensionless transport coefficients which are related to the corresponding dimensional coefficients in (1.16) in the following way:

$$\eta^{i,\,e} = \frac{\eta^i_{\dim}{}^e}{Lm^i n_0 v_0};\qquad \zeta^{i,\,e} = \frac{\zeta^{i,e}_{\dim}}{Lm^i n_0 v_0},\qquad (1.21)$$

where $\eta^{i,e}$ is the first viscosity coefficient and $\zeta^{i,e}$ is the second viscosity coefficient;

$$\varkappa^{i,\,e} = \frac{\varkappa^{i,\,e}_{\dim}}{Lm^i n_0 v_0 c_v},\qquad (1.22)$$

the thermal conductivities;

$$\nu = \frac{c^2}{4\pi\sigma L v_0} = \frac{1}{\mathrm{Re}_m},\qquad (1.23)$$

the dimensionless magnetic viscosity, which is the reciprocal of the characteristic value of the magnetic Reynolds number Re_m or of the dimensionless conductivity; and

$$B = \frac{LB_{\dim}}{m^i n_0 v_0 c_v},$$

which is the coefficient of heat transfer between ions and electrons.

These coefficients are expressed in terms of the unknown functions and are given below in the form required for the calculations.

Generally speaking, Eqs. (1.18) comprise a parabolic quasilinear system. With zero viscosity and heat conduction ($\Pi^{i,e} = 0$; $\varkappa^{i,e} = 0$) these equations form a hyperbolic system with three types of characteristics: ion, electron, and electromagnetic. The corresponding propagation velocities of small perturbations with respect to the medium are, respectively,

$$C^i = \sqrt{\frac{\gamma p^i}{n^i}}; \quad C^e = \sqrt{\frac{\gamma p^e}{\mu n^e}}; \quad C = \frac{1}{\sqrt{\varepsilon}} .$$

These velocities are very different. In the situations of interest the parameters μ and ε are very small, so that serious difficulties arise when one attempts to solve or analyze Eqs. (1.18). A numerical integration of these equations would require an extremely fine calculation grid. Nevertheless, Eqs. (1.18) with a finite value of μ are of interest in the theory of plasma flows in channels since they describe the flows quite thoroughly. They can play a role in the study of the processes which occur in the electrode sheaths, since an artificial increase in μ is equivalent to a model in which these sheaths are enlarged. To simulate certain other situations it is also meaningful to "make the electron heavier" or "slow down the velocity of light," i.e., to increase μ and ε. Accordingly, calculations carried out on the basis of the complete system of equations in (1.18) lead to interesting results, so that it is still of interest to devise an effective approximation method for solving these equations.

Magnetogasdynamic Equations. Incorporation of the Hall Effect. In a numerical solution of the problem we use equations derived from (1.18) by employing certain simplifications, the most important of which are associated with the difficulties mentioned above; the other simplifications correspond to neglecting effects which can be treated as secondary in certain particular problems. Let us examine the possible simplifications of system (1.18).

1. The most important is

$$\mu = 0; \quad \varepsilon = 0. \tag{1.24}$$

This simplification corresponds to the neglect of the electron inertia $m^e d^e \mathbf{v}^e/dt$ and the displacement current $\partial \mathbf{E}/c\partial t$ in (1.16) and (1.17), as noted in §1.1. Assumption (1.24) has several consequences.

a. In the case $\varepsilon = 0$, Eq. (1.18h) becomes

$$\mathbf{j} = \mathrm{curl}\ \mathbf{H}, \qquad (1.25)$$

and from (1.18a), (1.18d), and (1.25) we find

$$\partial (n^i - n^e)/\partial t = \xi\ \mathrm{div}\ \mathbf{j} = 0.$$

This condition corresponds to a plasma which is neutral everywhere ($n^i \equiv n^e$) if we specify neutrality at the initial time. Below we assume $n^i = n^e = \rho$, in accordance with the usual density notation in gasdynamics. When $\mu = 0$ the dimensionless ion number density is equal to the dimensionless mass density of the plasma (i.e., the mass density divided by the unit of mass density, $\rho_0 = m^i n_0$).

b. When the electron mass m^e, is neglected it is reasonable to assume that the electron viscosity satisfies $\mathbf{\Pi}^e = 0$. Further justification for this approach comes from the fact that the ratio of the electron and ion viscosity coefficients is of the order $\sqrt{\mu}$ [13, 59]. Writing $\mathbf{v}^i = \mathbf{v}$, $p^1 + p^e = p$, and $\mathbf{\Pi}^1 = \mathbf{\Pi}$, and combining Eqs. (1.18b) and (1.18e), we find the magnetogasdynamic equation of motion:

$$\rho \cdot \frac{dv}{dt} = -\nabla p + [\mathbf{j}, \mathbf{H}] + \mathrm{Div}\ \mathbf{\Pi}. \qquad (1.26)$$

c. When $\mu = 0$ and $\mathbf{\Pi}^e = 0$, Eq. (1.18e) becomes a generalized Ohm's law that relates the electric field $\mathbf{E}$ to the current $\mathbf{j}$. When $n^i = n^e$ we have

$$v^e = \mathbf{v} - \xi \mathbf{j}/\rho, \qquad (1.27)$$

so that this generalized Ohm's law becomes

$$\mathbf{E} = \nu\mathbf{j} - [\mathbf{v}, \mathbf{H}] + \frac{\xi}{\rho}([\mathbf{j}, \mathbf{H}] - \nabla\ p^e) \qquad (1.28)$$

and differs from the usual Ohm's law in magnetogasdynamics in that it incorporates a finite (rather than an infinitesimally small) difference between the ion and electron velocities, i.e., the Hall effect.

In the case $\xi = 0$, Eq. (1.28) becomes the usual Ohm's law.

d. Using Eqs. (1.18g) and (1.28), we can eliminate the electric field **E** from the basic system of equations:

$$\frac{\partial \mathbf{H}}{\partial t} = \mathrm{curl}\, [\mathbf{v}, \mathbf{H}] - \mathrm{curl}\, \nu \mathbf{j} - \xi\, \mathrm{curl} \frac{[\mathbf{j}, \mathbf{H}] - \nabla p^e}{\rho}, \qquad (1.29)$$

where $\mathbf{j} = \mathrm{curl}\, \mathbf{H}$. Equation (1.29) is the diffusion equation for the magnetic field. The last term on the right side, with the coefficient ξ, corresponds to the Hall effect.

e. Finally, when we use (1.27) and $\mathbf{\Pi}^e = 0$, Eq. (1.18f) becomes

$$\frac{\beta \rho}{2(\gamma-1)} \cdot \frac{dT^e}{dt} + p^e \,\mathrm{div}\, \mathbf{v} = \frac{\beta}{2(\gamma-1)} [\mathrm{div}\,(\varkappa^e \nabla T^e) - B(T^e - T^i)] +$$
$$+ \nu j^2 + \xi \left[\frac{\beta}{2(\gamma-1)} (\mathbf{j}, \nabla T^e) + p^e \,\mathrm{div}\, \frac{\mathbf{j}}{\rho} \right]. \qquad (1.30)$$

The last term, containing ξ, is again due to the Hall effect.

In summary, by setting the parameters μ and ε equal to zero, and consequently neglecting Π^e, we find that the problem is described by Eqs. (1.18a), (1.18c), (1.26), (1.29), and (1.30), with the six unknown functions ρ, T^i, T^e, v_z, v_r, and H. These are the magnetogasdynamic equations for a two-temperature plasma in which the Hall effect is taken into account. They represent the first step in the transformation from the complete two-fluid model in (1.18) to the case of classical magnetogasdynamics and can serve as the basis for calculations for particular problems. For example, calculations have been carried out for the pinch effect [31, 74] in the two-temperature formulation.

The problems dealt with in the present review do not incorporate two temperatures; further simplifications are used.

2. The next way to simplify the problem is to specify a relationship between T^i and T^e:

$$T^i = f_i(T); \quad T^e = f_e(T); \quad T = T^i + T^e,$$

where f_i and f_e are given functions. Combining (1.18c) and (1.30) we find the energy equation for the resultant temperature:

$$\frac{\beta \rho}{2(\gamma-1)} \cdot \frac{dT}{dt} + p \,\mathrm{div}\, \mathbf{v} = \frac{\beta}{2(\gamma-1)} \,\mathrm{div}\,(\varkappa \nabla T) + \sum_{\alpha\beta} \Pi_{\alpha\beta} \times$$

$$\times \frac{\partial v_\alpha}{\partial x_\beta} + v j^2 + \xi \left[\frac{\beta}{2(\gamma-1)} f_e'(T)(j, \nabla T) + f_e(T) \frac{p}{T} \operatorname{div} \frac{j}{\rho} \right], \qquad (1.31)$$

where $\varkappa = \varkappa^i f_i^{\,\prime}(T) + \varkappa^e f_e^{\,\prime}(T)$. In the discussion which follows and in the calculations which have been carried out, the simplest functional relation is adopted:

$$T^i = T^e = T/2. \qquad (1.32)$$

This assumption has the consequence $p^i = p^e = p/2$. This assumption is clearly justified in isothermal flows; in general, it is used solely to simplify the calculations. In this case the description of the problem reduces to the equations

$$\left.\begin{array}{l}
\dfrac{\partial \rho}{\partial t} + \operatorname{div} \rho \mathbf{v} = 0; \\[2mm]
\rho \cdot \dfrac{d\mathbf{v}}{dt} = -\nabla p + [\mathbf{j}, \mathbf{H}] + \operatorname{Div} \boldsymbol{\Pi}; \\[2mm]
\dfrac{\beta \rho}{2(\gamma-1)} \cdot \dfrac{dT}{dt} + p \operatorname{div} \mathbf{v} = \dfrac{\beta}{2(\gamma-1)} \operatorname{div}(\varkappa \nabla T) + \sum_{\alpha\beta} \Pi_{\alpha\beta} \dfrac{\partial v_\alpha}{\partial x_\beta} + \\[2mm]
\quad + v j^2 + \dfrac{\xi}{2} \left[\dfrac{\beta}{2(\gamma-1)} (\mathbf{j}, \nabla T) + p \operatorname{div} \dfrac{\mathbf{j}}{\rho} \right]; \\[2mm]
\dfrac{\partial \mathbf{H}}{\partial t} = \operatorname{curl}[\mathbf{v}, \mathbf{H}] - \operatorname{curl} v\mathbf{j} - \xi \operatorname{curl} \dfrac{[\mathbf{j},\mathbf{H}] - \dfrac{1}{2}\nabla p}{\rho},
\end{array}\right\} \qquad (1.33)$$

where $\mathbf{j} = \operatorname{curl} \mathbf{H}$, $p = (\beta/2)\,\rho T$, $\varkappa = \frac{1}{2}(\varkappa^i + \varkappa^e)$, and the $\Pi_{\alpha\beta}$ are defined in accordance with (1.18).

System (1.33) is used below in certain versions of the calculation (§2.6). These are the magnetogasdynamic equations incorporating three types of dissipative processes: viscosity, thermal conductivity, and electrical conductivity. These equations also incorporate the Hall effect. The Hall effect enters the energy and diffusion equations for the magnetic field in the terms with the coefficient ξ; it represents the least traditional element in Eqs. (1.33). Making use of the axial symmetry, we can rewrite these equations as

$$\begin{aligned}
\frac{\beta}{2(\gamma-1)} (\mathbf{j}, \nabla T) + p \operatorname{div} \frac{\mathbf{j}}{\rho} &= \left(\mathbf{j}, \frac{\beta}{2(\gamma-1)} \nabla T + p \nabla \frac{1}{\rho} \right) = \\
&= \frac{\beta T}{2(\gamma-1)} \left(\mathbf{j}, \frac{\nabla T}{T} - (\gamma-1)\frac{\nabla \rho}{\rho} \right) = \frac{\beta T}{2(\gamma-1)} (\mathbf{j}, \nabla s) = \\
&= \frac{\beta T}{2(\gamma-1)} \left(\frac{1}{r} \cdot \frac{\partial Hr}{\partial r} \frac{\partial s}{\partial z} - \frac{\partial H}{\partial z} \cdot \frac{\partial s}{\partial r} \right), \qquad (1.34)
\end{aligned}$$

where $s = \ln(T/\rho^{\gamma-1}) = \ln(2p/\beta\rho^{\gamma})$ is the dimensionless entropy, and

$$\operatorname{curl} \frac{[j, H] - \frac{1}{2}\nabla p}{\rho} = \frac{1}{\rho}\operatorname{curl}[j, H] + \left[\nabla\frac{1}{\rho}, [j, H] - \frac{1}{2}\nabla p\right] =$$

$$= \frac{1}{\rho}[(H, \nabla)j - (j, \nabla)H] + \left(\nabla\frac{1}{\rho}, H\right)j - \left(\nabla\frac{1}{\rho}, j\right)H + \frac{\beta}{4\rho}[\nabla\rho, \nabla T].$$

Generally speaking, this latter equation contains second derivatives with respect to the coordinates, but in the case of two-dimensional flow across a magnetic field under consideration here [Eq. (1.10)], this equation is simplified:

$$\operatorname{curl}\frac{[j, H] - \frac{1}{2}\nabla p}{\rho} = \frac{\partial}{\partial r}\left(\frac{H}{\rho r}\right)\frac{\partial Hr}{\partial z} - \frac{\partial}{\partial z}\left(\frac{H}{\rho r}\right)\frac{\partial Hr}{\partial r} +$$

$$+ \frac{\beta}{4\rho}\left(\frac{\partial\rho}{\partial z}\cdot\frac{\partial T}{\partial r} - \frac{\partial\rho}{\partial r}\cdot\frac{\partial T}{\partial z}\right). \qquad (1.35)$$

Equations (1.33) generally form a quasilinear parabolic (because of the dissipative terms) system. The Hall effect introduces an additional nonlinearity in (1.34) and (1.35) in less important terms of the equations. However, if we neglect the dissipative processes, these terms become the main terms and have a strong effect on the nature of the solution of the equations (see § 2.5 for more details).

3. Other obvious simplifications of the problem and of Eqs. (1.33) can be achieved by neglecting the Hall effect and the dissipative processes. Without assigning any strict priority,[†] we can cite the following possibilities: a) neglect of the Hall effect, i.e., the use of the assumption $\xi = 0$ in (1.33). The remaining equations then refer to the single-fluid model of the plasma, i.e., the usual magnetogasdynamic problem of a viscous, heat-conducting medium with a finite conductivity; b) neglect of the viscosity, i.e., $\Pi_{\alpha\beta} = 0$ $(\eta = \zeta = 0)$; c) neglect of the thermal conductivity, i.e., $\varkappa = 0$; d) the assumption of an infinite plasma conductivity, i.e., the neglect of the magnetic viscosity, $\nu = 0$.

[†]Each such simplification is used independently of the others on the basis of physical considerations and the particular purposes of the solution.

If we neglect all these factors, then Eq. (1.33) becomes the system of equations of ideal (dissipationless) magnetogasdynamics:

$$
\left.
\begin{aligned}
&\frac{\partial \rho}{\partial t} + \operatorname{div} \rho \mathbf{v} = 0; \\[4pt]
&\rho \cdot \frac{d\mathbf{v}}{dt} = -\nabla p + [\mathbf{j}, \mathbf{H}]; \\[4pt]
&\frac{\beta \rho}{2(\gamma - 1)} \cdot \frac{dT}{dt} + p \operatorname{div} \mathbf{v} = 0; \\[4pt]
&\frac{\partial \mathbf{H}}{\partial t} = \operatorname{curl}[\mathbf{v}, \mathbf{H}]; \\[4pt]
&p = \frac{\beta}{2}\rho T; \quad \mathbf{j} = \operatorname{curl} \mathbf{H}.
\end{aligned}
\right\}
\tag{1.36}
$$

This is again a hyperbolic system of equations. In the flow situation under consideration here, corresponding to (1.10), this system has (in addition to the trivial entropic and vortex characteristics) magnetosonic characteristics, which correspond to $C_m = \sqrt{C^2 + \frac{H^2}{\rho}}$, as the propagation velocity for small perturbations with respect to the medium. Here $C^2 = \gamma p / \rho$ is the square of the gasdynamic sound velocity.

4. The next possible simplification is to eliminate the energy equation from (1.33) or (1.36), replacing it by the isentropic assumption s = const, $p = \beta \rho^\gamma / 2$, or the isothermal assumption, T = const, $p = \beta \rho / 2$. This approach reduces the number of equations and the number of unknown functions by unity.

5. When the process is a weak function of radius the calculations can be carried out in the quasi-one-dimensional (hydraulic) approximation. In this approach the channel cross section is specified by a slowly varying function $f(x)$, where x is the coordinate along the channel. All quantities are averaged over the cross section and are functions of the variables t and x alone. The velocity $\mathbf{v}$ is parallel to the x axis, while the magnetic field $\mathbf{H}$ is again perpendicular to the flow.

In this case the plasma flow is described by the equations [4, 17, 24]

$$
\frac{\partial \rho f}{\partial t} + \frac{\partial \rho v f}{\partial x} = 0;
$$

$$\rho \frac{dv}{dt} + \frac{\partial}{\partial x}\left(p + \frac{H^2}{2}\right) = \frac{\partial}{\partial x}\left(\eta\,\frac{\partial v}{\partial x}\right);$$

$$\frac{\beta \rho f}{2(\gamma-1)}\frac{dT}{dt} + p\,\frac{\partial vf}{\partial x} = \frac{\beta}{2(\gamma-1)}\cdot\frac{\partial}{\partial x}\left(\varkappa f\,\frac{\partial T}{\partial x}\right) +$$

$$+ f\left[\eta\left(\frac{\partial v}{\partial x}\right)^2 + \nu\left(\frac{\partial H}{\partial x}\right)^2\right];$$

$$\frac{\partial Hf}{\partial t} + \frac{\partial Hvf}{\partial x} = \frac{\partial}{\partial x}\left(\nu f\,\frac{\partial H}{\partial x}\right);$$

$$p = \frac{\beta}{2}\rho T.$$

(1.37)

The quasi-one-dimensional description of the problem in (1.37) does not include the Hall effect, because this effect is inherently a two-dimensional effect. The dissipative terms and the energy equation can be either completely or partially neglected, depending on a given problem.

Transport Coefficients. The Parameter $\omega^e \tau^e$.
The transport coefficients η, ζ, $\varkappa$, and ν in Eqs. (1.33), (1.36), and (1.37) are specified as known functions of T, ρ, and H, after Braginskii [13]. These functions contain the dimensionless parameter $\chi = \omega^e \tau^e$, which is a measure of the "magnetization" of the electron fluid and is frequently used for quantitative estimates of the importance of the Hall effect. Here τ^e is the mean time between collisions, and ω^e is the electron gyrofrequency. In terms of dimensional quantities, we have

$$\omega^e = eH/m^e c; \quad \tau^e = 3\sqrt{m^e}\,(kT^e)^{3/2}/4\sqrt{2\pi}\lambda e^4 n^i,$$

where λ is the Coulomb logarithm.† Transforming to the units chosen above, we find the dimensionless equation

$$\chi = \omega^e \tau^e = X\frac{H(T^e)^{3/2}}{\lambda\rho},$$

(1.38)

where X is the dimensionless constant:

$$X = \frac{3}{4\sqrt{2\pi}}\cdot\frac{H_0(kT_0)^{3/2}}{\sqrt{m^e}\,n_0\,e^3 c}.$$

(1.39)

---

† See [13, 59]. In the problems under consideration here $\lambda \approx 10$.

The analogous parameter for the ions, $\omega^i \tau^i$, can be neglected, since

$$\omega^i \tau^i = \sqrt{2\mu} \, (T^i/T^e)^{3/2} \, \omega^e \tau^e \ll \omega^e \tau^e.$$

In flow across the magnetic field under assumption (1.32) the dimensionless transport coefficients in (1.21)–(1.23) can be written

$$\mathrm{Re}_m = \frac{1}{\nu} = \frac{X}{\xi \lambda \alpha_0 (\chi)} \left(\frac{T}{2} \right)^{3/2}; \tag{1.40}$$

$$\varkappa = \frac{1}{2} \varkappa^e = \frac{(\gamma-1) \, \beta \xi X \gamma_0 (\chi)}{4\lambda} \left(\frac{T}{2} \right)^{5/2}; \tag{1.41}$$

$$\eta = \frac{0.96 \beta \xi \sqrt{\mu} \, X}{\sqrt{2} \lambda} \left(\frac{T}{2} \right)^{5/2}. \tag{1.42}$$

In Eq. (1.41) we take account of the fact that the ion thermal conductivity satisfies

$$\varkappa^i \approx \sqrt{\mu} \varkappa^e \ll \varkappa^e,$$

so that it can be neglected. The functions $\alpha_0(\chi)$ and $\gamma_0(\chi)$ vary slowly over the temperature range under consideration here; the function $\alpha_0(\chi)$ never vanishes;

$$\alpha_0 (\chi) = 1 - \frac{1.837 + 6.416 \chi^2}{3.77 + 14.79 \chi^2 + \chi^4}; \quad \gamma_0 (\chi) = \frac{11.92 + 4.664 \chi^2}{3.77 + 14.79 \chi^2 + \chi^4}.$$

The second viscosity coefficient ζ is usually set equal to the first: $\zeta = \eta$. In the calculation program, the transport coefficients are used in a slightly simplified form:

$$\eta = \eta_1 + \eta_2 \, T^{5/2}; \tag{1.43}$$

$$\zeta = \zeta_1 + \zeta_2 \, T^{5/2}; \tag{1.44}$$

$$\varkappa = \varkappa_1 + \varkappa_2 \, T^{5/2}; \tag{1.45}$$

$$\mathrm{Re}_m = (1/\nu) = \sigma_1 + \sigma_2 \, T^{3/2}, \tag{1.46}$$

where all eight coefficients are constant.

If we approximate λ, α_0, and γ_0 by constants, the functions in (1.40)-(1.42) obviously transform into the particular case of Eqs. (1.43)-(1.46).

On the other hand, the transport coefficients can be assumed constant in the solution of the problem. This approach is meaningful in a study of qualitative questions regarding the roles of the dissipative processes.

Finally, the first terms in Eqs. (1.43)-(1.45) allow us to retain a certain minimum level of the dissipative effects at small values of T; this feature can be useful in shock-wave calculations.

Boundary Conditions and Initial Conditions. Under the assumption of axial symmetry, (1.10), Eqs. (1.33) are examined in that range of the cylindrical variables (z, r) corresponding to the channel geometry (Fig. 5). The shape of the electrodes that form the channel is specified by the curves

$$r = r_1(z); \; r = r_2(z). \tag{1.47}$$

We can draw up a complete list of the boundary conditions at the boundaries of the region in accordance with the discussion of these conditions in §1.1.

1. At the entrance to the channel, at z = 0, the conditions

$$\rho = 1; \; T = 1, \; H = r_0/r \tag{1.48}$$

are specified. Here r_0 is the scale value of the radius in the entrance cross section, e.g., $r_0 = (1/2)[r_1(0) + r_2(0)]$. If the viscosity is $\Pi = 0$, conditions (1.48) are sufficient, since the velocity at the entrance is subsonic.[†]

The presence of a viscosity also requires boundary conditions for the velocity. In the calculations these conditions are specified as

$$\partial (\rho v_z f)/\partial z = 0; \; v_r = (dr/dz)v_z, \tag{1.49}$$

where $f(z)$ is the cross-sectional area of the coordinate tube, and dr/dz is the slope of the coordinate line (§1.3) with respect to the z axis. The first condition in (1.49) means that the flow at the

---

[†]Otherwise it is necessary to supplement (1.48) with the specification of the velocity at z = 0.

entrance is approximately stationary, while the second means that the plasma enters the channel along characteristic directions, which are taken to be the coordinate lines.

2. The conditions specified at the electrodes, (1.47), are usually

$$\mathbf{v} = 0; \ E_\tau = 0; \ \partial T/\partial n = 0, \tag{1.50}$$

where $\mathbf{n}$ is the unit vector normal to the electrode surface and τ is tangent to this surface. In certain cases the third condition in (1.50) is replaced by a condition regarding heat transfer across the channel wall:

$$\partial T/\partial n = -kT. \tag{1.51}$$

We note that Eqs. (1.33) are required under conditions (1.50) only if there are dissipative terms. If $\Pi = 0$, the condition $\mathbf{v} = 0$ is replaced by the slip condition $v_n = 0$. If $\varkappa = 0$, the third boundary condition in (1.50) or condition (1.51) drops out. If $\xi = 0$ and $\nu = 0$, the second condition in (1.50) drops out.

Accordingly, if the problem is solved on the basis of the ideal equations in (1.36), then the boundary conditions at the walls in (1.50) are replaced by the single condition $v_n = 0$.

3. If the inner electrode is shorter than the outer electrode (Fig. 5), part of the boundary of the region lies on the channel axis. In this case the obvious conditions of axial symmetry are imposed:

$$v_r = 0; \ H = 0; \ \partial\rho/\partial r = \partial T/\partial r = \partial v_z/\partial r = 0. \tag{1.52}$$

4. The boundary conditions at the channel exit, at $z = z_1$, are less obvious; here we cannot predict any definite relations on the basis of physical considerations. For this reason, the conditions required for the calculations are somewhat artificial.

In this case the equations of ideal magnetogasdynamics in (1.36) require an additional condition if the exit velocity is lower than the propagation velocity of small perturbations with respect to the medium (the "signal" velocity), $v < C_m$, but they require nothing if $v > C_m$. If the problem includes viscosity, thermal conductivity, and a finite electrical conductivity, then Eq. (1.33) requires boundary conditions on v, T, and H, respectively.

One possible form of the boundary conditions,

$$\partial v_z/\partial z = \partial v_r/\partial z = 0, \ \partial T/\partial z = 0; \ \partial H/\partial z = 0, \tag{1.53}$$

corresponds to the traditional approach in gasdynamics [8]. If the calculation only includes certain dissipative processes, then the superfluous conditions in (1.53) can be discarded.

In the calculations associated with [17-20], without viscosity and without thermal conductivity, the third condition in (1.53) is replaced by

$$\partial Hf/\partial z = 0. \tag{1.54}$$

This condition is chosen on the basis of a series of calculations carried out in the quasi-one-dimensional version of the problem [17], in which the channel is artificially lengthened by means of tubes of various cross sections. From the results of these calculations we get the impression that the results are very insensitive to the boundary condition on H at the exit. This condition is imposed in connection with the magnetic viscosity ν, so that it should play a particularly important role at high values of ν. In this case, however, the calculations show that the magnetic field penetrates the channel slightly, being very weak at the exit; conditions (1.53) and (1.54) become equivalent and can be replaced by the condition H = 0. At small values of ν, on the other hand, the magnetic field penetrates over the entire length of the channel and its distribution is approximately the same as that of the density.†

One of the main purposes of the solution of these problems is to study the steady-state, i.e., time-independent plasma flow in channels. Formally, this study can be carried out by solving the problem in its steady-state version, i.e., by setting $\partial/\partial t \equiv 0$ in Eqs. (1.33). The boundary conditions listed above are consistent with this approach, since all are of a steady-state nature. However, this approach runs into serious difficulties: As a rule, steady-state flow in a nozzle is subsonic at the entrance to the channel and supersonic at the exit. The line at which the magnetosonic velocity C_m is reached lies near the minimum cross section of the nozzle. On this line the steady-state equations of ideal magnetogasdynamics

---

† The magnetic field is frozen in the plasma when ν = 0 [see (1.2b)]. In this case no boundary condition is necessary.

change in nature: To the left of the line they constitute an elliptic system; to the right they constitute a hyperbolic system. The position of the line itself must be determined in the calculations. Although the introduction of dissipative terms in the equations does result in equations of the same kind, it is not the answer to the problem, since the transport coefficients in real problems are frequently very small. Accordingly, calculations for the steady-state model are extremely complicated.[†]

Furthermore, the transient processes that precede the steady-state flow are of physical interest in themselves, and the study of these processes provides results in certain cases. For this reason the problem is solved on the basis of the nonstationary model in (1.33) by an asymptotic method. The initial conditions at $t = 0$ can be specified arbitrarily, provided that they lead to acceleration of the plasma from left to right. For example, it is sufficient to set $\mathbf{v} = 0$ and to require that ρ, T, and H be monotonically decreasing functions of z which are consistent with the boundary conditions in (1.48).

The problem is thus one of solving Eqs. (1.33) in that region of the variables (z, r) which is shown schematically in Fig. 5, under boundary conditions (1.48)-(1.54) and under the initial conditions just noted.

§1.3. Method for Numerical Solution of the Problem

Spatial Coordinate Systems. The cylindrical coordinates (z, r) used in formulating the problem in §1.2 are generally not convenient for the numerical solution. The boundaries of the calculation region are curvilinear, and this feature complicates the calculation procedure and the calculation equations at the boundaries. The calculations which we are reviewing here are carried out in curvilinear coordinates which correspond better to the shape of the region and which convert it into a rectangle with sides parallel to the axes. In various stages of these calculations we use three such coordinate systems.

[†] These calculations can be carried out under very stringent restrictions, e.g., in a study of supersonic flow.

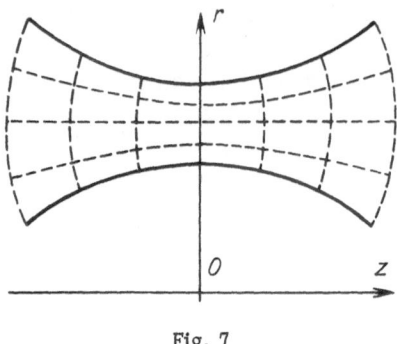

Fig. 7

1. The hyperbolic coordinates[†] (ξ, η), which are related to (z, r) by

$$r = a + \frac{1}{b}|\xi|\eta;$$

$$z = \frac{\text{sign}\,\xi}{b}\sqrt{(b^2 - \eta^2)(\xi^2 - b^2)}.$$

(1.55)

The coordinate grid consists of the hyperbolas $\eta = $ const and the ellipses $\xi = $ const, which are orthogonal to the hyperbolas (Fig. 7). The coordinates in (1.55) are useful when the electrodes can be incorporated in a system of coordinate hyperbolas through an appropriate choice of the constants a and b. In this case the left and right boundaries of the region should not be specified as the straight line segments z = const, but as the elliptic arcs $\xi = $ const; this requirement can easily be reflected in the initial formulation of the problem. The calculation region is transformed into the two rectangles $\xi_0 \leq \xi \leq -b$; $b \leq \xi \leq \xi_1$; $\eta_0 \leq \eta \leq \eta_1$, which should be thought of as rectangles joined to make a single rectangle along the lines $\xi = -b$ and $\xi = b$, which correspond to z = 0.

The advantages of using coordinate system (1.55) are its orthogonal nature and the resulting relative simplicity and natural form of the equations. On the other hand, this choice limits the choice of electrode shape. In particular, the interesting case of a shortened central electrode (Fig. 5) cannot be treated. The magnetogasdynamic equations are written out in hyperbolic coordinates[‡] in [18].

[†] The notation for the hyperbolic coordinates (ξ, η) is that of [18]. We do not change the notation in order to avoid confusion with the replacement factor ξ and the viscosity coefficient η, since these hyperbolic coordinates are not used anywhere except in this section.

[‡] In simplified form these equations are s = const, $\Pi = 0$, $\varkappa = 0$, $\xi = 0$.

2. The next coordinate system is based on a "straightening" of the boundaries (the electrodes) and the construction of a radially uniform grid between these boundaries. In addition, the grid can be made nonuniform along z, with a varying step Δz along the channel, if desired. The coordinates (α, λ) are defined by

$$z = z(\alpha); \; r = (1 - \lambda)r_1(\alpha) + \lambda r_2(\alpha); \; 0 \leqslant \alpha; \; \lambda \leqslant 1. \quad (1.56)$$

Here $z(\alpha)$ is a monotonic function of α. In the simplest case, a grid which is uniform along z, we have $z(\alpha) \equiv \alpha$. The functions $r_1(\alpha)$ and $r_2(\alpha)$ describe the shape of the electrodes in accordance with (1.47). If the inner electrode is shorter than the outer electrode (Fig. 5), we set $r_1(\alpha) = 0$ for $1 \leq z \leq z_1$. Both functions are arbitrary and cover a variety of electrode shapes. The only requirements are that they be single-valued and that they have no vertical tangents. In the actual calculations the functions $z(\alpha)$, $r_1(\alpha)$, and $r_2(\alpha)$ are specified in polynomial form.

The coordinate λ provides a uniform distribution of the calculation points along the radius. The lines of $\lambda = $ const are obviously found through a linear interpolation between the electrodes.

The results of a solution of the problem in terms of the coordinates in (1.56) are reported in [19-21] and in §§2.4-2.6 of the present paper.

3. The two coordinate systems discussed immediately above are Euler coordinates, i.e., fixed in space. The third system is of a mixed type, being an Euler system along the channel axis but a Lagrangian system along the radius. These coordinates must be used if the flow in the channel involves zones in which the plasma density exhibits a highly nonuniform radial profile. In this case it is useful to have a grid with a relatively uniform step in terms of the mass. Euler–Lagrange coordinates are used in calculations for two-dimensional gasdynamics [26]. In the work being reviewed here they are used in calculations for the magnetoplasma compressor (see § 2.6 and [23]).

We retain the Euler coordinate z as one of the coordinates.[†] We denote the other coordinate — the Lagrangian coordinate — by q and require that the coordinate line $q(t, z, r) = $ const move along

---

[†] It is possible to replace z by any monotonic function α, as in the preceding case.

with the plasma as time elapses; i.e., we require that the component of the velocity **v** normal to this line coincide with the velocity of the line itself along the normal to this line at each point at all times. We normalize q so that the value q = 0 corresponds to the central electrode and q = 1 corresponds to the outer electrode. Then the range of the coordinates (z, q) is the rectangle $0 \le z \le z_i$; $0 \le q \le 1$. It is not difficult to show that these requirements reduce to the requirement that the Euler coordinate r(t, z, q) be determined from the auxiliary equation[†]

$$(\partial r/\partial t) + u \, (\partial r/\partial z) = v, \qquad (1.57)$$

where $u = v_z$, $v = v_r$, and where the transformation of Eqs. (1.33) to the new coordinates is carried out on the basis of

$$\left. \begin{array}{l} \partial/\partial t \rightarrow \partial/\partial t - (r_t/r_q)(\partial/\partial q); \\ \partial/\partial z \rightarrow \partial/\partial z - (r_z/r_q)(\partial/\partial q); \\ \partial/\partial r \rightarrow (1/r_q)(\partial/\partial q), \end{array} \right\} \qquad (1.58)$$

where r_t, r_z, and r_q are the partial derivatives of r with respect to t, z, and q. In particular, it follows from (1.57) and (1.58) that

$$d/dt \equiv \partial/\partial t + u \, (\partial/\partial z) + v \, (\partial/\partial r) \rightarrow \partial/\partial t + u \, (\partial/\partial z). \qquad (1.59)$$

This simplification of the transport operator in (1.59) is a manifestation of the Lagrangian nature of the new coordinate q.

Difference Scheme. Turning now to the numerical solution of the problem, we note that Eq. (1.33) for the two-dimensional plasma flow across a magnetic field and (1.10) are similar to the gasdynamic equations; the magnetic field H acts formally as an additional pressure. Accordingly, in working out a method for numerical integration of the equations we can draw on the extensive experience which has been acquired in calculations in similar gasdynamic problems with thermal conductivity.

Let us review the calculation schemes used in the work being discussed here. The calculation scheme used for the simplified version of the equations is described in [15]. The difference scheme is constructed in the same manner for all the coordinate systems described above; in describing this scheme we temporarily adopt

---

[†] This equation must obviously be supplemented with the initial distribution of r and with a boundary condition at z = 0.

the notation (x, y) for the spatial coordinates. To illustrate the
process of approximating differential equations (1.33) by difference
equations, we distinguish three distinct elements of these equations
and discuss each element separately.

1. Equation (1.33) is based on (1.36), the magnetogasdynamic equations of an ideal medium. We rewrite these latter equations as

$$\begin{aligned}
d\mathbf{v}/dt &= -(1/\rho)\,\nabla\, p + (1/\rho)[\,\mathrm{curl}\ \mathbf{H},\ \mathbf{H}]; \\
d\rho/dt &= -\rho\,\mathrm{div}\,\mathbf{v}; \\
dT/dt &= -(\gamma-1)\,T\,\mathrm{div}\,\mathbf{v}; \\
dH/dt &= -\mathbf{H}\,[\mathrm{div}\,\mathbf{v}-(v/r)]; \\
d/dt &= \partial/\partial t + a\,(\partial/\partial x) + b\,(\partial/\partial y),
\end{aligned} \right\} \qquad (1.60)$$

where a and b are expressed in terms of z, r, u, and v; they are
the projections of the velocity **v** onto the directions orthogonal to
the (x, y) coordinate lines.

We introduce calculation points in two kinds, in a checker-
board arrangement (Fig. 8). The unknown functions are divided
into two groups. The values of the functions of the first group
(u, v, and, if necessary, z and r) are only calculated only at points
of the first kind (the circles), while the functions of the second
group (ρ, T, H, and thus p; also, if necessary, the entropy s) are
calculated at the points of the second kind (the crosses). The
crosses are also shifted by half a time step from the circles,
so that we denote the points corresponding to the circles by integer indices and those corresponding to the crosses by half-

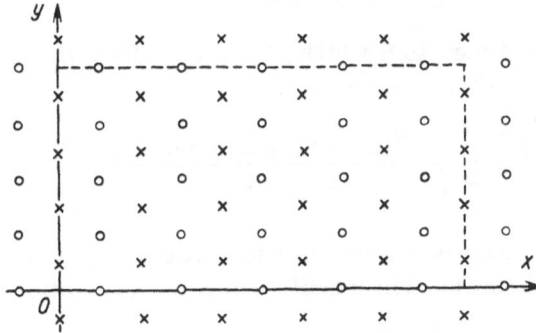

Fig. 8

integer indices. For example, u_{lm}^n is the calculated value of u at the points with coordinates $t = n\Delta t$, $x = l\Delta x$, $y = m\Delta y$, and $\rho_{l+1/2,\,m+1/2}^{n+1/2}$ is the value of $t = (n + 1/2)\Delta t$ at the point with coordinates $x = (l + 1/2)\Delta x$, $y = (m + 1/2)\Delta y$.

The operator d/dt on the left side of all the equations in (1.60) is approximated by the difference operator

$$\frac{du}{dt} \sim \frac{u_{lm}^{n+1} - u_{lm}^n}{\Delta t} + \frac{a + |a|}{2} \cdot \frac{u_{lm}^n - u_{l-1,\,m}^n}{\Delta x} + \frac{a - |a|}{2} \cdot \frac{u_{l+1,\,m}^n - u_{lm}^n}{\Delta x} +$$

$$+ \frac{b + |b|}{2} \cdot \frac{u_{lm}^n - u_{l,\,m-1}^n}{\Delta y} + \frac{b - |b|}{2} \cdot \frac{u_{l,\,m+1}^n - u_{lm}^n}{\Delta y}, \qquad (1.61)$$

where the values of the functions a and b are taken at the point (l, m, n). The meaning of Eq. (1.61) is that the derivatives with respect to x and y in the operator d/dt are replaced by the one-sided differences, with the particular side depending on the direction of the velocity. We approximate dv/dt in the same way and use the same procedure (but only for the half-integer points) for $d\rho/dt$, dT/dt, and dH/dt. In the latter case the values of a and b are taken to be the average values found from the four nearest integer points. The right sides of Eqs. (1.60) contain first derivatives with respect to x and y only of the functions of the opposite[†] group. We approximate them by differences with the help of the four nearest points of the intermediate layer; e.g., we approximate the derivative which appears in ∇p in the first equation by

$$\frac{\partial p}{\partial x} \sim \frac{p_{l+1/2,\,m+1/2}^{n+1/2} + p_{l+1/2,\,m-1/2}^{n+1/2} - p_{l-1/2,\,m+1/2}^{n+1/2} - p_{l-1/2,\,m-1/2}^{n+1/2}}{2\Delta x}, \qquad (1.62)$$

and we approximate that which appears in div **v** in the last three equations by

$$\frac{\partial u}{\partial x} \sim \frac{u_{l+1,\,m}^n + u_{l+1,\,m+1}^n - u_{lm}^n - u_{l,\,m+1}^n}{2\Delta x}. \qquad (1.63)$$

The difference scheme constructed for Eqs. (1.60) is a generalization of the scheme used in one-dimensional gasdynamic problems [56].

[†]With respect to the left sides of the equations.

2. The next element in Eqs. (1.33) consists of the second derivatives with respect to x and y in the dissipative terms:

$$\text{Div } \Pi; \quad \text{div}\,(\varkappa \nabla T); \quad -\text{curl}\,(\nu\,\text{curl}\,H). \tag{1.64}$$

We note that the second derivatives involved here are those of only that function which is differentiated with respect to t on the left side of the equation. In other words, each equation containing the expressions in (1.64) is an equation of the heat-conduction type. It is well known that the solution of an equation of this type by means of the simplest explicit difference scheme requires an extremely stringent restriction on the time step[†] Δt. Accordingly, an explicit scheme can be recommended only for very small values of the corresponding dissipation coefficient: η, ζ, $\varkappa$, or ν. Otherwise the second derivatives in (1.64) are better approximated by implicit difference expressions.

In the calculations we adopt the separation method discussed in detail by Yanenko [62]. In the transition from each time layer to the next, two "sweeps" are made — horizontal and vertical. The mixed second derivatives are approximated by explicit difference expressions. We use the refinement that in the viscous case the equations for each component of the velocity **v** contain second derivatives of the other component. They are also approximated by explicit expressions.

3. The third element of Eqs. (1.33) consists of the terms which are quadratic in first derivatives and are associated with dissipative processes,

$$\sum_{\alpha,\,\beta} \Pi_{\alpha,\,\beta} \frac{\partial v_\alpha}{\partial x_\beta} + \nu\,(\text{curl } H)^2,$$

and the Hall effect (1.34) and (1.35). These terms play a secondary role in the equations, since they are subordinate to the main terms — the second derivatives — in the case of viscosity and a finite conductivity; when $\Pi = 0$, $\nu = 0$ they do not appear.[‡] Accordingly, they can be approximated by differences in an arbitrary way. In the case under consideration here, all these first derivatives are

---

[†]Here $\Delta t \sim \min\,[(\Delta x)^2,\,(\Delta y)^2]$ [62].

[‡]It is shown in §2.5 that in the case $\nu = 0$ fundamental difficulties generally arise in attempts to incorporate the Hall effect.

replaced by central differences on the basis of the nearest points: for $\partial v_\alpha / \partial x_\beta$, four points are used, by analogy with (1.63), while for the derivatives of ρ, T, and H two points are used (for each function), e.g.,

$$\frac{\partial H}{\partial x} \sim \frac{1}{2\Delta x} (H_{l+3/2,\, m+1/2}^{n+1/2} - H_{l-1/2,\, m+1/2}^{n+1/2}).$$

They are taken from a known time layer; i.e., they constitute an explicit part of the difference scheme.

The boundary conditions listed at the end of § 1.2 can be written without difficulty for the difference equations. It is convenient to construct the upper and lower boundaries of the region from the integer points and to construct the left and right boundaries from the half-integer points (Fig. 8). For convenience in writing the boundary conditions at the left, the bottom, and the top, we also introduce fictitious points extending half a step outside the region. Then the boundary conditions written in their final form are easily transferred to the necessary points on the boundary, and the derivatives in them are replaced by difference ratios.

A few words are in order regarding the choice of steps for the difference grid. The spatial steps Δx and Δy are fixed for each calculation version in accordance with the accuracy level required in the given version. One criterion for the accuracy of the result is the agreement within the specified error with the result of calculations carried out with finer steps.

The time step Δt is determined by the requirement that the difference scheme be stable. If the equations contain dissipative terms, (1.64), and the solution is carried out by means of implicit schemes, then the scheme is stable for any value of Δt [62]. Otherwide, the problem is solved in terms of Eq. (1.60) by an explicit scheme; for this scheme to be stable, it is necessary to satisfy the Courant condition

$$\max_{(z,\, r)} (|\mathbf{v}| + C_m) \Delta t \leqslant \min (\Delta z, \Delta r), \tag{1.65}$$

which is a generalization of the corresponding condition in one-dimensional gasdynamic problems [56].

In certain versions of the problem, however, the dissipative effects are not all taken into account or are small. It is therefore desirable that the scheme remain stable in any case, so that condition (1.65) is applied in all the calculations. This condition is the rule used for choosing the time step Δt in each time layer of the calculation.

Chapter 2

FLOW OF FULLY IONIZED PLASMA

§ 2.1. Establishment of Steady State; Stability of the Flow

The calculation method described above makes it possible to follow the evolution of a flow beginning from some initial state. Depending on the particular model adopted for the calculations, three different cases are observed:

(1) The flow reaches a steady state in a certain time τ_{tra}; at $t \gg \tau_{tra}$ all functions become essentially independent of the time. This is always the situation in the case of the flow of a plasma which is a good conductor ($Re_m \gg 1$) with no Hall effect ($\xi = 0$).

(2) The flow does not reach a steady state. The functions describing the flow continue to vary with the time in some quasi-periodic manner. Fluctuating flows of this type are observed at a low plasma conductivity ($Re_m \ll 1$) at the channel entrance with no Hall effect (see Chap. 3 for details).

(3) A solution of the flow problem does not exist. This means that after a certain time the solution increases without bound near some point in the flow. This effect is observed in two-dimensional flows with the Hall effect if the replacement factor ξ is large at the specific plasma conductivity. In particular, in the case of ideal conductivity, flows incorporating the Hall effect "never exist." It is shown at the end of §2.5 that in this case the magnetogasdynamic equations are unstable against perturbations which grow arbitrarily rapidly; the equations are said to be "nonevolutionary."† Physically, the meaning of this result is that it is necessary to take account of

---

† The corresponding Cauchy problem is not well posed.

the electron inertia and, perhaps, kinetic effects. It is suggested
that the "nonexistence" of solutions discovered in the calculations
is due to the formation of conducting bridges at the anode (see §§1.1
and 2.5).

Let us examine the steady-state flow in detail. The steady state
is reached in a time τ_{tra} which is of the order of the transit time
through the system $\tau_0 = L/v$, where L is the length of the channel,
and v is the characteristic velocity of the steady-state flow.[†] The
situation is formulated precisely as follows: For arbitrarily initial
conditions, the solution of the nonstationary problem reaches a "quasi-
stationary state" in a time τ_{tra}; after this time (at $t > \tau_{\text{tra}}$) the
functions that describe the flow only vary over a few percent. The
ultimate steady state is reached later (i.e., all the properties involved in the calculation reach steady-state values in all the significant quantities); the time required for this steady state to be
reached varies over a broad range, depending on the particular version of the problem. How this process leading to the establishment
of the actual steady state depends on the flow parameters is described below.

Whether the flow does or does not reach a steady state in the
calculations bears directly on the very serious and difficult problem of the stability of plasma flow in a channel. The stability problem, i.e., the problem of the time evolution of various perturbations, is essentially a three-dimensional problem, regardless of
the symmetry of the main flow. Nevertheless, there is definite
interest in the simplified two-dimensional and one-dimensional
versions of this problem, which can be treated in the present calculations, since the two-dimensional perturbations which are
stretched out along the magnetic lines of force are generally the
most dangerous from the stability standpoint. This statement is
the magnetohydrodynamic analog of the Squire theorem [5, 55].
This feature adds substantially to the value of calculations for
nonstationary two-dimensional flows as a tool for detecting flow
instabilities.

The most common method for analyzing stability is the method
of linear perturbations. [‡] The problem reduces to the solution of

---

†In terms of the units chosen in §1.2, $\tau_0 \approx 1$.
‡The use of the linearization method is justified in [61].

linearized equations for small perturbations or, at least, to an estimate of the time evolution of the solutions of these linearized equations. This method yields results in certain relatively simple problems. In particular, an interesting case arises when the perturbation is a plane wave with wavelength $l \to 0$. In this case the problem reduces to linear equations with constant coefficients. One such problem is solved in § 2.5. Artyushkov and Morozov [4] prove the stability against fine-scale one-dimensional (longitudinal) perturbations in the flow of a plasma which is a good conductor in a channel.

In a study of the stability of a steady-state flow the coefficients of the linearized equations are independent of the time. In this case the time evolution of the solution is governed by a factor like $\exp(\lambda t)$, where the quantity λ is found by solving an eigenvalue problem.† A variational (energy) principle is widely used in analyzing equilibrium plasma configurations ($\partial / \partial t = 0$, $\mathbf{v} = 0$). This principle makes it possible to skip the step of finding the individual eigenvalues and to proceed immediately to the derivation of a stability condition in terms of a positive value of the potential energy of small oscillations [33, 58].

Numerical methods have been used frequently during the past few years to study the stability of hydrodynamic flows. The effectiveness of this approach is demonstrated in the many papers cited in [11]. We should expect this method to also be effective for studying plasma flows. Although the work cited below was carried out to study the overall flow pattern, rather than to study the stability in particular, the results are of use in an analysis of the stability.

The establishment of a steady state in the calculations means that the two-dimensional steady-state plasma flow in the channel is stable against two-dimensional perturbations of finite or low frequency. The calculations reveal no oscillations with wavelength comparable to or larger than the step of the calculation grid which would prevent a steady state from being reached. On the other hand, the fact that a steady state is reached tells us nothing about fine-scale oscillations; if these oscillations do exist they can be suppressed by the stabilizing dissipative properties of the calculation process itself.

---

† The use of this method for stability problems, regardless of the completeness of the eigenfunctions, is justified in [14].

Accordingly, although the numerical methods do not represent a universal tool for studying the stability of plasma flows, they do lead to definite results, including quantitative results, regarding the stability in a certain class of problems (see Fig. 20 in § 2.5).

§ 2.2.   Quasi-One-Dimensional Flow

Certain properties of the plasma flow in channels — the integrated characteristics and the simplest aspects of the behavior — can be found by solving the problem in the one-dimensional formulation. There are two classes of steady-state one-dimensional flow problems.

The first type is the laminar flow of a viscous, conducting fluid in an infinitely long channel of constant cross section. The functions describing the flow depend on only a single coordinate, in the transverse direction. The problem is described by ordinary differential equations; the integration of these equations yields the familiar solutions of the Poiseuille, Couette, Hartmann, etc., problems for an incompressible medium (see, e.g., [5, 24, 57]).

The other class of problem presupposes an averaging of the flow characteristics over the cross section. As a result of this averaging the functions become independent of the transverse coordinates and the only nontrivial spatial coordinate is that along the channel. This situation usually involves a narrow channel of finite length and corresponds reasonably well to the actual flow pattern when the transverse gradients of the properties are very small over most of the channel and the contribution from the discontinuities in the properties near the walls can be neglected. The narrow-channel approximation is widely used in gas dynamics, e.g., in rough calculations for nozzles, since the boundary layer is thin. Attempts to apply this model to plasma flows have been less successful (see § 1.1).†

The problems of this second class usually involve channels of varying cross section. The flow is described by Eqs. (1.37). The steady-state quasi-one-dimensional problem reduces to ordinary differential equations, which, in the dissipationless case, contain singularities at points at which the flow velocity becomes

---

†See [42] regarding the applicability of the one-dimensional approximation.

equal to the magnetosonic velocity. Various cases, with various ratios of the plasma and sound velocities, have been studied qualitatively by many investigators; these studies represent a magnetogasdynamic generalization of the theory of the Laval nozzle [5, 24, 34, 39, 57]. Solutions of certain problems in the quasi-one-dimensional formulation are available, with some of the functions assumed to be given. As an example we can cite the induction of currents in the known flow of a conducting fluid in a given magnetic field [39, 57]. Certain questions can be solved by using additional simplifying assumptions. As an illustration of this approach we cite the series of papers by Bam-Zelikovich [6]. Bam-Zelikovich neglects the pressure forces in comparison with the electromagnetic forces and assumes a constant, small conductivity. As a result it becomes possible to separate the system of equations into simpler equations and to calculate the flow.

In general the problem (steady-state or transient) must be solved numerically. A solution of this kind — the first such solutions in the series of papers being reviewed here — is reported in [17] for the flow of an inviscid, zero-thermal-conductivity plasma with finite electrical conductivity in a nozzle (Fig. 2):

$$f(x) = 0.3 - 0.8x(1-x); \quad (0 \leqslant x \leqslant 1). \tag{2.1}$$

The calculation reveals the establishment of a steady state and

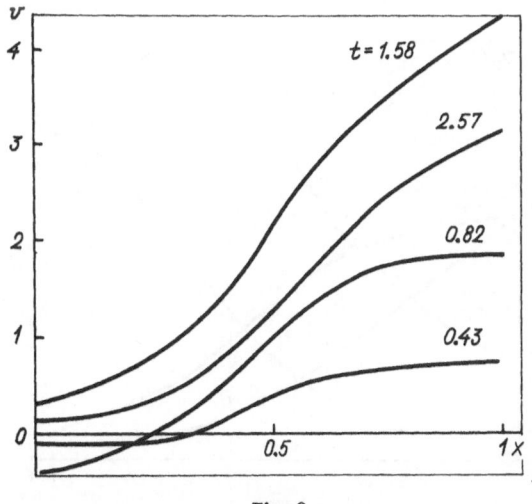

Fig. 9

makes it possible to study the dependence of the steady-state flow
parameters on the governing dimensionless parameters, speci-
fically β, and the coefficients σ_1 and σ_2 in Eq. (1.46) for the conductivity. These calculations show that for all values $Re_m > 0.1$
the flow reaches a steady state in a time of the order of the transit time. Figure 9 illustrates this process for the velocity profile. The curve corresponding to the largest time ($t = 2.57$) essentially represents steady-state flow.

The steady-state solutions have the following properties: The
flow is subsonic at the entrance and supersonic at the exit. The
magnetosonic velocity is reached near the minimum cross section
of the channel. The velocity and entropy increase monotonically,
while the density and magnetic field fall off along the channel. Figure 10 shows illustrative plots of these functions, corresponding
to the values $\beta = 0.84$; $\sigma_1 = \sigma_2 = 0.1$. Figure 11 shows the velocity
v at the exit from the channel as a function of the conductivity for
$\beta = 0.84$. The curve in Fig. 11a shows the velocity as a function
of the constant conductivity $Re_m = \sigma_1$ ($\sigma_2 = 0$). In Fig. 11b the conductivity is variable [see Eq. (1.46)] with equal coefficients $\sigma_1 = \sigma_2$
(whose values are plotted along the abscissa). It is evident from
this figure that as the conductivity Re_m is reduced the flow velocity
at the exit from the nozzle increases. This behavior is attributed to
the fact that decreasing Re_m at a given current increases the plasma

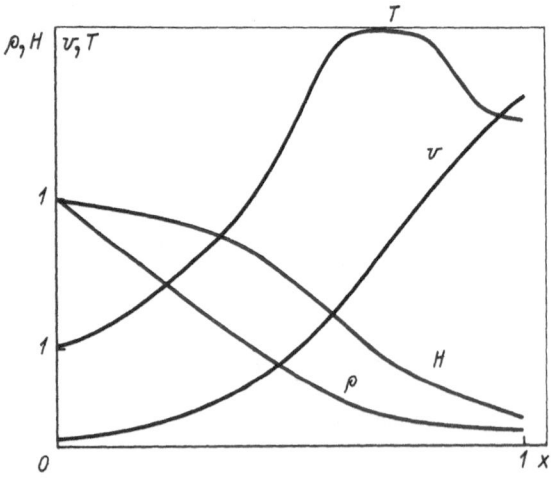

Fig. 10

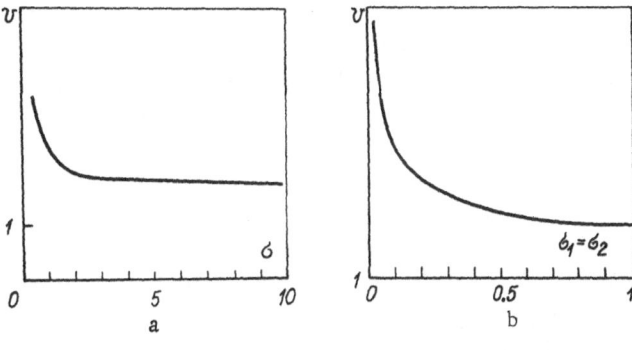

Fig. 11

heating. The exit velocity also increases with decreasing β, i.e., with increasing magnetic field at the entrance and with increasing energy contribution per unit mass.

At high values of Re_m the steady-state solution of the problem rapidly approaches the solution corresponding to a plasma of infinite conductivity. The system in (1.37) is readily integrated when $\eta = \varkappa = \nu = 0$, $\partial/\partial t = 0$, and reduces to the algebraic equations

$$\rho v f = \text{const}; \quad v^2/2 + \gamma p/(\gamma - 1)\rho + H^2/\rho = \text{const};$$
$$H/\rho = \text{const}; \quad s = \text{const}, \tag{2.2}$$

which are a dimensionless form of Eq. (1.2).

Another purpose of the one-dimensional calculations is to find an appropriate boundary condition for the magnetic field at the right end of the channel (see § 1.2). For this purpose the channel is lengthened at $x > 1$ by tubes of various lengths and various cross sections — a constant cross section (f = const) and a linearly increasing cross section (f = ax + b). It turns out that over most of the channel ($0 \le x \le 1$) the flow is very insensitive to the shape of the tube, while the magnetic field satisfies the condition $\partial H f/\partial x = 0$ quite accurately at $x = 1$. As a result of this study, boundary condition (1.54) was adopted for the two-dimensional calculations.

The events leading to the steady-state flow in a narrow channel can be described as follows: The quasisteady state is established in a transit time, as noted above, while the steady state proper is reached in a time which depends on the conductivity. If the conductivity is high, the steady state is reached rapidly; as Re_m is reduced the process is extended. Small oscillations, comparable in magnitude to the step of the calculation grid, then appear. At small values of Re_m (at $Re_m < 0.1$ in the case under consideration here), the flow never reaches a steady state: Pronounced temperature oscillations arise in the calculations. Accordingly, if the conductivity is low the flow becomes unstable. One reason for this effect may be the boundary condition at the right side, which must be examined very carefully at small values of Re_m. Further evidence for these arguments comes from the fact that the time required for the final steady state to be reached depends slightly on the nature of the tube attached to the right side of the channel; specifically, the steady state is reached more rapidly as the cross section of this tube is increased.

When $Re_m = 0$ the magnetic field becomes constant, the electric current vanishes, and the set in (1.37) becomes the gasdynamic equations (see § 3.1). The instability described above thus prevents an asymptotic approach to this state in the calculations.

§ 2.3. Two-Dimensional Flow of Ideal Plasma

Analytic methods are less effective in the two-dimensional case than in the one-dimensional case. At best, they yield solutions which are series in terms of some small parameter, e.g., $Re_m \ll 1$. Certain particular solutions for two-dimensional problems can be found under the assumption that certain quantities that describe the flow are given. Methods using this approach and illustrative solutions of various magnetogasdynamic problems are reported in [5, 24, 34, 39, 57]. In many cases an important smallness parameter is the ratio of the scale length of the inhomogeneities across and along the flow, $\delta = (l_\perp/l_\parallel) \ll 1$. By expanding the functions ρ, $\mathbf{v}$, etc., in power series in δ, it becomes possible to analyze several properties of the flow (this is the method used for a channel with a slowly varying cross section). This approach has been used to study the qualitative distribution of the density, velocity, and electric current in a coaxial channel [49], to study the influence of the Hall effect on the flow [50], and to reach certain estimates regarding the contraction of the plasma at the axis of a magnetoplasma compressor [43] (see also [52]).

The quasi-one-dimensional treatment of plasma flow in a channel (§ § 1.1 and 2.1) leads to the conclusion that acceleration of the plasma requires a nozzle-shaped channel. This approach also yields an estimate for the exit velocity in the ideal case. These results do not provide enough information for systems in which the distance between the electrodes is comparable to the length of the channel, in which the radial variations in the magnetic field generate interesting flow features. It is also important to learn how the effects which distinguish the plasma from an ideal plasma influence the two-dimensional flow.

The first calculation for two-dimensional plasma flow in a coaxial channel was carried out by N. I. Gerlakh and the present authors in 1962 [18]. The basic purpose of this study was to get some idea of the role played by the channel geometry and the associated two-dimensional details of the flow. The problem was posed in its simplest form: the isentropic flow of a single-com-

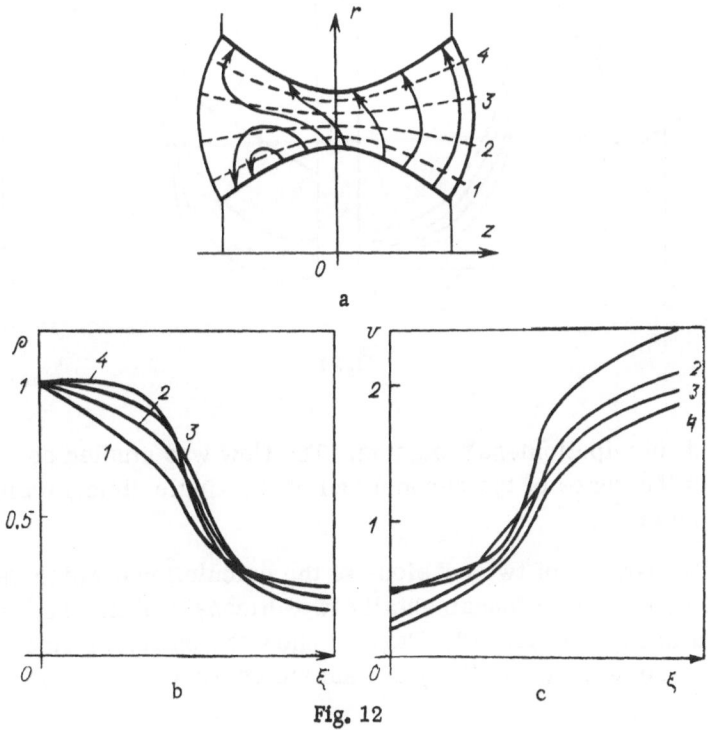

Fig. 12

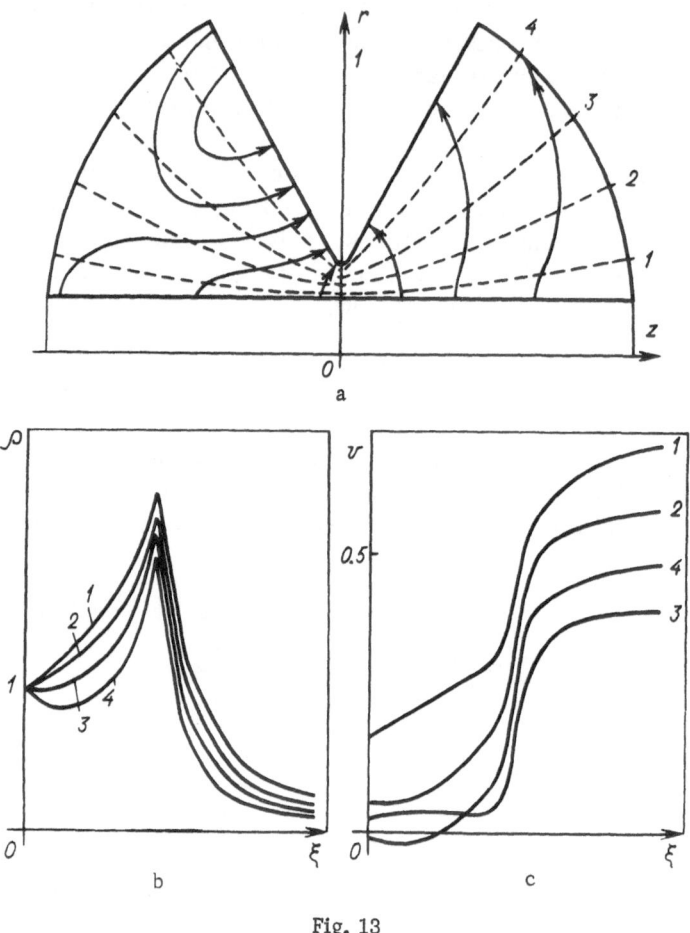

Fig. 13

ponent, dissipationless† plasma. The flow was studied as a function of the shape of the channel and of the single dimensionless
parameter β.

The results of two versions of the calculations, which revealed effects of a fundamentally two-dimensional nature, are
shown in Figs. 12 and 13. Parts a show the channel shape (the
solid arrows show the lines of electric current $\mathbf{j} = \operatorname{curl} \mathbf{H}$, while

---

†More precisely, the conductivity was assumed to be constant and high ($\operatorname{Re}_m \gg 1$).

the dashed lines are the coordinate lines). Parts b and c show
the profiles of the density ρ and the velocity v_τ along the coordinate lines in the steady state. Here v_τ is the projection of the
velocity on the coordinate lines. The plots of this velocity give
a complete description of the flow since the velocity component
perpendicular to this one is very small and the coordinate lines
then coincide approximately with the trajectories of the plasma
particles. The curves are labeled with the index of the coordinate
line as defined in parts a of these figures. The values of β for Figs.
12 and 13 are 1.84 and 0.034, respectively.

The results found in a series of calculations of this kind lead
to the following conclusions.

1. The flow reaches a steady state or a quasisteady state in a
stable manner in a transit time.

2. A radial variation of the magnetic field has the consequence
that the velocity becomes a function of the radius; in particular, the
velocity at the channel exit is not constant. In general, the magnitude of the velocity v_τ and its r dependence vary with β. If β is
large the flow is of a gasdynamic nature and approximately one-
dimensional. In this case the function $v(r)$ is very weak. Decreasing β, i.e., enhancing the role of the magnetic field, increases the
velocity spread at the exit. In very strong magnetic fields ($\beta < 0.3$)
the velocity distribution at the exit becomes essentially independent
of β in terms of qualitative features. Finally, at all values of β
the r dependence of the exit velocity is weaker than $1/r$, i.e.,
weaker than the variation $H(r)$ specified at the entrance.

In questions related to plasma accelerators, a knowledge of
$\mathbf{v}(r)$ at the channel exit is extremely important since a radial spread
of the velocity reduces the usefulness of the accelerator (in terms
of its energy characteristics). In certain calculations an attempt is
made to improve the $\mathbf{v}(r)$ profile at the exit (to make it flatter) by
using an inhomogeneous input of the density ρ at the channel entrance. It turns out that when $\rho(r)$ is a linearly increasing function
at the entrance to the channel the velocity at the exit is more uniform, but in part of the channel the velocity is then in the vertical
direction or even in the backward direction. Furthermore, pronounced electric-current eddies (see the discussion below) appear,
reducing the effectiveness of the accelerator.

3. In gasdynamic nozzles the gas is accelerated as the result of a pressure drop along the channel, so that the density falls off monotonically along the z axis (in the case of adiabatic flow). The calcuations show that for flow of a plasma in a strong magnetic field in a channel with a rapidly varying profile (Fig. 13) the density does not always vary monotonically along the flow trajectory (for the case of a monotonic variation in the velocity). In other words, plasma compression is observed near the channel entrance — an increase in ρ with increasing z. The explanation for this behavior runs as follows: According to the Euler equation in (1.26), which can be rewritten as

$$\rho \,(dv/dt) = - \nabla[p + H^2/2] + (\mathbf{H}, \nabla)\mathbf{H},$$

the plasma is not only subjected to the gradient of the total pressure, but also to a force $(\mathbf{H}, \nabla)\mathbf{H}$, due to the tendency of a line of force to contract. The discovery of this compression effect in the calculations was followed up by a study on the basis of the narrow-channel approximation [49, 52], and the conditions under which this compression occurs were found. For example, when the velocity increases linearly along a narrow tube of constant cross section with a shape described by $r = r_0(1 - z^2)^\alpha$, the density ρ and the magnetic field H increases from left to right if $\alpha > 1$. In this case the magnetic field is assumed to be strong, i.e., $\beta \ll 1$.

This nonmonotonic. variation in the plasma density in plasma flow in a channel subsequently led to the theory of compressional flows (see § 2.6).

4. Certain versions of the flow calculations reveal eddies in the electric current **j** (see Figs. 12 and 13; the current lines are shown by the solid arrows). These eddies occur in the subsonic part of the flow near the electrodes, so that at certain parts of the electrodes the currents flow in the backward direction. The current distribution over the electrodes turns out to be very irregular. As a result of these eddies the backward currents in the plasma generate a force directed opposite to the flow.

The calculations show that these current eddies arise only if the channel is sharply curved and if the magnetic field is weak. As the field is increased (as β is reduced) the eddies become weaker and ultimately vanish (in a given channel). The current lines in Fig. 14 correspond to the value $\beta = 0.0754$.

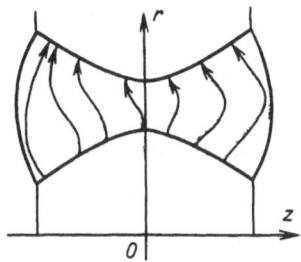

Fig. 14

The current eddies found in these calculations were also found later in the narrow-channel approximation [2, 49, 52]. It turns out that the appearance of these eddies is a consequence of the conversion of thermal energy into magnetic energy, so that the eddies can be observed only at large values of β.

Finally, both the nonmonotonic variation in the density and the current eddies have been subsequently verified experimentally, in agreement with the understanding reached on the basis of the calculations described above.

Accordingly, even the first, highly simplified calculations revealed several effects of a fundamentally two-dimensional nature. These new effects stimulated a further development of both theory and experiment.

§ 2.4.   Influence of a Finite Conductivity on the

Flow

In the subsequent calculations, various complications were introduced. In this section we examine the role of a finite plasma conductivity. The problem remains highly simplified: The energy equation is replaced by the isothermal condition (T = const), so that the conductivity is assumed to be constant (Re$_m$ =const). The viscosity, thermal conductivity, and Hall effect are all neglected, as in the earlier calculations. Accordingly, the characteristics of the flow are the two dimensionless parameters β and $\nu = 1/\text{Re}_m$ and the channel geometry.

These calculations show that the solution reaches a quasisteady state in a transit time. The ultimate steady state is reached after a time which depends on both these parameters; the time is shorter, the larger the value of β (the weaker the magnetic field) and the lower the value of β (the higher the conductivity). As β and Re_m are reduced the time required for the steady state to be reached becomes longer, as in the quasi-one-dimensional model (see §2.2).

The steady-state two-dimensional flow has the following properties.

1. In a weak magnetic field ($\beta \sim 10$) the gas pressure is higher than the magnetic pressure. The flow is of a gasdynamic nature, and the parameters, except for the magnetic field H, remain essentially the same over a broad range of the conductivity. The variation of the gasdynamic properties along the radius r is extremely weak so that the flow is nearly one-dimensional. Accordingly, at large values of β the plasma conductivity can be neglected unless we are interested in the behavior of the magnetic field or the electric current.

2. As the magnetic field is strengthened (as β is reduced), the two-dimensional nature of the flow becomes more apparent: All properties become functions of r. The nonmonotonic variation of the density in the converging part of the channel (mentioned in § 2.3 for an ideal plasma) becomes more pronounced in the case of a plasma of finite conductivity; specifically, this nonmonotonic variation can be observed even in a more smoothly curved channel. Figure 15 shows the lines of constant density in a channel with $\beta = 0.0754$ for three values of the conductivity. This figure indicates that this compression effect is extremely weak for a given channel geometry for the value $\nu = 0.001$. It becomes appreciable as ν is increased. The most clearly defined maximum in the density corresponds to the value $\nu = 1$ and lies halfway between the electrodes. The density falls off monotonically in the electrode boundary layers. At higher resistivity ν, the density becomes a monotonically decreasing function everywhere. It is a strong function of the radius, and its value far from the electrodes is much higher than that near the electrodes. The reason is that the current density in the channel is perpendicular to the electrodes, so that the medium near the electrodes is accelerated in the direction tangent to the electrodes, that is to say, toward the center of the gap. When $\nu \to \infty$ the contribution of the electromagnetic forces is weak; when $\nu \to 0$ the field and the plasma are both compressed, so that the pattern is changed.

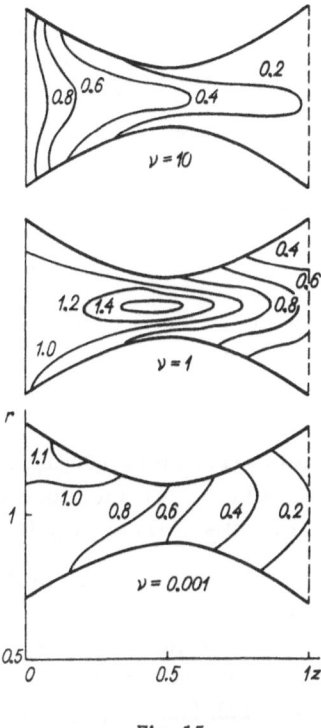

Fig. 15

3. The conductivity also affects the flow velocity, in particular, the nature of the velocity distribution at the channel exit. The fact that the exit velocity increases with increasing ν (with decreasing conductivity) is also revealed in calculations based on the quasi-one-dimensional model (see § 2.2). As in the case of an ideal conductivity, the two-dimensional analysis also shows that the exit velocity falls off monotonically along the radius for all values of ν; this velocity peaks at the inner electrode and is a minimum near the outer electrode. As ν is increased the flow becomes more uniform as it is accelerated, with the velocity of the central part increasing more rapidly than that near the periphery. This result can be seen in Fig. 16, which shows the velocity $u = v_z$ at the channel exit as a function of the resistivity ν. The upper curve corresponds to the maximum (central) velocity, while the lower curve corresponds to the minimum (outer) velocity. The channel shape is the same as that for Fig. 15, and the parameter is $\beta = 0.0754$.

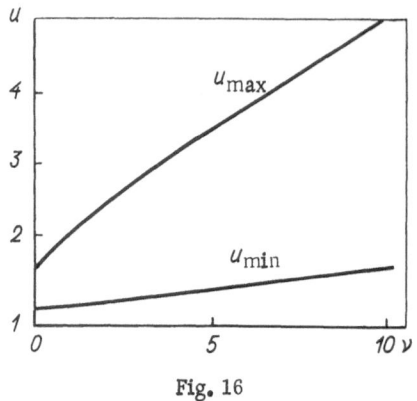

Fig. 16

Accordingly, as ν is varied from 0 to 10, the ratio of the maximum velocity to the minimum velocity — a measure of the nonuniformity at the flow of the exit — increases from about 1.5 to 3.

4. In the distribution of the magnetic field and thus the electric current in an ideal plasma we can distinguish two limiting cases: In strong fields ($\beta \ll 1$) the field shows a $1/r$ radial variation. In other words, the current flow between the electrodes is primarily in the vertical direction. In weak fields ($\beta \gg 1$), the nonuniformity of the r profile of the field becomes more pronounced, and the field is a nonmonotonic function of z, as can be seen from the current eddies.

A finite conductivity smooths the distribution of the magnetic field. As the resistivity ν is increased the nonuniformities become less pronounced, and the region occupied by a current eddy contracts, degenerates into a point at the electrode, and then vanishes completely. Figure 17 shows the electric–current lines Hr = const (the curves are labeled with the values of Hr) for the case $\beta = 7.54$ and for various values of the resistivity. In this channel geometry the eddy disappears at a value near $\nu = 0.1$. For values of ν near unity or higher the radial variation of the field is roughly $1/r$ and the current is perpendicular to the flow. Figure 17 also shows that at higher ν the depth to which the magnetic field penetrates into the channel is reduced: H falls off rapidly along z, and the lines of Hr = const for the case $\nu = 10$ group closer to the center of the channel† than in the case $\nu = 1$.

---

†The reason is that in the limit $\nu \to \infty$ we have $j = \sigma E$; i.e., the current density peaks at the critical cross section of the channel, because of the condition σ = const, and because the potential difference between the electrodes is constant.

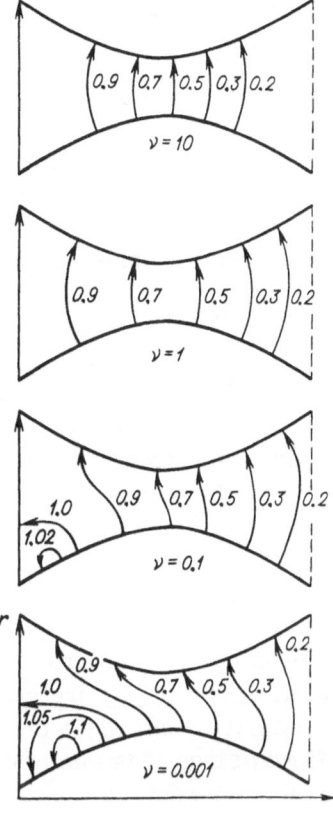

Fig. 17

In a strong magnetic rield (small values of β), the radial profile of H is described approximately by $1/r$, and there are no current eddies at any of the conductivity values considered here.
The conductivity only has an appreciable effect on the depth to
which the field penetrates into the channel; specifically, the depth
increases with increasing conductivity. The results discussed in
this section are reported in [20].

§ 2.5. Hall Effect

Replacement Factor. Preliminary Considera-
tions. When the Hall effect is taken into account the plasma
flow is described in Eq. (1.33). In this description the replacement
factor ξ is a measure of the Hall effect [Eq. (1.20)]. This param-

eter is a measure of the ratio of the third term on the right side of Ohm's law in (1.28) to the second term:

$$\xi \sim \xi \, [\mathbf{j}, \mathbf{H}]/\rho \, [\mathbf{v}, \mathbf{H}] \sim |\mathbf{v}^i - \mathbf{v}^e|/v^i.$$

According to (1.18), $\xi \mathbf{j} = \rho(\mathbf{v}^i - \mathbf{v}^e)$, and this relation follows from the fact that the field $\mathbf{H}$ is orthogonal to the vectors $\mathbf{v}$ and $\mathbf{j}$. This relation tells us that the replacement factor is also a measure of the angle between the electron and ion trajectories.

The parameter ξ can be furthermore regarded as the ratio $\xi \sim I_d/I_m$ of the current through the system, I_d, to the mass flow rate expressed in current units, $I_m = e\dot{m}/m^i$ [51].

This replacement factor can be related to other dimensionless parameters which are also measures of the Hall effect: the Hall parameter $\chi = \omega^e \tau^e$ and the "ion number per unit length" Π [13, 51]. It follows from Eqs. (1.38)-(1.40) that $\chi/\alpha_0(\chi) = \xi \, \mathrm{Re}_m H/\rho$, that is, in order-of-magnitude terms, $\chi \approx \xi \cdot \mathrm{Re}_m$; also $\Pi = 4\pi e^2 n L^2/m^i c^2 = \rho \xi^2$, where the right side of the first of these equations for Π consists of dimensional quantities, while the right side of the second equation consists of dimensionless quantities. Comparing Π with ξ, we note that the former is apparently more convenient for an analysis of equilibrium configurations, while the latter is more convenient for analyzing flows, since it is easily determined from measurements of the integrated characteristics.

The sign of ξ can be used to specify the electrode polarity. According to the formal definition, (1.20), we have $\xi \geq 0$. Reversal of the electrode polarity implies a change in the sign of the magnetic field H. It is evident, however, that the system in (1.33) is invariant to simultaneous changes in the sign of H and ξ. This assumes that the polarity $H \leq 0$ in the case $\xi \geq 0$ can formally be replaced by the polarity $H \geq 0$ with $\xi \leq 0$, which is more convenient for the calculations. Accordingly, we assume $H \geq 0$ everywhere below, assuming that the central electrode is the anode when $\xi > 0$ or the cathode when[†] $\xi < 0$. If $\xi = 0$, Eq. (1.33) is invariant against a change in the sign of H and a reversal of the electrode polarity, as noted in §1.1. In the same section, we discuss certain general considerations and report experimental data on the influence of the Hall effect in plasma flow in accelerators.

[†]According to (1.20) changing the sign of ξ means changing the sign of the elementary charge e; this change can also be treated as a change in the electrode polarity.

Certain properties of steady-state flows associated with the Hall effect have been established in work by Morozov, Solov'ev, and Shubin [50-53]. Of primary interest among these are the fact that the magnetic field is frozen in the electron fluid of the plasma and the fact that a potential drop occurs near the anode. This drop occurs because the electric potential is constant along the electron trajectories in the steady state so that at large values of the parameteter ξ (the angle between trajectories) the cathode potential is "carried" by the electrons to essentially all points in the channel volume.

It is more difficult to predict the properties of nonstationary flow, especially if the conductivity is finite and there are other deviations from an ideal plasma. It should be noted that the problem with the Hall effect is intrinsically two-dimensional, that is to say, no meaningful one-dimensional analog exists. This problem is pursued in numerical calculations, since the authors believe that this is the only way to avoid undesirable simplifications.

Flow Calculations with the Hall Effect. The first calculations for two-dimensional plasma flow in a channel with the Hall effect were carried out in 1966, and the results were published in [19, 46]. These calculations are based on (1.33) with extremely important simplifications: The plasma is assumed to be inviscid ($\Pi = 0$) with zero thermal conductivity ($\varkappa = 0$), and the flow is assumed to be isothermal, in agreement with the experimentally established constancy of the electron temperature (§1.1). The conductivity is constant because the flow is isothermal. Accordingly, the problem contains the three dimensionless parameters β, ξ, and ν, and the calculations show how the flow varies with these parameters. The basic result of this numerical solution is that the flow does not reach a steady state for arbitrary values of the replacement factor ξ, but only for values in the range

$$\xi_- (\beta, \nu) < \xi < \xi_+ (\beta, \nu), \tag{2.3}$$

where $\xi_- < 0$, $\xi_+ > 0$. This result means that there is a critical value of ξ for both electrode polarities (ξ_+ if the central electrode is the anode and $| \xi_- |$ in the opposite case) beyond which the flow does not reach a steady state. This critical value is a function of β and ν. Furthermore, at values of ξ outside the interval in (2.3) no solution of the problem exists, in the sense discussed above (§2.1). The calculation for nonstationary flow is ended at some

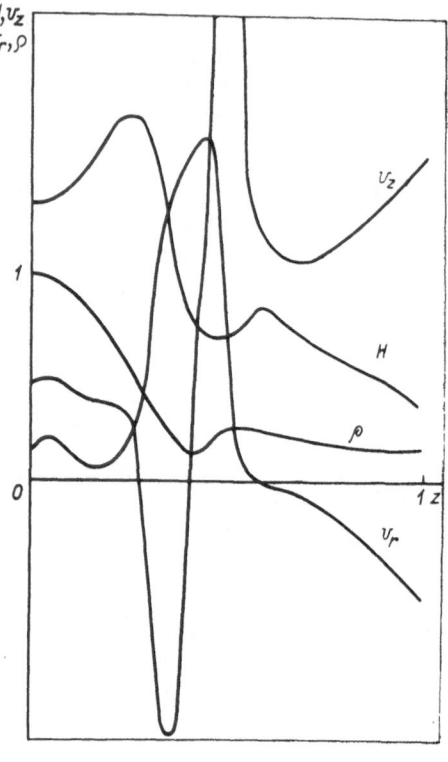

Fig. 18

point because the flow becomes unstable: the magnetic field and
the current density increase rapidly near some point on the anode.
In this region the density falls, and there is an increase in the
particle velocity, which is everywhere directed away from the
singularity. Figure 18 shows the ρ, H, v_z, and v_r as functions of
z near the central electrode (anode) at time t = 1.54 (shortly before the calculation is ended). These results refer to the channel
shown in Fig. 17 and to the parameter values β = 7.54, ν = 1, and
ξ = 0.5. The flow pattern becomes smoother with distance from
the anode. Figure 19 shows the behavior of the same quantities near
the cathode.

Thus, near the anode an "explosion" occurs which acts as a
source of fast particles. These features are found in the calculations
for various channel geometries, various parameter values, and
various electrode polarities, but they always occur at the anode.

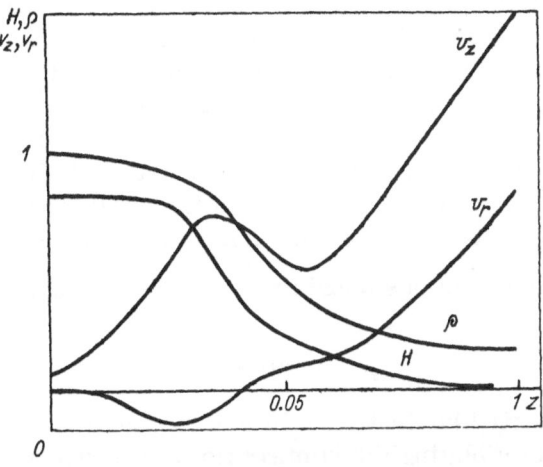

Fig. 19

The anode "explosions" found in these numerical calculations may be related to certain experimental results noted in §1.1: an instability of the flow near the anode, involving "current bridges" and pronounced anode erosion. It appears that the same phenomenon can lead to the formation of fast particles in rail devices and in Marshall guns [60].

The critical values ξ_+ and ξ_- are found in calculations for various values of β and ν. Figure 20 shows ξ_+ as a function of ν for three values of β (the channel geometry is the same as in Fig. 17).

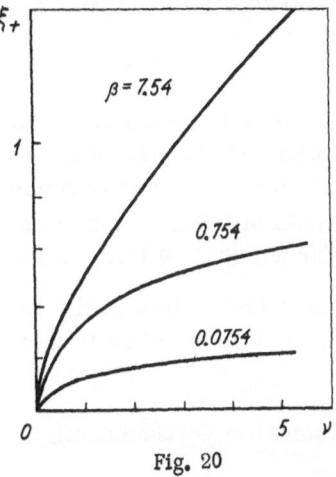

Fig. 20

The other critical value, $|\xi_-(\beta, \nu)|$ show the same behavior. We see from these curves that $\xi_+(\beta, 0) = 0$; thus the flow of an ideally conducting plasma cannot be calculated with the Hall effect taken into account. On the other hand, increasing the magnetic viscosity ν increases the range of permissible values of ξ. The behavior of the curves as functions of β shows that the Hall effect is more pronounced in stronger magnetic fields (at smaller values of β); specifically, the critical value of ξ_+ falls off with decreasing β.

The numerical values found for $\xi_+(\beta, \nu)$ can be approximated by

$$\xi_+ \approx 0.3\beta^{1/4}\nu^{1/2}. \tag{2.4}$$

It follows in particular that it is difficult to achieve high exit velocities without violating the laminar flow. The exit velocity[†] is $v_{max} = P/m$, where P is the momentum carried off by the flow per unit time, and m is the flow rate of working medium (per unit time). Obviously $P = aI^2$, so that

$$v_{max} = b\xi I. \tag{2.5}$$

Here, a and b are constants governed by the geometry of the system and the nature of the working medium. Under the assumption of isothermal flow (T = const, σ = const) we find

$$\left.\begin{array}{l} \beta \sim 1/v^2_{max}(\text{since } v^2_{max} \sim H^2/\rho); \\ 1/\nu = Re_m \sim v_{max}. \end{array}\right\} \tag{2.6}$$

Substituting (2.4) and (2.6) into (2.5) we find $v_{max} \sim I/v_{max}$; thus

$$v_{max} \sim I^{1/2}. \tag{2.7}$$

At the present time there is no experimental basis for this latter relation; on the other hand there is no contradiction, since no one has yet been able to achieve a pronounced increase in the velocity of the entire mass of the gas, rather than in the velocity of certain fast particles for which $\xi \sim 1$.

When the replacement factor lies in the range in (2.3), the flow reaches a steady state which is reached in a transit time. No os-

---

[†]All the quantities under discussion here are dimensional.

cillations are found in the calculation. Let us now examine the influence of the Hall effect on the steady-state conditions.

A general property of these flows is their two-dimensional nature. All quantities have radial profiles which are less uniform than in the corresponding magnetogasdynamic problems (§§2.3 and 2.4). The plasma density ρ is higher near the cathode and lower near the anode than in flows with $\xi = 0$. As a result of the Hall effect, the plasma is squeezed toward the cathode; this effect is more pronounced, the stronger the magnetic field, i.e., the lower the value of β (for given values of ξ and ν).

The Hall effect has a particularly noticeable influence on the nonuniformity of the magnetic-field distribution. As ξ increases, the field near the anode increases and the field near the cathode decreases. As a result the electric-current distribution of the channel is distorted. In particular, in cases in which the current is directed primarily toward the flow for $\xi = 0$, the current is deflected to the side when $\xi \neq 0$ and grazes the anode.

Figure 21 shows the current distribution in the channel for $\beta = 7.54$, $\nu = 1$, and three values of ξ: $\xi = 0$ (a flow with no Hall effect), $\xi = 1/3$ (the central electrode is the anode), and $\xi = -1/3$ (the central electrode is the cathode, and the replacement factor is the same). As ξ is increased and approaches the critical value,† the current distribution becomes even more distorted: A current eddy forms near the anode and leads to an anode explosion at the critical value of ξ.

Finally, we consider the role played by the ohmic resistivity ν in these calculations. In addition to expanding the range of permissible values of ξ, as noted above, the resistivity generally reduces the influence of the Hall effect in this region. Consider a calculation for a flow with fixed values of β and ξ; if the value of ν is raised, the nonuniformities in the steady-state distributions of the density and magnetic field due to the Hall effect (noted above) become less significant. For example, the lines of electric current (Fig. 21, $\xi = \pm 1/3$) tend to become more nearly vertical.

Accordingly, this series of calculations makes it possible to analyze the effect of all three parameters on the flow. Although

---

† In this case $\xi_+ \approx 0.5$ (Fig. 20) and $\xi \approx -0.5$.

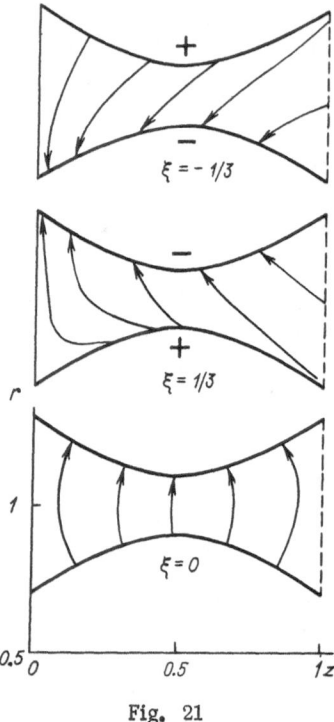

Fig. 21

some of the results can be predicted qualitatively on the basis of approximate theories, e.g., the deflection of the current or the compression of the plasma toward the cathode, these calculations furnish a picture of the overall flow in any given case. Finally, the calculations led to a new result: the anode explosions that follow from the solution of Eqs. (1.33). The corresponding analysis of the magnetogasdynamic equations with the Hall effect is given below.

Evolutionary Nature of the Magnetogasdynamic Equations with the Hall Effect. The unstable nature of the plasma flow with the Hall effect, which is found in the numerical solution of the problem and in experiment, raises the question of the stability of such a flow. A comprehensive analytic study of the stability requires the solution of a difficult problem, even in the linear approximation. Nevertheless, it is possible to establish the following important result: The Cauchy problem for the magnetogasdynamic equations with the Hall effect but without the dissipative terms is not a well-posed problem in the case of plane, two-dimensional flow across a magnetic field. This statement means that the solution of the Cauchy problem is unstable against fine-scale oscillations which grow arbitrarily rapidly in time. These equations are called "nonevolutionary" [27]. This result obviously also holds for the axisymmetric flows with azimuthal magnetic field under discussion here, since fine-scale oscillations are examined locally; from the local standpoint these flows are the same as plane flows across a magnetic field.

The nonevolutionary nature of the equations is a more profound result than an ordinary instability, since the question of stability for dissipationless Hall flow is thus permanently resolved.

We now consider the evolutionary nature of the equations. A dissipationless plasma Hall flow is described by Eq. (1.33), with $\Pi = 0$, $\varkappa = 0$, and $\nu = 0$. For simplicity, we discard the energy equation, assuming adiabatic flow ($p = \beta\rho^\gamma/2$) or isothermal flow ($\gamma = 1$). Let us examine plane two-dimensional flows in the plane of the Cartesian coordinates (x, y), assuming the magnetic field to be directed along the z axis. By analogy with (1.10), $v_z = 0$, $H_x = H_y = 0$, and $\partial/\partial z = 0$, and Eq. (1.33) becomes (under these assumptions)

$$\left.\begin{array}{l} \dfrac{\partial\rho}{\partial t}+\dfrac{\partial\rho u}{\partial x}+\dfrac{\partial\rho v}{\partial y}=0; \\[2mm] \rho\,\dfrac{du}{dt}=-\dfrac{\partial}{\partial x}\left(p+\dfrac{H^2}{2}\right); \\[2mm] \rho\,\dfrac{dv}{dt}=-\dfrac{\partial}{\partial y}\left(p+\dfrac{H^2}{2}\right); \\[2mm] \dfrac{\partial H}{\partial t}+\dfrac{\partial Hu}{\partial x}+\dfrac{\partial Hv}{\partial y}+\xi\dfrac{H}{\rho^2}\left(\dfrac{\partial\rho}{\partial x}\cdot\dfrac{\partial H}{\partial y}-\dfrac{\partial\rho}{\partial y}\cdot\dfrac{\partial H}{\partial x}\right)=0, \end{array}\right\} \tag{2.8}$$

where

$$u=v_x;\ \ v=v_y;\ \ H=H_z;\ \ \frac{d}{dt}=\frac{\partial}{\partial t}+u\frac{\partial}{\partial x}+v\frac{\partial}{\partial y}.$$

A particular feature of two-dimensional flow across a magnetic field is the degenerate nature of the Hall effect (§ 1.2). In the equations written here, the terms due to the Hall effect contain only first derivatives with respect to the coordinates; in the general three-dimensional case they contain second derivatives. We note further that the Hall terms are always nonlinear in terms of the first derivatives, in contrast with all other terms in Eq. (2.8). Both these features play a role in the discussion below: The simplicity of the basic result is a consequence of the degeneracy while the nonlinearity is apparently responsible for the nonevolutionary nature of the equations.

To study the flow stability we linearize Eqs. (2.8). The coefficients of the resulting system of linear equations are set equal to constants, since we are treating fine-scale perturbations; such pertrubations can be examined locally, in a small neighborhood of any point, with the values of the coefficients being "frozen" at this point. In the linearization of the nonlinear Hall terms in (2.8) the derivatives of the unperturbed solution serve as coefficients and are treated as constants.

Denoting the small perturbations by the subscript "1" (ρ_1, u_1, v_1, H_1), we find, after linearizing (2.8),

$$
\left.
\begin{aligned}
&\frac{d\rho_1}{dt} + \rho\left(\frac{\partial u_1}{\partial x} + \frac{\partial v_1}{\partial y}\right) = 0; \\[4pt]
&\frac{du_1}{dt} + \frac{C^2}{\rho}\cdot\frac{\partial\rho_1}{\partial x} + \frac{H}{\rho}\cdot\frac{\partial H_1}{\partial x} = 0; \quad C^2 = \gamma p/\rho; \\[4pt]
&\frac{dv_1}{dt} + \frac{C^2}{\rho}\cdot\frac{\partial\rho_1}{\partial y} + \frac{H}{\rho}\cdot\frac{\partial H_1}{\partial y} = 0; \\[4pt]
&\frac{dH_1}{dt} + H\left(\frac{\partial u_1}{\partial x} + \frac{\partial v_1}{\partial y}\right) + \xi\,\frac{H}{\rho^2}\left(\frac{\partial\rho}{\partial x}\cdot\frac{\partial H_1}{\partial y} - \frac{\partial\rho}{\partial y}\cdot\frac{\partial H_1}{\partial x} + \right. \\[4pt]
&\qquad\left. + \frac{\partial H}{\partial y}\cdot\frac{\partial\rho_1}{\partial x} - \frac{\partial H}{\partial x}\cdot\frac{\partial\rho_1}{\partial y}\right) = 0.
\end{aligned}
\right\}
\tag{2.9}
$$

Here all the coefficients, including $\partial\rho/\partial x$, $\partial\rho/\partial y$, $\partial H/\partial x$, and $\partial H/\partial y$, are assumed to be constant.

We seek a plane-wave solution of system (2.9), exp ($i\omega t + ik_1x + ik_2y$), with constant coefficients. Here k_1 and k_2 are arbitrary real numbers [27]. We write $k^2 = k_1^2 + k_2^2$, $\omega = k\lambda$, $k_1 = kl_1$, and $k_2 = kl_2$. Then the direction of the plane wave is specified by the unit vector $l = (l_1, l_2)$, and

$$
l_1u + l_2v = v_l; \quad l_1\partial/\partial y - l_2\partial/\partial x = \partial/\partial\tau,
$$

where τ is the direction along the wave front, which is orthogonal to l. Finally, setting $\lambda + v_l = z$, we find that the characteristic equation, which relates ω, k_1, and k_2, is

$$z\left(z^3 - \frac{\xi H}{\rho^2} \cdot \frac{\partial \rho}{\partial \tau} z^2 - C_m^2 z + \frac{\xi H}{\rho^2} \cdot \frac{\partial \mathscr{P}}{\partial \tau}\right) = 0, \qquad (2.10)$$

where $C_m^2 = C^2 + H^2/\rho$ and $\mathscr{P} = p + H^2/2$. Obviously, the condition under which the Cauchy problem for systems (2.8) and (2.9) is a well-posed problem is the inequality

$$\text{Im}\,\omega = k\,\text{Im}\,\lambda = k\,\text{Im}\,z \geqslant \text{const} \qquad (2.11)$$

for any real values of l_1, l_2 $(l_1^2 + l_2^2 = 1)$ and $k > 0$. For finite values of k the solutions z are always finite; hence in practice we require that conditions (2.11) hold in the limit $k \to \infty$. The coefficients in Eq. (2.10) are real and independent of k. If the equation has two complex-conjugate roots, for one of them $\text{Im}\,z < 0$ and $\text{Im}\,\omega \to -\infty$ in the limit $k \to \infty$. In other words, condition (2.11) does not hold. It follows that in the well-posed Cauchy problem all three roots of Eq. (2.10) must be real.

All the roots of the equation $z^3 + pz^2 + qz + r = 0$ with real coefficients are real if, and only if, the discriminant satisfies

$$D \equiv p^2q^2 + 18pqr - (4p^3r + 4q^3 + 27r^2) \geqslant 0.$$

For Eq. (2.10) we have

$$D = 4gY^2 + (C_m^4 + 18C_m^2 g - 27g^2)Y + C_m^6, \qquad (2.12)$$

where

$$Y = [(\xi H/\rho^2)(\partial \rho/\partial \tau)]^2; \quad g = (\partial \mathscr{P}/\partial \tau)/(\partial \rho/\partial \tau).$$

An elementary analysis of the quadratic trinomial in (2.12) shows that $D \geq 0$ for all $Y \geq 0$ if, and only if,

$$0 \leqslant g \leqslant C_m^2. \qquad (2.13)$$

Thus the system of equations in (2.8) is evolutionary if condition (2.13) holds for all possible directions of the vector τ. If this condition does hold, Eq. (2.10) has complex roots in the following cases: 1) $g < 0$, $Y > Y_2$; 2) $g > C_m^2$, $Y_1 < Y < Y_2$, where Y_1 and Y_2 are the roots of the trinomial D.

The condition in (2.13) usually does not hold in real plasma flows. Specifically, if the gradients of ρ and $\mathscr{P} = p + H^2/2$ are not collinear (i.e., if ρ and H are functionally independent), it is always possible to find a wave direction l such that g < 0. Thus, Eq. (2.8) is nonevolutionary. This result is consistent with the fact that it is not possible to compute Hall flow with an ideal conductivity ($\nu = 0$). This result is also apparently responsible for the instability which appears in the flow calculations — to a greater extent, the smaller ν and the larger ξ.

The equations are evolutionary if the magnetic viscosity is finite, $\nu > 0$. This conclusion can easily be reached through arguments analogous to those above and is not surprising: When $\nu \neq 0$, the fourth equation in (2.8) is of second order, and the Hall terms in this equation play a secondary role in the fine-scale perturbations [16].

In conclusion we emphasize again that these results are due to the degenerate nature of the Hall effect in Eq. (2.8) for two-dimensional flow across a magnetic field. In the general three-dimensional cases the equations cannot be studied by the plane-wave method. A repetition of these arguments in this case (or even for the three-dimensional perturbations of the flow under consideration here) leads to the following results.

1. If the conductivity is finite ($\nu > 0$) the equations are again evolutionary.

2. With an ideal conductivity ($\nu = 0$), the characteristic equation which is the generalization of (2.10) has some roots like $z = a\mathrm{k} + O(1)$, i.e., $\omega = a\mathrm{k}^2 + O(\mathrm{k})$ with real a. The leading term of this expression is "neutral," and whether condition (2.11) holds in the limit k $\rightarrow \infty$ depends on the next term. On the other hand, in the plane-wave method, in which the coefficients are assumed to be constant, only the leading terms of the asymptotic series are meaningful.

§2.6. Compressional Plasma Flows

Simplest Compressional Flow. The density increase caused by the self-magnetic field of the plasma, noted in §2.3, is much more pronounced in flow in a channel in which the inner electrode is shorter than the outer eletrode (Fig. 5). The earliest calculations for this case showed that a region is formed on the channel axis, near the end of the inner electrode, in which the plasma density is extremely high and increases with increasing magnetic field. A particularly high density is observed in calculations carried out for isothermal flow [21]. These results attracted attention in connection with the successful experiments by Filippov and Filippova [66], Mather [67], and others with a highly compressed plasma focus.[†] The basic distinction between the calculations and the experiments is that the calculations deal with steady-state flows in a channel while the plasma-focus system generates a single-shot compression of a fixed plasma mass.

An analysis of this pronounced compression carried out for an ideal plasma in the narrow-channel approximation [43] provides an explanation of this phenomenon and a proof of the existence of a qualitatively new class of steady-state flows — "compressional flows." This analysis led A. I. Morozov to the concept of the magnetoplasma compressor [44, 45].

For the steady-state flow of an ideal plasma the generalized Bernoulli theorem

$$U \equiv i + v^2/2 + H^2/\rho = \text{const} \quad [i = \int (dp/\rho)] \qquad (2.14)$$

holds along the trajectory of the plasma particles. To show this we note that the magnetogasdynamic equations in (1.36) yield

$$\rho \left[\partial U/\partial t + (v, \nabla) U \right] = (\partial/\partial t)[p + H^2/2].$$

If the flow is steady, we have $(v, \nabla)U = 0$, which means that U is constant along a trajectory.[‡]

If the scale value of the magnetic field is high (if β is small), and if the entrance velocity is low, then in the initial part of each trajectory the main term in the Bernoulli integral U is the "magnetic enthalpy" H^2/ρ. During the flow process this enthalpy is converted to $v^2/2$, in which case the problem is accelerated by electromagnetic forces, or to $i(\rho)$, corresponding to compressional flow. Systems with compressional flows of this kind are called magnetoplasma compressors.

If either regime is to be achieved, the magnetic enthalpy H^2/ρ on the corresponding part of the trajectory must be small in comparison with the other terms in (2.14). In a channel with a relative-

---

[†]See the two-dimensional calculations for the plasma focus in [32, 74].
[‡] See the analogous integrals in (1.2c) and (2.2) for the one-dimensional case.

ly short inner electrode (Fig. 5), this condition is clearly satisfied at the axis, where H = 0. With regard to the trajectory leaving the system at the channel axis, the integral in (2.14) means that

$$i + v^2/2 = i(0) + v^2(0)/2 + H^2(0)/\rho(0), \qquad (2.15)$$

where the argument "0" on the right side corresponds to the entrance cross section of the channel, z = 0. Hence we can easily make simple estimates for the compression at the axis† if we assume the flow to be isentropic or isothermal.‡ In the case of isentropic flow we have

$$i(\rho) = \frac{\gamma}{\gamma-1} \cdot \frac{p}{\rho} = \frac{\gamma\beta}{2(\gamma-1)} \rho^{\gamma-1}. \qquad (2.16)$$

By virtue of the boundary conditions in (1.48) we have $\rho(0) = 1$, and $H(0) = r_0/r$ is of the order of unity. Let us assume for simplicity that $H(0) = 1$. Then (2.15) and (2.16) yield

$$\frac{\gamma\beta}{2(\gamma-1)} \rho^{\gamma-1} = 1 + \frac{\gamma\beta}{2(\gamma-1)} + \frac{v^2(0)-v^2}{2}.$$

Evidently the highest values of the density at the axis are reached where the velocity is low, i.e., $v \approx v(0)$. In this case

$$\rho_{max} = \left(\frac{2(\gamma-1)}{\gamma\beta} + 1 \right)^{1/(\gamma-1)}. \qquad (2.17)$$

Since the density unit used here is the density at the entrance, this value of ρ_{max} represents the maximum relative plasma compression.

In isothermal flow we have $i(\rho) = (\beta/2) \ln \rho$, and similar arguments lead to the equation

$$\rho_{max} = e^{2/\beta}. \qquad (2.18)$$

Thus, in isentropic (or isothermal) flow of an ideal plasma at the

---

†The velocity at the accelerator exit is evaluated in an a similar way [(1.4) in §1.1].
‡This is a crude assumption since plasma compression can, in principle, be caused by a shock wave, in which the entropy is generally not conserved. Accordingly, these estimates as well as the calculations of [21], based on the same assumptions, should be taken as a first approximation of the study of the compression phenomenon. Nevertheless, the calculations which have been carried out to date do confirm the general features of this scheme, even with a shock wave in the channel.

axis of the channel with a relatively short inner electrode (Fig. 5) a pronounced plasma compression can occur, provided that the flow velocity is not high. The estimates in (2.17) and (2.18) show that as the magnetic field is strengthened the maximum possible compression increases (along with the ratio $1/\beta$) according to a power law (exponential).

This analysis indicates the possibility of compressional flows, but leaves a number of questions open. Nevertheless, the analysis shows that attempts to achieve solutions can reduce to a search for flow regimes in which the velocity at the end of the central electrode is not excessive as compared with the entrance velocity. Finally, the analysis is carried out for the case of an ideal plasma; when the results are extended to real flows it is necessary to take account of effects which have been ignored up to this point.

We now turn to these calculations [21]. They are carried out for the isentropic flow ($\gamma = 5/3$) and isothermal flow ($\gamma = 1$) of an inviscid, zero-thermal-conductivity plasma in the channel shown in Fig. 22. The flow is characterized by the parameter β, the conductivity Re_m (which is constant), and the replacement factor ξ.

These calculations show that a compression zone is formed at the axis with no Hall effect and an ideal conductivity; the compres-

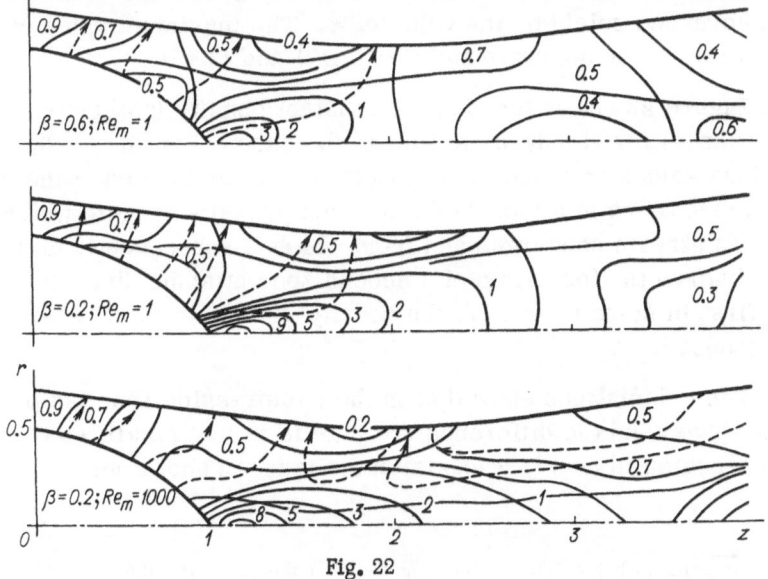

Fig. 22

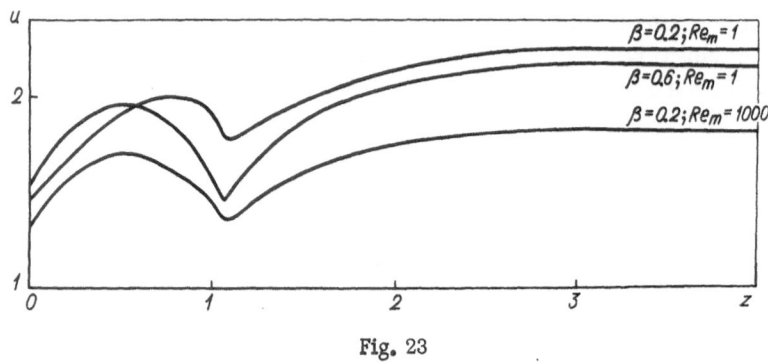

Fig. 23

sion agrees with the estimates in (2.17) and (2.18) for various values of β and for both values of γ.

The plasma compression at the axis is also found for finite values of the conductivity. The maximum compression and the point at which it occurs are nearly independent of the conductivity over a broad range, $0.1 \le \text{Re}_m \le 1000$. Incorporation of the Hall effect also has no effect on the extent of the compression, at least in that range of ξ in which the steady-state flow is established.[†] Accordingly, the compression at the channel axis and the magnitude of the compression in the case of an inviscid, zero-thermal-conductivity plasma, for adiabatic (or isothermal) flow, are governed solely by the value of β. The maximum value of the density is given approximately by (2.17) and (2.18).

Figure 22 shows the lines of constant density (solid curves) in the channel for steady-state flow in the case $\gamma = 5/3$, $\xi = 0$, with various values of β and the conductivity. Near the end of the central electrode there is a clearly defined compression zone to which the plasma stream converges and from which it subsequently diverges. We observe the formation of a second zone at some distance from the first in some cases, but the density maximum is not nearly as pronounced.

The calculations show that in the compression zone the plasma flow slows, and the difference $v - v^2(0)$ is in fact small, so that the compression is nearly a maximum. Figure 23 shows curves of the

---

[†]For example, with $\gamma = 5/3$, $\beta = 0.6$, and $\text{Re}_m = 1$ this range is $-0.5 < \xi < 0.5$ (§2.5).

velocity u(z) along the central electrode and the channel axis. The
influence of the conductivity and the Hall effect on the flow velocity
is described qualitatively in § § 2.4 and 2.5; the quantitative effect
is quite weak for the given geometry (the upper and lower curves
in Fig. 23). The conductivity only has a strong effect on the distribution of the magnetic field and the electric current in the channel. With an ideal conductivity the field is frozen in the plasma
and is distributed in the channel in the same way as the density.
As the ohmic resistivity ν increases, the penetration of the field
into the channel becomes progressively smaller, and at high values
of ν the plasma in the compression zone is almost completely "de-
energized." Figure 22 shows the lines of the electric current
(dashed curves). We note that current eddies form near the outer
electrode at high values of Re_m.

The Hall effect also has an important influence on the field and
the electric current, deflecting the current lines in one direction or
the other, depending on the electrode polarity.

After the calculations were carried out and analyzed, Morozov,
Kovrov, and Vinogradova verified the existence of compressional
flows experimentally. These experiments were carried out in 1967
with nitrogen as a medium at currents from 200 to 400 kA. It was
found [47] that at currents of 300 kA and higher it is possible to
achieve a compressional regime with $\rho_{max} \approx 100$ and $T_e \approx 2\text{-}3$ eV,
which exists through the discharge (70 μsec). This flow is extremely
stable. Experiments with hydrogen were begun in 1968. Here again
it is possible to achieve a steady-state compressional flow which
lasts for a time comparable to the bank discharge time (100 μsec).
At currents of 300 kA the values $\rho \approx 30$ and $T_i \approx T_e \approx 10$ eV are
achieved [45]. These experimental results are described in more
detail in [48].

Effect of Viscosity and Thermal Conductivity
on the Flow. The results discussed above should be treated
as preliminary since they are based on a simplified model. A
further study of compressional flows requires refinements in the
formulation of the problem and in the calculations themselves. Experiment indicates that the plasma compression at the axis generally heats the plasma. Furthermore, shock waves can be generated in the channel and either assist or oppose the compression.
For example, in the figures showing the flow between cylindrical

electrodes in the published calculations of Morse [68, 71] a shock
wave that bonds a region of hot but low–density plasma at the channel axis is clearly evident. This situation usually corresponds to
flow around a blunt object.

Accordingly, the assumption of an isentropic flow cannot be
supported, nor can the subsequent calculations based on the complete system of equations in (1.33), including the energy equation
and all the dissipative processes — the viscosity, thermal conductivity, and electrical conductivity.

The numerical values of the transport coefficients are estimated, as mentioned in §1.2, on the basis of the experimental values of the dimensional quantities specified at the channel entrance:
$n_0 \approx 2 \cdot 10^{16}$ cm^{-3}, $T_0 \approx 5$ eV, $H_0 \approx 10^4$ Oe, and $L \approx 30$ cm. Using
Eqs. (1.40)–(1.46) we find

$$\eta_2 \approx 10^{-5} - 10^{-6}; \quad \varkappa_2 \approx 10^{-3} - 10^{-4}; \quad \sigma_2 \approx 10^8. \qquad (2.19)$$

These estimates show that the assumptions $\Pi = 0$ and $\varkappa = 0$ used
earlier are not too crude. The viscosity and thermal conductivity
can only affect the flow substantially in the hot region. In the calculation program, however, they are overestimated slightly because
of the introduction of the first terms η_1, ζ_1, and $\varkappa_1$ in Eqs. (1.43)–
(1.45). This approach makes it possible to carry out calculations
for flow with shock waves, which can arise at small and moderate
values of T, without resorting to special calculation methods.†
Furthermore, to some extent this approach relieves the contradiction which arises between the low viscosity, (2.19), and the boundary condition that the plasma "sticks" to the channel walls‡ ($v = 0$).

Let us examine the results of the two calculations.

First, a series of calculations has been carried out with constant values of the transport coefficients to determine the qualitative influence of dissipative effects on the flow. The channel shape
and the basic parameters ($\beta = 0.2$, $\xi = 0$, and $\nu = 0.001$) are the
same as at the bottom of Fig. 22; the viscosity coefficients $\eta = \zeta$
and the thermal conductivity $\varkappa$ are varied over broad ranges.

---

†One possibility is to introduce another, artificial viscosity [56].

‡An optimum calculation program would ignore the viscosity below a certain minimum
value and replace the sticking boundary condition with a slip condition.

The results show that in a first approximate a steady state is reached (in a transit time). The properties of this steady-state flow depend on the dissipative processes.

The introduction of viscosity sharply alters the nature of the flow, even at small values of the coefficients η and ζ. The most important role here is played by the first viscosity, η; the second viscosity has almost no effect. The flow is retarded and heated as a result of friction at the channel wall. At values $\eta \sim (0.01-0.05)$ and above the viscous friction turns out to be the governing factor for the flow. The flow velocity is subsonic, and the characteristic value of this velocity is approximately a linear function of Re = $1/\eta$, falling off rapidly with increasing η. The plasma temperature increase along the walls, reaching values 30-40 times the values at the entrance. The strongly heated electrode layers show a low density, so that there is a narrow zone of low-density plasma along the channel axis. The compression region discussed above spreads out, acquiring the shape of a long, broad tube, within which the plasma density is low. In this case the maximum compression is much lower and, like the tube dimensions, changes very slowly as a function of η.

When $\eta < 0.01$ the velocity at the channel exit is supersonic. It increases with increasing Re, but more slowly than in the preceding case: Specifically, as η is reduced by a factor of ten, from 0.01 to 0.001, the velocity increases by less than a factor of three. Part of the energy is still converted into heat: At $\eta = 0.001$ the maximum temperature at the inner electrode is 20 times that at the entrance, and the maximum temperature at the outer electrode is ten times that at the entrance. The low-density sheaths near the electrodes become thinner. The compression zone is tubular, but its transverse dimensions are smaller; it is contracted toward the axis.

As expected, the thermal conductivity smooths the temperature distribution in the channel, modifying the flow properties established by the nonuniform plasma heating. Even when $\varkappa$ is comparable to η, the hot zones near the electrodes vanish. The compression zone again lies at the axis, but fills a volume larger than that found in calculations on the basis of the ideal model, and the maximum compression is half as pronounced.

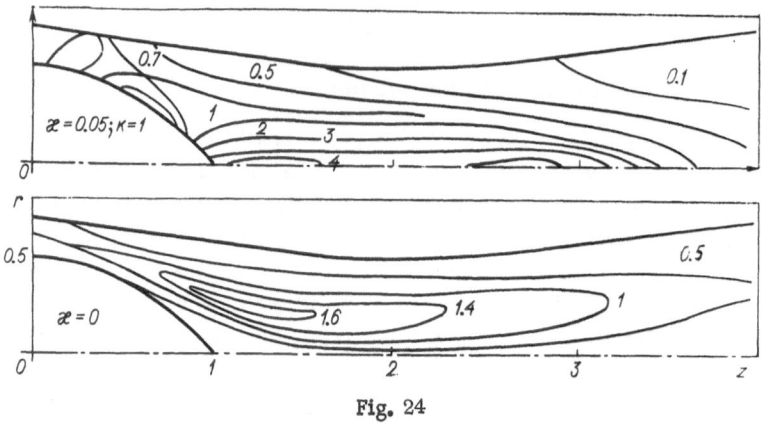

Fig. 24

This pattern can be constructed more rapidly by incorporating in the problem (along with the thermal conductivity) a heat sink at the channel walls which obeys the law $\partial T/\partial n = -kT$ instead of using the condition that the walls are thermal insulators. The velocity of the plasma flow in the channel is nearly independent of the thermal conductivity, being governed primarily by the values of β, ν, and η. Figure 24 shows lines of constant density for the case $\beta = 0.2$, $\nu = 0.001$, and $\eta = \zeta = 0.05$ in the cases with an without the thermal conductivity. Figure 25 shows the maximum flow velocity at the axis at the exit from the channel as a function of the viscosity coefficient η (for the same values of β and ν as in Fig. 24).

The second series of calculations is carried out for the flow of a plasma with variable transport coefficients, approximately the

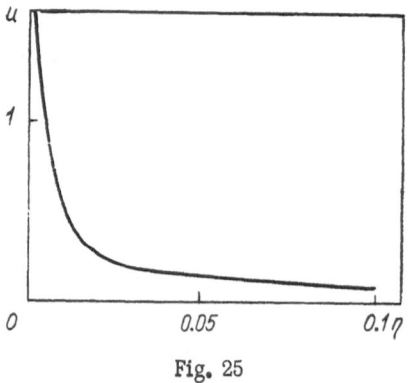

Fig. 25

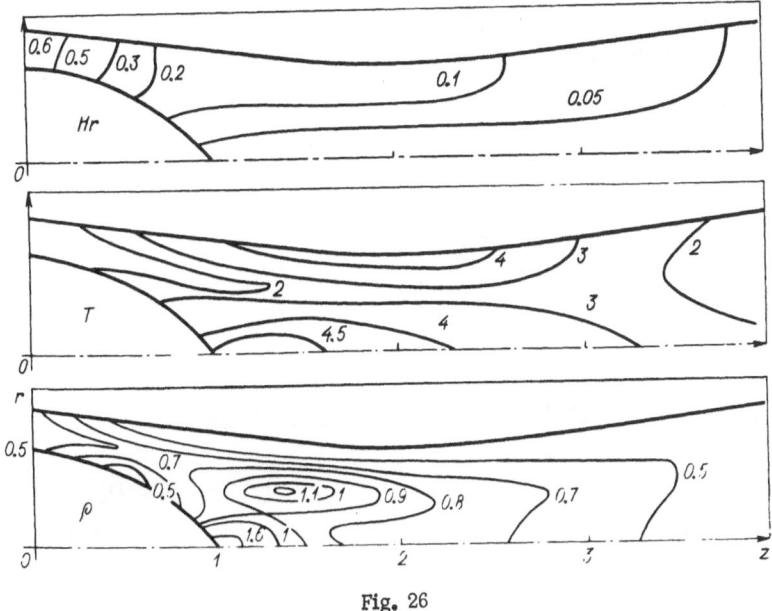

Fig. 26

same as those given by the estimates in (2.19); specifically, the values are $\beta = 0.2$, $\xi = 0$, $\eta = \zeta = 0.001 + 10^{-5} T^{5/2}$, $\varkappa = 0.001 + 10^{-4} T^{5/2}$, and $Re_m = 100\, T^{3/2}$. The boundary condition at the electrodes corresponds to a heat sink that follows (1.5) with $k = 1$. Figure 26 shows lines of constant density ρ, constant temperature T, and constant values of the function Hr (i.e., the electric-current lines). The density distribution shows both tendencies mentioned earlier: The region of slight compression is tubular, while compression is observed at the axis, directly adjacent to the central electrode. The temperature in this latter compression zone is higher than that at the surface of the inner electrode; i.e., this temperature is governed not only by friction but also by the compression. With a longer outer electrode we see clearly a layer of relatively low-density plasma, heated to $T \approx 4$.

In reviewing these results we should emphasize the difficulty in comparing them with experiment. In the first chapter we noted that the boundary condition $\mathbf{v} = 0$ does not correspond to actual conditions since there is nothing to prevent the ions from reaching the electrode and disappearing there. Under these conditions the flow pattern near the cathode and in the compression zone can be very different from that found in these calculations.

Consequently, the results found in calculations that ignore the viscosity are apparently a better approximation to the actual conditions. Nevertheless, the calculations with viscosity are clearly of methodological interest and are useful for discussing the features of the real physical processes.

Chapter 3

FLOW OF A GAS WHICH IS IONIZED IN THE CHANNEL

§ 3.1.   Ionization Process in the Flow Model

In this chapter we include ionization in the channel in calculations of the plasma flow. As noted in § 1.1, in an actual experiment a neutral gas enters the channel and then is ionized almost completely as it is accelerated and heated in the converging part of the nozzle. In steady-state flow it is generally possible to observe a clearly defined ionization front; to the left of this surface there is a flowing gas and to the right there is a flowing plasma.

The gas flow at the left of the front is described by the gasdynamic equations, while the plasma flow at the right of the front is described by Eq. (1.33). The problem can be reduced to a numerical integration of these equations if the interface is known. However, the usual conservation laws generally do not form a complete set of boundary conditions at the ionization shock wave [40]. The other conditions required must be chosen on the basis of the shock structure.[†] The number of other conditions that are required is ultimately governed by the requirement that the shock be of an evolutionary nature, i.e., by the ratios of the gas, magnetosonic, and shock-wave velocities [7]. Accordingly, it is desirable to incorporate the ionization in this methematical model for the flow in a way which allows us to avoid a direct calculation of the position of the ionization front. It can be assumed that ionization occurs when the gas reaches a certain critical temperature $T*$; thus, the interface is the surface $T = T*$.

[†]The physical meaning is that the ionization front is "thick."

The neutral gas has no electrical conductivity; thus in this gas $\text{Re}_m = 0$ ($\nu = \infty$). Under this condition we find $\mathbf{j} = 0$ from Eq. (1.33), and these equations automatically become the gasdynamic equations. In the ionized gas the conductivity is nonvanishing, being governed by Eq. (1.46), while the flow is described by Eq. (1.33).

Thus, the ionization of the gas in the channel can be incorporated into the model of Chap. 1 and in the calculation scheme if the conductivity Re_m is assumed to change abruptly at a certain temperature; specifically, for temperatures $T < T^*$ we set $\text{Re}_m = 0$ and for temperatures $T > T^*$ we use Eq. (1.46). This procedure leaves certain calculation difficulties associated with the unbounded increase in the magnetic viscosity ν. In principle it is possible to circumvent these difficulties, say, by solving the diffusion equation for the magnetic field by the tridiagonal method [30], which is sometimes used to solve problems of this kind in their one-dimensional formulation [10], and by introducing an additional condition in the program for the solution of the energy equation. This condition is that the Joule heat vanishes when $\text{Re}_m = 0$.

We use a different approach in the present model. The problem is formulated slightly less exactly but in a more consistent way. The neutral gas is assigned a small but nonvanishing conductivity, so that the flow of the entire medium can be described by the magnetogasdynamic equation, (1.33). This formulation of the problem is completely consistent with the experimental conditions: The high-current discharge generates intense UV radiation, which causes an appreciable ionization of the gas entering the system, thus imparting a finite conductivity to the gas.

The incorporation of ionization into this model of the problem means that Eq. (1.46) is replaced by the following temperature dependence for the conductivity:

$$\text{Re}_m = \frac{1}{\nu} = \begin{cases} \sigma_0, & T < T^*; \\ \sigma_1 + \sigma_2 T^{3/2}, & T > T^*, \end{cases} \tag{3.1}$$

where $\sigma_0 \ll 1$. We can proceed in a similar way with regard to the thermal conductivity in (1.45). We note further that heat is dissipated during the ionization process. To reflect this dissipation in the calculations, to the right side of the energy equation in

(1.33) or (1.37) we add a term

$$-I_0\, \rho \exp\left[-\alpha\,(T - T^*)^2\right], \qquad\qquad (3.2)$$

which corresponds to the heat loss at T = T*. At large values of α the exponential function in (3.2) has the property that it is only important near the value T = T*, corresponding to the ionization. Equations (3.1) and (3.2) constitute the refinement of the formulation of the problem associated with ionization.

The calculations described below (§ 3.3) are carried out in the quasi-one-dimensional formulation of the problem, i.e., Eq. (1.37) without viscosity (η = 0).

§ 3.2. Steady-State Flow with a Conductivity Discontinuity in a Channel of Constant Cross Section

Before taking up the calculated results we examine the following simplified problem.

We assume one-dimensional, steady-state flow of an inviscid, zero-thermal-conductivity plasma in an infinite channel of constant cross section. The plasma conductivity varies in some way from a constant value σ_L on the left to a constant value σ_R on the right; here $\sigma_L < \sigma_R$. We assume that the flow properties are constant, being independent of the coordinate along the channel (x) at points sufficiently far from the region in which σ changes. The x axis is in the direction of flow so that the velocity v > 0. Finally, we assume that the flow is subsonic at points sufficiently far to the left.

The quantities ρ_0, T_0, p_0, H_0, and v_0 are the constant values of the density, temperature, etc., which are given at the left (at x → $-\infty$), and we adopt these values and some scale length L (e.g., the length of the region over which the change in σ occurs) as the units for the corresponding quantities by analogy with the procedure used in § 1.2. Then the dimensionless equations describing the flow are Eq.[†] (1.37) in which we set $f(x)$ = const, $\eta = \varkappa = 0$. Evaluating the integrals, we find

$$\rho v = C_1, \qquad\qquad (3.3a)$$

---

†For simplicity here we omit the final term in (3.2).

$$\rho v^2 + p + H^2/2 = C_2, \tag{3.3b}$$

$$C_1 (i + v^2/2) + EH = C_3, \tag{3.3c}$$

$$-v \, dH/dx + Hv = C_4, \tag{3.3d}$$

where

$$\left.\begin{array}{l} p = (\beta/2)\,\rho T; \quad i = \int (dp/\rho) = (\gamma\beta/2\,(\gamma-1))\,T; \\ C_1 = v_L; \ C_2 = v_L^2 + (\beta+1)/2; \ C_3 = (v_L/2)[\gamma\beta/(\gamma-1) + v_L^2]; \\ C_4 = E, \end{array}\right\} \tag{3.4}$$

and $v_L = v_0(4\pi\rho_0)^{1/2}/H_0$ is the velocity at the left in the coordinate system we have chosen. Since the flow on the left is constant $(dH/dx = 0)$, it follows from (3.3d) that $E = v_L$.

After ρ and T are eliminated, the first three equations in (3.3) reduce to a single equation between H and v:

$$v^2 + [\gamma v/(\gamma + 1)v_L](H^2 - 2C_2) - 2[(\gamma - 1)/(\gamma + 1)](H - 1) = 0. \tag{3.5}$$

Thus, we can write v in terms of H and substitute the result into (3.3d), which must be solved with the initial condition $H(-\infty) = 1$. This approach is a familiar approach in the study of shock waves and ionization waves in magnetogasdynamics [3, 40].

Equation (3.5) describes a curve in the (H, v) plane, as shown schematically in Fig. 27. The exact position of this curve depends

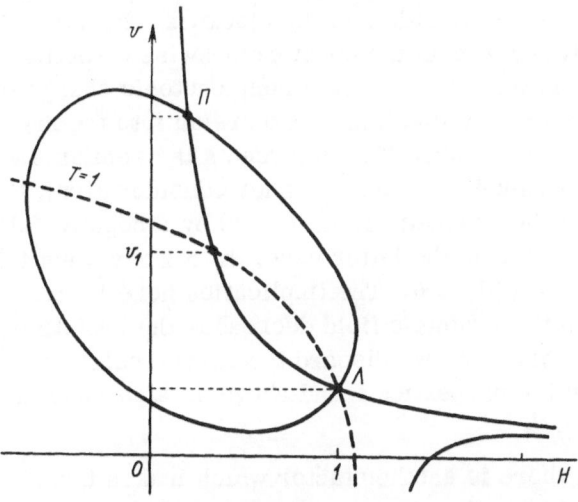

Fig. 27

on the two parameters β and v_L. According to Eq. (3.3d) the constant flow regions at the left and at the right correspond to the intersections of curve (3.5) with the hyperbola

$$Hv = E = v_L. \tag{3.6}$$

The coordinates of these points are given by

$$F(v) \equiv (v - v_L) F_1(v) \equiv (v - v_L) \left[v^2 - \left(\frac{2\gamma}{\gamma + 1} C_2 - v_L^2 \right) \frac{v}{v_L} - \frac{2 - \gamma}{\gamma + 1} \right] = 0. \tag{3.7}$$

In the half-plane $v > 0$ curves (3.5) and (3.6) intersect at two points: $v = v_L$ and $v = v_R$, where v_R is the positive root of the quadratic trinomial $F_1(v)$. The point L obviously corresponds to the flow at the left ($v = v_L$, $H = 1$), while the point R represents the sole possibility for constant flow at the right ($v = v_R$, $H = v_L/v_R$). Point R lies above the point L (Fig. 27) by virtue of the assumption that the velocity at the left is subsonic: if $v_L^2 < [C^2 + H^2/\rho]_L = \gamma \beta/2 + 1$, $F_1(v_L) = [2/(\gamma + 1)][v_L^2 - \gamma\beta/2 - 1] < 0$, so that $v_L < v_R$. This means that there is an acceleration from v_L to v_R in the flow and a corresponding reduction of the plasma density from $\rho = 1$ at the left to $\rho = v_L/v_R < 1$ at the right.

If the assumed type of continuous flow does exist, then the solution of the problem is represented by that part of curve (3.5) which connects the points L and R (Fig. 27). The transition from L to R can be made formally either in the clockwise direction (below the hyperbola $Hv = v_L$) or in the counterclockwise direction (above the hyperbola). In the former case a nonmonotonic change occurs in H, which necessarily involves an excursion into the region of negative H [we see that curve (3.5) intersects the semiaxis $v > 0$ at two points for all values of β and v_L under consideration]; flow implies a reversal of the magnetic field ($H < 0$) by a negative electric current ($dH/dx > 0$). In the latter case, $Hv > E = v_L$, but this is possible only if $\nu dH/dx > 0$. The implication here is that in the region in which the magnetic field decreases the resistivity ν must be negative; on this basis we discard the second solution. The first solution with the properties found above is also not reasonable in this problem.

Finally, there is another factor which makes these solutions unsuitable. Increasing the conductivity involves an increase in the

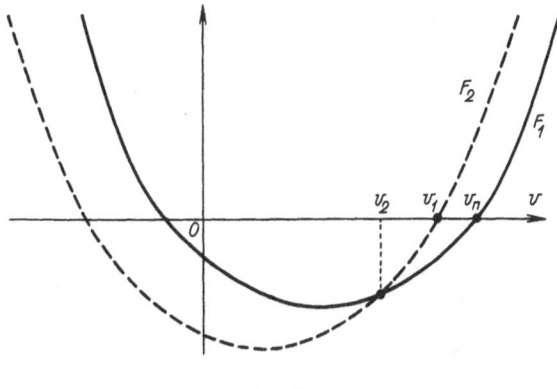

Fig. 28

temperature, so that we can try to assign a physical meaning to the solution if the temperature at the right is higher than that at the left, i.e., if $T_R > 1$. However, we always find that the opposite inequality holds, as can be shown in the following way.

It follows from (3.3c) that the parabolas

$$v^2/2 + H = C_3/v_L - [\gamma\beta/2(\gamma-1)]T$$

in the (H, v) plane correspond to constant temperatures T; as T increases, the entire parabola shifts to the left. At T = 1, the parabola passes the point L (Fig. 27), and the other intersection of this parabola with the hyperbola in (3.6) is given by the positive root v_1 of the equation

$$F_2(v) \equiv v^2 + v_L v - 2 = 0. \qquad (3.8)$$

Comparing (3.7) and (3.8), we see that the plots of $F_1(v)$ and $V_2(v)$ (Fig. 28) intersect at the point $v_2 = 3v_L/2C_2$ and that $F_2(v_2) < 0$. It follows that $v_1 < v_R$; thus, the curve corresponding to T = 1 intersects the hyperbola $Hv = v_L$ a second time below the point R, and the point R lies in the region† T < 1 (Fig. 27). This feature also rules out a third possibility — that of allowing a discontinuity in the conductivity and constructing a continuous solution containing the points L and R.

---

†Furthermore, at certain values of β and ν_L, we cannot rule out the possibility that the point R will fall in the region T < 0, in which the solution would have no meaning.

These results show that there are no steady-state physical plasma flows in a channel of constant cross section in which the conductivity increases continuously or discontinuously and in which there is acceleration and heating. This result can be associated with the established fact that rarefaction shock waves do not occur in the usual gasdynamics and magnetogasdynamics of an ideal medium.

Strictly speaking, this analysis does not extend to channels of variable cross section. Nevertheless, these results indicate that even in this case there may be no steady-state continuous solutions with acceleration and heating.

Experimentally, however, as noted above, an abrupt transition occurs from a gas to a plasma; this is accompanied by acceleration, heating, and a density decrease. The discrepancy between these experimental results and our calculated result means only that the ionization process is much more complicated than the simplified model adopted here. The difficulties involved in a more detailed theoretical approach again show the need for a numerical solution of this problem.

§3.3. Calculations for the Flow of an Ionizing Gas

Flow calculations incorporating the ionization process in accordance with the scheme specified in §3.1 are carried out in the quasi-one-dimensional approximation in a nozzle channel. To facilitate the comparison of the results with the available data on plasma flow (§2.2), we carry out these calculations for a channel with the same cross-sectional shape [(2.1)]:

$$f(x) = \begin{cases} 0.3 - 0.8x(1-x), & 0 \leqslant x \leqslant 1; \\ 1.5x - 1.2, & 1 \leqslant x \leqslant 5. \end{cases} \qquad (3.9)$$

Two series of calculations have been carried out, in connection with the two possible interpretations of the temperature dependence of the conductivity. Equation (3.1), taken literally, means that when a gas — already ionized — is cooled to a temperature below T^* there is another transition, to a nonionized state, $\mathrm{Re}_m = \sigma_0$. This model can correspond to the acceleration of a dense plasma, which is of interest in the development of wind tunnels.

In the second interpretation of Eq. (3.1), it is assumed that after ionization, the gas remains ionized, regardless of the subsequent temperature; in this case we introduce the additional requirement that recombination does not occur. In other words, the change in the state (Re_m) for each particle occurs in accordance with Eq. (3.1) only when the temperature increases.

The results found in the calculations are qualitatively very similar so that we restrict the discussion of the quantitative results below to the first series of calculations.

As a result of all these calculations it is established that the flow does not reach a steady state with a continuous solution. This result is in agreement with that found at the end of § 3.2. Let us examine the nature of the solution in detail.

If the conductivity of the ionized gas is finite, that is, if the coefficients σ_1 and σ_2 in Eq. (3.1) are of order unity, the flow reaches a steady state in a time of the order of the transit time. The steady state is characterized by smooth, rather than discontinuous, functions: Specifically, T and v increase monotonically, while ρ and H decrease monotonically along the channel. The temperature reaches the critical value T* in a continuous way near the minimum cross section of the channel. The most pronounced reaction to the ionization process is found in the magnetic field H, which drops sharply immediately after T reaches T*; correspond-

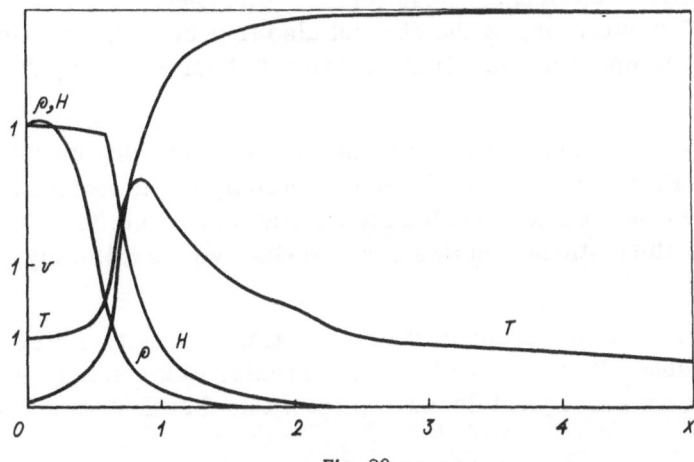

Fig. 29

ingly, the electric current in this region increases abruptly.
Figure 29 shows ρ, T, v, and H as functions of distance along the
channel [with the shape in (3.9)] in steady-state flow for the following parameter values:

$$\left.\begin{array}{l} \beta = 0.2; \ \sigma_0 = 0.05; \ \sigma_1 = \sigma_2 = 1; \ \varkappa = 0; \ T^* = 1.5; \\ I_0 = 0.3. \end{array}\right\} \qquad (3.10)$$

The flow regime can be described as a function of the conductivity ahead of the ionization front and as a function of the channel
shape: As σ_0 is reduced (from 0.05 to 0.01 in these calculations)
there is a slight (up to 20-30%) increase in the temperature T and
in the velocity v, a decrease in the density ρ (corresponding to the
velocity increase), and a shift of the position of the ionization front
(the point corresponding to T = T*) to the right (from x = 0.55 to
x = 0.75). Furthermore, the flow becomes unstable with decreasing†
σ_0: At $\sigma_0 = 0.01$ we observe undamped temperature oscillations
(with an amplitude of up to 20%) in the exit part of the channel, on
the descending branch of the T(x) curve.

If the channel shape is changed so that the minimum cross section is reduced in comparison with the maximum cross section [in
the calculations, it is reduced to a size smaller by a factor of four
than that of the cross section in (3.9)], then the flow with the parameters in (3.10) changes in the following way: The velocity decreases (by up to 30%), the density increases, and the ionization
front shifts to the left (x = 0.45 for $\sigma_0 = 0.05$ and x = 0.55 for $\sigma_0 =$
0.01). The narrowing of the channel also improves the stability of
the flow; temperature oscillations are not observed at any value
of σ_0.

This is the flow pattern for small but nonvanishing values of σ_1
and σ_2. If, on the other hand, the conductivity of the ionized gas is
high, the flow does not reach a steady state, but exhibits quasi-
periodic fluctuations. In this case we observe the following phenomenon.

In a time of the order of the transit time a density distribution
is established in the channel which is similar to the steady-state
distribution; subsequent deviations from this distribution are very

---

†It is understood that the current I remains constant.

weak. The density is 30% lower than that found in the calculations
for the flow of a fully ionized plasma (§ 2.2).

On the other hand, there are sharp temperature oscillations.
The position of the ionization front oscillates in time near the minimum cross section of the nozzle (x ≈ 0.5). Ahead of the front the
flow is nearly in a steady state. Behind the front there is a narrow
hot region in which the temperature maximum oscillates. Hot conducting layers of plasma periodically separate from the front and
move to the right, cool down, and become wider. The magnetic
field falls off from left to right, essentially to zero; the entire decrease occurs in the fluctuating heated region behind the ionization
front. This behavior corresponds to a process in which the electric current "punctures" the channel in a hot region, with a relatively
high conductivity. The velocity in the channel generally increases
monotonically from left to right, and the velocity oscillations are
correlated with the temperature oscillations. The amplitude of these
oscillations does not exceed 25-30% of the average velocity beyond
the ionization front.

The flow pattern can be described as follows: A parcel of gas,
rapidly accelerated, tends to increase the voltage

$$U = Ef = (-v\partial H/\partial x + vH)f$$

at the electrodes that form the channel walls. The result is a
breakdown of the gas by the electric current.

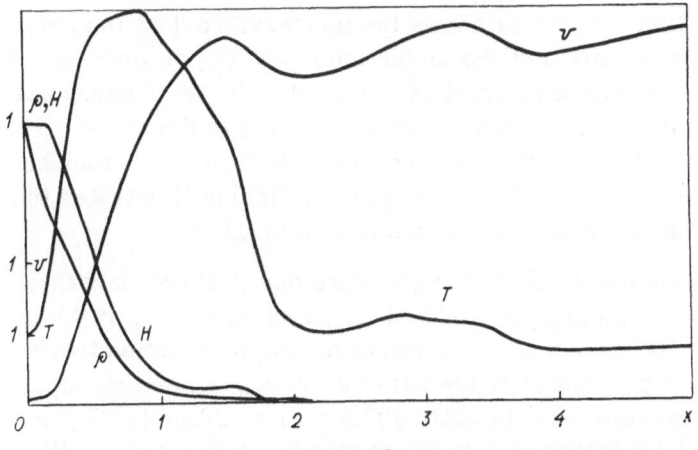

Fig. 30

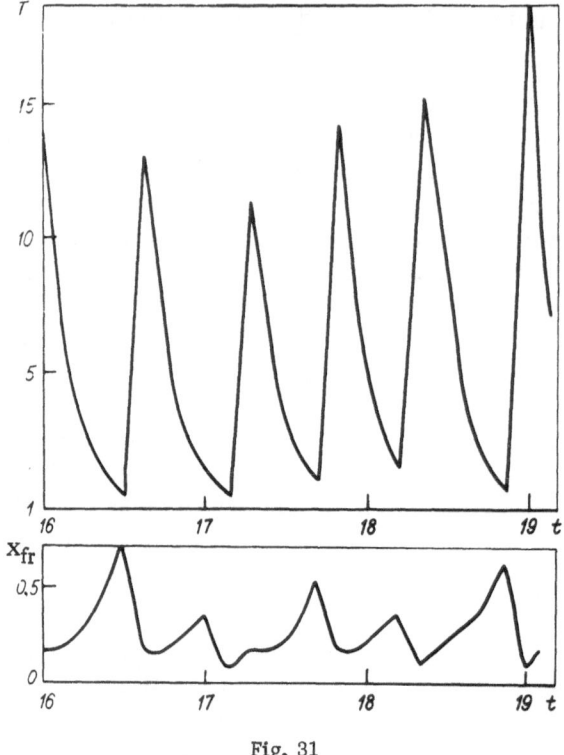

Fig. 31

Figure 30 shows ρ, T, v, and H as functions of the distance along the channel for a typical time, t = 16.75, longer than the transit time. Figure 31 shows the temperature T at the point x = 0.7 and the position of the ionization front, x_{fr}, as functions of the time over the time interval 16 < t < 19. The two figures correspond to the same channel shape, (3.9), and to the same parameter values, (3.10), except for the conductivity beyond the ionization front: Here $\sigma_1 = \sigma_2 = 100$. Figure 31 clearly illustrates the fluctuating nature of the flow with a period of $\Delta t \approx 0.6$.

The amplitude of the temperature oscillations increases with increasing σ_1 and σ_2 or at fixed values of σ_1 and σ_2, if the conductivity at the entrance, σ_0, is reduced. As β is varied there are changes in the values of the parameters σ_0, σ_1, and σ_2 separating the steady-state and fluctuating flows. For example, if β is increased by a factor of four in the steady-state version with the parameters in (3.10), the flow becomes quasiperiodic, although the oscillation amplitude and frequency are small. The flow calculations for this case are reported in [22]. At higher β these oscillations vanish, as expected, and the steady-state flow regimes approach the gasdynamic regimes.

The absence of a steady-state flow regime in the case $Re_m \to \infty$ with $T > T^*$ can also be explained on the basis of the results of § 3.2. If the flow is steady, Eqs. (3.3) hold in a small neighborhood of the ionization front. In particular, when $Re_m = \infty$, Eq. (3.3d) means that

$$-vdH/dx + vH = E, \quad x < x_{fr}; \quad \rbrace \qquad (3.11)$$
$$vH = E, \qquad \qquad x > x_{fr} . \quad \rbrace$$

At small values of σ_0 the quantity $vj = -vdH/dx$ is nonvanishing at the left of the front (as confirmed by the calculations); it follows from (3.11) that the product vH is not continuous at $x = x_{fr}$ (§ 3.2).

The existence of an oscillating ionization front implies that inertial effects play an important role and again confirms that the front has a finite width.

We emphasize that the temperature oscillations found in these calculations correspond to the zero-thermal-conductivity model for the plasma. This point cannot be ignored in the analysis since the thermal conductivity undoubtedly plays a role in real flows with heating and ionization. A nonvanishing (even small) thermal conductivity stabilizes the flow. For example, the flow with $\beta = 0.2$, $\sigma_0 = 0.01$, and $\sigma_1 = \sigma_2 = 100$ oscillates in the case $\varkappa = 0$ with an amplitude $T = 100$, but the flow becomes steady if a constant thermal conductivity of $\varkappa = 0.2$ is incorporated in the calculation. The temperature becomes a smooth, monotonically increasing function of x.

We note in conclusion that Belyaev et al. [10] have also observed the flow of an ionizing gas to be fluctuational. They carried out calculations (in terms of Lagrangian coordinates) for the one-dimensional flow of a plasma in a pulsed rail accelerator and found that quasiperiodic narrow hot layers develop from an initial temperature perturbation in a nonconducting gas.

REFERENCES

1. V. A. Abramov et al., Report No. 3.1.11.8, in: Eighth International Conference on Phenomena in Ionized Gases, Vienna, 1967.
2. L. M. Alekseeva and 1. S. Solov'ev, Prikl. Matem. i Mekh., 28:987-995 (1964).
3. J. E. Anderson, Magnetohydrodynamic Shock Waves, MIT Press, Cambridge, Mass. (1963).
4. E. V. Artyushkov and A. I. Morozov, Teplofiz. Vys. Temp., 2:525-534 (1964).
5. S.-I. Pai, Magnetogasdynamics and Plasma Dynamics, Springer-Verlag, New York (1962).
6. G. M. Bam-Zelikovich, Izv. Akad. Nauk SSSR, Ser. Mekhan. i Mashinostr., No. 2:3-10 (1963); No. 5:9-15 (1964); Prikl. Matem. i Mekh., 28:664-669 (1964); Zh. Prikl. Mekhan. i Tekh. Fiz., No. 6:36-40 (1968).
7. A. A. Barmin and A. G. Kulikovskii, Dokl. Akad. Nauk SSSR, 178:55-58 (1968) [Sov. Phys. Dokl. 13:4 (1968)].
8. T. D. Butler, Mekhanika, No. 4:91-108 (1968).
9. R. B. Basharov et al., in: Physics of Gas-Discharge Plasmas [in Russina], No. 2, Atomizdat, Moscow (1969), pp. 161-184.
10. S. A. Belyaev et al., "Calculation of unsteady plasma acceleration in the one-dimensional approximation," Preprint, Institute of Applied Mathematics, Academy of Sciences of the USSR, Moscow, 1969.
11. R. Betchov and W. O. Criminale, Stability of Parallel Flows, Academic Press, New York (1967).
12. S. I. Braginskii, I. M. Gel'fand, and R. P. Fedorenko, in: Plasma Physics and the Problem of Controlled Thermonuclear Reactions [in Russian], Vol. 4, Izd. Akad. Nauk SSSR, Moscow (1958), pp. 201-221.
13. S. I. Braginskii, Reviews of Plasma Physics, Vol. 1 (ed. M. A. Leontovich), Consultants Bureau, New York (1965), p. 205.
14. K. V. Brushlinskii, Izv. Akad. Nauk SSSR, Ser. Matem., 23:893-912 (1959).
15. K. V. Brushlinskii, Zh. Vychislit. Matem. i Matem. Fiz., 8:1039-1048 (1968).
16. K. V. Brushlinskii and A. I. Morozov, Prikl. Matem. i Mekh., 32:957-959 (1968).
17. K. V. Brushlinskii, N. M. Zueva, and A. I. Morozov, Izv. Akad. Nauk SSSR, Ser. Mekhanika, No. 5:3-6 (1965).
18. K. V. Brushlinskii, N. I. Gerlakh, and A. I. Morozov, in: Fourth Riga Conference on MHD, June 1964. Abstracts of Reports [in Russian], Izd. Akad. Nauk Latv. SSR, Riga (1964); Izv. Akad. Nauk SSSR. Mekhan. Zhidk. i Gaza, No. 2:189-192 (1966).
19. K. V. Brushlinskii, N. I. Gerlakh, and A. I. Morozov, Magnitn. Gidrodinam., No. 1:3-8 (1967).
20. K. V. Brushlinskii, N. I. Gerlakh, and A. I. Morozov, Magnitn. Gidrodinam., No. 2:31-34 (1967).
21. K. V. Brushlinskii, N. I. Gerlakh, and A. I. Morozov, in: Third All-Union Congress on Theoretical and Applied Mechanics. Abstracts of Reports [in Russian], Izd. Akad. Nauk SSSR, Moscow (1968), p., 50; Dokl. Akad. Nauk SSSR, 180:1327-1330 (1968) [Sov. Phys. Dokl. 13:588 (1968)].

22. K. V. Brushlinskii, A. I. Morozov, and V. V. Paleichik, Izv. Akad. Nauk SSSR Mekhan. Zhidak. i Gaza, No. 5:29-32 (1970).

23. K. V. Brushlinskii, A. I. Morozov, and V. V. Paleichik, Plasma Accelerators (ed. L. A. Artsimovich) [in Russian], Mashinostroenie, Moscow (1973), pp. 251-254.

24. A. B. Vatazhin, G. A. Lyubimov, and S. A. Regirer, Magnetohydrodyanmic Flows in Channels [in Russian], Nauka, Moscow (1970).

25. E. P. Velikhov et al., Dokl. Akad. Nauk SSSR, 184:578-581 (1969) [Sov. Phys. Dokl. 14:68 (1969)].

26. B. Alder, S. Fernbach, and M. Rotenberg, (editors) Methods in Computational Physics. Vol. 4. Applications in Hydrodynamics, Academic, New York.

27. I. M. Gel'fand, Usp. Matem. Nauk, 14:87-158 (1959).

28. I. M. Gel'fand et al., Zh. Vychislit. Matem. i Matem. Fiz., 7:322-347 (1967).

29. A. V. Gubarev, L. M. Degtyarev, A. A. Samarskii, and A. P. Favorskii, Dokl. Akad. Nauk SSSR, 192:520-524 (1970) [Sov. Phys. Dokl. 15:445 (1970)]. Proceedings of the Fifth International Conference on MHD Energy Conversion, München, 1971.

30. L. M. Degtyarev and A. P. Favorskii, Zh. Vychislit. Matem. i Matem. Fiz., 8:679-684 (1968); 9:211-218 (1969).

31. V. F. D'yachenko and V. S. Imshennik, Reviews of Plasma Physics, Vol. 5 (ed. M. A. Leontovich), Consultants Bureau, New York (1970), p. 447.

32. V. F. D'yachenko and V. S. Imshennik, this volume, p. 199.

33. B. B. Kadomtsev, Reviews of Plasma Physics, Vol. 2 (ed. M. A. Leontovich), Consultants Bureau, New York (1966), p. 153.

34. L. E. Kalikhman, Fundamentals of Magnetohydrodynamics [in Russian], Atomizdat, Moscow (1964).

35. I. G. Kesaev, Cathode Processes in an Electric Arc [in Russian], Nauka, Moscow (1968).

36. A. Ya. Kislov, A. I. Morozov, and G. N. Tilinin, Zh. Tekh. Fiz., 38:975-978 (1968) [Sov. Phys. Tech. Phys. 13:736 (1968)].

37. A. Ya. Kislov et al., Report No. 3.1.11.9, in: Eighth International Conference on Phenomena in Ionized Gases, Vienna (1967).

38. P. E. Kovrov et al., Dokl. Akad. Nauk SSSR, 172:1305-1308 (1967) [Sov. Phys. Dokl. 12:155 (1967)].

39. A. G. Kulikovskii and G. A. Lyubimov, Magnetohydrodynamics [in Russian], Fizmatgiz, Moscow (1962).

40. A. G. Kulikovskii and G. A. Lyubimov, Dokl. Akad. Nauk SSSR, 129:52-55 (1959) [Sov. Phys. Dokl. 4:1185 (1960)].

41. L. D. Landau and E. M. Lifshits (Lifshitz), Fluid Mechanics, Addison-Wesley, Reading, Mass. (1959).

42. G. A. Lyubimov, Izv. Akad. Nauk SSSR. Mekhan. Zhidk. i Gaza, No. 3:3-11 (1966).

43. A. I. Morozov, Zh. Tekh. Fiz. 37:2147-2159 (1967) [Sov. Phys. Tech. Phys. 12:1580 (1968)].

44. A. I. Morozov, Nuclear Fusion. Special Application [in Russian], Vienna (1969), pp. 111-120.

45. A. I. Morozov, Vestn. Akad. Nauk SSSR, No. 6:28-36 (1969).
46. A. I. Morozov, K. V. Brushlinskii, N. I. Gerlakh, and A. P. Shubin, Report No. 3.1.11.7, in: Eighth International Conference on Phenomena in Ionized Gases [in Russian], Vienna (1967).
47. A. I. Morozov, P. E. Kovrov, and A. K. Vinogradova, Pis'ma Zh. Eksp. Teor. Fiz., 7:257-260 (1968) [JETP Lett. 7:199 (1968)].
48. A. I. Morozov et al., "Experimental study of flow in a magnetoplasma compressor," Preprint No. 2275, I. V. Kurchatov Institute of Atomic Energy, Moscow, 1973.
49. A. I. Morozov and L. S. Solov'ev, Zh. Tekh. Fiz., 34:429-443 (1964) [Sov. Phys. Tech. Phys. 9:337 (1964)].
50. A. I. Morozov and L. S. Solov'ev, Zh. Tekh. Fiz., 34:1141-1153 (1964) [Sov. Phys. Tech. Phys. 9:889 (1965)].
51. A. I. Morozov and L. S. Solov'ev, Dokl. Akad. Nauk SSSR, 164:80-83 (1965) [Sov. Phys. Dokl. 10:834 (1966)].
52. A. I. Morozov and L. S. Solov'ev, this volume, p. 1.
53. A. I. Morozov and A. P. Shubin, Teplofiz. Vys. Temp., 3:827-837 (1965), Zh. Prikl. Mekhan. i Tekh. Fiz., No. 5:14-20 (1967).
54. I. N. Ostretsov, V. A. Petrosov, A. A. Porotnikov, I. B. Safonov, and S. D. Tseitlin, in: Plasma Accelerators (ed. L. A. Artsimovich) [in Russian], (1973), pp. 254-257.
55. S. A. Regirer, Magnetohydrodynamic Flows in Channels and Tubes, Scientific Progress, Mechanics Series [in Russian], VINITI, Moscow (1966).
56. R. D. Richtmyer, Difference Methods for Initial-Value Problems, Interscience, New York (1958).
57. G. W. Sutton and A. Sherman, Engineering Magnetohydrodynamics, McGraw-Hill, New York (1965).
58. L. S. Solov'ev, Zh. Eksp. Teor. Fiz., 53:2063-2069 (1967) [Sov. Phys. JETP 26:1161 (1968)].
59. L. Spitzer, Physics of Fully Ionized Gases, Interscience, New York (1956).
60. V. M. Fedorov, Magnitn. Gidrodinam., No. 2:44-54 (1965).
61. V. I. Yudovich, Dokl. Akad. Nauk SSSR, 161:1037-1040 (1965) [Sov. Phys. Dokl.].
62. N. N. Yanenko, Fractional-Step Method for Solving Multidimensional Problems of Mathematical Physics [in Russian], Nauka, Novosibirsk (1967).
63. T. D. Butler, J. L. Cook, and R. L. Morse, Proceedings of the APS Topical Conference on Numerical Simulation of Plasma, Los Alamos, 1968, N C6.
64. D. Düchs, Phys. Fluids, 11:2010-2018 (1968).
65. J. R. Freeman and F. O. Lane, Proceedings of the APS Topical Conference on Numerical Simulation of Plasma, Los Alamos, 1968, N C7.
66. N. V. Filippov and T. J. Filippova, Plasma Physics and Controlled Nuclear Fusion Research (Proceedings of the Culham Conference, 1965), Vol. 2, IAEA, Vienna (1966), p. 405.
67. J. W. Mather, Plasma Physics and Controlled Nuclear Fusion Research (Proceedings of the Culham Conference, 1965), Vol. 2, IAEA, Vienna (1966), p. 389.
68. J. G. Linhart, Nucl. Fusion, 10:211-234 (1970).
69. B. Alder, S. Fernbach, and M. Rotenberg, (editors), Methods in Computational Physics, Vol. 9, Plasma Physics, Academic Press, New York (1970).

70. P. D. Morgan, N. J. Peacock, and D. E. Potter, Third European Conference on Plasma Physics and Controlled Fusion, Utrecht, June 1969.

71. R. L. Morse, in: Methods in Computational Physics (ed. B. Alder, S. Fernbach, and M. Rotenberg), Vol. 9. Plasma Physics, Academic Press, New York (1970), pp. 213-239.

72. D. E. Potter, Phys. Fluids, 14:1911-1924 (1971).

73. Proceedings of the APS Topical Conference on Numerical Simulation of Plasma, Los Alamos, September 18-20, 1968.

74. K. V. Roberts and D. E. Potter, in: Methods in Computational Physics, Vol. 9, Plasma Physics (ed. B. Alder, S. Fernbach, and M. Rotenberg), Academic Press, New York (1970), pp. 339-420.

TWO-DIMENSIONAL MAGNETOHYDRODYNAMIC MODEL FOR THE DENSE PLASMA FOCUS OF A Z PINCH

V. F. D'yachenko and V. S. Imshennik

INTRODUCTION

Earlier analysis of the transient plasma compression in a Z pinch makes use of the one-dimensional approximation [1]. All the MHD properties are assumed to be functions of the cylindrical radius r and the time t. The incoporation of dissipative processes on the basis of the classical transport theory yields a detailed description of the focusing effects of the collapse of the shock wave and the current sheet and also makes it possible to find the maximum plasma temperature and density near the pinch axis. A comparison of the quantiative results of the one-dimensional theory with the experimental results in a noncylindrical Z pinch leads to an important conclusion: The high velocity with which the plasma collapses to the axis, and the associated high temperature, are reached experimentally as a result of a mechanism which ejects plasma from the region in which the plasma parameters reach their maximum values. This comparison shows that the typical ejection involves at least 90% of the mass per unit length of the plasma. This conclusion implies substantial deviations from the cylindrical symmetry assumed in the formulation of the one-dimensional problem. These deviations were observed in the very earliest experiments and were in fact deliberately produced by Filippov et al. [2] in order to raise the parameters of the compressed plasma and

199

to increase the neutron yield. The term "noncylindrical Z pinch" was applied to these experiments even before theoretical work on the problem began. The conclusion noted above — that a substantial fraction of the plasma mass is ejected from the region in which the current sheet approaches the axis most closely — is interpreted more accurately as a reflection of the essential nature of the process rather than simply as evidence that the pinch is noncylindrical. Experimentally, it is difficult to measure the ejected mass, although indirect conclusions regarding this effect (the observation of an axial plasma jet) were reached a long time ago [3]. Accordingly, a better theoretical description of the actual processes that occur in a noncylindrical Z pinch means that it is necessary to go beyond the one-dimensional MHD theory.

The next important question is to determine whether it is more important to incorporate the dependence on z or on φ (the two remaining spatial coordinates in the cylindrical coordinate system) in the formulation of the problem. At one point it appears that a three-dimensional model would be needed if both these coordinates turned out to be important. Fortunately, the experimental data show that it is sufficient to restrict the analysis to the z dependence of all quantities, assuming that they are independent of φ; in other words, the axisymmetric approximation can be used. In this connection we note that any improvement in the azimuthal symmetry of the pinch increases the neutron yield [4]. Much care has been taken to ensure an axisymmetric pinch in the initial stage of the discharge [5], but experimentally there is a certain "optimum" deviation of the initial conditions from cylindrical symmetry in the z direction. There are also independent theoretical considerations regarding the various roles of the coordinates z and φ in the pinch. A question which might be asked is which perturbation grow in a cylindrically symmetric Z pinch. In the linear approximation we are concerned with the stability of the plasma interface with a magnetic field when this interface is accelerated toward the plasma. By making use of the old results of Kruskal and Schwarzschild [6], we conclude that in practice the perturbations of the boundary in the φ direction, which occur along the magnetic field, are always suppressed [7]. The perturbations in the z direction, across the magnetic field, grow freely as long as their wavelength is large enough so that dissipative processes — primarily the plasma viscosity — do not have a damping effect [8].

On the basis of these considerations we develop a formulation of the two-dimensional MHD problem in which all quantities are functions of the coordinates r and z, and the time t. This formulation of the problem and certain results found in a numerical solution of this problem are reported in [4] and, in greater detail, in [9]. More recently, Roberts and Potter [11] have solved a similar formulation of the problem for the coaxial version of a noncylindrical Z pinch developed by Mather [10]. This problem has also been formulated and solved by Watteau et al. [12] and, later, by Robuoch and DiCola [13], in the simplified snowplow model.†

In the idealized theoretical model for the noncylindrical Z pinch examined below we assume at the outset that 1) the MHD description of the plasma is applicable; 2) it is permissible to ignore ionization, recombination, radiation, and the deviation from the ideal equation of state (all of which are characteristic of a low-temperature plasma); and 3) the pinch effect is axisymmetric. These three basic assumptions are supplemented with several other, secondary, simplifications related to the particular initial and boundary conditions and the dissipative processes in the system of MHD equations. In contrast with the one-dimensional theory [1], we use the Kirchhoff equation, with certain approximations, in order to determine the time dependence of the total current through the plasma. All of these additional assumptions are specifically stated at the appropriate points.

In Chap. 1 we formulate the two-dimensional MHD problem for the case of an ideal, fully ionized deuterium plasma. In Chap. 2 we describe the numerical method used to solve this problem. Finally, in Chap. 3 we report the most important calculated results, draw certain physical conclusions, and compare the results with the results of related calculations by other investigators.

---

†It can be shown that the snowplow model follows from the MHD equations when the flow has the appropriate profile in the direction perpendicular to the plasma boundary.

Chapter 1

MHD ANALYSIS OF A NONCYLINDRICAL Z PINCH

§1. Description of Dissipative Processes

a. Structure of the Current Sheet; Plasma
Conductivity. The classical conductivity of a plasma across
a strong magnetic field is lower than that along the field, but the
quantiative difference is insignificant because of compensating effects [14, 15] ($\sigma_\perp$ is smaller than the isotropic plasma conductivity
by only a factor of two). However, there is serious reason not to
use classical conductivity theory for the particular conditions under consideration here. An analysis of the experimental results
for collisionless Z pinches carried out by Dubovoi et al. [16, 17]
has shown that the directed electron velocity in the plasma does
not exceed the ion-acoustic velocity $c_S = (kT_e/m_i)^{1/2}$ multiplied
by a coefficient k_T, which is a function of the temperature ratio
T_e/T_i. In good agreement with the theory for the ion-acoustic instability in a collisionless plasma, the coefficient k_T increases
with decreasing ratio T_e/T_i. Dubovoi et al. [17] use the approximation $k_T \approx 75(T_e/T_i)^{-3/2}$ for this function. Thus, when the temperatures are equal the directed electron velocity can reach the electron thermal velocity.

No experimental work has been reported on the directed electron velocities or the current distributions for hot, collision-dom-
inated plasmas, but it has been shown theoretically [18] (see also
[19]) that the limitations on the directed velocity in this case may
be even more stringent than in the absence of electron collisions.
It is first necessary to show that the collisionless approximation
is not applicable under these conditions. The corresponding necessary conditions are given in [19]. For the ion-acoustic instability
the necessary condition is

$$d_e^3\, n_e\, m_e/m_i \ll 1,$$

where d_e and n_e are the electron Debye length and density. Noting
that the number of electrons within a Debye sphere is typically
$n_e d_e^3 \approx 10^3$ (e.g., with $T_e = 10^{6\circ}$K and $n_e = 6 \cdot 10^{16}$ cm^{-3}, we have
$d_e = 2 \cdot 10^{-5}$ cm; with $T_e = 10^{7\circ}$K and $n_e = 6 \cdot 10^{18}$ cm^{-3}, we have

$d_e = 6 \cdot 10^{-6}$ cm), we see that this condition is satisfied. Furthermore, when the electron and ion temperatures are equal, ion-acoustic oscillations grow even if the directed electron velocity is only a few times higher than the ion-acoustic velocity ($k_T \sim 1$) [18]. Even if the electron and ion temperatures are different (and this is a better description of the situation; see the discussion below), it can be assumed that the directed electron velocity $v_e = j/n_e e$ nowhere exceeds $k_T c_S = k_T (kT/m_i)^{1/2}$, where $k_T \gtrsim 1$. The necessary inequality $v_e \lesssim k_T c_S$ can be satisfied by introducing an effective plasma conductivity $\sigma_e = k_e \sigma_\perp$ with a coefficient k_e determined by experiment.

Let us estimate the value of k_e for a deuterium plasma with $m_i = m_D = 2m_p$. The directed velocity is $v_e = \dfrac{j}{n_e e} = \dfrac{Jm_D}{2\pi R \rho e d}$, where the sheet thickness is $d = \left(\dfrac{c^2}{4\pi k_e \sigma_\perp} \cdot \dfrac{R}{v} \right)^{1/2}$ The characteristic time for the growth of the magnetic field used to estimate d is assumed to be R/v; this assumption is strictly correct if the velocity of the plasma boundary, v, is constant. For the conditions in which we are interested, the boundary velocity changes relatively little [1]. Equating v_e to $k_T c_S$, and substituting in the equation for $\sigma_\perp$ from [14], we find†

$$k_e = 14 k_T^2 \cdot \frac{c^2 e^4 m_e^{1/2}}{m_p^3} \cdot \frac{R^3 \rho^2}{J^2 (kT)^{1/2} v} . \tag{1.1}$$

We substitute into (1.1) the values of the constants and the following parameters, which are characteristic of the plasma focus [7]: J = 0.7 mA, T = 1.2 keV, R = 0.65 mm, ρ = 2.4 $\cdot 10^{-4}$ g/cm³, and v = 3.5 $\cdot 10^7$ cm/sec. Then

$$k_e = 10^{-2} k_T^2. \tag{1.2}$$

On the right side of (1.1) there is a quantity proportional to $(\rho R^2)^3 \cdot R^{-1} J^{-4}$, since the order-of-magnitude conditions $T \sim J^2 (\rho R^2)^{-1}$ and $v \sim (T)^{1/2}$ are satisfied in the pinch. Here ρR^2 is the plasma density per unit length, which we can find if we determine the mass ejected from the focus. Furthermore, the equation for k_e contains a relatively insensitive function of the plasma radius, $J^4 R$, since

---

†The Coulomb logarithm is assumed to be constant: $\lambda_c = 10$.

the current J at maximum compression decreases with increasing inductive reactance when the radius R is reduced. From (1.2) we conclude that for classical plasma conductivity ($k_e = 1$) the directed electron velocity at the focus should be an order of magnitude higher than the ion-acoustic velocity ($k_e \approx 10$); this can hardly be the actual case, in view of the instabilities mentioned above [18, 19].[†]

As long as the current sheet is far from the axis and the shock wave has not reached the axis, it is not appropriate to introduce the coefficient k_e. The directed velocity is well below the critical value. Substituting typical plasma parameters into Eq. (1.2) ($J = 1$ mA, $T = 10$ eV, $R = 10$ cm, $\rho = 1.2 \cdot 10^{-6}$ g/cm^3, and $v = 3.5 \cdot 10^6$ cm/sec [7]), and using $k_e = 1$, we find $k_T \approx 0.10$. Nevertheless, we are primarily interested in the stage of maximum plasma compression, in which the plasma focus is formed. It is in this stage that the sheet thickness becomes (more precisely, we should say "may become" at this point — before we have examined the calculated results) comparable to the plasma radius. As the sheet collapses to the axis this thickness definitely becomes small in comparison with the distance from the outer boundary to the shock front, even if the conductivity is reduced by two orders of magnitude, as follows from (1.2). We also note that the energy associated with Joule heating is insignificant in comparison with the work performed by the ponderomotive forces (Joule heating in the sheet depends on the coefficient k_e according to $j^2 d/k_e \sigma_\perp \sim k_e^{-1/2}$). It can therefore be argued that the introduction of an effective plasma conductivity σ_e, with the coefficient k_e taken from (1.2), has little effect on the collapse of the current sheet to the axis of the system when $k_T \gtrsim 1$. On the other hand, in the most important stage — that in which the plasma focus is formed — the use of an effective plasma conductivity σ_e limits the value of the directed electron velocity to the ion-acoustic velocity, multiplied by k_T. In this way the role played by possible plasma instabilities is taken into account. The calculations will show how well the condition involving a limitation on the directed electron velocity is satisfied.

[†]We might also add that there is the possibility of a gradient instability, by analogy with the collisionless situation examined in [20].

b. Influence of a Magnetic Field on the Viscosity and Thermal Conductivity of a Plasma. Difference between Electron and Ion Temperatures.

We begin with the equation for the effective plasma conductivity σ_e:

$$\sigma_e = k_e \sigma_\perp. \tag{1.3}$$

Substituting the constant factor k_e from (1.2), and assuming $k_T = 1$, we find

$$\sigma_e = 10^{-2}\sigma_\perp. \tag{1.3'}$$

It should not be concluded that the conductivity reduction in (1.3), which is justified on the basis of physical considerations, has a strong effect on the basic calculated results, in particular, the maximum values of the parameters at the plasma focus. The range of physical conditions under consideration here is such that even the conductivity σ_e given in (1.3') is extremely high. This statement can be verified by direct comparison of the results of the calculations below with earlier data [9] obtained under the assumption of an infinite plasma conductivity. Until the second compression begins the basic parameters obtained by the two calculations are the same. The finite conductivity plays a most important role in the second (and subsequent) plasma compressions, in which the dissipationless model is meaningless [9]. We must not forget, however, that somewhere in this process the analysis goes beyond the range of applicability of the MHD description.

How does the sheet thickness compare with the ion gyroradius $r_{\Lambda i} = v_{Ti}/\omega_{\Lambda i}$? By using the expression for $\sigma_\perp$ used in deriving (1.1) in (1.3') we can write the ratio of these distances for deuterons:

$$\frac{r_{\Lambda i}^2}{d^2} = \frac{v_{Ti}^2}{\omega_{\Lambda i}^2} \cdot \frac{4\pi\sigma_e}{c^2} \cdot \frac{v}{R} = 5.4 \cdot 10^{-3} \frac{c^2 m_p}{m_e^{1/2} e^4} \cdot \frac{(kT)^{5/2} vR}{J^2} \tag{1.4}$$

where $v_{Ti} = (3kT/2m_D)^{1/2}$ is the deuteron thermal velocity, $H_\varphi = 2J/cR$ is the magnetic field at the boundary, and $\omega_{\Lambda i} = eH_\varphi/m_D c$ is the deuteron gyrofrequency. Substituting the parameters given above for the plasma focus into Eq. (1.4), we find $r_{\Lambda i}^2/d^2 = 0.4$. For the stage in which the current sheet collapses, again using these parameters we find $r_{\Lambda i}^2/d^2 = 3 \cdot 10^{-5}$. Initially the sheet thickness is much larger than the ion gyroradius, but the two quantities eventually become comparable in magnitude. This is a reasonable result since the magnetic pressure and gas pressure are equal: $(H_\varphi^2/8\pi) = (2\rho kT/m_i)$. This equality holds approximately throughout all stages of the process. Evidently the equality $d = Jm_i/2\pi\mathrm{Re}\rho v_{Ti}$, where $v_e = v_{Ti}$ and $r_{\Lambda i} = v_{Ti}/\omega_{\Lambda i}$, is essentially equivalent to a pressure balance. With $v \sim (T)^{1/2}$ and $T \sim J^2(\rho R^2)^{-1}$ the right side of (1.4) is proportional to $RJ^4(\rho R^2)^{-3}$ and thus varies in inverse proportion to the right side of (1.1).

The basic measure of the effect of the magnetic field on the transport process is the product $\omega_{\Lambda i}\tau_i$, where $\tau_i = l_i/v_{Ti}$ is the mean free time, and $l_i = (kT)^2/2n_e e^4 \lambda_C$ is the mean free path. If a deuterium plasma in which the magnetic and gas pressures are equal exhibits a temperature that satisfies the inequality $T > T_{cr}$, where

$$kT_{cr} = 2.2\, c^{1/2}e^{3/2}\rho^{1/4}. \tag{1.5}$$

The effect of the magnetic field on ion transport processes is important since $\omega_{\Lambda i}\tau_i > 1$. With a density $\rho = 2.4 \cdot 10^{-4}$ g/cm^3 at the plasma focus, (1.5) shows the value $T_{cr} = 0.31$ keV. In the stage in which the sheet collapses with $\rho = 1.2 \cdot 10^{-6}$ g/cm^3, we have $T_{cr} = 80$ eV.

It follows that the magnetic field can affect the ion transport processes in the later stage of the process, in which $\omega_{\Lambda i}\tau_i \gg 1$. It has been shown, however [in the discussion of Eq. (1.4)], that in this case $r_{\Lambda i} \sim d$. Strictly speaking, a kinetic description is needed for the ions in the sheet: The plasma pressure and thus the density and/or temperature change substantially in a distance d, which satisfies the condition

$$d \sim r_{\Lambda i} \ll l_i. \tag{1.6}$$

Within the framework of the hydrodynamic approximation it is then consistent to ignore the ion magnetization in the transport coefficients. Our only remaining task is to discuss a possible intermediate stage of the process in which the conditions $d \gg r_{\Lambda i}$ and $r_{\Lambda i} \ll l_i$ hold simultaneously. Formally, this case occurs for the conductivity from (1.3'), but it is sufficient to recall that in the early stages, before the focus forms, the introduction of σ_e results in an artificial increase in the sheet thickness. We can make the following estimate for a deuterium plasma with a classical conductivity: $d > r_{\Lambda i}$ if $T < T'_{cr}$, where [by analogy with (1.5), we assume $p_{gas} = p_{mag}$ and $v = v_{Ti}$]

$$kT'_{cr} = 3.1 m_e^{1/4} m_p^{-3/4} e^2 \rho^{1/2} R^{1/2}. \tag{1.7}$$

If T'_{cr} is less than the value of T_{cr} given in (1.5), this intermediate stage does not occur. Comparing (1.5) and (1.7), we find the condition under which $T'_{cr} < T_{cr}$:

$$(\rho R^2)^{1/4} < 0.7 c^{1/2} m_e^{-1/4} m_p^{3/4} e^{-1/2}. \tag{1.8}$$

The left side of inequality (1.8) tends to decrease as the sheet collapses to the axis as a result of the ejection of mass. The product ρR^2 is thus largest early in the process. Substitution of the corresponding parameter values $\rho = 1.2 \cdot 10^{-6}$ g/cm³ and $R = 10$ cm into equality (1.8) shows that the two sides are roughly equal. We conclude that there is no point in taking account of the dependence of the ion transport coefficients on the magnetic field or in taking account of any possible intermediate stage. This stage does not occur in a plasma with a classical conductivity.

Pursuing analogous arguments for the electron component, we see that the entire range in which we are interested falls in the interval defined by the inequalities

$$r_{\Lambda e} \ll l_e; \quad r_{\Lambda e} \ll d. \tag{1.9}$$

These inequalities ensure the applicability of the classical theory of dissipative processes in a strong magnetic field. Accordingly, the electronic thermal conductivity (the contribution of electrons to the ion viscosity is negligible) is given by the expressions in [21]. However, the electronic thermal conductivity is also neglected in this problem. This neglect can be justified to a certain extent. In the current sheet the product $\omega_{\Lambda e} \tau_e$ is so large that the electronic thermal conductivity essentially vanishes. The length δ_e, defined as the distance from the outer boundary of the plasma to the point at which $\omega_{\Lambda e} \tau_e \sim 1$, is essentially equal to the plasma radius $(R - \delta_e \ll R)$, at least in the later stage of the process.

The case with different electron and ion temperatures has been studied on the one-dimensional model [1]. For the axial region, of primary interest here, this difference is not of fundamental importance. The difference decreases as the conditions for the ap-

plicability of the MHD description of the plasma improve. There
is no point in taking account of this difference at extremely small
values of the parameter $\alpha = p_0^2/\rho_0^3 R_0$, which is a measure of the
"hydrodynamic" nature of the plasma [1] in the problem studied
below. These is also no point in including this temperature difference in a problem with zero electronic thermal conductivity.
In general, it appears that the use of the single-temperature approximation does not lead to any significant errors in the determination of the plasma temperature outside the sheet. Within this sheet
the single-temperature approximation is a rough approximation.
Nevertheless, the one-dimensional calculations [1] with Joule heating show no substantial difference between the electron and ion temperatures in the current sheet. Accordingly, in the problem below
we restrict the discussion to a model in which the two temperatures
are equal:

$$T_i = T_e = T. \tag{1.10}$$

It is possible to formulate the problem more accurately by taking
account of dissipative processes in the electron component. In a
more accurate model it would be necessary to take account of the
electronic thermal conductivity and the difference between the
electron and ion temperatures.

Let us summarize the results of the discussion to this point.
First a comparison with experiment leads to an effective plasma
conductivity which is two orders of magnitude lower than the classical conductivity. This choice leads to the restriction that the electron directed velocity cannot exceed the ion-acoustic velocity in
the most important stage — that in which the plasma focus is
formed. Second, it has been shown that there is no point in incorporating the effect of the magnetic field in the classical ion
viscosity and thermal conductivity. Third, we have presented arguments for neglecting the electronic thermal conductivity and
the difference between the ion and electron temperatures in the
first step of the solution. Finally, to avoid any confusion, we emphasize that there is no rigid constraint on the conclusions reached
here in terms of the particular set of parameters used for the
plasma in the pinch; hence, these conclusions hold over the entire
range of typical experimental conditions.

§ 2.   Two-Dimensional Magnetohydrodynamic Equations (MHD Model)

a.   Equation for the Magnetic Field in the Plasma.

We begin with Maxwell's equations in the quasi-stationary approximation,

$$\text{curl }\mathbf{E} = -\frac{1}{c} \cdot \frac{\partial \mathbf{H}}{\partial t} \; ; \tag{2.1}$$

$$\mathbf{j} = \frac{c}{4\pi} \text{curl }\mathbf{H} \, , \tag{2.2}$$

and the generalized Ohm's law in the form [14, 15]

$$\mathbf{j} = \sigma_e \left(\mathbf{E} + \frac{1}{c} [\mathbf{vH}] - \chi [\mathbf{jH}] + \chi c \, \text{grad } p_e \right), \tag{2.3}$$

where σ_e is the effective plasma conductivity in the strong magnetic field [see Eqs. (1.3) and (1.3')], $\chi = 1/n_e ec = 2m_p/\rho ec$ is the Hall coefficient, and $p_e = n_e kT = \rho kT/2m_p$ is the electron pressure.[†]

We take account of the coordinate and the time dependences of σ_e and χ in the subsequent transformations, and we adopt the notation

$$\lambda = c^2/4\pi\sigma_e; \; \varkappa_c = c^2\chi/4\pi. \tag{2.4}$$

Substituting $\mathbf{E}$ from (2.3) into Eq. (2.1), and using (2.2) to eliminate the current density $\mathbf{j}$, we find a vector equation for the magnetic field:

$$\frac{\partial \mathbf{H}}{\partial t} = - \text{curl } (\lambda \text{curl} \mathbf{H}) + \text{curl}[\mathbf{vH}] - \text{curl}\{\varkappa_c \, [\text{curl} \mathbf{H} \cdot \mathbf{H}]\} + \tag{2.5}$$
$$+ 4\pi \text{curl}(\varkappa_c \, \text{grad } p_e).$$

[†]The crude approximation to the dissipative processes in the electron component used in §1 is continued when also used in the generalized Ohm's law (2.3). Comparison of (2.3) with the corresponding expression derived in [21] shows that some of the numerial coefficients in (2.3) are rounded off as compared with the rigorous expressions.

Using the following identities from vector analysis,

$$\mathrm{curl}\,(\lambda\,\mathrm{curl}\,\mathbf{H}) = \lambda\,\mathrm{curl}\,\mathrm{curl}\,\mathbf{H} + [\mathrm{grad}\,\lambda \cdot \mathrm{curl}\,\mathbf{H}];$$

$$\mathrm{curl}\,(\varkappa_c\,\mathrm{grad}\,p_e) = [\mathrm{grad}\,\varkappa_c \cdot \mathrm{grad}\,p_e];$$

$$\mathrm{curl}\,\{\varkappa_c\,[\mathrm{curl}\,\mathbf{H}\cdot\mathbf{H}]\} = \varkappa_c\mathrm{curl}[\mathrm{curl}\,\mathbf{H}\cdot\mathbf{H}] + [\mathrm{grad}\,\varkappa_c\,[\mathrm{curl}\,\mathbf{H}\cdot\mathbf{H}]] =$$

$$= \varkappa_c\mathrm{curl}[\mathrm{curl}\,\mathbf{H}\cdot\mathbf{H}] + \mathrm{curl}\,\mathbf{H}\,(\mathrm{grad}\,\varkappa_c\cdot\mathbf{H}) - \mathbf{H}(\mathrm{curl}\,\mathbf{H}\cdot\mathrm{grad}\,\varkappa_c),$$

we can write (2.5) in the more convenient form

$$\frac{\partial \mathbf{H}}{\partial t} = -\lambda\,\mathrm{curl}\,\mathrm{curl}\,\mathbf{H} - [\mathrm{grad}\,\lambda \cdot \mathrm{curl}\,\mathbf{H}] + \mathrm{curl}\,[\mathbf{v}\mathbf{H}] - \varkappa_e\mathrm{curl}[\mathrm{curl}\,\mathbf{H}\cdot\mathbf{H}] -$$

$$-\mathrm{curl}\,\mathbf{H}\,(\mathrm{grad}\,\varkappa_c\cdot\mathbf{H}) + \mathbf{H}\,(\mathrm{curl}\,\mathbf{H}\cdot\mathrm{grad}\,\varkappa_c) + 4\pi\,[\mathrm{grad}\,\varkappa_c\cdot\mathrm{grad}\,p_e]. \qquad (2.6)$$

The last four terms in (2.6) are called "Hall" terms, since they vanish when $\varkappa_e = 0$.

The field equations written in this cylindrical coordinate system have an interesting property. Let us assume that at some time t the field components H_r and H_z vanish everywhere. Then it follows immediately from (2.2) that $j_\varphi = 0$. According to (2.3), the azimuthal component of the electric field, E_φ, also vanishes, provided that we make the further assumption that the problem is axisymmetric, $\partial/\partial\varphi = 0$ (so that the pressure gradient vanishes). Finally, examining Eq. (2.1), we conclude that $(\partial H_r/\partial t) = (\partial H_z/\partial t) = 0$. Accordingly, if $H_r = H_z = 0$ in the specification of the initial conditions at t = 0, in the axisymmetric problem these field components (H_r, H_z) vanish everywhere at any later time t > 0 (also $j_\varphi \equiv E_\varphi \equiv 0$).

Restricting the analysis to those solutions of the axisymmetric problem for which H_r and H_z vanish, we write the azimuthal component of Eq. (2.6) as

$$\frac{\partial H_\varphi}{\partial t} = \frac{\partial}{\partial r}\left[\frac{\lambda}{r}\cdot\frac{\partial}{\partial r}(rH_\varphi)\right] + \frac{\partial}{\partial z}\left(\lambda\frac{\partial H_\varphi}{\partial z}\right) - \frac{\partial}{\partial r}(v_r H_\varphi) - \frac{\partial}{\partial z}(v_z H_\varphi) +$$

$$+ H_\varphi\left(\frac{2\varkappa_c}{r} - \frac{\partial\varkappa_c}{\partial r}\right)\frac{\partial H_\varphi}{\partial z} + H_\varphi\frac{\partial\varkappa_c}{\partial z}\left(\frac{H_\varphi}{r} + \frac{\partial H_\varphi}{\partial r}\right) +$$

$$+ 4\pi\left(\frac{\partial\varkappa_c}{\partial z}\cdot\frac{\partial p_e}{\partial r} - \frac{\partial\varkappa_c}{\partial r}\cdot\frac{\partial p_e}{\partial z}\right). \qquad (2.7)$$

We note that the Hall terms remain in (2.7) solely as the result of incorporating the variation in the coefficient $\varkappa_c$ and the cylindrical correction; the other terms, including the leading second-order field derivatives [see Eq. (2.6)], drop out of the final equation. This case can be treated as degenerate with respect to the Hall effect, while the secondary terms remaining in Eq. (2.7) can be completely ignored without introducing any error of physical importance. Accordingly ($\varkappa_c \equiv 0$), we can go from Eq. (2.7) to the final equation for the single nonvanishing component of the magnetic field, H_φ:

$$\frac{\partial H_\varphi}{\partial t} = \frac{\partial}{\partial r}\left[\frac{\lambda}{r}\cdot\frac{\partial}{\partial r}(rH_\varphi)\right] + \frac{\partial}{\partial z}\left(\lambda\frac{\partial H_\varphi}{\partial z}\right) - \frac{\partial}{\partial r}(v_r H_\varphi) - \frac{\partial}{\partial z}(v_z H_\varphi). \qquad (2.8)$$

Finally, we write the other nonvanishing properties of the electromagnetic field:

$$\left.\begin{array}{l} E_r = \dfrac{1}{c}\left(v_z H_\varphi - \lambda\dfrac{\partial H_\varphi}{\partial z}\right); \quad E_z = \dfrac{1}{c}\left[\dfrac{\lambda}{r}\cdot\dfrac{\partial}{\partial r}(rH_\varphi) - v_r H_\varphi\right]; \\[2mm] j_r = -\dfrac{c}{4\pi}\cdot\dfrac{\partial H_\varphi}{\partial z}; \quad j_z = \dfrac{c}{4\pi}\cdot\dfrac{1}{r}\cdot\dfrac{\partial}{\partial r}(rH_\varphi). \end{array}\right\} \qquad (2.9)$$

We now consider the equations of motion.

b. Equations of Motion of the Plasma. When the ponderomotive and viscous forces are taken into account, the equation of motion of the plasma becomes

$$\rho\frac{dv}{dt} = -\operatorname{grad} p + \frac{1}{4\pi}[\operatorname{curl}\mathbf{H}\cdot\mathbf{H}] + \operatorname{Div}\sigma. \qquad (2.10)$$

The divergence of the viscous tensor σ is given in terms of the orthogonal curvilinear coordinates ξ^i by [22]

$$(\operatorname{Div}\sigma)_k = \frac{1}{h_k}\left[\frac{1}{h_1 h_2 h_3}\cdot\frac{\partial}{\partial\xi^i}\left(\frac{h_1 h_2 h_3 h_k}{h_i}\sigma_{ik}\right) - \sigma_{ii}\frac{\partial\ln h_i}{\partial\xi^k}\right], \qquad (2.11)$$

that is, it is written in terms of the Lamé coefficients, whose squares are

$$h_i^2 = (\partial x/\partial\xi^i)^2 + (\partial y/\partial\xi^i)^2 + (\partial z/\partial\xi^i)^2. \qquad (2.12)$$

Here, x, y, z are the Cartesian coordinates. The right side of (2.11) is summed over the index i. In turn, the components of the symmetric viscous tensor σ with the isotropic ion-viscosity coefficient η are expressed in terms of the velocity components $v_{\xi i}$ and their derivatives [22, 23]:

$$\sigma_{ik} = \eta \left[\frac{1}{h_k} \cdot \frac{\partial v_\xi^i}{\partial \xi^k} + \frac{1}{h_i} \cdot \frac{\partial v_\xi^k}{\partial \xi^i} - \frac{1}{h_i h_k} \left(v_\xi^i \frac{\partial h_i}{\partial \xi^k} + v_\xi^k \frac{\partial h_k}{\partial \xi^i} \right) + \right.$$

$$\left. + 2\delta_{ik} \frac{v_\xi^\lambda}{h_\lambda} \cdot \frac{\partial \ln h_\lambda}{\partial \xi^\lambda} \right] - \frac{2}{3} \eta \delta_{ik} \left[\frac{v_\xi^\lambda}{h_\lambda} \cdot \frac{\partial}{\partial \xi^\lambda} \ln (h_1 h_2 h_3) + \frac{\partial}{\partial \xi^\lambda} \left(\frac{v_\xi^\lambda}{h_\lambda} \right) \right], \quad (2.13)$$

where a summation must be carried out over λ. The quantity in the second pair of brackets in (2.13) appears when the plasma compressibility is taken into account, while the expression in the first pair of brackets corresponds to the incompressible case. In terms of the Cartesian coordinates, from (2.13) we find the familiar expression for the viscosity tensor of an ideal gas with a vanishing second viscosity coefficient [24].

We now apply general equations (2.11)-(2.13) to the cylindrical coordinate system, $\xi^1 = r$, $\xi^2 = \varphi$, $\xi^3 = z$. From (2.12), $h_1 = 1$, $h_2 = r$, and $h_3 = 1$. Using (2.11), we can write the components of the viscous force for the axisymmetric case:

$$(\mathrm{Div}\ \sigma)_r = \frac{\partial \sigma_{rr}}{\partial r} + \frac{\partial \sigma_{rz}}{\partial z} + \frac{\sigma_{rr} - \sigma_{\varphi\varphi}}{r}; \quad (2.14)$$

$$(\mathrm{Div}\ \sigma)_\varphi = \frac{\partial \sigma_{r\varphi}}{\partial r} + \frac{\partial \sigma_{z\varphi}}{\partial z} + \frac{2\sigma_{r\varphi}}{r}; \quad (2.15)$$

$$(\mathrm{Div}\ \sigma)_z = \frac{\partial \sigma_{rz}}{\partial r} + \frac{\partial \sigma_{zz}}{\partial z} + \frac{\sigma_{rz}}{r}. \quad (2.16)$$

From (2.13) we find the components of the viscous tensor to be

$$\sigma_{rr} = \eta \left[2 \frac{\partial v_r}{\partial r} - \frac{2}{3} \left(\frac{\partial v_r}{\partial r} + \frac{\partial v_z}{\partial z} + \frac{v_r}{r} \right) \right];$$

$$\sigma_{zz} = \eta \left[2 \frac{\partial v_z}{\partial z} - \frac{2}{3} \left(\frac{\partial v_r}{\partial r} + \frac{\partial v_z}{\partial z} + \frac{v_r}{r} \right) \right];$$

$$\sigma_{rz} = \eta \left(\frac{\partial v_r}{\partial z} + \frac{\partial v_z}{\partial r} \right); \quad (2.17)$$

$$\sigma_{\varphi\varphi} = \eta \left[2 \frac{v_r}{r} - \frac{2}{3} \left(\frac{\partial v_r}{\partial r} + \frac{\partial v_z}{\partial z} + \frac{v_r}{r} \right) \right];$$

$$\sigma_{r\varphi} = \eta \left(\frac{\partial v_\varphi}{\partial r} - \frac{v_\varphi}{r} \right); \quad \sigma_{z\varphi} = \eta \frac{\partial v_\varphi}{\partial z}. \quad (2.18)$$

The azimuthal component of the ponderomotive force vanishes, so that the field components H_r and H_z vanish everywhere. Accordingly, the plasma rotation velocity v_φ satisfies an equation obtained by expanding the total time derivative on the left side of (2.10) and substituting the viscous force from (2.15) and (2.18) into the right side:

$$\frac{\partial v_\varphi}{\partial t} + v_r \frac{\partial v_\varphi}{\partial r} + v_z \frac{\partial v_\varphi}{\partial z} + \frac{v_r v_\varphi}{r} = \frac{1}{r} \left(\frac{\partial}{\partial t} + v_r \frac{\partial}{\partial r} + v_z \frac{\partial}{\partial z} \right) (r v_\varphi) =$$

$$= \left(\frac{\partial}{\partial r} + \frac{2}{r} \right) \left[\eta \left(\frac{\partial v_\varphi}{\partial r} - \frac{v_\varphi}{r} \right) \right] + \frac{\partial}{\partial z} \left(\eta \frac{\partial v_\varphi}{\partial z} \right). \qquad (2.19)$$

The viscous forces vanish in rigid rotation, $v_\varphi = \omega_0 r$; in all other cases they redistribute the initial angular momentum $r v_\varphi$. However, if we assume that the plasma is not rotating at $t = 0$ ($v_\varphi = 0$), then (2.19) indicates that the velocity component v_φ also vanishes everywhere at any later time $t > 0$. The situation is similar to the situation for the field components H_r and H_z. When $v_\varphi \equiv 0$ the centrifugal force vanishes in the radial component of the equation of motion. We now write the remaining equations of motion in (2.10), expanding the total time derivatives and the ponderomotive force, using (2.9), and substituting the viscous force from (2.14) and (2.16):

$$\rho \left(\frac{\partial v_r}{\partial t} + v_r \frac{\partial v_r}{\partial r} + v_z \frac{\partial v_r}{\partial z} \right) = - \frac{\partial p}{\partial r} - \frac{H_\varphi}{4\pi r} \cdot \frac{\partial}{\partial r} (r H_\varphi) + \frac{\partial \sigma_{rr}}{\partial r} +$$

$$+ \frac{\partial \sigma_{rz}}{\partial z} + \frac{\sigma_{rr} - \sigma_{\varphi\varphi}}{r} ; \qquad (2.20)$$

$$\rho \left(\frac{\partial v_z}{\partial t} + v_r \frac{\partial v_z}{\partial r} + v_z \frac{\partial v_z}{\partial z} \right) =$$

$$= - \frac{\partial p}{\partial z} - \frac{H_\varphi}{4\pi} \cdot \frac{\partial H_\varphi}{\partial z} + \frac{\partial \sigma_{rz}}{\partial r} + \frac{\partial \sigma_{zz}}{\partial z} + \frac{\sigma_{rz}}{r} . \qquad (2.21)$$

In substituting the four components of the viscous tensor σ_{rr}, $\sigma_{\varphi\varphi}$, σ_{zz}, and σ_{rz} from (2.17) into (2.20) and (2.21), we must take account of the dependence of the ion-viscosity coefficient on the coordinates r and z, which is extremely important for a fully ionized plasma. To avoid the cumbersome equations we do not make this substitution; we treat this latter equation as the entropy equation for the plasma (the usual continuity equations and the equations of state of the plasma also appear at this step).

   c.   Entropy Equation for the Plasma.   To determine the common ion and electron temperature T (taking account
of the ion viscosity and thermal conductivity and the Joule heating
due to the effective conductivity), we write the entropy equation

$$\rho T \frac{dS}{dt} = \text{div} (\sigma v) - v \, \text{Div} \, \sigma - \text{div} \, F + \frac{c^2}{16\pi^2 \sigma_e} (\text{curl} H)^2. \qquad (2.22)$$

The energy flux due to the ion thermal conductivity, **F**, figures in
this equation; with an isotropic thermal conductivity v, this energy flux is

$$\mathbf{F} = -v \, \text{grad} \, T. \qquad (2.23)$$

The right side of (2.22) must be written in the cylindrical coordinate system with the axial symmetry taken into account. Following the rules for tensor differentiation in orthogonal curvilinear
coordinates, we write the energy dissipation due to the viscosity
as [22, 23]

$$\text{div} (\sigma v) - v \, \text{Div} \, \sigma = \sigma_{ik} \frac{h_k}{h_i} \cdot \frac{\partial}{\partial \xi^i} \left(\frac{v_\xi^k}{h_k} \right) + \frac{v_\xi^k}{h_k} \sigma_{ii} \frac{\partial \ln h_i}{\partial \xi^k}, \qquad (2.24)$$

with summations over both i and k. In the cylindrical coordinate
system, from (2.24) we find

$$\text{div} (\sigma v) - v \, \text{Div} \, \sigma = \sigma_{rr} \frac{\partial v_r}{\partial r} + \sigma_{rz} \left(\frac{\partial v_r}{\partial z} + \frac{\partial v_z}{\partial r} \right) + \sigma_{zz} \frac{\partial v_z}{\partial z} + \sigma_{\varphi\varphi} \frac{v_r}{r},$$

$$(2.25)$$

where the tensor components σ_{ik} are given by (2.17).

   We use the thermodynamic identity

$$T dS = d\mathscr{E} + p d (1/\rho), \qquad (2.26)$$

where $\mathscr{E}$ is the internal energy per unit mass of the plasma and we
use the continuity equation

$$\frac{d\rho}{dt} + \rho \, \text{div} \, v = \frac{d\rho}{dt} + \rho \left(\frac{\partial v_r}{\partial r} + \frac{\partial v_z}{\partial z} + \frac{v_r}{r} \right) = 0. \qquad (2.27)$$

Then the left side of (2.22) can be rewritten

$$\rho T \frac{dS}{dt} = \rho \frac{d\mathscr{E}}{dt} - \frac{p}{\rho} \cdot \frac{d\rho}{dt} = \rho \frac{d\mathscr{E}}{dt} + p \left(\frac{\partial v_r}{\partial r} + \frac{\partial v_z}{\partial z} + \frac{v_r}{r} \right). \qquad (2.28)$$

As a result of transformation (2.28) the entropy equation in (2.22) becomes and equation for the temperature T if we also introduce the equation of state for the fully ionized ideal deuterium plasma:

$$p = (k/m_p)\rho T; \quad \mathscr{E} = (3/2)(k/m_p)T. \tag{2.29}$$

Substituting (2.23), (2.25), and (2.28) into (2.22), and using definition (2.24), we find an equation for the temperature in coordinate form:

$$\rho \frac{d\mathscr{E}}{dt} = -p\left(\frac{\partial v_r}{\partial r} + \frac{\partial v_z}{\partial z} + \frac{v_r}{r}\right) + \sigma_{rr}\frac{\partial v_r}{\partial r} + \sigma_{rz}\left(\frac{\partial v_r}{\partial z} + \frac{\partial v_z}{\partial r}\right) +$$

$$+ \sigma_{zz}\frac{\partial v_z}{\partial z} + \sigma_{\varphi\varphi}\frac{v_r}{r} + \frac{1}{r}\cdot\frac{\partial}{\partial r}\left(rv\frac{\partial T}{\partial r}\right) + \frac{\partial}{\partial z}\left(v\frac{\partial T}{\partial z}\right) +$$

$$+ \frac{\lambda}{4\pi}\left[\frac{1}{r^2}\left(\frac{\partial}{\partial r}rH_\varphi\right)^2 + \left(\frac{\partial H_\varphi}{\partial z}\right)^2\right], \tag{2.30}$$

where the operator is $d/dt = \partial/\partial t + v_r\,\partial/\partial r + v_z\,\partial/\partial z$ (this operates on any scalar function).

In summary, in this section we have formulated the complete system of differential equations of the MHD model for the present problem. This system includes the equation for the magnetic field, (2.8) (the degenerate Hall effect is ignored, and an effective plasma conductivity is introduced); the equations of motion, (2.20) and (2.21), along with the viscous-tensor components (2.17) (the isotropic viscosity of the plasma ion is taken into account); the entropy equation, (2.30) (the isotropic ion viscosity and thermal conductivity as well as the Joule heating contribute to the nonadiabatic terms of this equation); and, finally, the continuity equation, (2.27), and the equation of state, (2.29). We have made the important assumption that the field components H_r and H_z and the velocity component v_φ are initially zero.

§3. Energy Conservation Law and Kirchhoff Equation

a. Energy Conservation Law in Integral Form. Entropy equation (2.22) is equivalent to a local energy conservation law [25]. In the numerical solution of the problem the energy conservation law in integral form is very important when the integration is carried out over the entire plasma volume. In a moving plasma, this volume changes with time. Below we derive the integral law in a plasma volume bounded by moving external boundaries. This law is derived in [26] for the more general case in which the ion and electron temperatures are different. From Eqs. (2.22), (2.26), (2.27), and (2.10) we derive the local energy conservation law:

$$\rho \frac{d}{dt} \left(\mathscr{E} + \frac{v^2}{2} \right) = \operatorname{div} [(\sigma - p)\, \mathbf{v} - \mathbf{F}] +$$

$$+ \frac{1}{4\pi} (\mathbf{v} \,[\operatorname{curl}\mathbf{H} \cdot \mathbf{H}]) + \frac{c^2}{16\pi^2 \sigma_e} (\operatorname{curl}\mathbf{H})^2. \tag{3.1}$$

To integrate (3.1) over the entire plasma volume we introduce the Lagrangian variable dm = ρdV and assume that the total plasma mass is conserved:

$$\int \frac{d}{dt} \left(\mathscr{E} + \frac{v^2}{2} \right) \rho dV = \int \frac{\partial}{\partial t} \left(\mathscr{E} + \frac{v^2}{2} \right) dm = \frac{d}{dt} \int \left(\mathscr{E} + \frac{v^2}{2} \right) dm =$$

$$= - \oint [(p - \sigma)\, \mathbf{v} + \mathbf{F}]\, d\mathbf{S} + \frac{1}{4\pi} \int (\mathbf{v}\,[\operatorname{curl}\, \mathbf{H} \cdot \mathbf{H}])\, dV +$$

$$+ \frac{c^2}{16\pi^2} \int \frac{(\operatorname{curl}\mathbf{H})^2}{\sigma_e}\, dV. \tag{3.2}$$

The change in the magnetic field energy in the quasistationary case is described by the following equation, as follows from Eqs. (2.1) and (2.2):

$$\frac{\partial}{\partial t} \left(\frac{H^2}{8\pi} \right) = -(\mathbf{j}\mathbf{E}) - \operatorname{div} \frac{c}{4\pi}\, [\mathbf{E}\mathbf{H}]. \tag{3.3}$$

We integrate (3.3) over the variable plasma volume and add the term

$$\int \left(\mathbf{v} \operatorname{grad} \frac{H^2}{8\pi} \right) dV$$

to both sides in order to find the total time derivative. Using continuity equation (2.27), and introducing the Lagrangian variable m, we find

$$\int \frac{d}{dt} \left(\frac{H^2}{8\pi\rho} \right) \rho dV - \int \frac{H^2}{8\pi} \operatorname{div} v dV = \frac{d}{dt} \int \frac{H^2}{8\pi\rho} dm - \int \frac{H^2}{8\pi} \operatorname{div} v dV =$$

$$= - \int (jE) \, dV - \oint \frac{c}{4\pi} [EH] \, dS + \int \left(v \operatorname{grad} \frac{H^2}{8\pi} \right) dV. \qquad (3.4)$$

The total energy of the magnetic field in the variable plasma-filled volume is then

$$\frac{d}{dt} \int \frac{H^2}{8\pi\rho} \, dm = - \int (jE) \, dV - \oint \frac{c}{4\pi} [EH] \, dS + \oint v \, \frac{H^2}{8\pi} \, dS. \qquad (3.5)$$

Using Ohm's law, (2.3) (with $\chi = 0$), we eliminate the electric field E from (3.5) and introduce (2.4). Integral equations (3.2) and (3.5) are then combined. In this case the sum of all the volume integrals vanishes[†]:

$$\frac{1}{4\pi} \int (v \, [\operatorname{curl}H \cdot H]) \, d V + \frac{c^2}{16\pi^2} \int \frac{(\operatorname{curl}H)^2}{\sigma_e} \, dV - \int (jE) \, dV = 0. \qquad (3.6)$$

Simplifying (3.6) we find that the right side of the resultant equation only contains surface integrals over the outer boundary of the plasma:

$$\frac{d}{dt} \int \left(\mathscr{E} + \frac{v^2}{2} + \frac{H^2}{8\pi\rho} \right) dm = - \oint [p - \sigma) v + F] \, dS -$$

$$- \frac{1}{4\pi} \oint \lambda \, [\operatorname{curl}H \cdot H] \, dS + \frac{1}{4\pi} \oint (vH) \, HdS - \oint v \, \frac{H^2}{8\pi} \, dS. \qquad (3.7)$$

In the present case the plasma boundary is the axisymmetric surface $R(x, t)$ and $v_\varphi \equiv H_r \equiv H_z \equiv 0$ (taking account of the initial conditions). We now write integral energy conservation law (3.7) in terms of the cylindrical coordinates, noting that the vector dS has the components

$$dS = \left\{ Rd\varphi dz, \ 0, \ -Rd\varphi \frac{\partial R}{\partial z} \, dz \right\}. \qquad (3.8)$$

[†] The additional work of the electric field in the moving conductor, due to the presence of the term $(1/c) \, [\nu \, H]$, is equal to the negative value of the work performed by the ponderomotive force [25].

Integrating over the angle φ, we have

$$\frac{d}{dt} \int \left(\mathscr{E} + \frac{v_r^2 + v_z^2}{2} + \frac{H_\varphi^2}{8\pi\rho} \right) \rho r\,dr\,dz = -\int\limits_{(z)} \left[(\rho v_r - \sigma_{rr}\,v_r - \sigma_{rz}\,v_z + F_r) - \right.$$

$$-\frac{\lambda}{4\pi} \cdot \frac{H_\varphi}{R} \cdot \frac{\partial}{\partial r}(rH_\varphi)_{r=R} + v_r \frac{H_\varphi^2}{8\pi} \bigg] R\,dz + \int\limits_{(z)} \bigg[(\rho v_z - \sigma_{rz}\,v_r - $$

$$- \sigma_{zz}\,v_z + F_z) - \frac{\lambda}{4\pi} H_\varphi \frac{\partial H_\varphi}{\partial z} + v_z \frac{H_\varphi^2}{8\pi} \bigg] R\,\frac{\partial R}{\partial z}\,dz. \qquad (3.9)$$

The curvilinear integral along z extends along the entire plasma boundary, from $z_{\min}$ to $z_{\max}$. We note that the third surface integral in (3.7) vanishes, since the field lines are parallel to the plasma surface.

Finally, we note that the electromagnetic energy fluxes across the plasma boundary in (3.7) cannot be separated into two parts corresponding to Joule heating and the work performed by the ponderomotive forces, although one effect or the other appears alone in limiting cases ($\lambda \to 0$ or $v \to 0$). Equation (3.5) shows that only the sum of the Joule heat and the work performed by the ponderomotive forces within the plasma, which is equal to the integral $\int (jE)dV$, as follows from (3.6), can be represented as the sum of fluxes in (3.7).

    b.   Complete Energy Conservation Law for the Electromagnetic Field; Kirchhoff Equation.   The Kirchhoff equation can be derived on the basis of energy considerations extended to the entire system, including the gaseous conductor and the external electric circuit. For a chamber volume originally filled with gaseous deuterium with a constant density bounded by metal electrodes and insulators, we can write the following integral relation, which follows from (3.3):

$$\int\limits_{(V_0)} \frac{\partial}{\partial t} \left(\frac{H^2}{8\pi} \right) dV = \frac{d}{dt} \int\limits_{(V_0)} \frac{H^2}{8\pi}\,dV = -\int\limits_{(V_0)} (jE)\,dV - \oint\limits_{(S_0)} \Pi d\mathbf{S}, \qquad (3.10)$$

where the volume integrals are evaluated over the entire constant volume (V_0), and the surface integral (Π is the flux density of electromagnetic energy) is extended to all the solid surfaces (the electrodes and the insulator) that bound the chamber volume (S_0). If it is assumed that all the energy of the magnetic field produced by the current in the plasma is confined in the chamber volume, we can express the integral $\oint\limits_{(S_0)} \Pi dS$ in terms of the change in the energy in the external electric circuit. In the quasistationary approximation, we have the following equation for the circuit [25]:

$$\oint\limits_{(S_0)} \Pi\, dS = \frac{d}{dt}\left(\frac{L_0 J^2}{2c^2}\right) + \frac{d}{dt}\left(\frac{Q^2}{2C_0}\right) + R_{\Omega 0}\, J^2. \tag{3.11}$$

Here, L_0 is the self-inductance, C_0 is the capacitance, and $R_{\Omega 0}$ is the resistance. The charge on the capacitors, Q, is related to the total current J by

$$J = -\frac{dQ}{dt};\quad Q = Q_0 - \int\limits_0^t J\,dt' = C_0 U_0 - \int\limits_0^t J\,dt', \tag{3.12}$$

since the initial capacitor charge Q_0 can be expressed in terms of the initial voltage on the capacitors U_0. Substituting Q from (3.12) into (3.11), we find the rate of energy loss from the chamber volume to be

$$\oint\limits_{(S_0)} \Pi dS = \frac{L_0 J}{c^2}\cdot\frac{dJ}{dt} - J\left(U_0 - \frac{1}{C_0}\int\limits_0^t J\,dt'\right) + R_{\Omega 0} J^2. \tag{3.13}$$

Using Ohm's law with $\chi = 0$, we then eliminate the electric field $\mathbf{E}$ and transform from (3.10) to

$$\frac{d}{dt}\left(\int\limits_{(V_p)} \frac{H^2}{8\pi}\,dV + \int\limits_{(V_V)} \frac{H^2}{8\pi}\,dV\right) = -\int\limits_{(V_p)} \frac{j^2}{\sigma_e}\,dV + \frac{1}{c}\int\limits_{(V_p)} (\mathbf{j}\,[\mathbf{v}\mathbf{H}])\,dV - \oint\limits_{(S_0)} \Pi dS =$$

$$= -\int\limits_{(V_p)} \frac{j^2}{\sigma_e}\,dV - \frac{1}{c}\int\limits_{(V_p)} (\mathbf{v}\,[\mathbf{j}\mathbf{H}])\,dV - \oint\limits_{(S_0)} \Pi dS. \tag{3.14}$$

The integral on the left side of (3.14) is broken up into two parts: The first is the integral over the plasma-filled volume (V_p), while the second is that over the remaining, vacuum part of the chamber volume (V_V). In the hydrodynamic model the plasma is "plowed" completely by the current sheet. The two integrals over the volume V_p on the right side of (3.14) represent the Joule heating and the negative of the work performed by the ponderomotive

forces [see (3.6) and the associated footnote regarding the transformation of the second integral]. The integral over the vacuum
part of the volume can be evaluated over r, since the magnitude of
the field in the vacuum is $H = H_\varphi = 2J/cr$:

$$\int\limits_{(V_V)} \frac{H_\Phi^2}{8\pi} 2\pi r\, dr\, dz = \frac{J^2}{c^2} \int\limits_{z_{\min}}^{z_{\max}} \sum_i \ln \frac{R_{ex}^j(z)}{R_{in}^j(z)}\, dz, \qquad (3.15)$$

where R_{ex}^j is the radius of the external boundary of the vacuum gap
for a given value of z, and R_{in}^j is the radius of the internal boundary.
The summation over j in the integral is carried out over all parts
of the vacuum, separated by the parts of the plasma, for a given z.
In the outermost part the outer boundary coincides with R_{ex}^0, the
chamber radius. The summation over j in the integral in (3.15)
allows for a plasma configuration so complicated that there can be

points on the surface at which $\left|\dfrac{dR}{dz}\right| \to \infty$ (with $\dfrac{d^2R}{dz^2} \neq 0$). We can

say that the integration in (3.15) is carried out over the entire volume in which the equation for the magnetic field H_φ is valid. In
accordance with (3.15), we introduce the external self-inductance
of the plasma conductor:

$$L_p = 2 \int\limits_{z_{\min}}^{z_{\max}} \sum_i \ln \frac{R_{ex}^j(z)}{R_{in}^j(z)}\, dz. \qquad (3.16)$$

Actually, the electromagnetic energy introduced into the plasma
is still small in comparison with the field energy in the vacuum
part of the volume and in the elements of the external circuit. Consequently, in a first approximation, in (3.14) we can neglect all
integrals over the plasma volume (V_p). Using (3.14), substituting
in the surface integral from (3.13), and using (3.15) and (3.16), we
then write the Kirchhoff equation in the very simple form

$$\frac{L_p + L_0}{c^2} \cdot \frac{dJ}{dt} + \frac{J}{2c^2} \cdot \frac{dL_p}{dt} - \left(U_0 - \frac{1}{C_0}\int\limits_0^t J\, dt'\right) + R_{\Omega 0}J = 0. \qquad (3.17)$$

This equation takes account of the time dependence of the external
self-inductance of the plasma conductor, L_p.

c. Incorporation of the Energy Supplied to the Plasma in the Kirchhoff Equation. Taking account of the electromagnetic energy supplied to the plasma, we write the Kirchhoff equation as

$$\frac{L_p+L_0}{c^2}\cdot\frac{dJ}{dt}+\frac{J}{2c^2}\cdot\frac{dL_p}{dt}-\left(U_0-\frac{1}{C_0}\int_0^t J\,dt'\right)+R_{\Omega 0}\,J=-\frac{W_p}{J}, \qquad (3.18)$$

where the rate of change of the energy due to the plasma, W_p, is given by the following equation, as follows from (3.14) and (3.17):

$$W_p=\frac{d}{dt}\int_{(V_p)}\frac{H^2}{8\pi}\,dV+\int_{(V_p)}\frac{j^2}{\sigma_e}\,dV+\frac{1}{c}\int_{(V_p)}(v\,[jH])\,dV. \qquad (3.19)$$

In cylindrical coordinates, the equations for W_p for this problem take the following form, when we use (2.4) and (2.9):

$$W_p=\frac{1}{4}\cdot\frac{d}{dt}\int_{(V_p)}H_\varphi^2\,r\,dr\,dz+\frac{1}{2}\int_{(V_p)}\lambda\left[\left(\frac{\partial H_\varphi}{\partial z}\right)^2+\left(\frac{\partial H_\varphi}{\partial r}\right)^2+\right.$$

$$\left.+\frac{H_\varphi}{r}\left(2\frac{\partial H_\varphi}{\partial r}+\frac{H_\varphi}{r}\right)\right]r\,dr\,dz-\frac{1}{4}\int_{(V_p)}\left[v_z\frac{\partial H_\varphi^2}{\partial z}+v_r\left(\frac{\partial H_\varphi^2}{\partial r}+2\frac{H_\varphi^2}{r}\right)\right]r\,dr\,dz. \quad (3.20)$$

The right side of (3.19) can be written in a different form. Using (3.2) to introduce the plasma characteristics, from (3.19) we find an alternate equation for W_p:

$$W_p=\frac{d}{dt}\int_{(V_p)}\left(\frac{H^2}{8\pi}+\mathscr{E}\rho+\frac{\rho v^2}{2}\right)dV+\int_{(S_p)}[(p-\sigma)\,v+F]\,dS. \qquad (3.21)$$

We note that in substituting (3.21) into (3.18) we find a trivial energy conservation law for the entire system. Using the integral energy conservation law in the plasma, (3.7), we can now express the rate of change of the energy W_p from (3.21) solely in terms of integrals over the surface S_p:

$$W_p=-\frac{1}{4\pi}\oint_{(S_p)}\lambda\,[\text{curl }H\cdot H]\,dS+\frac{1}{4\pi}\oint_{(S_p)}(vH)\,H\,dS-\oint_{(S_p)}v\,\frac{H^2}{8\pi}\,dS. \qquad (3.22)$$

By analogy with (3.20), we can write (3.22) in coordinate form as a curvilinear integral along the plasma boundary z_p:

$$W_p = \int\limits_{(z_p)} \left[\frac{\lambda}{4\pi} \cdot \frac{H_\varphi}{R} \cdot \frac{\partial}{\partial r} (rH_\varphi)_{r=R} - v_r \frac{H_\varphi^2}{8\pi} \right] R\,dz -$$

$$- \int\limits_{(z_p)} \left(\frac{\lambda}{4\pi} H_\varphi \frac{\partial H_\varphi}{\partial z} - v_z \frac{H_\varphi^2}{8\pi} \right) R \frac{\partial R}{\partial z}\,dz. \qquad (3.23)$$

This equation is the same as (3.9) if the gaseous components of the fluxes are omitted from the latter.

Up to this point, in deriving the total Kirchhoff equation, (3.18), we have assumed that all the current from the external circuit flows through the plasma. Actually, however, the contact between the external circuit and the plasma can be broken. This can occur as the result of a secondary breakdown of the gas-filled gap [27] or as a result of the capture of magnetic field when the separate moving parts of the current sheet come together. Field capture leads to a configurations in which the vacuum part of the chamber volume is multiply connected. The magnetic field in the plasma bridge is dissipated in a certain time, so that a new, toroidal, current loop appears in the plasma after a transient process. This current loop has no electromagnetic coupling with the external circuit. The same result is found in secondary breakdown of the gas-filled gap. Only part of this new current loop can flow at the surface of the metal electrodes or in the ionized gas that remains in the volume. A total current J' flows along the boundary of the resulting toroid; when the circuit is broken this current is equal to the total current J. Subsequently, the current J' must be determined from Eq. (3.10) with $\Pi = 0$. Carrying out the necessary manipulations in accordance with (3.18), we write the following equation for the current J':

$$\frac{L_p'}{c^2} \cdot \frac{dJ'}{dt} + \frac{J'}{2c^2} \cdot \frac{dL_p'}{dt} = -\frac{W_p'}{J'}. \qquad (3.24)$$

The self-inductance of the toroid outside the conductor, L_p', is governed by Eq. (3.16), while the rate of change of energy due to loss in the plasma is governed by (3.19) [in coordinate form, by (3.20)]. In this case the integration in (3.16) is extended to the vacuum volume enclosed by the toroidal surface and the integration in (3.19) is carried out over the plasma volume V_p' directly adjacent to the same surface.† Equation (3.24) describes the damping of the current in the disconnected circuit as the result of Joule heating and the work performed by the ponderomotive forces (this second factor can also lead to current generation). Those transformations of Eq. (3.19) which make use of energy conservation throughout the plasma volume, i.e., Eqs. (3.21)-(3.23), cannot be applied in this case.

In §3 we have derived an integral law for energy conservation within the plasma, (3.9). This law is useful in the numerical solution of the problem. The total conservation law for the energy of the magnetic field of the entire system, (3.10), is used to derive the Kirchhoff equation for the total current J in various forms — ranging from the simplest form in (3.17), which ignores the flux of electromagnetic energy into the plasma, to Eq. (3.24), which takes account of the current loop that has been disconnected from the external circuit. The flux of electromagnetic energy into the plasma W_p [see the right side of the total Kirchhoff equation, (3.18)], is written in several equivalent forms; in particular, in (3.22) it is expressed in terms of the surface integrals alone. The development of a disconnected current loop is extremely complicated; here we have only considered current in the disconnected loop.

§4. Initial Conditions and Boundary Conditions for the MHD Problem

a. Choice of Initial Conditions. In choosing the initial conditions and boundary conditions we are guided by considerations of simplicity and convenience; choose conditions that are compatible with the actual conditions in the noncylindrical Z pinch and the philosophy of the MHD model (§§1-3). The conditions for the applicability of the MHD model — in particular, the assumption of a fully ionized plasma and complete plowing of this plasma by the current sheet — are not satisfied in the initial stage of the discharge, during breakdown, in the formation of the current sheet near the insulator, and in the detachment of this sheet from the insulator [28]. Although the motion of the sheet into the upper part of the chamber does satisfy the conditions of the MHD model, this motion is affected by the complicated chamber geometry. To develop an analysis for this process it is necessary to incorporate the

---

†The definition of the volume V'_p that corresponds to the new current loop is relative. Actually, this volume is bounded by some surface at which the magnetic field H_ϕ vanishes nearly everywhere.

lower part of the chamber (around the insulator) in the calculation region; it is also necessary to treat the motion of the diverging part of the current sheet, which is not of fundamental importance to the final stage of the discharge. Furthermore, the detachment of the sheet from the insulator is undoubtedly described in a satisfactory manner by the simplified snowplow model [12], which was recently used in calculations for this geometry [7]. The stage in which the current sheet appears in the upper part of the chamber is, of course, very important, since it is in this stage that the initial conditions for the formation of the plasma focus are established. We know from experiment that the deviation of the current sheet from cylindrical symmetry is governed by the properties of the plasma focus. Fortunately, the position of the sheet above the anode (while it is still far from the chamber axis) can be determined reliably by magnetic and optical measurements (in addition, this position can be found with the help of the snowplow model). Thus, we conclude that a tedious calculation of the MHD problem is rather ineffective and not fully justified in the initial stage of the discharge, in which the current sheet moves into the upper part of the chamber. The initial conditions is, instead, taken to be a position for the current sheet determined from experiment or from an approximate calculation. We ignore completely the diverging part of the sheet and the presence of the lower part of the chamber. Correspondingly, in calculating the self-inductance L_p we assume that the external boundaries of the vacuum are the planes $z_{min} = 0$ and $z_{max} = z_0$, where z_0 is the fixed distance from the anode to the upper metal cover of the chamber, and R_{ex}^o is the outer radius of the chamber. If the initial location of the current sheet is far from the axis of the system (if, say, the sheet radius is half the anode radius), we can ignore the energy stored in the plasma at this time ($t = 0$) in comparison with the energy of the final stage of the discharge. Neglecting the diverging part of the sheet and the lower part of the chamber is to be understood in the same sense (their contributions to the energy in the final stage are very small).

In this way, formulate the initial conditions of the MHD problem in order to calculate the final stage of the process. It is reasonable to generalize the corresponding conditions of the one-dimensional theory [1]. Assume that at $t = 0$ we are given an arbitrary external plasma boundary $R = R(z, 0)$, so that the volume filled by the plasma is

$$z_0 \geqslant z \geqslant 0; \quad R(z, 0) \geqslant r \geqslant 0. \qquad (4.1)$$

In this region the plasma is assumed to be at rest, unmagnetized, and cold and to have a constant density at t = 0:

$$v_r = v_z = 0; \quad H_\varphi = 0; \quad T = 0; \quad \rho = \rho_0. \qquad (4.2)$$

The density ρ_0 is the same as the initial gas density in the chamber if the mass captured from the external volume is neglected; this mass can be comparable in magnitude to that specified on the basis of conditions (4.1) and (4.2). Plowing of the mass in the earlier stages of the discharge is not very important; even so, it is possible, in principle, to take plowing into account through an appropriate increment in ρ_0. To solve the Kirchhoff equation we must also specify the current and the decrease in the capacitor charge at t = 0:

$$J = J_0; \quad \Delta Q = \int_{-\infty}^{0} J dt. \qquad (4.3)$$

The initial self-inductance L_p must be calculated from Eq. (3.16) and account must be taken of the comments above regarding the size and shape of the vacuum part of the chamber volume (t = 0); hence $L = L_0 + L_p$:

$$L = L_0 + 2 \int_{0}^{z_0} \ln \frac{R_{ex}^0}{R(z, 0)} \, dz, \qquad (4.4)$$

if the analysis is restricted to initial configurations R(z, 0) for which $\left| \frac{dR}{dz} \right| < \infty$ everywhere.

b. **Choice of Boundary Conditions.** Certain boundary conditions are generalizations from the one-dimensional problem, but the boundary conditions also include conditions at the anode and the upper cover of the chamber, which coincide with the end planes of the cylindrical region under consideration (z = 0, z = z_0). The conditions at these planes can obviously be chosen in several ways. Complicated physical processes occur at the interface between the metal and the hot plasma. The metal is melted and evaporated as its surface is subjected to the Joule heating, ion and electron thermal conduction, fast-electron beams, radiation, etc. It is clear, however, that these boundary-layer effects have very little effect on the plasma far from the electrodes, since we are dealing with a rapid process. Experiment confirms [27] that the propagation velocity of the metal vapor is roughly an order of magnitude lower than the characteristic plasma velocities. Thus, the vapor only fills a small fraction of the chamber volume, lagging behind the current sheet. The thickness of the boundary layer in the plasma itself must be of the order of the particle mean free path, which is extremely short when the MHD approximation applies. A complete description of the actual plasma – metal boundary layer goes far beyond the scope of the present review. It is therefore best to formulate the boundary conditions in such a way that we can prevent the appearance of any boundary layer near the electrodes. In particular, this can be done with the MHD equations with dissipation by specifying mirror symmetry. Specifically, at z = 0 and z = z$_0$ we set

$$\left. \begin{array}{l} v_z = 0; \quad \partial v_r/\partial z = 0; \quad \partial T/\partial z = 0; \\ \partial \rho/\partial z = 0; \quad \partial H_\varphi/\partial z = 0. \end{array} \right\} \qquad (4.5)$$

The symmetry of the pressure follows from the equation of state when (4.5) is taken into account. We note that the usual condition for a viscous fluid **v** = 0 leads to an illusory refinement of the problem; in practice this condition is extremely inconvenient because it leads to the appearance of a narrow boundary layer.

At the external boundary of the plasma the most important conditions are that both components of the momentum flux density vanish. In the coordinate system that moves with the mass these conditions are equivalent to the condition that no force is exerted [24]. This condition should hold at the free external interface of the plasma with the vacuum [1]:

$$pn_r - \sigma_{ri}n_i = 0; \quad pn_z - \sigma_{zi}n_i = 0. \qquad (4.6)$$

Equation (4.6) contains the unit outward normal to the plasma surface, **n**, with components n$_r$ and n$_z$, which can obviously be written in terms of the components of d**S** from (3.8):

$$\mathbf{n} = \frac{d\mathbf{S}}{|d\mathbf{S}|} = \frac{d\mathbf{S}}{Rd\varphi dz \sqrt{1 + \left(\dfrac{\partial R}{\partial z}\right)^2}} = \frac{1}{\sqrt{1 + \left(\dfrac{\partial R}{\partial z}\right)^2}} \left\{1, 0, -\dfrac{\partial R}{\partial z}\right\}. \qquad (4.7)$$

Substituting **n** from (4.7) into (4.6), we find two conditions at r = R(z, t):

$$(p - \sigma_{rr}) + \sigma_{rz}\frac{\partial R}{\partial z} = 0; \quad (p - \sigma_{zz})\frac{\partial R}{\partial z} - \sigma_{rz} = 0. \qquad (4.8)$$

Conditions (4.8) at the external plasma boundary must be supplemented by conditions on the temperature and the magnetic field. Using (4.7), we have

$$
\left.
\begin{aligned}
n_r \frac{\partial T}{\partial r} + n_z \frac{\partial T}{\partial z} &= \frac{\partial T}{\partial r} - \frac{\partial R}{\partial z} \cdot \frac{\partial T}{\partial z} = 0; \\
H_\varphi &= \frac{2J}{cR(z,t)}.
\end{aligned}
\right\}
\tag{4.9}
$$

Making use of (4.8) and (4.9) and the energy conservation law in (3.9), we find that we are only left with terms due to the electromagnetic energy flux across the plasma boundary. At the plasma boundaries at $z = 0$ and $z = z_0$ all the energy fluxes vanish completely, if (4.5) is used. Accordingly the energy conservation relation (3.9) assumes the following simple form under boundary conditions (4.5), (4.8), and (4.9):

$$
\begin{aligned}
\frac{d}{dt} \int \left(\mathcal{E} + \frac{v_r^2 + v_z^2}{2} + \frac{H_\varphi^2}{8\pi\rho} \right) \rho r\, dr\, dz &= \\
= -\int_{(z_p)'} \left[-\frac{\lambda}{4\pi} \cdot \frac{H_\varphi}{R} \cdot \frac{\partial}{\partial r} (rH_\varphi)_{r=R} + v_r \frac{H_\varphi^2}{8\pi} \right] R\, dz &+ \\
+ \int_{(z_p)'} \left(-\frac{\lambda}{4\pi} H_\varphi \frac{\partial H_\varphi}{\partial z} + v_z \frac{H_\varphi^2}{8\pi} \right) R \frac{\partial R}{\partial z}\, dz,
\end{aligned}
\tag{4.10}
$$

where the integral is only evaluated along the free interface of the plasma with the vacuum [over the region $(z_p)'$]. This choice of boundary conditions does not affect the equations for W_p in (3.20) or (3.23) [the only difference is that $(z_p)'$ and (z_p) are identified].

We must still write boundary conditions at $r = 0$; these conditions follow from the boundedness of all quantities at the axis of the system:

$$
v_r = 0; \quad \partial v_z/\partial r = 0; \quad \partial T/\partial r = 0; \quad \partial \rho/\partial r = 0; \quad H_\varphi = 0.
\tag{4.11}
$$

In contrast with the one-dimensional case [1], the condition $\partial v_z/\partial r = 0$ must also be satisfied here, so that the quantity σ_{rz} from (2.17) will vanish at $r = 0$. This condition is sufficient to ensure the boundedness of the right side of the equation of motion in (2.21).

We are primarily interested in applying the MHD model to the final stage of the discharge; the initial stage can be described by the simplified snowplow model, if necessary. Accordingly, at the initial time for the MHD problem we assume that the position of the current sheet in the upper part of the chamber is known, e.g., from experiment. In formulating the initial conditions we introduce an important simplification: We ignore the plasma energy and the distribution of the energy density within the current sheet. The boundary conditions of this problem ignore the complicated physical processes that occur at the metal – plasma interface. This comment applies to the mirror-symmetry condition, which prevents the appearance of a boundary layer in the plasma. The external plasma boundary is assumed to be a free interface with the vacuum, so that the externally applied force and the conduction heat flux vanish at this boundary. The magnetic field at the external plasma boundary is related to the total current; to determine this current we solve the Kirchhoff equation with the appropriate initial data. In general, these initial conditions and boundary conditions provide a correct description of the formation of the plasma focus.

§5. Dimensionless Form of the Equations; Complete Formulation of the MHD Problem

a. Dimensionless Form of the MHD Equations.

We write the viscosity, thermal conductivity, and diffusion coefficient for the magnetic field for a deuterium plasma in the following form [1] (see also §1):

$$\eta = B(kT)^{5/2}; \quad \nu = D(kT)^{5/2}; \quad \lambda = E(kT)^{-3/2}. \tag{5.1}$$

The constants B and D are determined from the classical ion viscosity and thermal conductivity for the isotropic case. According to (1.3'), the constant E is given by $E = 10^2 E_{cl}$, where E_{cl} corresponds to the classical conductivity $\sigma_\perp$. Thus

$$B = 1.4 \cdot 10^{24}; \quad D = 2.4 \cdot 10^{32}; \quad E = 1.7 \cdot 10^{-9}. \tag{5.2}$$

All the physical quantities are written in dimensionless form:

$$\left.\begin{aligned}
&\tilde{r} = r/R_0; \quad \tilde{z} = z/R_0; \quad \tilde{t} = t/t_0; \quad \tilde{v}_r = v_r/v_0; \\
&\tilde{v}_z = v_z/v_0; \quad \tilde{\rho} = \rho/\rho_0; \\
&\tilde{p} = p/p_0; \quad \tilde{T} = T/T_0; \quad \tilde{\mathcal{E}} = \mathcal{E}/\mathcal{E}_0; \quad \tilde{H}_\varphi = H_\varphi/H_0; \\
&\tilde{J} = J/J_0.
\end{aligned}\right\} \tag{5.3}$$

The only independent units here are R_0, ρ_0, and J_0; the other units are related to these by

$$H_0 = 2J_0/cR_0; \quad p_0 = H_0^2/4\pi;$$
$$v_0 = (p_0/\rho_0)^{1/2}; \quad t_0 = R_0/v_0; \quad T_0 = (m_p/k) \cdot (p_0/\rho_0); \tag{5.4}$$
$$\mathscr{E}_0 = p_0/\rho_0.$$

Using (5.3) and (5.4), we can now write the entire system of MHD equations, including (according to §2), the equation for the magnetic field, (2.8); the equations of motion, (2.20) and (2.21), along with definitions (2.17); the entropy equation, (2.30); the continuity equation, (2.7); and the equation of state, (2.29). Since we will be using the dimensionless quantities from (5.3) for the most part, we omit the tilde for simplicity. In the order given above, the MHD equations are

$$\frac{dH_\varphi}{dt} + H_\varphi \left(\frac{\partial v_r}{\partial r} + \frac{\partial v_z}{\partial z} \right) = \frac{\partial}{\partial r} \left[\frac{\lambda}{r} \cdot \frac{\partial}{\partial r} (rH_\varphi) \right] + \frac{\partial}{\partial z} \left(\lambda \frac{\partial H_\varphi}{\partial z} \right); \tag{5.5}$$

$$\rho \frac{dv_r}{dt} + \frac{H_\varphi}{r} \cdot \frac{\partial}{\partial r} (rH_\varphi) = -\frac{\partial p}{\partial r} + \frac{\partial \sigma_{rr}}{\partial r} + \frac{\partial \sigma_{rz}}{\partial z} + \frac{\sigma_{rr} - \sigma_{\varphi\varphi}}{r}; \tag{5.6}$$

$$\rho \frac{dv_z}{dt} + H_\varphi \frac{\partial H_\varphi}{\partial z} = -\frac{\partial p}{\partial z} + \frac{\partial \sigma_{rz}}{\partial r} + \frac{\partial \sigma_{zz}}{\partial z} + \frac{\sigma_{rz}}{r}; \tag{5.7}$$

$$\rho \frac{d\mathscr{E}}{dt} = -p \left(\frac{\partial v_r}{\partial r} + \frac{\partial v_z}{\partial z} + \frac{v_r}{r} \right) + \sigma_{rr} \frac{\partial v_r}{\partial r} + \sigma_{rz} \left(\frac{\partial v_r}{\partial z} + \frac{\partial v_z}{\partial r} \right) +$$
$$+ \sigma_{zz} \frac{\partial v_z}{\partial z} + \sigma_{\varphi\varphi} \frac{v_r}{r} + \frac{1}{r} \cdot \frac{\partial}{\partial r} \left(r\nu \frac{\partial T}{\partial r} \right) + \frac{\partial}{\partial z} \left(\nu \frac{\partial T}{\partial z} \right) +$$
$$+ \lambda \left[\frac{1}{r^2} \left(\frac{\partial}{\partial r} rH_\varphi \right)^2 + \left(\frac{\partial H_\varphi}{\partial z} \right)^2 \right]; \tag{5.8}$$

$$\frac{d\rho}{dt} + \rho \left(\frac{\partial v_r}{\partial r} + \frac{\partial v_z}{\partial z} + \frac{v_r}{r} \right) = 0; \tag{5.9}$$

$$p = \rho T; \quad \mathscr{E} = 3T/2. \tag{5.10}$$

The components of the viscous tensor are [in agreement with (2.17)]

$$\sigma_{rr} = \frac{4}{3} \eta \left(\frac{\partial v_r}{\partial r} - \frac{1}{2} \cdot \frac{v_r}{r} - \frac{1}{2} \cdot \frac{\partial v_z}{\partial z} \right);$$
$$\sigma_{zz} = \frac{4}{3} \eta \left(\frac{\partial v_z}{\partial z} - \frac{1}{2} \cdot \frac{v_r}{r} - \frac{1}{2} \cdot \frac{\partial v_r}{\partial r} \right);$$
$$\sigma_{\varphi\varphi} = \frac{4}{3} \eta \left(\frac{v_r}{r} - \frac{1}{2} \cdot \frac{\partial v_r}{\partial r} - \frac{1}{2} \cdot \frac{\partial v_z}{\partial z} \right); \tag{5.11}$$
$$\sigma_{rz} = \eta \left(\frac{\partial v_r}{\partial z} + \frac{\partial v_z}{\partial r} \right).$$

The dimensionless dissipation coefficients λ, ν, and η in Eqs. (5.5), (5.8), and (5.11) are determined from (5.1):

$$\lambda = E^* T^{-3/2}; \quad \nu = D^* T^{5/2}; \quad \eta = B^* T^{5/2}, \tag{5.12}$$

where the constant numerical factors are

$$E^* = (E/m_p^{3/2})(1/\alpha\beta); \quad D^* = (Dm_p^{7/2}/k)\,\alpha; \quad B^* = Bm_p^{5/2}\alpha. \tag{5.13}$$

The latter equations include two characteristic combinations of the basic units:

$$\alpha = p_0^2/\rho_0^3 R_0; \quad \beta = \rho_0 R_0^2, \tag{5.14}$$

whose physical meaning has been pointed out earlier [1] ($\alpha \sim l_{i,e}/R_0$,

$$\alpha\beta \sim \frac{R_0}{r_{\Lambda i,\Lambda e}} \cdot \frac{l_{i,e}}{r_{\Lambda i,\Lambda e}},$$

where $l_{i,e}$ and $r_{\Lambda i,\Lambda e}$ are the ion and electron mean free path and gyroradius, respectively).

b. Initial Conditions and Boundary Conditions in Dimensionless Form. In dimensionless form the initial conditions in (4.2) are

$$v_r = v_z = 0; \; H_\varphi = 0; \; T = 0; \; \rho = 1; \; R = R(z, 0)(t = 0) \tag{5.15}$$

Correspondingly, the initial conditions for the Kirchhoff equation, (4.3) and (4.4), are

$$J = 1; \; \Delta Q = \int_{-\infty}^{0} J dt; \; L = \frac{L_0}{R_0} + 2\int_{0}^{z_0} \ln \frac{R_{ex}^0}{R(z, 0)} dz. \tag{5.16}$$

The problem is solved in the region $z_0 \geq z \geq 0$, $R(z, t) \geq r \geq 0$, with the following boundary conditions at the external boundary [(4.8) and (4.9)]:

$$\left.\begin{array}{l}
(p - \sigma_{rr}) + \dfrac{\partial R}{\partial z}\,\sigma_{rz} = 0; \quad (p - \sigma_{zz})\dfrac{\partial R}{\partial z} - \sigma_{rz} = 0; \\[2mm]
\dfrac{\partial T}{\partial r} - \dfrac{\partial T}{\partial z} \cdot \dfrac{\partial R}{\partial z} = 0; \quad H_\varphi = \dfrac{J}{R(z, t)} \\[2mm]
[r = R(z, t)].
\end{array}\right\} \tag{5.17}$$

The functions σ_{rf}, σ_{zz}, and σ_{rz} are taken from (5.11). At the electrode boundaries we have the symmetry conditions from (4.5):

$$v_z = 0; \quad \frac{\partial v_r}{\partial z} = 0; \quad \frac{\partial T}{\partial z} = 0; \quad \frac{\partial \rho}{\partial z} = 0; \quad \frac{\partial H_\varphi}{\partial z} = 0 \, (z = 0, \, z_0). \qquad (5.18)$$

At the axis of the system from (4.11) we have

$$v_r = 0; \quad \frac{\partial v_z}{\partial r} = 0; \quad \frac{\partial T}{\partial r} = 0; \quad \frac{\partial \rho}{\partial r} = 0; \quad H_\varphi = 0 \, (r = 0). \qquad (5.19)$$

In terms of the units chosen here, the energy conservation law in (4.10) becomes

$$\frac{d}{dt} \int \left(\mathscr{E} + \frac{v_r^2 + v_z^2}{2} + \frac{H_\varphi^2}{2\rho} \right) \rho r dr dz = - \int_{(z_p)'} \left[- \lambda \frac{H_\varphi}{R} \cdot \frac{\partial}{\partial r} (r H_\varphi)_{r=R} + \right.$$

$$\left. + v_r \frac{H_\varphi^2}{2} \right] R dz + \int_{(z_p)'} \left(-\lambda H_\varphi \frac{\partial H_\varphi}{\partial z} + v_z \frac{H_\varphi^2}{2} \right) R \frac{\partial R}{\partial z} \, dz, \qquad (5.20)$$

where the dimensionless diffusion coefficient for the magnetic field, λ, is given in (5.12).

c. Dimensionless Form of the Kirchhoff Equation. First ignoring the energy loss in the plasma, we write the current equation from (3.17) in the following dimensionless form:

$$L^{1/2} \frac{d}{dt} (L^{1/2} J) + \Omega J + \Phi + \Psi \left(\int_0^t J dt' + \Delta Q \right) = 0, \qquad (5.21)$$

where the total dimensionless inductance L is, according to (3.16), (4.4), and (5.16),

$$L = \frac{L_0}{R_0} + 2 \int_0^{z_0} \sum_j \ln \frac{R_{ex}^j (z)}{R_{in}^j (z)} \, dz, \qquad (5.22)$$

and the new constant coefficients Ω, Φ, and Ψ are given as follows in terms of the basic units and the circuit constants (all are expressed in electrostatic units):

$$\Omega = c^3 R_{\Omega 0} \frac{R_0}{J_0} (\pi \rho_0)^{1/2}; \quad \Phi = -c^3 U_0 \frac{R_0}{J_0^2} (\pi \rho_0)^{1/2}; \quad \Psi = \frac{c^4}{C_0} \cdot \frac{R_0^3}{J_0^2} \pi \rho_0. \qquad (5.23)$$

The constant ΔQ is given in the initial conditions in (5.16).

When the energy loss in the plasma is taken into account we must use the complete Kirchhoff equation in (3.18); hence, on the right side of dimensionless equation (5.21) the zero is replaced by the following quantity, as follows from (3.20):

$$-\frac{W_p}{J} = -\frac{1}{J} \cdot \frac{d}{dt} \int\limits_{(V_p)} H_\varphi^2\, rd\,rdz - \frac{2}{J} \int\limits_{(V_p)} \lambda \left[\left(\frac{\partial H_\varphi}{\partial r} \right)^2 + \left(\frac{\partial H_\varphi}{\partial z} \right)^2 + \right.$$

$$\left. + \frac{H_\varphi}{r} \left(2 \frac{\partial H_\varphi}{\partial r} + \frac{H_\varphi}{r} \right) \right] rdrdz + \frac{1}{J} \int\limits_{(V_p)} \left[v_z \frac{\partial H_\varphi^2}{\partial z} + v_r \left(\frac{\partial H_\varphi^2}{\partial r} + 2 \frac{H_\varphi^2}{r} \right) \right] rdrdz.$$

$$(5.24)$$

The mathematical problem is thus one of solving system of equations (5.5)-(5.10), (5.21), with initial conditions (5.15) and (5.16) and boundary conditions (5.17)-(5.19) in the region $z_0 \geq z \geq 0$ and $R(z, t) \geq r \geq 0$.

d. Concrete Version of the MHD Problem. In order to pursue the determination of the parameters of the system of dimensionless equations, we must specify the units R_0, ρ_0, and J_0 and then use Eq. (5.4) to determine the other units. We assume $R_0 = 13$ cm, $\rho_0 = 2.4 \cdot 10^{-7}$ g/cm^3, $J_0 = 1$ mA, $H_0 = 1.54 \cdot 10^4$ G, $p_0 = 1.88 \cdot 10^7$ dyn/cm^2, $v_0 = 8.8 \cdot 10^6$ cm/sec, $t_0 = 1.47 \cdot 10^{-6}$ sec, $T_0 = 82$ eV, and $\mathscr{E}_0 = 7.7 \cdot 10^{13}$ ers/g. We find the parameters α and β from (5.14), and we find E*, D*, and B* from (5.13): $\alpha = 1.98 \cdot 10^{33}$, $\beta = 4.06 \cdot 10^{-5}$, E* $= 1.0 \cdot 10^{-2}$, D* $= 2.0 \cdot 10^{-2}$, and B* $= 1.0 \cdot 10^{-2}$.

We specify the parameters of the external circuit to be $U_0 = 24$ kV, $R_{\Omega 0} = 0$, $C_0 = 180$ μF, and $L_0 = 50$ cm. Then from (5.23) we find $\Omega = 0$, $\Phi = -2.72$, $\Psi = 0.92$, and $L_0/R_0 = 3.84$. To determine ΔQ from (5.21), we use experimental data and Eq. (5.16). It is assumed that the current reaches a maximum of 1.05 mA at $t_{max} = 4 \cdot 10^{-6}$ sec and that the initial time for the calculation is $t_{init} = 5 \cdot 10^{-6}$ sec. The typical oscilloscope trace of the current can be approximated by the function $J = 4J_{max}t/t_1(1 - t/t_1)$, where $t_1 = 8 \cdot 10^{-6}$ sec (e.g., with $t = t_{init}$, $J = 0.985$ mA) and $J_{max} = 1.05$ mA. The decrease in capacitor charge is

$$\Delta Q = \int\limits_0^{t_{init}} 4 \frac{J_{max}\,t}{J_0\,t_1} \left(1 - \frac{t}{t_1} \right) \frac{dt}{t_0} = \int\limits_0^{t_{init}/t_0} 4 \frac{J_{max}\,t_0}{J_0\,t_1} \times$$

$$\times \tau \left(1 - \frac{t_0}{t_1} \tau \right) d\tau = 0.775 \int\limits_0^{3.40} \tau (1 - 0.184\tau)\, d\tau = 2.59,$$

where we have substituted the values of the current J_0 and the time t_0.

The initial conditions for the problem, (5.15), also include the boundary shape $R(z, 0)$, which can be specified graphically. Here, $R(0, 0) = 1$ and $R(z, 0) \geq 1$. The distance from the anode to the upper cover of the chamber is $z_0 = 0.54$.

These concrete values for all the parameters of the system of equations, the boundary conditions, and the initial conditions can be taken as an illustration of the method of formulating the problem for numerical calculation. In this section, we have shown that the solution of the dimensionless problem depends on two parameters in the system of MHD equations (α and β) and five parameters in the Kirchhoff equation (Ω, Φ, Ψ, ΔQ, and L_0/R_0). Furthermore, the initial conditions and boundary conditions include another parameter (z_0) and the sheet shape, $R(z, 0)$. It should be noted, however, that in the calculation of the parameters of the plasma focus the parameter α plays the leading role; the net effect of all the circuit parameters is simply to cause some deviation of the current from the value $J = 1$. The parameter β only has an effect on the structure of the current sheet. It can be assumed that variations in the shape of the current sheet $R(z, 0)$ and the distance between the electrodes, z_0, do not affect the maximum values of the properties in the plasma focus.

Chapter 2

NUMERICAL SOLUTION OF THE TWO-DIMENSIONAL MHD PROBLEM

§6. General Comments

Approximate solutions of problems involving the integration of systems of partial differential equations are usually found by finite-difference methods. These methods are based on the replacement, in some way or other, of continuous quantities of the original problem by discretized quantities. The ranges of the independent variables are replaced by sets of calculations points, the unknown functions are replaced by tables, and the differential equations are replaced by finite-difference equations, i.e., algebraic relations between the values of the functions at adjacent calculation points. The problem obtained in this manner contains certain discreteness parameters, which characterize the set of calculation points, i.e., the number and closeness of these points.

This transformation from a functional problem to an arithmetic problem is only meaningful if the corresponding calculation algorithm is practical and only if the calculated results provide adequate information about the exact solution. An analysis of the agreement of the solutions of these two problems reduces to an analysis of the asymptotic behavior of the numerical solution, i.e., its convergence to the exact solution as the density of the calculation points is increased without bound. Although it is rarely possible to find a rigorous mathematical proof of convergence, there is usually an adequate basis for confidence in this convergence. In principle, it is possible to approach the exact solution arbitrarily closely and to estimate the deviation from the exact solution by comparing the results obtained with different numbers of calculation points.

Despite the seemingly formidable capabilities of modern computers the amount of information to be processed in the solution of complicated multidimensional problems is so large that it is not possible to reach values of the algorithm parameters for which asymptotic behavior is established. Nevertheless, finite-difference methods are still used in this situation. The usual justification is that a really "accurate" solution is not required and that a small number of calculation points is adequate for evaluating the effects of interest. This completely reasonable argument neglects one important point: Reducing the density of calculation points does more than only degrade the calculation accuracy in the sense that important local details of the solution are smoothed over; it can also lead to qualitative changes in the properties of the difference problem itself. The only justification for the use of any numerical method lies in its asymptotic behavior. In this sense, all these methods are equivalent to the differential problem and to each other, and all are good if enough points are used. If, however, there are not enough calculation points, any one of these methods departs from the original differential problem and begins to display features of its own, reacting in a unique way to the particular features of the problems involved, essentially becoming a discrete model of the physical process. This circumstance is responsible, in particular, for the continuing efforts of mathematicians to develop new calculation algorithms that are applicable to broad classes of problems.

Let us briefly examine those particular features of the problem involved in the present work which are pertinent to the choice of a difference method. Foremost among these features are the extreme deformations of the plasma volume as the plasma moves toward the axis of the system; these can change certain dimensions by two or three orders of magnitude.

Under these conditions it is natural to use calculation grids based on a Lagrangian coordinate system, and thus to move the calculation points of the finite-difference scheme together with the moving mass elements of the plasma. The approach has acquitted itself very well in the one-dimensional problems, but in two-dimensional problems the use of Lagrangian grids runs into serious difficulties: In the one-dimensional problem, the elementary calculation cell, which joins points at which the values of the properties directly affect each other, is a line segment; in the two-dimensional problem, however, this cell is a plane figure, generally a quadrangle. As the medium moves the cell is deformed. A line segment can only contract or lengthen, but the number of degrees of freedom of a quadrangle is much greater. Even with comparatively regular flow, a quadrangle which, for example, increases in size along one of its diagonals can actually become a "one-dimensional" figure. Howerver, this quadrangle is the calculation cell, and its shape strongly affects the accuracy with which the differential equations are approximated by the finite-difference equations. Deformations of the cell degrade the accuracy of the approximation; in turn we find the generation of irregular, parasitic deformations, etc. As a result, the calculation can become misleading.

The reasons for this defect are not clear. In the absence of a satisfactory mathematical theory of the nonlinear problem we cannot identify the essential questions involved nor discover the pertinent shortcomings of the hydrodynamic description of the continuous medium. Here it is sufficient to note that in finite-difference schemes based on Lagrangian calculation grids the approximation losses lead quite rapidly to a "computational catastrophe." These schemes are usually provided with a variety of tools for correcting the grids to reflect the particular features of the problem and the particular "taste" of the investigator.

A second important feature of this problem is that it involves different scales. Although all parts of the plasma are of course related, the most important parameters are those of the plasma focus, which is a region of negligible volume at the axis of the system. The processes occurring at the periphery are essentially of no interest. In constructing a calculation grid, however, we cannot arbitrarily make it finer in certain regions without introducing "extra" coordinate lines and calculation points in other regions, since the grid must be regular to some extent. The net result is that the number of necessary calculations is increased severalfold.

The finite-difference method which we use, the free-point method, makes it possible to treat these features of the problem while avoiding associated difficulties.

§7. The Free-Point Method

As noted above, the catastrophic distortions of the elementary Lagrangian calculation cell are caused by the deformations of this cell that arise from the relative motion of different mass elements of the medium. These deformations are a natural consequence of the problem; however, it is not necessary to use the deformed Lagrangian cells as the calculation cells. They simply reflect a relative arrangement of points which prevails at the initial time but which ceases to be useful as time elapses. One way to avoid the difficulty is to abandon the fixed relative arrangement of points dictated by the coordinate grid and to use, instead, natural calculation cells that reflect the actual arrangement of points at each time. This is the basic idea behind the free-point method. In order to implement this method it is necessary to a) specify a method for choosing the neighboring points, i.e., construct a calculation grid, and b) to write out finite-difference equations that apply for a calculation cell of arbitrary shape.

There is no particular difficulty in specifying which points are adjacent to a given point and thereby constructing the calculation cell for any particular situation. The problem is to transform our common-sense understanding of proximity into an algorithm which is universal but, at the same time, reasonably efficient and logically simple, so that it can be used on a computer.

Information on the relative arrangement of the calculation points is embodied in a table of coordinates of these points. The algorithm must obviously involve a comparison of the relative coordinates of the points and the choice of the "nearest" points. The term nearest must be defined exactly. We place the origin of coordinates (polar coordinates, R, φ) at the point M_0, for which the calculation grid is constructed. If any two points $M_1(R_1, \varphi_1)$ and $M_2(R_2, \varphi_2)$ lie on a single ray emerging from M_0, i.e., if $\varphi_1 = \varphi_2$, then the choice of the nearest point is trivial. If φ_1 and φ_2 are approximately the same, while R_1 and R_2 are different, the problem is resolved in a similar way; one of the points puts the other in its "shadow." Let us give these arguments an exact meaning.

We define the "shadow" of a given point $M_1(R_1, \varphi_1)$ to be that part of the plane which lies (from the standpoint of M_0) outside a certain curve of

$$ F\left(\frac{R}{R_1},\ |\varphi - \varphi_1|\right) = 0. \tag{7.1} $$

These particular combinations of arguments must be used because of the requirements that the curve be symmetric with respect to the ray $\varphi = \varphi_1$ and invariant with respect to the choice of the zero of the angle φ and with respect to the choice of the linear scale. The selection process reduces to the construction of the shadow of each point, the rejection of those points which fall in the shadow of any other point, and thus the identification of the sequence of points which are neighbors of M_0. Before taking up the choice of the function F we discuss certain general requirements which the selection must satisfy. Obviously, the neighboring points cannot include points which are too remote, since this is equivalent to increasing the step size of the difference scheme and, thus, to a degradation of the calculation accuracy; it also makes unnecessary demands on the computer. On the other hand, we do not want to minimize the total number of neighboring points since this approach makes it necessary to choose between points of approximately equivalent position, increases the weight of each point in the calculation scheme, and therefore leads to larger perturbations on neighbors due to the motion. These requirements are not definite enough for our purposes. Actually we have a single criterion for judging the quality of the selection method (which again is not very definite): In particular cases involving regular arrangements of the calculation points, the method must result in the selection of the natural calculation cell.

One characteristic of a natural cell is a rather dense distribution of neighboring points along the angle φ. We extend this idea to the general case: Any angle of some specified size (of the order of $\pi/2$) must contain at least one neighboring point.

The second requirement also follows naturally: The result of the selection must not depend on random factors (the order in which the points are tested, etc.).

It is clear that the most important characteristic of the shadow is its effective angular size. If the shadow is narrow the neighboring points will include remote points; if it is broad there will be few remaining points and certain directions φ will be poorly represented. In general, it is not possible to avoid both these extreme situations, so we consider the following iterative selection procedure.

The selection is carried out by using a shadow with an angular size of the order of a specified value $\Delta\varphi$ (this approach guarantees that remote points will be covered); we test the angular density of the sequence of points chosen. If certain directions φ are unoccupied, we repeat the selection procedure at points lying at these unoccupied angles, reducing the angular size of the shadow. After several iterations the desired sequence of neighboring points can be identified (provided that this sequence exists). When this method is used to construct the calculation cell there is some nonequivalence in directions φ, but this simply reflects the features of the particular distribution of calculation points.

From the standpoint of their effects on the result, the other characteristics of the shadow are relatively unimportant. However, it is not simply the result which is important; the method by which it is reached, i.e., the number of calculations required to construct the difference cell, is also important. The number of such calculations is governed primarily by the number of pairs of points whose coordinates are to be compared. Regardless of the comparison mechanism it would be wasteful to examine all the calculation points available. A simple sorting procedure requires a number of operations of the order of the square of the number of points (per cell). It is clear, therefore, that it is of primary importance to examine the efficiency of the selection algorithm.

§ 8. The Selection of Points

As noted above, the set of points to be examined must be as small as possible and still contain the desired neighboring points. The selection algorithm is thus broken up into two steps: the selection of the set (vicinity) and the selection of the neighboring points.

If the coordinates of the points are not supplemented with any additional information about the relative arrangement of these points, a general sorting procedure is required. This additional information is usually specified as the order of the points, e.g., by writing their coordinates in a table in which position corresponds to the actual position of the points. In a one-dimensional array of points this process resolves the problem. With an arbitrary two-dimensional array this ordering is completely devoid of meaning — even with the position of each point available. It is possible, however, to carry out a partial ordering; then in the choice of vicinities it is possible to restrict the examination to a comparatively small number of points, whose number is, at any rate, independent of the total size of the array.

We denote the Cartesian coordinates of the calculation points by r, z. We partition the r, z plane into strips using lines with $z =$ const. Those points which fall in a given strip are ordered along the coordinate r, while the strips themselves are ordered along z. The strip width is the parameter in this ordering, and it must be matched to the actual density of calculation points. If the strips are too narrow or too wide, the ordering degenerates into a one-dimensional ordering along z or along r, respectively, and the efficiency is degraded.

The dimensions of a neighborhood are defined in terms of Δr, Δz; in making the choice we only evaluate the differences between coordinates. The dimension Δz gives the necessary number of strips while Δr gives the number of points on the strips which must be included in a neighborhood. We note that when we approach the choice of the vicinity of a given point we already have information about it by virtue of the order of the points — the vicinity of the calculation point treated previously. Accordingly, the region of a given calculation point is well defined, and initial data for the choice is available. The strategy used for choosing the values of Δr and Δz

is important. We wish to minimize the chosen set but keeping it
adequately large, that is, we wish to achieve equivalence to the entire array of calculation points. If we also wish to simplify the
techniques for making the choice, a compromise is necessary. The
following procedure is used.

We first consider only that strip which includes given calculation point $M_0(r_0, z_0)$. The quantity Δr is taken to be of the order
of the width of this strip, being set equal to half this width. Of the
points of this one strip we choose those for which the coordinate r
differs from r_0 by less than Δr. We find the order of this set of
points along the angle φ around M_0 and check to see whether the
angular density of points is adequate. If not, the set that has been
chosen clearly does not contain the desired sequence of neighboring
points. In this case we determine (roughly) the direction of the unoccupied angle and make an additional choice, either by considering another strip or by changing the parameter Δr for one side
of the vicinity. The angular-density condition is then evaluated
and the process repeated, if necessary.

In each step the chosen vicinity is characterized by only four
parameters; along z, the number of strips on one side and the
other; along r, the values of Δr^+ and Δr^-, respectively. In each
step, a set containing the sequence of neighboring points is found.

It is evident that this procedure is dictated by the desire to
maximize the efficiency of the algorithm; in certain cases it can
lead to a set of points that are not equivalent to the entire array
(primarily because the choice is carried out in terms of r and z,
while the subsequent choice of neighboring points is carried out
in terms of R and φ). This situation can be avoided by improving
the algorithm or by increasing the dimensions of the vicinity substantially.

In the choice it is important to use the order of the points in the
r, z plane. Since the calculation points move as time elapses (they
move from one strip to another and change places in the same
strip), it is also necessary to take account of the work which must
be carried out on the systematic reordering of the points in estimating the calculation volume. We minimize the work required
for this reordering by permitting an arbitrary arrangement of points
in the table (in the computer memory) and by specifying the posi-

tions (addresses) of the two neighbors along r for each point. In particular, this ordering procedure makes it a simple matter to add or eliminate calculation points without reconstructing the entire information array.

We now turn to the second step of the algorithm — that of selecting the neighboring points. The fundamental aspects of this process (the comparison of the relative coordinates at the points through the construction of the shadow of each point) are described in the preceding section. Here we shall discuss the rational design of the corresponding selection algorithm.

We first note that the initial data for the selection are a sequence of points which are already ordered along the angle φ around the central point. Accordingly, taking account of the angular dimensions of the shadow, we can easily distinguish those points which can, in general, cover each other, thereby avoiding the comparison of the coordinates of all pairs of points. In this way, the number of necessary calculations is reduced substantially.

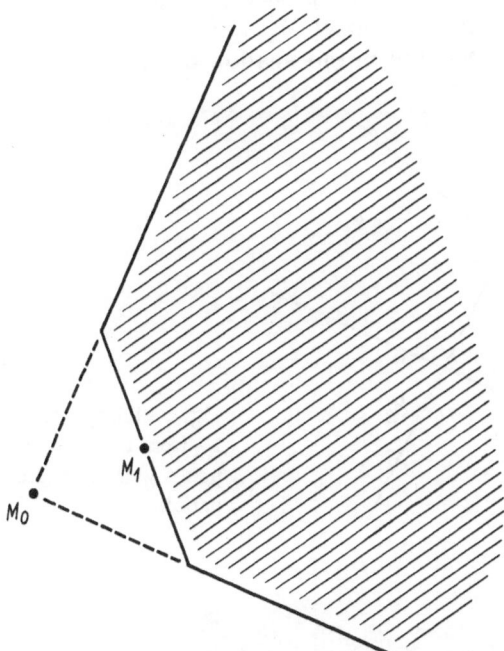

Fig. 1. Shadow of the point M_1.

The comparison involves determining the sign of the function F from (7.1) when the coordinates of a given pair of points are substituted. The nature of the function F must of course be as simple as possible; specifically, we use the function (Fig. 1)

$$F\left(\frac{R}{R_1}, \ |\varphi-\varphi_1|\right) = \begin{cases} \dfrac{R}{R_1}\cos(\varphi-\varphi_1)-1, & |\varphi-\varphi_1|<\dfrac{\pi}{4}; \\[2mm] |\varphi-\varphi_1|-\dfrac{\pi}{4}, & \dfrac{R}{R_1}\cos(\varphi-\varphi_1)>1. \end{cases} \qquad (8.1)$$

This choice of F is convenient because it only includes the quantities R, $\cos\varphi$, and $\sin\varphi$, which can easily be calculated from the differences $r-r_0$ and $z-z_0$. If other functions of φ are used it is necessary to carry out the operations of taking the arc sine, etc., which are very demanding of a computer. To order the selected set along the angle φ we only use these quantities, without calculating the value of φ itself.

The selection process is clear: We have a pair of points $M_1(R_1, \varphi_1)$ and $M_2(R_2, \varphi_2)$ such that $|\varphi_2 - \varphi_1| < \pi/4$ [actually, $\cos(\varphi_2 - \varphi_1) > 0.7$]. If $(R_2/R_1)\cos(\varphi_2 - \varphi_1) > 1$, then, according to (8.1), M_2 lies in the shadow of M_1; if $(R_1/R_2)\cos(\varphi_2 - \varphi_1) > 1$, then, on the contrary, M_1 lies in the shadow of M_2.

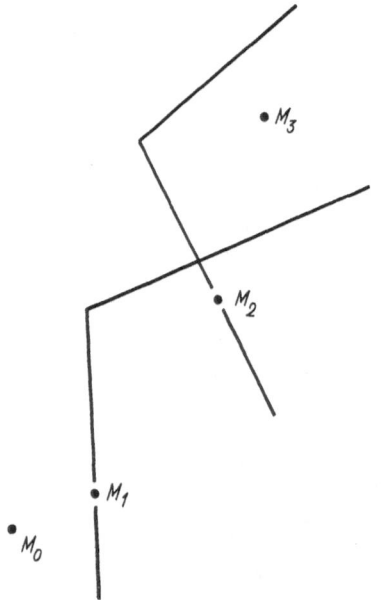

Fig. 2. Elimination of the point M_2 from the set of neighboring points while retaining it for comparison.

It is important to note that the fact that a point lies in a shadow does not mean that it is eliminated from subsequent comparisons, although it is not included among the neighboring points. If it were eliminated from subsequent comparisons, one of the selection conditions might be violated — the condition that the result be independent of the order in which the points are processed. This feature can be demonstrated by considering the situation in Fig. 2. Obviously, only M_1 of these three points should be included as neighbors of the indicated point. However, if M_2 is eliminated from the algorithm because it lies in the shadow of M_1, then point M_3 might also turn out to be identified as a neighboring point. To handle this situation properly it is necessary to treat M_3 after M_2 is compared with M_1, and M_2 is eliminated.

If the neglect of a point from the algorithm permissible (this reduces substantially the number of calculations required), the shadows must satisfy the following conditions: if the point M_2 belongs to the shadow of point M_1, then the shadow of M_2 is completely contained in the shadow of M_1. The construction of the corresponding functions F is feasible, but leads either to algorithms involving complicated calculations or unnatural calculation cells. The main point, however, is that it can be shown that in this case the boundary of the shadow of the given point M_1 [the curve $R(\varphi)$] satisfies the condition

$$R_1 \leqslant R(\varphi) \leqslant R_1 \exp k |\varphi - \varphi_1|;$$

that is, the angular dimension of the shadow is unbounded. This circumstance deprives us of the possible use of the order of points along φ to reduce the number of calculations.

To complete this summary of the selection algorithm we consider the construction of the calculation cell for points lying near the boundary of the region in which the calculations are being carried out. This question is significantly related to the nature of the boundary condition; this boundary condition dictates certain requirements on the shape of the calculation cell. At the same time, the choice of a vicinity and the selection of neighboring points in this case differ from those described above only in details: Unoccupied φ sectors are allowed, certain points must be incorporated among the neighboring points, etc.

§9. Method for Finding the Difference Equations

We start from the principle that the solution of the difference problem converges asymptotically to the solution of the differential problems, so that the construction of the difference method must be consistent with the approximation and stability requirements. Without going into the exact mathematical meaning of these terms, we note the following: The approximation requirement means that the difference equations transform into the differential equations of the original problem asymptotically as the calculation points become denser. There is usually no difficulty in checking this requirement. The stability requirement results from the need to take proper account of the range of the solution; this requirement is essentially a requirement on the correctness of the difference problem. A complete mathematical study of this question cannot be carried out for real problems, so that it is customary to simply check the stability of certain linear models of the difference problems. Although these models are simplifications, they provide reliable conclusions concerning the properties of these problems. A calculation cannot be carried out by an unstable difference method since the results become meaningless rapidly; thus, the calculation always includes a check of the stability requirement.

Accordingly, at a given time t we have a certain calculation cell, i.e., a point M_0 surrounded by the neighboring points M_1, $M_2, \ldots, M_N$. The value of each of the functions $f\{=u, v, T, \ldots\}$ at the point M_n is denoted by f_n. We denote by f^0 this value of f at the time $t + \tau$ at the point to which M_0 moves in the time τ. The quantity f^0 is to be calculated.

We write the system of differential equations in the compact form

$$d\mathcal{F}/dt = \mathcal{D}(\mathcal{F}), \qquad (9.1)$$

where $\mathcal{F}$ is the set of functions f, i.e., a vector function, and $\mathcal{D}$ is a differential operator (with respect to r and z), i.e., a combination of the derivatives $\partial f/\partial r$, $\partial f/\partial z$, $\partial^2 f/\partial r^2$, etc.

The most convenient way to find the difference equations is the following. We can obviously set

$$df/dt \approx (f^0 - f_0)/\tau. \qquad (9.2)$$

We then write each function $f(t, r, z)$ in the vicinity of M_0 in the form

$$f = f_0 + f_r (r - r_0) + f_z (z - z_0) + f_{rr} \frac{(r-r_0)^2}{2} + f_{rz} (r - r_0)(z - z_0) + \ldots =$$

$$= f_0 + R (f_r \cos \varphi + f_z \sin \varphi) + \frac{R^2}{2} f_{rr} \cos^2 \varphi + R^2 f_{rz} \cos \varphi \sin \varphi + \ldots, \qquad (9.3)$$

where R, φ are the polar coordinates ($r - r_0 = R \cos \varphi$, $z - z_0 = R \sin \varphi$), and f_r, f_z, f_{rr},.... are undetermined coefficients. In expansion (9.3) we only retain a finite number of terms and require that the resulting polynomial (in powers of $r - r_0$, $z - z_0$) assume the proper values f_k at several neighboring points M_k; i.e., we require that this polynomial be an interpolation polynomial. This approach leads us to a system of linear equations for f_r, f_z, f_{rr},.... like

$$f_r \cos \varphi_k + f_z \sin \varphi_k + \frac{R_k}{2} f_{rr} \cos^2 \varphi_k +$$

$$+ R_k f_{rz} \cos \varphi_k \sin \varphi_k + \ldots = \frac{f_k - f_0}{R_k}. \qquad (9.4)$$

If the number of neighboring points M_k is the same as the number of unknowns, the unknowns are determined unambiguously and we write all the coefficients f_r, f_z, f_{rr}, etc., in terms of R_k, φ_k, f_k. Obviously

$$\frac{\partial f}{\partial r} \sim f_r; \quad \frac{\partial f}{\partial z} \sim f_z; \quad \frac{\partial^2 f}{\partial r^2} \sim f_{rr}, \ldots. \qquad (9.5)$$

Hence, by replacing the derivatives in (9.1) by the corresponding expressions from (9.4) and (9.5) we find certain difference equations. In this approach, the problem reduces to the choice of the degree of the interpolation polynomial and to a determination of the role played by each of the neighboring points in the construction of this polynomial, since there are usually more of these points than are necessary. For example, to approximate first derivatives it is obviously sufficient to construct an interpolation function which is linear in terms of $r - r_0$, $z - z_0$:

$$f = f_0 + R (f_r \cos \varphi + f_z \sin \varphi). \qquad (9.6)$$

Using any pair of neighboring points $M_1(R_1, \varphi_1)$, $M_2(R_2, \varphi_2)$ and solving the system of two equations, we find the coefficients f_r and

f_z from

$$f_r = \frac{\begin{vmatrix} \delta f_1 & \sin \varphi_1 \\ \delta f_2 & \sin \varphi_2 \end{vmatrix}}{\sin (\varphi_2 - \varphi_1)}; \quad f_z = \frac{\begin{vmatrix} \cos \varphi_1 & \delta f_1 \\ \cos \varphi_2 & \delta f_2 \end{vmatrix}}{\sin (\varphi_2 - \varphi_1)}, \tag{9.7}$$

where

$$\delta f_{1,2} = (f_{1,2} - f_0)/R_{1,2}. \tag{9.8}$$

Each pair M_1, M_2 leads to particular values of f_r, f_z. There is no reason for assigning a preference to any of these values, in accordance with the principle of the a priori equivalence of neighboring points, which is used in constructing the calculation cell. Accordingly, in order to take account of all the points, we must either increase the degree of the polynomial or use some method to average the values of f_r and f_z obtained for the different pairs M_1, M_2.

The first approach increases the order of the approximation, but the increase is uncontrolled and random, since the number of neighboring points is arbitrary. Furthermore, we are then faced with the meaningless question of what powers of $r - r_0$, $z - z_0$ should be present in expansion (9.3). Finally, the need to solve high-order systems of linear equations greatly complicates the algorithm.

The choice of a method for averaging the functions f_r and f_z is not always a trivial problem. It is necessary to determine which points should be combined into pairs, what weight to assign each pair, etc. The particular solution adopted for these questions affects the stability of the method, although it does not affect the accuracy of the approximation. The stability is related to the effective incorporation of the range of the solution and is governed by precisely these factors.

It is evident that similar problems arise in the approximation of second derivatives.

In order to have a common method for resolving these questions and to simplify the problem considerably, we use the following approach in constructing the difference equations.

Returning to system (9.1), we approximate it locally (within a single calculation cell) by a system of linear equations with constant coefficients like

$$\partial \mathcal{F} / \partial t = D \mathcal{F} + C, \tag{9.9}$$

where D is a linear differential operator. The values of the functions f and their derivatives are contained in the coefficients D and C. We take all these values at the central point of the cell, M_0, calculating them in any way (e.g., that described above), concerning ourselves only with the closeness of the approximation. Accordingly, the values of C and the coefficients of the operator D can be assumed to be known.

The transformation from (9.1) to (9.9) is not single-valued. The partitioning of $\mathcal{D}(\mathcal{F})$ can be carried out in several ways, by distinguishing from $\mathcal{D}(\mathcal{F})$ various differential expressions $D\mathcal{F}$. The purpose of this partitioning and the suitability of the various methods for carrying it out will become clear from the discussion below.

From the theory for systems of linear differential equations with constant coefficients, as in (9.9), we know that it is possible to find equations that give solution in terms of the initial data. In particular, the solution at time $t + \tau$ at point M^0 can be expressed in terms of the values of the solution at time t, i.e., $\mathcal{F}(R, \varphi)$, by means of

$$\mathcal{F}^0 = \iint Q(\tau, R, \varphi) \mathcal{F}(R, \varphi) \, dR d\varphi + C\tau, \tag{9.10}$$

where Q (a matrix of functions) is completely defined by the operator D.

This integral equation cannot be used directly since the functions $\mathcal{F}(R, \varphi)$ are not known. We only have a discrete set of values of this function at the points M_0, M_1, ..., M_N, which form the given calculation cell. It is natural to use $\mathcal{F}(R, \varphi)$ as the interpolation function constructed from these points. If this function is written as

$$P(R, \varphi) = \sum_{i=0}^{N} a_i(R, \varphi) \mathcal{F}_i, \tag{9.11}$$

where $\mathcal{F}_i$ is the value of $\mathcal{F}$ at the point M_1, and this expression is used in (9.10), we find the difference equation

$$\mathcal{F}^0 = \sum_{i=0}^{N} A_i \mathcal{F}_i + C\tau, \tag{9.12}$$

where

$$A_i = \iint a_i (R, \varphi) \, Q \, (\tau, R, \varphi) \, dR d\varphi. \qquad (9.13)$$

Here again we are confronted with the question of constructing the interpolation function. However, the requirements on this function and the method for using it are completely different in this case. It is not necessarily a differentiable or even continuous function; it must simply be an interpolation function which can be integrated. This circumstance means that simple interpolation methods can be used.

The separation of the linear operator D from $\mathscr{D}(\mathscr{F})$ greatly simplifies the solution of many questions, primarily because Eq. (9.10) gives explicitly the range of the solution and the "weights" of the various parts of the R, φ plane. It is obvious, however, that this information is only useful when the linear operator D is a good model of the original operator $\mathscr{D}$. By $D\mathscr{F}$ here we understand the leading terms in $\mathscr{D}(\mathscr{F})$, i.e., the main linear part of the operator.

§10. Calculation Equations

We turn now to the derivation of the calculation equations for a particular problem (§5). Rather than taking the approach of pursuing the principle outlined in the preceding section, we proceed in the following manner: The system is partitioned into parts corresponding to a given physical process (the magnetohydrodynamics or the various dissipative processes). We construct calculation equations for each part and then combine them.

We begin with the magnetohydrodynamics by examining the system of equations

$$\left. \begin{array}{l} \rho \, \dfrac{du}{dt} + H \, \dfrac{\partial H}{\partial r} + \dfrac{\partial p}{\partial r} = 0; \\[2mm] \rho \, \dfrac{dv}{dt} + H \, \dfrac{\partial H}{\partial z} + \dfrac{\partial p}{\partial z} = 0; \\[2mm] \rho \, \dfrac{d\mathscr{E}}{dt} + p \left(\dfrac{\partial u}{\partial r} + \dfrac{\partial v}{\partial z} \right) = 0; \\[2mm] \dfrac{dH}{dt} + H \left(\dfrac{\partial u}{\partial r} + \dfrac{\partial v}{\partial z} \right) = 0; \\[2mm] \dfrac{d\rho}{dt} + \rho \left(\dfrac{\partial u}{\partial r} + \dfrac{\partial v}{\partial z} \right) = 0. \end{array} \right\} \qquad (10.1)$$

It can be assumed that the other terms of original system (5.5)-(5.9) which have been omitted here are incorporated in the right side (C) in (9.9); and it can also be assumed that the terms which are written out correspond to the operator † $\mathscr{D}$. The presence of the term C simply adds a term C_T to difference equation (9.12). The method actually used for calculating that part of C which contains leading (second) derivatives is determined after an examination of all parts of the original system. Separation of (10.1) can be justified by the fact that situations can arise in which dissipative effects are negligible, so that the rank of the corresponding terms is only formal.

We now write

$$w = p + \frac{H^2}{2}; \quad c^2 = \frac{5}{3} T + \frac{H^2}{\rho} , \tag{10.2}$$

and transform system (10.1) to the following form (using $p = \frac{2}{3}\rho\mathscr{E} = \rho T$):

$$\left. \begin{aligned} &\frac{du}{dt} + \frac{1}{\rho} \cdot \frac{\partial w}{\partial r} = 0; \\ &\frac{dv}{dt} + \frac{1}{\rho} \cdot \frac{\partial w}{\partial z} = 0; \\ &\frac{dw}{dt} + \rho c^2 \left(\frac{\partial u}{\partial r} + \frac{\partial v}{\partial z} \right) = 0; \\ &\frac{1}{H} \cdot \frac{dH}{dt} = \frac{3}{2T} \cdot \frac{dT}{dt} = \frac{1}{\rho c^2} \cdot \frac{dw}{dt} ; \\ &\frac{d\rho}{dt} + \rho \left(\frac{\partial u}{\partial r} + \frac{\partial v}{\partial z} \right) = 0. \end{aligned} \right\} \tag{10.3}$$

The coefficients of this system are assumed to be constant and equal to their values at the central point of the calculation cell, M_0. Then the first three equations in (10.3) form a closed system with constant coefficients like (9.9); here $\mathscr{F} = \{u, v, w\}$ and C = 0. The matrix $Q(\tau, R, \varphi)$ for this system has been derived in [29]. We do not reproduce it here, since we use Eq. (9.10), after a slight conversion involving an integration by parts. Specifically, the in-

---

†For convenience in the discussion below we modify the notation slightly: $\nu_r \rightarrow u$, $\nu_z \rightarrow v$, $H_\varphi \rightarrow H$.

tegral equations corresponding to this case are written in the form

$$
\begin{aligned}
u^0 = u_0 &- \frac{1}{2\pi\rho_0 c_0} \int_0^{2\pi} d\varphi \int_0^{\tau c_0} \frac{\partial w}{\partial r} \cdot \frac{R\,dR}{\sqrt{\tau^2 c_0^2 - R^2}} + \\
&+ \frac{\tau c_0}{2\pi} \int_0^{2\pi} \cos\varphi\, d\varphi \int_0^{\tau c_0} \left(\frac{\partial u}{\partial r} + \frac{\partial v}{\partial z} \right) \frac{dR}{\sqrt{\tau^2 c_0^2 - R^2}}; \\
v^0 = v_0 &- \frac{1}{2\pi\rho_0 c_0} \int_0^{2\pi} d\varphi \int_0^{\tau c_0} \frac{\partial w}{\partial z} \cdot \frac{R\,dR}{\sqrt{\tau^2 c_0^2 - R^2}} + \\
&+ \frac{\tau c_0}{2\pi} \int_0^{2\pi} \sin\varphi\, d\varphi \int_0^{\tau c_0} \left(\frac{\partial u}{\partial r} + \frac{\partial v}{\partial z} \right) \frac{dR}{\sqrt{\tau^2 c_0^2 - R^2}}; \\
w^0 = w_0 &- \frac{\rho_0 c_0}{2\pi} \int_0^{2\pi} d\varphi \int_0^{\tau c_0} \left(\frac{\partial u}{\partial r} + \frac{\partial v}{\partial z} \right) \frac{R\,dR}{\sqrt{\tau^2 c_0^2 - R^2}} + \\
&+ \frac{1}{2\pi c_0} \cdot \frac{\partial}{\partial \tau} \int_0^{2\pi} d\varphi \int_0^{\tau c_0} (w - w_0) \frac{R\,dR}{\sqrt{\tau^2 c_0^2 - R^2}},
\end{aligned} \qquad (10.4)
$$

where R, φ are again the polar coordinates with origin at M_0; ρ_0, c_0, u_0, v_0, w_0 are the values at the point M_0; and u^0, v^0, w^0 are the values of these functions at the time $t + \tau$ at the point M^0 with coordinates

$$
r^0 = r_0 + u_0\tau; \quad z^0 = z_0 + v_0\tau, \qquad (10.5)
$$

to which M_0 moves over time step τ. The difference between (10.4) and (9.10) lies in the fact that here the integrals do not contain the functions themselves, but their derivatives with respect to r and z; this difference is obviously unimportant.

We must now choose an interpolation method in order to calculate the right side of (10.4). The simplest method is the following: From each pair of neighboring points M_n and M_{n+1} with sequential indices and from the central point, M_0, we construct linear functions like that in (9.6) (where f represents u, v, or w); these are only used in the sector $M_n M_0 M_{n+1}$ (Fig. 3). Equations for determining the coefficients of these functions — Eqs. (9.7), (9.8) (the subscripts "1" and "2" must obviously be written "n" and "n + 1")— are already available. Substituting (9.6) into (10.4) and integrating,

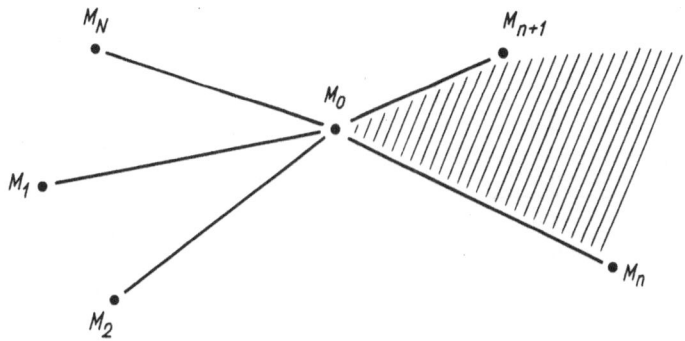

Fig. 3. Calculation cell for the point M_0.

we find the difference equations

$$u^0 = u_0 - \frac{\tau}{2\pi\rho_0} \sum w_r \,\Delta\varphi + \frac{\tau c_0}{4} \sum (u_r + v_z) \,\Delta \sin\varphi;$$

$$v^0 = v_0 - \frac{\tau}{2\pi\rho_0} \sum w_z \,\Delta\varphi - \frac{\tau c_0}{4} \sum (u_r + v_z) \,\Delta \cos\varphi;$$

$$w^0 = w_0 - \frac{\tau\rho_0 c_0^2}{2\pi} \sum (u_r + v_z) \,\Delta\varphi +$$

$$+ \frac{\tau c_0}{4} \sum (w_r \,\Delta \sin\varphi - w_z \,\Delta \cos\varphi),$$

(10.6)

where the summation is carried out over all sectors; $\Delta\varphi$, $\Delta \cos\varphi$, and $\Delta \sin\varphi$ denote the increments in φ, $\cos\varphi$, and $\sin\varphi$, respectively, for each sector; f_r, f_z are calculated with the help of (9.7), (9.8) (they vary from sector to sector).

Analysis of Eqs. (10.6) shows that the first sum in each equation is the obvious approximation of the corresponding differential equations of system (10.3). Averaging of f_r, f_z is carried out with "weights" equal to the angular intervals $\Delta\varphi$ of the corresponding sectors. The second sums in (10.6) do not have differential counterparts in (10.3). However, it is premature to conclude that these sums are extraneous. In the first place, as the dimensions of the cell and τ approach zero these terms vanish before the others do; i.e., they are of higher order in τ, R_n and do not affect the accuracy of the approximation. In the second place, it turns out that these sums have an important effect on the stability of difference equations (10.6). This feature can be shown by examining the particular case in (10.6) in which the array of cal-

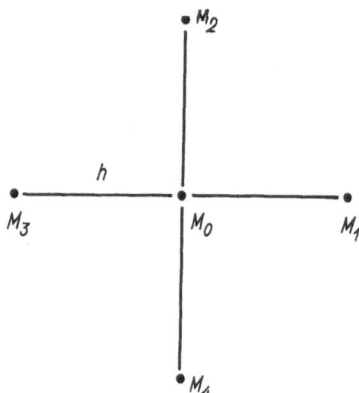

Fig. 4. Calculation cell for the case of a regular grid.

culation points forms a regular grid with steps h along r and z and in which the calculation cell is a set of five points that form the configuration shown in Fig. 4. In this case Eq. (10.6) degenerates into a difference system whose stability can be studied by the usual methods, and which leads to the condition

$$2\tau c_0 \leqslant h. \tag{10.7}$$

On the other hand, if the second sums are discarded, a similar analysis leads to the condition $\tau < \text{const } h^2$, which limits the applicability of this scheme.

Difference equations (10.6) can also be found through a direct approximation of (10.3), if we guess at the extraneous second sums. When the approach described in § 9 is taken this result is achieved automatically.

The other equations of system (10.3), which have not been discussed up to this point, can obviously be replaced by the difference equations

$$\left. \begin{aligned}
H^0 &= H_0 + \frac{H_0}{\rho_0 c_0^2} (w^0 - w_0); \\
T^0 &= T_0 + \frac{2}{3} \cdot \frac{T_0}{\rho_0 c_0^2} (w^0 - w_0); \\
\rho^0 &= \rho_0 - \frac{\tau \rho_0}{2\pi} \sum (u_r + v_z) \Delta\varphi.
\end{aligned} \right\} \tag{10.8}$$

Accordingly, the difference equations for the hydrodynamic part of (10.3) of this system have been constructed. The fact that Eqs.

(10.6) and (10.8), together with (10.5), approximate the system of differential equations in (10.3) is almost obvious; this point is easily checked.

On the other hand, we cannot specify any analytic methods for studying the stability of difference problem (10.6). It is necessary to restrict the study to particular cases in which the calculation points are in a regular arrangement. The simplest of these cases gives us condition (10.7); the others lead to analogous conditions. Our approach is as follows: Since the method used to construct the calculation cell leads to (by virtue of the condition $\Delta\varphi \lessgtr \pi/2$) cells which are "no worse" than that shown in Fig. 4, from the standpoint of taking account of the ranges of the solution, we expect that the stability condition for difference problem (10.6) can be found by generalizing condition (10.7), specifically, by imposing the inequality

$$2\tau c_0 \leqslant R_n \tag{10.9}$$

for each neighboring point.

We turn now to original system (5.5)-(5.9) and examine that part of the system which describes dissipative effects associated with the viscosity. Retaining the leading terms, we find the system of equations

$$\left.\begin{aligned}
\rho\,\frac{\partial u}{\partial t} &= \frac{4}{3}\,\eta\,\frac{\partial^2 u}{\partial r^2} + \eta\,\frac{\partial^2 u}{\partial z^2} + \frac{1}{3}\,\eta\,\frac{\partial^2 v}{\partial r\,\partial z}\,; \\
\rho\,\frac{\partial v}{\partial t} &= \frac{1}{3}\,\eta\,\frac{\partial^2 u}{\partial r\,\partial z} + \eta\,\frac{\partial^2 v}{\partial r^2} + \frac{4}{3}\,\eta\,\frac{\partial^2 v}{\partial z^2},
\end{aligned}\right\} \tag{10.10}$$

where, as usual, $\rho = \rho_0$ and $\eta = \eta_0$ are constants. It can be shown that the integral equation of the type in (9.10) that corresponds to this system is

$$\left.\begin{aligned}
u^0 &= u_0 + \frac{1}{2\pi}\int_0^{2\pi} d\varphi \int_0^\infty \left(\cos\varphi \exp\left(-\frac{3\rho R^2}{16\,\eta\tau}\right)\left(\frac{\partial u}{\partial r} + \frac{\partial v}{\partial z}\right) + \right. \\
&\quad \left. + \sin\varphi \exp\left(-\frac{\rho R^2}{4\eta\tau}\right)\left(\frac{\partial u}{\partial z} - \frac{\partial v}{\partial r}\right)\right) dR; \\
v^0 &= v_0 + \frac{1}{2\pi}\int_0^{2\pi} d\varphi \int_0^\infty \left(\sin\varphi \exp\left(-\frac{3\rho R^2}{16\eta\tau}\right)\left(\frac{\partial u}{\partial r} + \frac{\partial v}{\partial z}\right) - \right. \\
&\quad \left. - \cos\varphi \exp\left(-\frac{\rho R^2}{4\eta\tau}\right)\left(\frac{\partial u}{\partial z} - \frac{\partial v}{\partial r}\right)\right) dR.
\end{aligned}\right\} \tag{10.11}$$

The interpolation method used in the study of the hydrodynamic part of the system cannot be used here, since the difference equations found in this manner do not approximate the system of differential equations in (10.10). This result is not surprising since it is necessary to approximate second derivatives, and when a linear interpolation function is used these derivatives are included in the neglected interpolation error. Accordingly, we must use at least a quadratic interpolation, constructing functions like

$$f = f_0 + f_r \, (r - r_0) + f_z \, (z - z_0) + f_{rr} \frac{(r - r_0)^2}{2} +$$
$$+ f_{rz} \, (r - r_0) \, (z - z_0) + f_{zz} \frac{(z - z_0)^2}{2} \tag{10.12}$$

with five undetermined coefficients f_r, f_z, f_{rr}, f_{rz}, f_{zz}. To determine these coefficients we must know the values of f at five neighboring points, which give us five equations like

$$f_r \cos \varphi_n + f_z \sin \varphi_n + \frac{R_n}{2} \, (\cos^2 \varphi_n \, f_{rr} + \sin 2\varphi_n \, f_{rz} +$$
$$+ \sin^2 \varphi_n f_{zz}) = \frac{f_n - f_0}{R_n}, \tag{10.13}$$

these are equations for f_r, f_z, f_{rr}, f_{rz}, f_{zz}. Clearly, the number of neighboring points N that form the given calculation cell cannot be smaller than five. We assume at first that this number is precisely five. Then polynomial (10.12) is then determined unambiguously. Substituting (10.12) into (10.11), and carrying out the integration, we find

$$\left. \begin{array}{l} u^0 = u_0 + \dfrac{1}{\rho} \, \tau \left(\dfrac{4}{3} \, \eta u_{rr} + \eta u_{zz} + \dfrac{1}{3} \, \eta v_{rz} \right); \\[3mm] v^0 = v_0 + \dfrac{1}{\rho} \, \tau \left(\dfrac{1}{3} \, \eta u_{rz} + \eta v_{rr} + \dfrac{4}{3} \, \eta v_{zz} \right), \end{array} \right\} \tag{10.14}$$

which is an obvious and trivial result. Each derivative of system (10.10) is replaced by a corresponding coefficient of the interpolation polynomial. The motivation here is to make use of the same expression, (10.12), over the entire φ range from 0 to 2π. The quantities f_r and f_z do not appear in Eq. (10.14); in the solution of the system of five linear equations in (10.13) we can then first eliminate f_r and f_z, thereby simplifying the situation.

The difference equations in (10.14) obviously approximate the system of differential equations in (10.10). Study of the stability for the degenerate case (a regular rectangular grid with a step h) leads to the condition

$$\frac{16}{3} \frac{\eta}{\rho} \tau \leqslant h^2.$$

For a calculation cell of general shape it is reasonable to require that

$$\frac{16}{3} \frac{\eta}{\rho} \tau \leqslant R_n^2 \qquad (10.15)$$

and that the angular density of neighboring points be sufficiently high,

$$\Delta\varphi < \pi/2. \qquad (10.16)$$

This last condition also ensures that the condition $N \geq 5$ is satisfied.

We now turn to the general case of an arbitrary number of neighboring points (this number is of course equal to or larger than five). In order to break up the set of neighboring points into sequences of five points we must choose N to be a multiple of four (or equal to five). We cannot work from this basis. Nor is there a basis for a nonequivalent incorporation of the points involving the construction of special equations for certain of these points. Without going into all the possibilities here, we proceed immediately to the final algorithm that has been selected.

A set of five neighboring points is needed to construct the interpolation function; it is also sufficient if these five points form a figure that surrounds the central point of the cell. We break up all the sets of neighboring points into such groups of five in such a way that each point is only included in one of these groups. On the basis of each group of five points we calculate the corresponding values of f_{rr}, f_{rz}, f_{zz} and then average these values. The average values of f_{rr}, f_{rz}, f_{zz} obtained in this manner are used in Eqs. (10.14).

There will of course be some nonequivalence of the neighboring points, but it will be governed by the particular arrangement of these points. For example, the "deficit" points which, taken separately, represent a particular direction φ are included in a larger number of groups of five and thus have a higher relative weight. This is to be expected and is valid.

Finally, we consider the heat-conduction equation

$$\frac{3}{2} \rho \frac{\partial T}{\partial t} = \nu \left(\frac{\partial^2 T}{\partial r^2} + \frac{\partial^2 T}{\partial z^2} \right)$$

and the diffusion equation for the magnetic field,

$$\frac{\partial H}{\partial t} = \lambda \left(\frac{\partial^2 H}{\partial r^2} + \frac{\partial^2 H}{\partial z^2} \right).$$

The corresponding integral equations (9.10) are given by the familiar Poisson integral. The interpolation questions are resolved by a method like that just described, which leads to the difference equations

$$T^0 = T_0 + \tau \frac{2\nu}{3\rho} \left(\overline{T}_{rr} + \overline{T}_{zz} \right);$$

$$H^0 = H_0 + \tau\lambda \left(\overline{H}_{rr} + \overline{H}_{zz} \right),$$

where $\overline{f}_{rr}$ and $\overline{f}_{zz}$ are the average values of the coefficients of the interpolation polynomials; we also obtain the stability conditions

$$\frac{8}{3} \cdot \frac{\nu}{\rho} \tau \leqslant R_n^2; \quad 4\lambda\tau \leqslant R_n^2. \tag{10.17}$$

Accordingly, all parts of the original system of differential equations (5.5)-(5.9) which can be considered the most important parts have been considered. All the other parts, containing minor terms, are approximated by a single method. The values of the functions are taken at the point M_0 and the values of the derivatives are replaced by the averages over the sectors,

$$\left. \begin{aligned} \frac{\partial f}{\partial r} &\sim \overline{f}_r = \frac{1}{2\pi} \sum f_r \, \Delta\varphi; \\ \frac{\partial f}{\partial z} &\sim \overline{f}_z = \frac{1}{2\pi} \sum f_z \, \Delta\varphi, \end{aligned} \right\} \tag{10.18}$$

where f_r, f_z are calculated for each sector on the basis of Eqs. (9.7) and (9.8).

Combining these calculation equations, we have finally

$$
\left.
\begin{aligned}
u^0 &= u + \frac{\tau}{\rho}\Big[-w_r + \rho\Sigma_1 + \eta\Big(\frac{4}{3}u_{rr} + u_{zz} + \frac{1}{3}v_{rz}\Big) - \\
&\quad - \frac{H^2}{r} + \eta_r\Big(\frac{4}{3}u_r - \frac{2}{3}v_z - \frac{2}{3}\cdot\frac{u}{r}\Big) + \eta_z(u_z + v_r) - \\
&\quad - \frac{4}{3}\cdot\frac{\eta}{r}\Big(u_r - \frac{u}{r}\Big)\Big]; \\[4pt]
v^0 &= v + \frac{\tau}{\rho}\Big[-w_z + \rho\Sigma_2 + \eta\Big(\frac{1}{3}u_{rz} + v_{rr} + \frac{4}{3}v_{zz}\Big) + \\
&\quad + \eta_z\Big(\frac{4}{3}v_z - \frac{2}{3}u_r - \frac{2}{3}\cdot\frac{u}{r}\Big) + \eta_r(u_z + v_r) + \\
&\quad + \frac{\eta}{r}\Big(v_r + \frac{1}{3}u_z\Big)\Big]; \\[4pt]
T_0 &= T + \tau\frac{2}{3\rho}\Big[-\rho T(u_r + v_z) + \frac{T}{c^2}\Sigma_3 + \nu(T_{rr} + T_{zz}) - \\
&\quad - \frac{\rho u}{r} + T_r\Big(v_r + \frac{\nu}{r}\Big) + T_z v_z + \lambda\Big(\Big(H_r + \frac{H}{r}\Big)^2 + H_z^2\Big) + \\
&\quad + 2\eta\Big(u_r^2 + v_z^2 + \Big(\frac{u}{r}\Big)^2\Big) - \\
&\quad - \frac{2}{3}\eta\Big(u_r + v_z + \frac{u}{r}\Big)^2 + \eta(u_z + v_r)^2\Big]; \\[4pt]
H^0 &= H + \tau\Big[-H(u_r + v_z) + \frac{H}{\rho c^2}\Sigma_3 + \lambda(H_{rr} + H_{zz}) + \\
&\quad + \lambda_r\Big(H_r + \frac{H}{r}\Big) + \lambda_z H_z + \frac{\lambda}{r}\Big(H_r - \frac{H}{r}\Big)\Big]; \\[4pt]
\rho^0 &= \rho - \tau\rho\Big(u_r + v_z + \frac{u}{r}\Big); \\[4pt]
r^0 &= r + \tau u; \\[2pt]
z^0 &= z + \tau v.
\end{aligned}
\right\} \qquad (10.19)
$$

To simplify the notation we have omitted indices and averaging signs. Here $f^0(=u^0, v^0, \ldots)$ represents the value at the point $M^0(r^0, z^0)$ at time $t + \tau$. On the right sides, f represents f_0, i.e., the value at point M_0; f_r and f_z represent $\bar{f}_r$ and $\bar{f}_z$, the averages over the sectors [(10.18)]; and f_{rr}, f_{rz}, and f_{zz} represent $\bar{f}_{rr}$, $\bar{f}_{rz}$, and $\bar{f}_{zz}$, the averages over series of five points. Finally, the quantities $\Sigma_{1,2,3}$ are calculated as the sums over the sectors, in accordance with (10.6):

$$
\left.
\begin{aligned}
\Sigma_1 &= \frac{c}{4} \sum (u_r + v_z) \, \Delta \sin \varphi; \\[2mm]
\Sigma_2 &= -\frac{c}{4} \sum (u_r + v_z) \, \Delta \cos \varphi; \\[2mm]
\Sigma_3 &= \frac{c}{4} \sum (w_r \, \Delta \sin \varphi - w_z \, \Delta \cos \varphi).
\end{aligned}
\right\}
\qquad (10.20)
$$

Also, the quantities w and c are determined from (10.2).

§11. Difference Formulation of the Boundary Conditions

If the boundary condition gives the desired function directly, no problem arises. In most cases, however, the boundary condition simply provides additional information to supplement the differential equations of the problem. From the standpoint of the principle used here to find the calculation equations, the difference between the interior points and the boundary points reduces to the fact that the boundary points do not have a set of neighboring points sufficient for a correct construction of an interpolation polynomial and for taking account of the region of dependence. We make up for this insufficiency by making use of the boundary conditions.

The boundary conditions in (4.5) at $z = 0$ and $z = z_0$ are symmetry conditions. We introduce fictitious points M' which are the images of the real calculation points M across these boundaries. At a fictitious point we have

$$
u' = u; \; v' = -v; \; T' = T; \; \rho' = \rho; \; H' = H, \qquad (11.1)
$$

and the coordinates of the point are

$$
r' = r, \; z' = -z \qquad (11.2)
$$

in mapping across $z = 0$ or

$$
r' = r, \; z' = 2z_0 - z \qquad (11.3)
$$

in mapping across $z = z_0$. Constructing the corresponding fictitious point for each calculation point near the boundary we can convert all the real points into points which are formally interior points. The same approach makes it a simple matter to formulate the boundary conditions at the z axis, (4.11). In this case the mapping is carried out on the basis of the equations

$$r' = -r; \ z' = z; \ u' = -u; \ v' = v; \ T' = T; \ \rho' = \rho; \ H' = -H. \quad (11.4)$$

If a point falls directly on the z axis it is necessary to calculate the ratios f/r at $f = 0$, $r = 0$. An indeterminate form of this type can be resolved by setting

$$\left. \frac{f}{r} \right|_{r=0} = \bar{f}_r, \quad (11.5)$$

where $\bar{f}_r$ is calculated in the standard way (by averaging over sectors).

Accordingly, the introduction of fictitious points resolves the question of the boundary conditions at $r = 0$, $z = 0$, $z = z_0$. We turn now to the conditions at the outer boundary, $r = R(z, t)$, in (4.8) and (4.9). To formulate these conditions we actually make use of the same principle, but in a different form.

We first note that the outer boundary moves and is represented at any instant by a sequence of calculation points. Obviously, this sequence must be dense enough to make the interior calculation points neighboring points and also dense enough for a satisfactory description of the boundary curve.

Figure 5 shows an example of a violation of the first condition. Points M_1 and M_2 are boundary points, while M_0 is an interior point. Obviously, we must insert another boundary point M_{12} between M_1 and M_2 (by interpolation on the basis of M_1 and M_2).

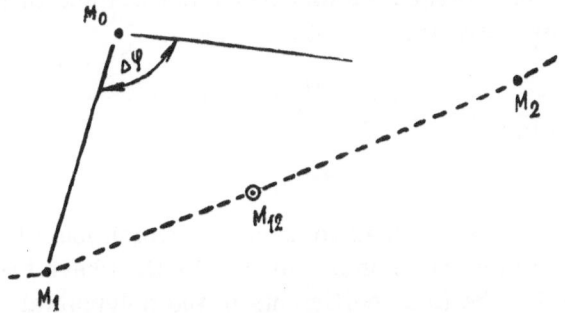

Fig. 5. Insertion of a boundary point.

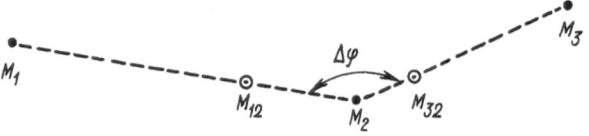

Fig. 6. Smoothing of the boundary.

The second condition is formulated as a restriction of the "difference curvature" of the boundary. Consider three sequential boundary points M_1, M_2, M_3 (Fig. 6), which form an angle $\Delta\varphi$. We impose the condition

$$|\Delta\varphi - \pi| < \Delta_\Gamma\varphi, \qquad (11.6)$$

where $\Delta_\Gamma\varphi$ is given (in the calculation we assume $\Delta_\Gamma\varphi \sim \pi/10$). If condition (11.6) does not hold, we assume that the difference curvature is large; to reduce it we "smooth" the boundary, replacing point M_2 by the pair of points M_{12} and M_{32}, which are found through interpolations in terms of M_1, M_2, and M_2, M_3, respectively. As a result, $\Delta\varphi$ is always approximately equal to π.

The method for constructing a calculation cell for a boundary point only differs from the standard method in the details. Specifically, the selection is carried out only from those points which lie within the angle $M_1M_2M_3$; M_1 and M_3 are necessarily among the neighbors. Since $\Delta\varphi \sim \pi$, the difference cell contains at least three neighboring points.

The boundary conditions are formulated by imposing the appropriate requirements on the interpolation polynomials. The quantities n and s denote the directions normal and tangent to the boundary, respectively. Condition (4.9),

$$\partial T/\partial n = 0, \qquad (11.7)$$

and the condition

$$\partial^2 T/\partial n\partial s = 0, \qquad (11.8)$$

which follows from it, give two equations which must be satisfied by the interpolation polynomial for T. We thus have two equations for determining the five coefficients of the polynomial: T_r, T_z, T_{rr}, T_{rz}, T_{zz}. The neighboring boundary points M_1 and M_3 give two more equations. Using one of the interior points we can close the system. Accordingly, in this case the group of five points is governed by an interior point. The final values of the coefficients in the interpolation polynomial are chosen to be the averages over all the groups of five used.

The same principle is used in constructing the interpolation polynomials for the functions u and v. Boundary conditions (4.8) (and the conditions which follow from them) give us four equations,

$$\sigma_{nn} = \sigma_{ns} = 0; \tag{11.9}$$

$$\partial \sigma_{nn}/\partial s = \partial \sigma_{ns}/\partial s = 0, \tag{11.10}$$

with ten coefficients u_r, v_r, u_z,..., v_{zz}. The pair of neighboring boundary points M_1, M_3 gives us four more equations and any interior point can be used to close the system.

The calculation of the values of $\bar{f}_r$, $\bar{f}_z$ corresponding to the less important terms of the system is carried out by averaging over sectors (as for the interior calculation points). The quantities f_r, f_z for the exterior sector $M_1 M_2 M_3$ are taken to be the average values of these quantities over the interior sectors.

In the solution of the systems of five and ten equations associated with (11.7)-(11.10) we first eliminate f_r, f_z and then calculate f_{rr}, f_{rz}, f_{zz}; this approach simplifies the situation.

Accordingly, to calculate the values of f at the boundary points (except for H, which is given), we make use of the same calculations (10.19). The only difference is in the method for finding the values of $\bar{f}_{rr}$, $\bar{f}_{rz}$, $\bar{f}_{zz}$.

§12. Formulation of the Calculation Problem

In carrying out calculations for a shock wave moving through a zero-temperature plasma we use an artificial viscosity; specifically, in the viscosity coefficient η from (5.12) we introduce a constant correction:

$$\eta = B^* T^{5/2} + B_1^*. \tag{12.1}$$

Furthermore, in order to eliminate the singularity $\lambda \to \infty$ in the limit $T \to 0$, the following equation is used for λ in place of (5.12):

$$\lambda = E^*/(T^{3/2} + E_1^*). \tag{12.2}$$

The stability conditions in (10.9), (10.15), and (10.17) impose restrictions on the relation between the time step τ and the distance to the neighboring points that form the calculation cell, R_n. The formal method we use makes it possible to carry out the calculation with any specified value of τ, if we eliminate from consideration those points which are "too close." To do this it is sufficient to introduce an additional requirement in the selection condition:

$$R_n \geqslant h, \tag{12.3}$$

where

$$h = 2 \max \left(c\tau, \ \sqrt{\frac{2\nu}{3\rho}\tau}, \ \sqrt{\frac{4\eta}{3\rho}\tau}, \ \sqrt{\lambda\tau} \right). \tag{12.4}$$

Points which do not satisfy (12.3) are not included among the neighboring points.

The presence of a "blind" h-vicinity for each calculation point can only lead to ambiguities in two cases.

First, it may turn out that the calculation cell of an interior point is not closed because a boundary point falls in its h-vicinity. If the point at which this occurs is not of any particular interest, such an interior point is simply discarded; otherwise, the value of h (i.e., τ) is reduced.

Second, it may happen that two boundary points are separated by a distance smaller than h in (12.4). If we allow a boundary point to freely cross the h-vicinity of the other boundary point, then the various questions associated with the determination of the difference curvature of the boundary are greatly complicated. We therefore eliminate such point pairs like these, replacing them by a single point (the average) (or by reducing h and thus τ). In certain cases enlargement of the boundary step comes into conflict with the reduction of this step which is required in order to smooth the boundary (Fig. 6). The effective result of this conflict is the cutoff of higher harmonics in the boundary curve.

In all other cases the presence of h-vicinities can actually make certain calculation points extraneous, but this does not necessarily present a problem in the calculations.

The time step τ is chosen so that the value of h in (12.4) does not exceed a given value h_{max}. Specifically, in each time layer we calculate the quantities

$$m_1 = 2 \max_M c; \qquad\qquad (12.5)$$

$$m_2 = 4 \max_M (\lambda, \; 2\nu/3\rho, \; 4\eta/3\rho), \qquad\qquad (12.6)$$

where the maximum value is determined from all the calculation points. In the calculation of the next layer we use

$$\tau = \min \left(\frac{h_{max}^2}{m_1}, \; \frac{h_{max}^2}{m_2} \right). \qquad\qquad (12.7)$$

This method makes it possible to keep the step τ under control and to choose it on the basis of the parameters in the region of most importance.

Chapter 3

MHD ANALYSIS OF THE Z PINCH

§13. A Few Comments

In this chapter we analyze formation of the plasma focus in a pinch on the basis of the MHD approximation which contains the dissipative processes that are treated in §1. The numerical method used for the solution is described in Chap. 2. A simplified, essentially dissipationless, model for the plasma focus is given in [9], and we refer to this model for comparison at certain points. We recall that the problem being solved here, which is formulated in §§4 and 5, describes the main stage of the discharge. In the initial stage this theoretical model is of no particular interest.

We also discuss certain related physical questions which go beyond the scope of the MHD problem formulated here. We discuss the conclusions reached on the basis of the linear theory for the instability of the plasma—vacuum boundary, including questions dealing with the preservation of the axial symmetry of the pinch. It turns out that dissipative effects — primarily the plasma viscosity —

play an important role in this problem. We also study the radiation emitted from the plasma, taking account of specified concentrations of multiply charged impurities such as C and Ar. The estimates found here imply a substantial total energy loss for the parameter values typical of the plasma focus. We offer certain arguments (mentioned in part in [9]) regarding the mechanism responsible for neutron production. Although the quantitative estimates of the neutron yield obtained on the basis of the thermonuclear mechanism correspond approximately to the experimental data, this complicated questions requires further study, especially since the MHD model is not valid in the plasma-focus stage.

The calculations show that the formation of the plasma focus does not lead to an equilibrium MHD configuration. The continuous outflow of plasma in the axial direction is accompanied by a progressive reduction in the radius of the plasma cloud and the calculated parameters of the focus ultimately reach values such that the MHD model becomes meaningless. It is concluded that even with dissipative effects the MHD approximation is not capable of a complete theoretical descrption of the plasma focus, although it does apply in stages other than the focus stage. For this reason, it is not possible to calculate the lifetime of the focus. Other important characteristics of the process, on the other hand, are described quite well by the MHD theory. Among these are the dynamics of the focusing of the shock wave and the current sheet, the axial outflow of plasma from the pinch, the repeated, cascade nature of the compression of the plasma near the axis, the formation of the focus, and the properties of the focus in this stage.

The series of numerical calculations which have been carried out, including those of [9, 30], essentially exhaust the possibilities of the MHD model for describing this phenomenon. Further refinement of theory for the focus requires the development of an appropriate kinetic model for the plasma, primarily for the plasma ions. No such calculation has yet been reported. The only work in this direction which can be cited is the interesting recent attempt by Potter and Haines [31] to develop a one-dimensional kinetic model for the process; we note, however, that the axial outflow of plasma from the focus cannot be treated correctly with a one-dimensional model.

In discussing the calculated results we will not make a systematic comparison with experiment; this comparison is made in [4, 7, 27, 28].

We will discuss two calculations which differ primarily in the definition of the effective plasma conductivity. In the first approach (version A, summarized in [7]) we take $\sigma_e = 10^{-2}\sigma_\perp$, in accordance with the analysis in §1. In the second version (B) we take $\sigma_e = \sigma_\perp$; i.e., we set the plasma conductivity equal to the classical transverse conductivity in a strong magnetic field. The results found in version B turn out to agree with those found on the basis of the dissipationless model [9], while the results found on the basis of version A are very different.

All the constants involved in the formulation of the physical problem and the units of measurement are given in §5. We also take (§12) $B_1^* = 0.003$, $E_{1A}^* = 0.1$, and $E_{1B}^* = 0.01$. In accordance with the experimental data, the initial shape of the current shell is given by

$$R = 1 + 0.312 \sin^2 (\pi z/2z_0). \tag{13.1}$$

The total number of calculation points is about 1000. The points are first arranged uniformly, becoming denser near the lower electrode. Toward the end of the calculation the number of calculation points increases to about 1500, because the shell assumes a more complicated shape and because of the need to carry out the calculations in more detail for the focus region.

§14. Collapse of the Shock Wave and the Current Sheet to the Axis of the System

The first stage of this collapse is a regular process — the contraction of the plasma, primarily near the lower electrode, involving the formation of a narrow constriction in this region. Figure 7 gives a general idea of this process; it shows the positions of the outer boundary at the initial time and at later times for both calculation versions (A and B).

Figure 8 shows distributions of the parameters ρ, v_r, T, and H_φ at $z \approx 0$ for three times. Only the results for version B are

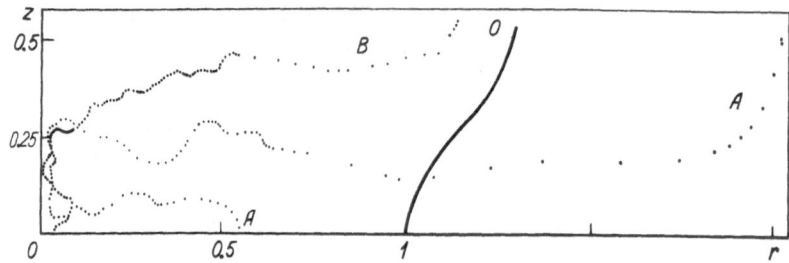

Fig. 7. Configuration of the boundary at various times. 0) Initial configuration
(t = 0); A) near the end of the process, for version A (t_A = 1.29); B) near the end
of the process, for version B (t_B = 0.83).

shown here, since these results are more realistic in this stage of
the process. On the whole, the calculated results agree with the
one-dimensional theory [1]. The shock wave forms at r ≳ 0.76 and
has an average velocity of 1.3 (i.e., $1.1 \cdot 10^7$ cm/sec), which is
nearly constant. The initial magnetic field penetrates far into the
plasma, but its contribution is small; at the shock front the magnetic pressure is about 0.2 of the gas pressure (at t = 0.2475).

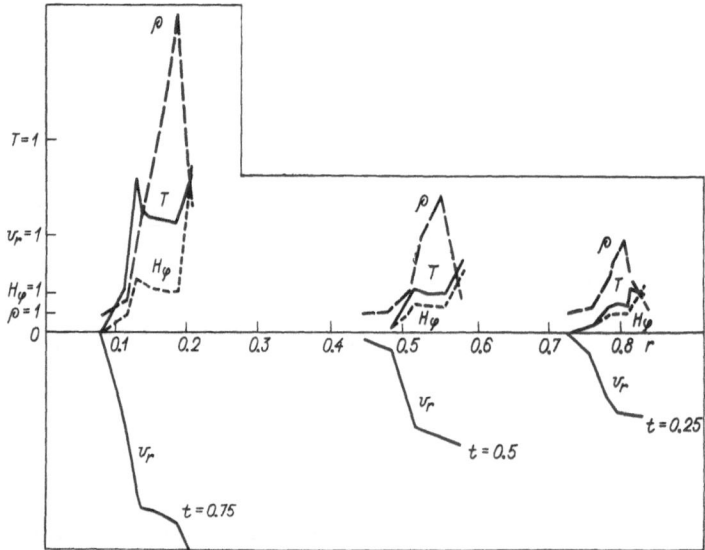

Fig. 8. Collapse of the shock wave and the current sheet to the axis.
Profiles of ρ, v_r, T, and H_φ at three times (version B).

There are two clearly defined peaks in the derivative of the magnetic field. The distance between the shock front and the current sheet remains very nearly constant, never exceeding 0.1 (i.e., 1.3 cm). The thickness of the shock front and the sheet are smaller than this distance by at least a factor of five.

In version A we find the following characteristic differences: The sheet thickness is larger by a factor of three or four (not by a factor of ten, since the Joule heating of the sheet is stronger). The magnetic field cannot be assumed to be frozen in. The collapse is more sluggish, especially toward the end.

There is a qualitative difference between these results and those found by Potter [30]; Potter does not introduce a steady-state shock front before the focus stage itself. The outward manifestations are a slight compression of the plasma within the current sheet (the compression factor is only a few units) and a "supersonic" motion of the sheet (the ratio of the plasma velocity to the sound velocity in [30] is six, while the present results for $t = 0.7457$ yield a ratio $v/c_S = v(5T/3)^{-1/2} = 1.56$; for an extremely strong shock wave this ratio would be $v/c_S \approx 1.34$). The differences between these results are apparently due primarily to the different geometrical arrangements.

The first stage of the process lasts different times in versions A and B ($1.1 - 0.8 = 0.3$). To determine the reason for this difference we examine Fig. 9, which shows the time dependence of the radius of the external boundary, $R(t)$, for $z = 0$ for the two versions. Also shown here is the total current. Evidently the difference between the collapse times is due primarily to a difference in the initial times. At the lower plasma conductivity (version A), the current sheet only begins to move inward after a time $\Delta t \approx 0.2$, which is expended in the formation of the sheet and the shock wave. Accordingly, the difference can be attributed, to a large extent, to the simplified choice of initial conditions, in particular, the assumption that the current sheet is at rest at $t = 0$. From the physical standpoint, this delay before the current sheet begins to move into the plasma is due to the reduced conductivity; in this formulation of the problem (with an "intermediate start") this delay should not be taken literally (roughly speaking, it should be taken into account in an analysis of the detachment of the current sheet from the insulator at the lower part of the chamber; see [1]).

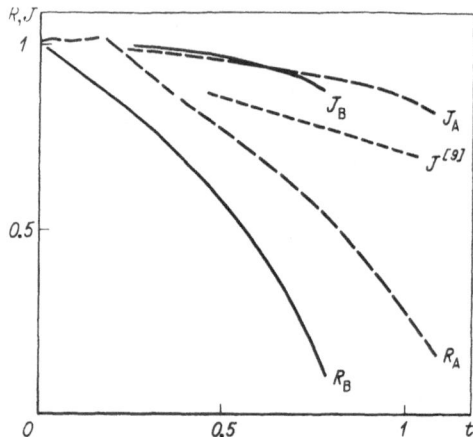

Fig. 9. The time dependence R(t) at z = 0 according to versions A and B. Also shown here are the corresponding curves of J(t), including the approximation of the current in the dissipationless calculation of J [9].

The collapse time in the dissipationless model is $t_{[9]} = 1.05$ in the present units, that is, it is essentially the same as $t_A = 1.1$, although physically it should be closer to $t_B = 0.8$. This agreement is fortuituous, due to the reduction of the current in the calculations of [9] (see Fig. 9), the cruder procedure of these calculations, etc.

The formation of the sheet and of the shock wave is accompanied by an outward motion of the sheet boundary, as shown in Fig. 10, which corresponds to the results found in version A at earlier times. In this version, for z > 0.3 the external boundary also moves outward at later times (Fig. 7). The waviness of the boundary evident in Figs. 7 and 10 is evidence of an instability of this boundary. In the first stage of the process the waves grow most rapidly along the boundary, with a typical wavelength of the order of 0.1 (1.3 cm). There are important differences in the shapes of the boundary in the two calculation versions.

§ 15. First Plasma Compression

This stage of the process has been discussed in detail on the dissipationless model of the plasma [9]; hence, in the present paper we simply report additional data obtained in versions A and B.

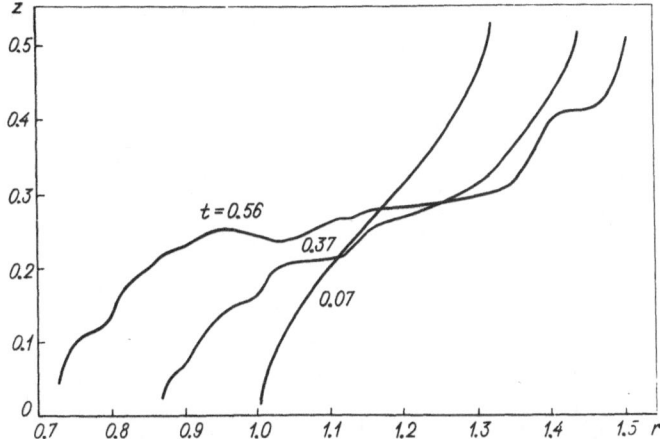

Fig. 10. Boundary configuration at several early times according to version A.

Let us trace the process by which the shock wave collapses. Figure 11 shows profiles of ρ, v_r, and T at z = 0 for three times (version B). The first instant, $t_1 = 0.784$, corresponds to the time at which the shock front reaches the axis of the system. Since $\partial v_z / \partial z > 0$ (Fig. 12), there is an upward outflow of plasma, even at z = 0 (in [30] this outflow is erroneously described by v_z). As a result, the calculations differ from those based on the one-dimensional theory. It is understandable that the plasma ejection, a two-dimensional effect, intensifies the collapse. In Fig. 11 we see a compression of the plasma to $\rho \gtrsim 100$; here T $\approx$ 5 and the sheet radius is R $\lesssim$ 0.05. In the calculations on the basis of the one-dimensional theory [1] we have $\rho \lesssim 80$, T $\lesssim$ 2, and R $\approx$ 0.14. The typical values of the parameters found in this stage (the first compression) for all three of the two-dimensional calculations are listed in Table 1.

The temporal difference is due (§ 14) to the simplified choice of initial conditions and the rough approximation of the time dependence of the total current in [9]. Apparently one should not assign any importance to the difference between the positions of the maximum plasma compression (the second line in the table), which is due largely to the nature of the boundary instability (Figs. 7 and 10). The differences in R_{min}, T_{max}, and ρ_{max}, on the other hand, are well substantiated.

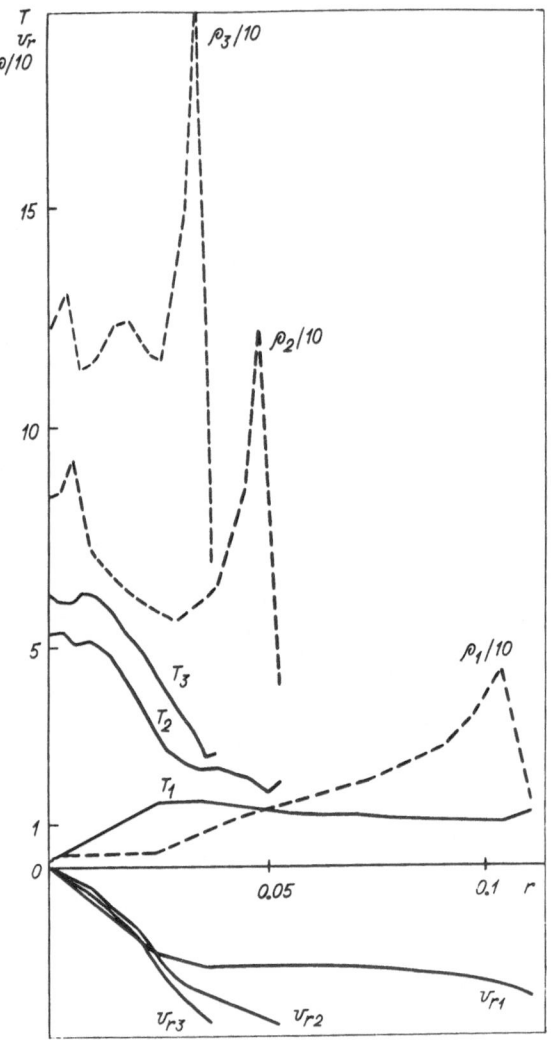

Fig. 11. Collapse of the shock wave to the axis. Pro-
files of ρ, ν_r, and T for z = 0 at three times: t_1 = 0.784;
t_2 = 0.802, and t_3 = 0.806 (version B).

In version A the sheet thickness is comparable to the radius of
the constriction. This thickness is added to the radius of the dense
core (the plasma density in the sheet is an order of magnitude
lower than the axial density), so that R_{min} is increased. Correspondingly, in version A the temperatures and densities at the axis
are lower.

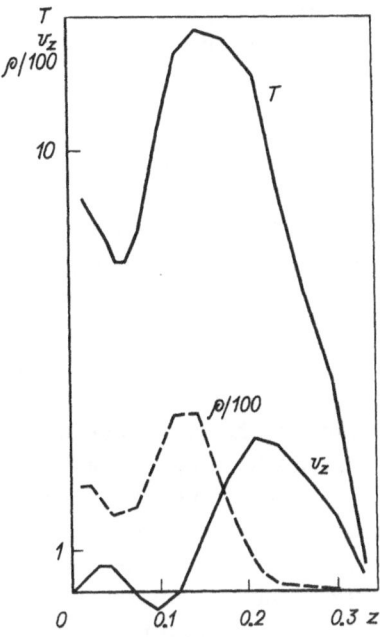

Fig. 12. Axial profiles of various properties: ρ, T, and v_z at r = 0 at time t = 0.8108 (version B).

In this stage of the process the viscous forces are unimportant, and the condition of a dynamic equilibrium of the magnetic and gas pressures is satisfied very accurately. The ratio of the gas pressures is $(\rho T)_A / (\rho T)_B \approx 17$, while $R^2_A / R^2_B \approx 25$. Also, noting that $J_B < I_A$ (Fig. 9), we find a nearly exact correspondence between changes in the magnetic and gas pressures.

In [9] we have treated the value of the temperature averaged over the constriction cross section as more realistic, since the ion

TABLE 1. Properties during the First Plasma Compression

Property	A	B	[9]
Time of first compression, t	1.12	0.81	1.05
Height above anode, z	0.15	0.15	0.06
Minimum sheet radius, R_{min}	0.075	0.015	0.02
Maximum axial temperature, T_{max}	3.4	13.0	15.0
Maximum axial density, ρ_{max}	100	450	180

thermal conductivity is neglected there. However, version B shows that the temperatures are not equalized by the thermal conductivity. Accordingly, Table 1 lists the axial temperature as T_{max} for the calculations of [9].

The parameters found for the first compression in calculation version B and in the calculations of [9] (which are more realistic in this stage) agree quite well. A pinch with radius of 2-3 mm shows a density several hundred times the initial density and a temperature of about 1 keV. These values are similar to those reported by Potter [30], but it is difficult to agree with Potter's simple interpretation of the relative role of the viscous heating of plasma ions as increasing with increasing viscosity coefficient. With nonlinear dissipation in the stage in which the shock wave collapses to the axis the relative amount of kinetic energy which is converted into thermal energy depends not only on the magnitude of the viscosity coefficient, but also on the nature of the flow behind the shock front (the plasma heating near the axis for the self-similar case [32] is determined, e.g., in [33]; for a given shock velocity, the maximum temperature at the axis, $T \sim \alpha^{-0.237}$, falls off with increasing viscosity coefficient, $\eta \sim \alpha T^{5/2}$; the parameter α is determined in § 5).

Let us evaluate the directed electron velocity v_e according to versions A and B. We write the ratio of v_e to the ion-acoustic velocity c_S in terms of dimensionless quantities (§ 5):

$$\frac{v_e}{c_S} \approx \frac{J}{ne\pi [R^2 - (R-d)^2]} \left(\frac{m_i}{kT}\right)^{1/2} \approx \frac{2.16 \cdot 10^{-2}}{\rho T^{1/2} d (2R - d)}. \qquad (15.1)$$

Substituting $\rho = 20$, $T = 3.4$, $d = 0.025$, and $R = 0.075$ in version A, we find $v_e/c_S \approx 0.2$. In version B the structure of the sheet is described very roughly, even at early times (Fig. 8). We adopt the corresponding values $\rho \approx 80$, $T \approx 4$, $d \approx 0.003$, and $R \approx 0.015$ and find the ratio $v_e/c_S \approx 1$. It is concluded that the model with the classical conductivity (version B) is still applicable for the first compression.

Figure 12 shows axial profiles of ρ, T, and v_z at r = 0 and at a time near the first compression, t = 0.8108. Here the shock front reaches the axis at the point z $\approx$ 0.33; it is meaningful to speak in terms of an axial wave rather than a converging wave. The width of the front is large in this case (about 0.1), corresponding to the high temperatures behind the front (of order 10). We see from this figure how the region of the dense pinch ($\rho \approx 400$) gives way at $z > 0.2$ to a region of an axial plasma jet ($\rho \lesssim 10$), which follows the shock front. The axial velocity at the time of the first compression at $z = 0.15$ is $v_z \approx 1.3$. These results agree with the data of [9].

§16. Second Plasma Compression

By "second" plasma compression we mean the complicated series of events which occur in the axial part of the plasma after the stage described above (the first stage) ends in the halt or retrograde motion of the current sheet. This second compression is less regular than the first; indeed one should speak in terms of a series of repeated compressions which occur at various points in the axial region. As long as we are interested in a broad area along the axis it is reasonable to use the results of version B. Accordingly, we discuss this second compression primarily on the basis of the data obtained in version B. Figure 13 illustrates the geometry of this process, with the positions of the calculation points in the final step of the calculations for both versions (most of these points appeared at the end of the calculation). In Fig. 13, as in Fig. 7, we see the boundary instability mentioned earlier, with a typical wavelength which has fallen to 0.05. The result is an axial contraction of the focus regions by more than a factor of two in this second compression.

The observed plasma focus corresponds to this second compression. This statement, which has been reported several times earlier [4, 7, 9], distinguishes our interpretation from that of Potter [30], who simply refers to a "repinch" (which apparently corresponds to the second compression being described here), but who uses the properties of the first compression as the characteristics of the plasma focus. Experiments [7, 28, 34] definitely indicate that this second compression is the predominant event.

Figure 14 shows the profiles of ρ, T, and v_z along the z axis for version B for the same time as in Fig. 13b. At $z \approx 0.17$ there are sharp peaks in ρ and T, which grow rapidly. At a point slightly higher ($z \approx 0.21-0.24$) there is also a second, shorter peak, but its growth comes to a halt as the main peak grows. It is evident that

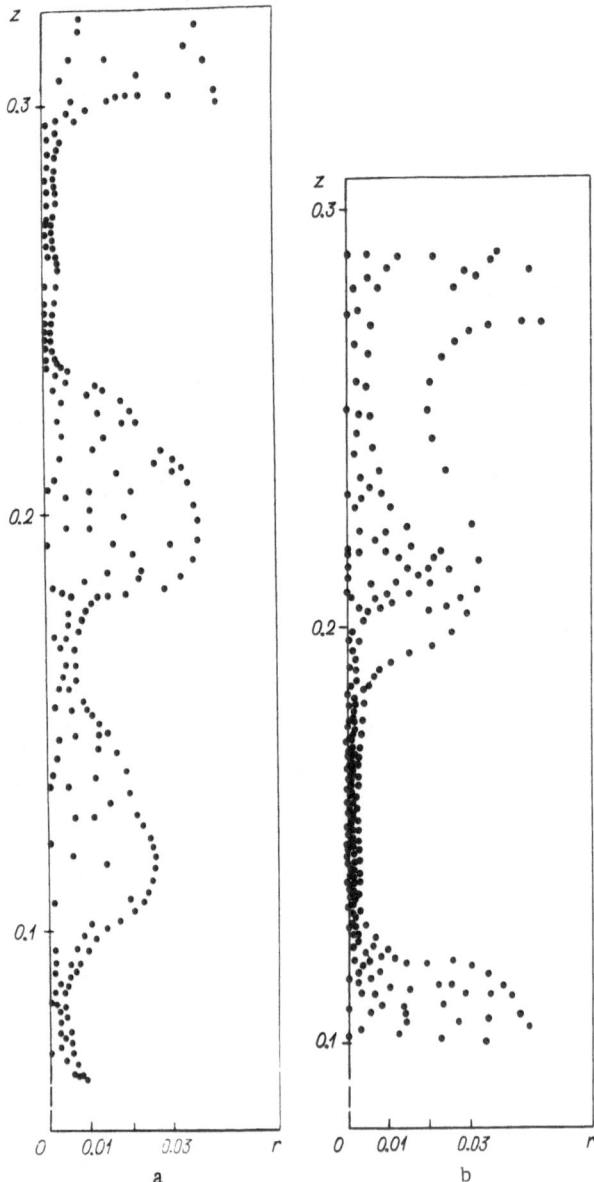

Fig. 13. Plasma configuration during the second compression. Arrangement of calculation points at the final step of the calculation. a) Version A at time $t_A = 1.297$; b) version B at time $t_B = 0.83$.

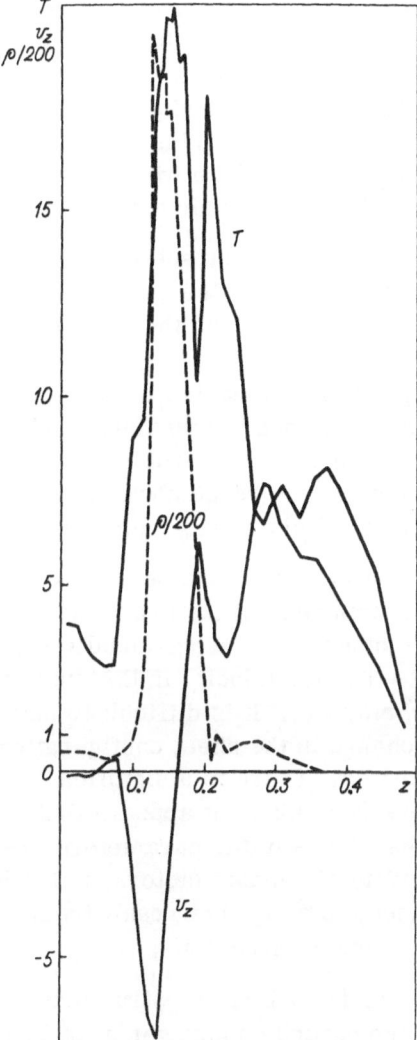

Fig. 14. Axial profiles of various properties during the second plasma compression; ρ, T, and vz at r $\approx$ 0 at time t = 0.83 (version B).

the plasma jet from this main peak plays a dominant role. The configuration of the current sheet in Fig. 13b corresponds to this distribution. At this main peak the sheet radius reaches R = 0.003; at the second peak it only reaches 0.02.

The axial velocity v_z passes through zero at this peak as soon as this peak dominates (Fig. 14). The plasma jet is not only expelled upward along the z axis but also downward. The outward velocity of the plasma from beneath the main constriction shows a tendency to increase since the gradient $\partial v_z / \partial z$ increases. We cannot be sure that the maxima of ρ and T coincide with the stopping point, $v_z = 0$. It can be assumed, however, that the maximum shifts with the velocity v_z from 1 to 3. Figure 14 shows clearly the high velocity of the axial plasma jet, which reaches values $v_z = 7$-8 (in comparison with the values 4-5 in the first compression; Fig. 12), which nearly corresponds to $v_z \approx 7 \cdot 10^7$ cm/sec [9, 30].

The other characteristics of the jet (its temperature, effective radius, etc.) are the same as those in [9]. As in [9], the calculated temperature behind the front of the axial shock wave is half that which follows from the Hugoniot condition ($T = v_z^2 / 3$). It cannot be argued in [9] that this difference is due to random factors. It is apparently due to transients and other effects. This question bears importantly on the problem of the neutron yield. The shock front is outside the range shown in Fig. 14. This front reaches the upper electrode ($z = z_0 = 0.54$) at $t = 0.83$.

In Version A the second compression is delayed slightly, but the main peak of the properties develops in a similar way. The two other peaks are gradually suppressed as a result of the axial jet. The jet velocity in this version is smaller by about 1.5 [(4-5) · 10^7 cm/sec]. It is difficult to say whether this difference is due to a change in the sheet configuration or a reduction of the collapse velocity. There is a significant upward displacement of the main peak ($z \approx 0.28$). It appears that the times and places at which these peaks in the parameters are observed are governed by a variety of random factors, which lead to various details in the sheet geometry that result from the competition of several developing constrictions.

In describing the structure of the main peak in the parameters of the second compression we will be referring to the calculations of version A (because of the considerations discussed in §1). Figure 15 shows the time dependence of ρ, T, and R corresponding to the two peaks, the first of which lies at about $z \approx 0.18$ (subscript "1"); the second lies at $z \approx 0.28$ (subscript "2"). As a result of the competition between constrictions, the second constriction replaces the first at time $t \approx 1.283$. This irregularity in the behavior of the parameters is typical of this second compression and is ultimately a consequence of the contraction of the typical wavelength of the perturbation of the external plasma boundary.

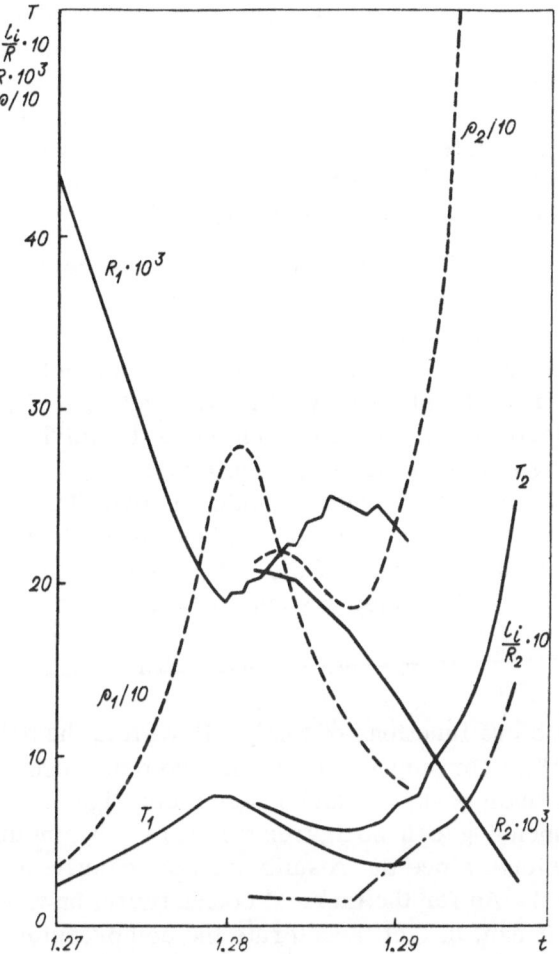

Fig. 15. Time dependence of the maximum plasma properties during the second compression. Subscript "1") Values at $z \approx$ 0.18; subscript "2") at $z \approx 0.28$ (version A). Also shown here is the ratio l_i/R.

Figure 15 shows that the shell radius R_2 continues to decrease even after the ratio of this radius to the mean free path ($l_i \approx 1.7 \cdot 10^{-2}T^2/\rho$) falls below unity. Accordingly, the calculations reveal a violation of the conditions for the applicability of the MHD approximation for the plasma in the region of the main peak. [The other ratio pertinent to the MHD approximation is $(r_{\Lambda i 2}/R_2) \sim 1$, since $d_2 \sim R_2$; § 1.] The uninterrupted plasma compression at a constriction is treated in [9]. On the basis of the calculations described here we conclude that the incorporation of dissipative processes and a change in the current does not terminate the plasma compression in the region of the main peak. The question of the subsequent fate of the plasma constriction, with the properties determined in the MHD approximation, cannot be fully resolved without incorporating kinetic effects.

In connection with the role of dissipative processes in the retardation of the compression, we note the following: Joule heating increases the electron pressure according to $p_e \sim R^{-2}$, as follows from the estimate in [9], but this estimate neglects the heat transferred away from the sheet as well as the exchange of thermal energy with ions. Accordingly, in the present case, in which the indicated effects leading to a decrease in p_e are important, the compression cannot be stopped as the result of Joule heating. The viscous forces retard the plasma compression (they play a progressively more important role in the later part of the compression), but they cannot stop it completely since they are automatically "cut off" at

$$v_r = ar; \; v_z = b \, (z - z_{pj}), \qquad (16.1)$$

where a and b are functions of t only. However, the behavior described by (16.1) corresponds to the second compression in the region of the main peak, as can be seen from Figs. 14 and 16. The times corresponding to these figures refer to the beginning of one of the compressions, since the results are not accurate at the peak itself (Fig. 13). As for the thermal conductivity, on the other hand, we note that it can, in fact, accelerate the compression (with the given plasma outflow). As follows from Fig. 16, the temperature becomes equalized along a radius. Finally, the increase in the inductive reactance of the plasma turns out to have a minor quantitative effect, in particular, because of the small axial dimensions of the region of the main peak. Between t = 1.280 and t = 1.297 the current only falls from 0.76 mA to 0.70 mA.

In summary, we cannot identify any factor which is effective in stopping the plasma compression in this formulation of the problem. This conclusion is consistent with the direct results of the calculations.

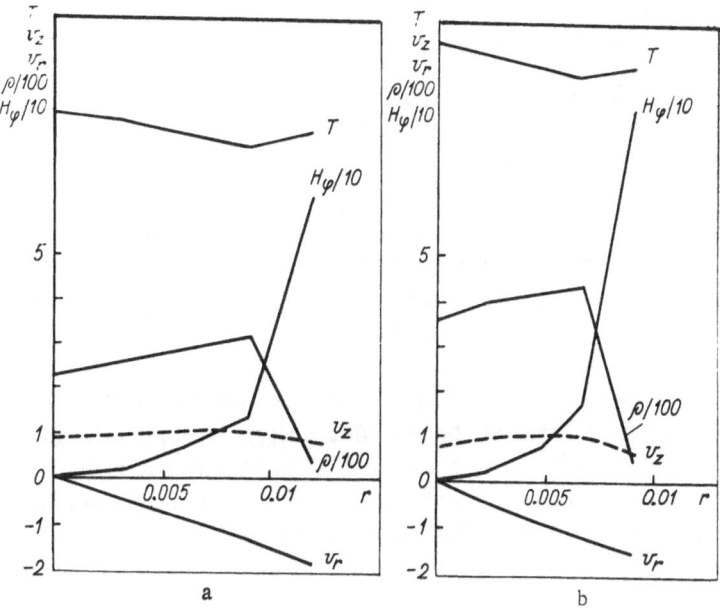

Fig. 16. Radial profiles of various properties during the second compression; T, ν_z, ν_r, ρ, and $H\varphi$ at z = 0.25. a) t_1 = 1.2905; t_2 = 1.2923 (version A).

This uninterrupted plasma compression is probably not found in the calculations of Potter [30], because of the Eulerian nature of his calculation. Nevertheless, it is noted in [30] that when the calculation grid is made finer a severalfold increase in the plasma density is observed; this implies directly that the quantitative data on the plasma properties in the focus stage are unreliable. For this reason, Potter's interpretations that the pinch radius is stabilized by the axial plasma flow through the focus region is not satisfactory. This flow is also observed in the present calculations (Fig. 12), but the plasma compression begins again shortly thereafter. The role played by the viscosity in the stability of the pinch is discussed below.

To end this discussion of the structure of the plasma focus (Fig. 16) we note that the tubular structure of the pinch is not as well defined in this second compression, while the skinning of the magnetic field is moderate: The skin thickness is slightly less than half the radius. In other words, the estimates in §1 are confirmed by these calculations.

Comparison of the results calculated in verions A and B also reveals a large difference in the time intervals between the first and second compressions. While this interval is 0.175 (250 nsec) in version A, that in version B is the same as in the dissipation-less model [9], 0.02 (30 nsec). Possible reasons for this difference have already been discussed. The results found in version B, which give a better description of the stage of the first compression, should be assigned more weight in this regard. The extreme small-ness of this time interval is one of the primary manifestations of the two-dimensional nature of the process (in the one-dimensional theory, this time interval is of order unity). Despite the large differences between the times and positions at which the main peak forms, the parameters T and R in this peak found in the two calculation versions agree well. The agreement in density ρ is slightly poorer, because of the different contributions of viscous forces.

Within the framework of the MHD model it is reasonable to take the parameters of the plasma focus to be the values calculated for the time at which $(l_i/R) \sim 1$. Accordingly (Fig. 15), we have T = 18, ρ = 1050, R = 0.004 (in version B, the value T = 18 corresponds to ρ = 3000 and R = 0.004). Furthermore, Δz = 0.06 (Fig. 13). Thus T = 1.5 keV, n = 7.5 · 10^{19} cm^{-3}, R = 0.5 mm, and Δz = 8 mm. Similar values are reported in [7].

§17. Instability of the Plasma Interface with the Magnetic Field

a. Linear Theory of the Instability. The calculations show that the plasma boundary becomes wavy, with a random, irregular structure characteristic of the onset of an instability. The influence of an instability of the boundary on the most important characteristics of the process, especially during the second compression, has already been noted (§§14-16). In this section we examine a concrete mechanism for the instability which can be responsible for the effect observed in the calculations. We first examine the linear theory of the instability and then compare it with the MHD calculations. At the end of this section we discuss the axial symmetry of the pinch from the standpoint of the linear theory of the instability.

During rapid compression of the plasma the plasma interface with the magnetic field is accelerated. If the acceleration is toward the plasma, this situation is equivalent, in a noninertial coordinate system, to the presence of a gravitational force acting outside the plasma. A Rayleigh—Taylor instability can occur under these similar conditions. Kurskal and Schwarzschild [6] have analyzed this instability in the MHD approximation for an ideal, incompressible plasma. This analysis is reported in an elegant form with certain generalizations by Stix [35]. For perturbations with wavevector **k** directed across the magnetic field, the growth rate of the instability increases without bound with increasing $|\overline{k}|$, $\gamma \sim \sqrt{k}$, in complete agreement with the classical Rayleigh—Taylor result for an incompressible fluid [36]. This unbounded increase in the growth rate is a consequence of an incomplete physical formulation of the problem. In analyzing the growth of short-wavelength perturbations it is natural to extend the original model of an ideal MHD fluid by incorporating dissipative effects, primarily, the plasma viscosity. Chandrasekhar [37] has reported an exhaustive analysis of the role played by the viscosity in an ordinary fluid. Imshennik [8] has studied the influence of the viscosity on the stability of the plasma interface with a magnetic field in the MHD case. The plasma is assumed to be homogeneous and incompressible, the interface is assumed to be planar, the viscosity is assumed to be isotropic, and other dissipative effects are ignored. We note that the role played by the viscosity in a steady-state cylindrical geometry is studied in [38], but a general dispersion relation is derived there.†

We will not repeat the derivation of the dispersion relation here [8]. We restrict the present discussion to the particular case most pertinent to the plasma-focus conditions, in which there is no magnetic field within the plasma $(H_0^p \equiv 0)$. In this case the dispersion relation becomes

$$\omega^2 - \varkappa + \varkappa^2\mu^2 - 4\varkappa^3 (\varkappa^2 + \omega)^{1/2} + 4\omega\varkappa^2 + 4\varkappa^4 = 0, \quad (17.1)$$

where we have introduced the dimensionless quantitites $\varkappa$ and ω and

---

†It is well known that a gravitational field can be used to simulate the effect of curvature of magnetic lines of force, so that an analysis of steady-state cylindrical pinch configurations is analogous to the present problem in many aspects [6, 36, 39].

the dimensionless parameter μ:

$$k = \varkappa \left(\frac{g\rho^2}{\eta^2}\right)^{1/3}; \quad \gamma = \omega \left(\frac{g^2\rho}{\eta}\right)^{1/3}; \quad \mu = \frac{k_{\parallel}}{k} \cdot \frac{H^v}{(4\pi\rho)^{1/2}} \left(\frac{\rho}{\eta g}\right)^{1/3}. \quad (17.2)$$

Perturbations along the plasma boundary are characterized by wave vector $\mathbf{k}$ with a component $k_{\parallel}$ along the field (the only component of $\mathbf{H}^V$ is H^v; the problem is in a plane geometry) and a growth rate γ (the perturbations grow as exp γt). Equation (17.1) is a single-parameter equation; the single parameter μ depends on a combination of the parameters H^V, ρ, η, and g and on the cosine of the angle between $\mathbf{H}^V$ and $\mathbf{k}$. When $\mu = 0$ (17.1) gives the equation for a viscous fluid [37]; in the limit $\eta \to 0$ the last three terms in (17.1) vanish and we are left with the familiar equation of the ideal MHD model [35]:

$$\omega' = (\varkappa - \mu^2\varkappa^2)^{1/2}. \quad (17.3)$$

Figure 17 shows the results of a numerical solution of (17.1), i.e., $\omega = \omega(\varkappa, \mu)$, for various values of the parameter μ. Shown for comparison are curves of ω' from (17.3) corresponding to the zero-viscosity case. The point at which the growth rate vanishes is independent of the viscosity [as can be seen directly by comparing Eqs. (17.1) and (17.3) in the case $\omega = \omega' = 0$]. We note from Fig. 17 that the viscosity reduces the growth rate until the parameter μ becomes too large (comparing the curves of ω and ω' in pairs), and the magnetic field suppresses the instability (comparing the curves of ω for increasing values of the parameter μ).

b. Nature of the Perturbations of the Plasma Interface in the MHD Model. Let us compare the conclusions reached on the basis of the linear theory for the instability with the numerical results found from the MHD model. Since the MHD problem is axisymmetric, we must set $k_{\parallel} = 0$ in the linear theory, i.e., $\mu = 0$ according to (17.2). The upper solid curve in Fig. 17 shows the growth rate in this case; it is in complete agreement with that found in [37]. We will first formulate several conditions under which the conclusions of the linear theory can be applied to the MHD problem. Since the comparison which will be made below is of a qualitative nature, we can write these conditions as simple inequalities.

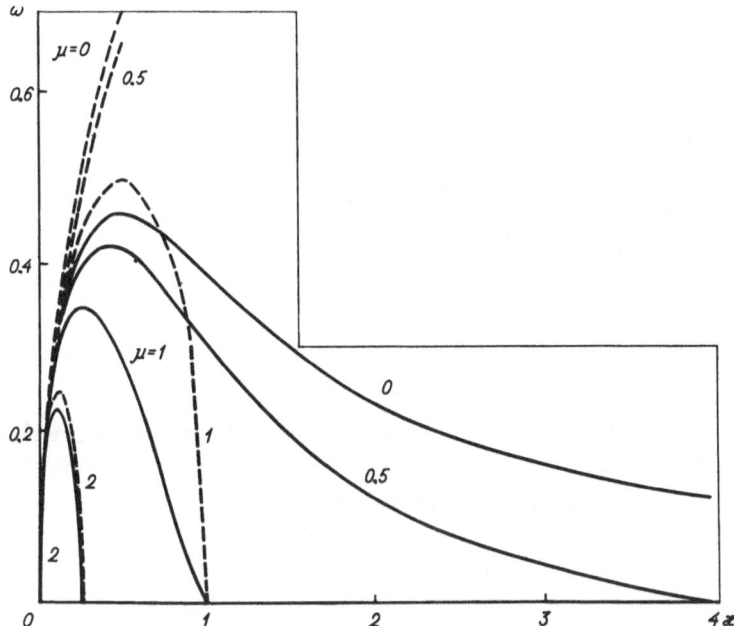

Fig. 17. Growth rate of the hydrodynamic instability, $\omega = \omega(\varkappa)$, for various values of the parameter μ (dimensionless quantities). Solid curves) Calculated, with effect of viscosity on the magnetic field; dashed curves) ideal MHD model.

The cylindrical nature of the problem and the magnetic field within the plasma can be neglected if the wave number k satisfies the conditions

$$R > k^{-1} > d, \tag{17.4}$$

where d is the sheet thickness. In the linear theory the perturbations fall off experimentally with distance into the plasma with exponent $k^p \geq k$; the exponent outside the plasma is $k^v = k$ [8]. We write the plasma–incompressibility condition as

$$c_S^2 > (\gamma/k)^2. \tag{17.5}$$

This condition follows from the entropy equation, which is not part of the original system in the linear theory. It is not difficult to write the conditions under which the thermal conductivity and electrical conductivity of the plasma are negligible:

$$t_i \gamma > 1 (i = \overline{T}, \overline{\Pi}); \quad t_j \approx C_v \rho / \nu k^2; \quad t_p \approx 1/k^2 \lambda, \qquad (17.6)$$

where we use the standard notation (Chap. 1) for the thermal conductivity (ν) and the diffusion coefficient of the magnetic field (λ). We also use trivial restrictions on the spatial scale (r_j) and temporal scale (t_j) for the changes in the density, velocity, and acceleration in the MHD problem:

$$r_\rho, \; r_v, \; r_g > k^{-1}, \; t_\rho, \; t_v, \; t_g > \gamma^{-1}. \qquad (17.7)$$

Inequalities (17.4)-(17.7) are the conditions for the applicability of the linear theory for the instability in the MHD problem.

In the comparisons below we will need the particular values of the acceleration g and the viscosity coefficient η from the MHD calculation. The unit of acceleration is (§ 5)

$$g_0 = \frac{v_0}{t_0} = \frac{v_0^2}{R_0} = \frac{p_0}{\rho_0 R_0} = \frac{H_0^2}{4\pi \rho_0 R_0} = \frac{J_0^2}{\pi c^2 \rho_0 R_0^3}. \qquad (17.8)$$

The best value of g is the sum

$$g \approx g_y + g_c = g_y + H^2/4\pi \rho R, \qquad (17.9)$$

where g_y is the calculated acceleration of the plasma boundary, and the second term gives a qualitative description of the dependence of the magnetic field on the radius of the plasma boundary. It is evident that a change in the magnetic field of a constant current in a perturbation of the plasma boundary is equivalent to an additional acceleration g_c [see (17.9)]. Equation (17.9) is changed slightly when we convert to dimensionless units. In accordance with (12.1) and the choice of the constant factors B* in § 5 and B_1^* in § 13, in dimensionless units the viscosity coefficient is

$$\tilde{\eta} = 10^{-2} \tilde{T}^{5/2} + 3 \cdot 10^{-3}. \qquad (17.10)$$

Let us consider the perturbation with the highest growth rate; clearly, if there are no special excitation conditions [7], it is these perturbations which grow preferentially. On the basis of Eqs. (17.2) (see also Fig. 17) we conclude that

$$\gamma_{max} = 0.46 \, (g^2 \rho / \eta)^{1/3}; \quad k_{max} = 0.50 \, (g \rho^2 / \eta^2)^{1/3}. \qquad (17.11)$$

Transforming to the dimensionless units of §5 and using (17.8)
and (17.11) we have

$$\tilde{\gamma}_{max} = 0.46 \left(\tilde{g}^2 \, \tilde{\rho}/\tilde{\eta}\right)^{1/3}; \quad \tilde{k}_{max} = 0.50 \left(\tilde{g} \, \tilde{\rho}^2/\tilde{\eta}^2\right)^{1/3}. \quad (17.12)$$

In the early part of the compression we set $\tilde{\rho} \approx 4$, $\tilde{T} \approx 0.2$, and $\tilde{g} \approx 3$
(§14) in (17.12) and find $\tilde{\gamma}_{max} \approx 10$ and $k_{max} \approx 10^2$. These values
and — most important — the perturbation wavelength agree qualita-
tively with the calculations (§14, Fig. 10). In the stage in which the
plasma focus forms, during which a significant acceleration of the
plasma boundary begins between the first and second compression
of the plasma, we can assume $\tilde{\rho} \approx 10^2$, $\tilde{T} \approx 6.5$, and $\tilde{g} \approx 5 \cdot 10^2$
(Fig. 15). Equation (17.12) then yields $\tilde{\gamma}_{max} \approx 1.5 \cdot 10^2$ and $\tilde{k}_{max} \approx$
10^2. The value $\tilde{k}_{max} \approx 10^2$ corresponds to a wavelength shorter than
0.1, as noted earlier in the discussion of the calculations (§16,
Fig. 13). A direct check of conditions (17.4)-(17.7) yields a generally satisfactory result. Conditions (17.4), (17.5), and (17.7) are
violated for wavelengths long in comparison with k_{max}^{-1}; for shorter
waves conditions (17.4) and (17.6) are violated.

We can conclude that the wave perturbations of the plasma
boundary found in the MHD calculations are a consequence of a
Rayleigh–Taylor instability with the stabilizing role of the viscosity taken into account [8, 37]. It is not appropriate to make a
quantitative comparison between the results of the linear theory and
the MHD problem because of 1) the limitations contained in (17.4)-
(17.7); 2) the uncertainty regarding the choice of values for ρ, T,
and g in (17.9) and (17.11); and 3) the important role of the nonlinear effects in the MHD model. We also note that the short wavelengths found in the linear theory (wavelength of the order of the
spatial step in the calculations) do not play a role in the onset of the
boundary instability. This result means that the difference method
does not introduce distortions in the description of the instability.

c. Axial Symmetry of the Pinch. Let us use the
linear theory of the instability to analyze the evolution of a perturbation of the plasma boundary with an azimuthal component
$k_{\parallel} \neq 0$. Filippov et al. [7] have shown why an instability that grows
in the rz plane does not show much effect in the azimuthal direc-
tion. If $k_{\parallel} \neq 0$, then $\mu \neq 0$, so that the general solution of dispersion relation (17.1) should be used. In a cylindrical geometry, the

component $k_{\parallel}$ can obviously only take on the discrete values

$$k_{\parallel}^m = m/R, \qquad (17.13)$$

where m is the mode index. Strictly speaking, from the standpoint of (17.4), the linear theory for the instability is only valid if m is large. However, qualitative results can be obtained if we apply the theory to cases $m \geq 2$. Perturbation modes with m = 2 pose a danger to the symmetry (the mode m = 1 simply shifts the focus axis). The magnetic field stabilizes all perturbation modes with $m' \geq m$, which are precisely in the azimuthal direction ($k \equiv k_{\parallel}$) if $k_{\parallel}^{ub} \leq k_{\parallel}^m$, where $k_{\parallel}^{ub}$ is the instability boundary on the short-wave side, which follows from (17.1) or (17.3):

$$k_{\parallel}^{ub} = 4\pi\rho g/(H^v)^2. \qquad (17.14)$$

From (17.13) and (17.14) we find the condition for azimuthal stability of the perturbations by setting m = 2 in (17.13):

$$g \leq (H^v)^2/2\pi\rho R. \qquad (17.15)$$

A rough check of condition (17.15) for an actual experiment is carried out in [7], but this check yields a negative result for the beginning of the process [especially if the effective acceleration g is determined from (17.9)]. Nevertheless, we should not conclude from this comparison that a real instability is possible in the azimuthal direction, since the growth rate is also important. Let us evaluate γ_{max}^{μ} for $\mu \neq 0$. In order-of-magnitude terms we have $g \sim H^2 \cdot (4\pi\rho R)^{-1}$, $\eta \sim l_i v_T \rho$, and $v_T \sim H/(4\pi\rho)^{1/2}$; hence using (17.2) we have

$$\mu \sim \frac{k_{\parallel}}{k}\left(\frac{R}{l_i}\right)^{1/3}, \qquad (17.16)$$

where l_i is the ion mean free path. When the MHD approximation is valid we have $R \gg l_i$; thus (17.16) shows that $\mu \gg 1$ for perturbations in the azimuthal direction. In this case the viscosity plays a very minor role and (17.3) yields $\gamma_{max}^{\mu} \approx \gamma_{max}^{\mu}\mu^{-1} \ll \gamma_{max}$, where γ_{max} is given in (17.11). Accordingly, the maximum instability growth rate is very small ($\gamma_{max}^{\mu} < 1$ with $\gamma_{max} \approx 10$, $R > 10^3 l_i$) and no instability is manifested on scale times of order unity.

We must still examine a possible disruption of axial symmetry due to mixed perturbations in which the component $k_{\parallel}$ satisfies

(17.13) but k_r and k_z are nonzero, but are so large that μ from (17.16) reaches a value of order unity ($\mu \lesssim 1$). Obviously, a necessary condition is $k_\parallel/k \sim (l_i/R)^{1/3} \ll 1$. The growth rate for these perturbations is again given roughly by the upper curve in Fig. 17; and with certain combinations of conditions it is possible that $\gamma \approx \gamma_{max}$. Stabilization by the magnetic field is not important. However, if account is taken of the limitation of the linear theory due to nonlinear effects, again it is not necessarily true that the axial symmetry is violated. Nonlinear effects come into play when the amplitude of the boundary perturbation is $\Delta R \sim k^{-1}$. Accordingly, the order of magnitude of the relative deviation from axial symmetry is $\Delta R/R \sim k_\parallel/k \sim (l_i/R)^{1/3} \ll 1$, since $k_\parallel \sim R^{-1}$. We conclude that with mixed perturbations the linear theory does not predict significant deviations from axial symmetry.

§18. Energy Balance and Neutron Yield
of the Plasma Focus

a. Radiation Loss in the Plasma. The radiative energy loss of the plasma has not been taken into account in the calculations. We can estimate this loss on the basis of the typical values of the plasma parameters, including those found from the calculations. Under actual conditions the radiative loss increases when heavy impurities are added deliberately or as the result of uncontrolled influx of impurities into the discharge. It is important to note at the outset that experiments have established that an admixture of heavy impurities (the typical concentrations of the N_2 or Xe impurities are about 1%) strongly affects the plasma compression; in particular, the impurities intensify or attenuate the neutron yield [7]. Although the role played by the heavy impurities is still not fully understood, it can be assumed that this role is important at the beginning of the discharge, when the initial ionization occurs, the current skin effect is important, and the current sheet is being formed. We will simply estimate the radiative energy loss in the final stage of the noncylindrical Z pinch, taking account of the contribution of heavy impurities to the radiation.

Galushkin [40] (cf. [41]) has studied in detail the radiation from typical heavy impurities under conditions characteristic of the noncylindrical Z pinch. The radiation is assumed to originate in

the volume, and the generalized radiation-collision model of
[42] is used. The distributions of multiply charged ions over
ionization state are calculated (for ionization states ranging from
the bare nuclei to beryllium-like ions), as are the distributions over
excited states (with principal quantum numbers $n \lesssim n_{max} \gtrsim 10$; above
n_{max}, the level population is assumed to be a Boltzmann population).
Because of the high electron densities ($n_e \approx 10^{16}$-10^{20} cm^{-3}), the
simple coronal approximation cannot be used. In a pinch, the conditions correspond to a case intermediate between the coronal and
Boltzmann cases and this complicates the calculations. Figure 18
shows the radiative energy loss of the plasma, q_A, due to heavy
impurities, specifically, carbon, C, and argon, Ar (for a concentration of 1% in terms of the number of atoms), calculated per electron and per impurity atom. Shown for comparison is the total

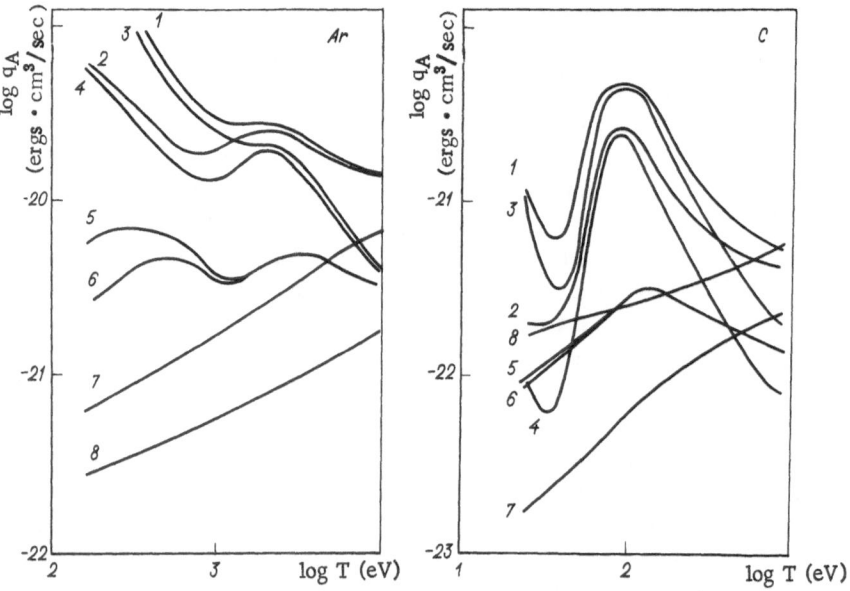

Fig. 18. Radiative energy loss of the plasma, q_A, due to heavy impurities of carbon
and argon in a concentration of 1% (in terms of the number of atoms) as a function of
the plasma temperature (calculated per electron and per impurity atom). 1,3,5) Elec-
tron density of $n_e = 10^{16}$ cm^{-3}; 2,4,6) $n_e = 10^{20}$ cm^{-3}; 7,8) arbitrary of n_e. 1,2) Total
radiation from all mechanisms; 3,4) line radiation; 5,6) recombination; 7) bremsstrah-
lung; 8) hydrogen emission, x10^2.

(bremsstrahlung plus recombination) emission of hydrogen, $10^2 q_H$. In argon $10^2 q_H \ll q_A$ over the entire temperature range and over the entire range of electron densities in the pinch; for carbon we have $10^2 q_H \gtrsim q_A$ at sufficiently high temperatures. Accordingly, under the pinch conditions the radiative loss is effectively governed by the heavy impurities. Figure 18 also shows the contributions of the various radiation mechanisms: line, recombination, and bremsstrahlung. Under the conditions that prevail in the pinch, even in carbon the line radiation dominates, becoming weaker than the bremmstrahlung only at high temperatures. Galushkin [40] has also reported detailed information on the spectral composition of the radiation.

Let us estimate the contribution of the radiative loss to the energy balance of the pinch (Fig. 18). We first write the ratio of the loss to the electron thermal energy:

$$\delta = \frac{q_A \langle t \rangle \langle n_A \rangle}{1.5 \langle T \rangle} = \frac{q_A \langle t \rangle \langle n_D \rangle}{1.5 \cdot 10^2 \langle T \rangle}. \tag{18.1}$$

This ratio involves the characteristic parameters of the plasma for a certain stage of the process, the time $\langle t \rangle$, the impurity and deuterium densities $\langle n_A \rangle$ and $\langle n_D \rangle$, and the temperature $\langle T \rangle$. Evaluation of (18.1) for the plasma-focus stage ($\langle n_D \rangle_{pf} \approx 10^{20}$ cm^{-3}, $\langle T \rangle_{pf} \approx 1.2$ keV, $\langle t \rangle_{pf} \approx 10^{-7}$ sec) yields $\delta_{pf} \approx 1$ (for argon) or $\delta_{pf} \approx 0.03$ (carbon). In other words, the relative importance of the radiative loss is almost the same as with typical parameters for the plasma focus. It should be noted that the conclusions reached regarding the radiative loss can be affected by several factors which have not been taken into account here [40]: the variations in the ion distribution over ionization states (the relaxation times for the populations of the pertinent ion levels are comparable to the scale time for the formation of the plasma focus), ionization loss (especially in nitrogen), and the uncertainty regarding the rates of the elementary processes. Nevertheless, this estimate shows that taking account of radiative loss in the MHD model does not introduce qualitative changes in the calculated results, since the ratio δ remains much larger than unity. It can be assumed that the radiative loss would increase the relative plasma density in the calculations. Therefore, it is not useful to incorporate the radiative loss in the MHD model.

b. Energy Balance of the Plasma. Of the various
characteristics of the plasma focus, the energy balance is very
important. In the MHD calculations, the values of the various forms
of energy in the plasma are calculated. The following integrals
are calculated (in dimensionless units) as functions of the time (§5):

$$E_K = \int\limits_{(V_p)} \frac{v_r^2 + v_z^2}{2}\, \rho r\, dr\, dz; \quad E_T = \int\limits_{(V_p)} \frac{3}{2} T \rho r\, d r\, dz;$$

$$E_M = \int\limits_{(V_p)} \frac{H_\varphi^2}{2}\, r\, d r\, dz; \quad E_\Sigma = E_K + E_T + E_M,$$

(18.2)

where E_Σ is the total energy in the plasma at time t. Figure 19
shows plots of all the quantities in (18.2) for calculation version B.
Version B seems to be preferable for describing the energy integrals since the contribution to these integrals from the immediate vicinity of the plasma focus, for which version A is more
appropriate, is quite small (see the discussion below). The curve
of the total energy, E_Σ, shows that after the first compression (the

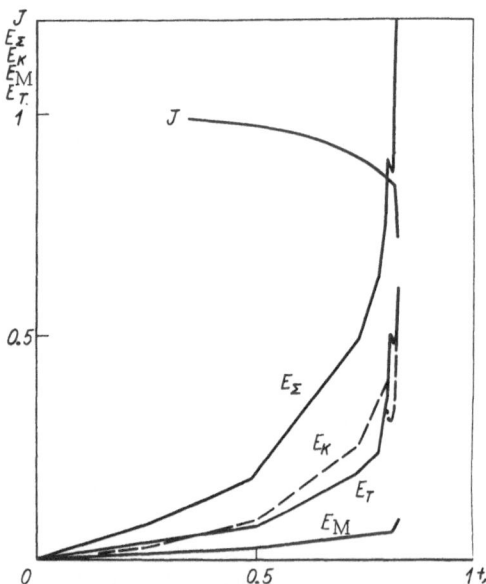

Fig. 19. Time evolution of the energy and total-
current integrals (version B).

weak energy peak at t = 0.813) there is a second compression, involving a rapid increase in the energy. Throughout the process the thermal (E_T) and kinetic (E_K) energies remain approximately equal. During the first compression, with $E_T \approx 1.5E_K$, the state with the lowest fraction of kinetic energy is reached. Then, as the second compression begins, the plasma begins to move toward the axis again, and the axial jet develops; then $E_T \approx 1.2E_K$. The relative magnitude of the magnetic energy, E_M, in version B is extremely small, except at the very beginning of the compression stage, at which $E_M \leq 0.1E_\Sigma$. In version A, the relative magnitude of E_M naturally increases; in accordance with the change in the sheet thickness (§ 14), it increases by a factor of about three. To convert the numbers in Fig. 19 into dimensional values, they must be multiplied by the coefficient (§ 5) $2\pi \mathcal{E}_0 \rho_0 R_0^3 = 2.55 \cdot 10^{11}$ ergs = 25.5 kJ. These results show that during the formation of the plasma focus more than half the bank energy $(\frac{1}{2}C_0 U_0^2 \approx 52$ kJ) is deposited in the plasma. Figure 19 also shows the current J calculated from the simplified Kirchhoff equation, which neglects the energy loss in the plasma, W_p (§§3 and 5). It follows from the calculations $(E_\Sigma \gtrsim 0.5C_0U_0^2/2)$ that the current J is overestimated, since the condition $W_p = 0$ is too crude.

It is interesting to estimate the fraction of the total plasma energy represented by the energy within the plasma focus in the strict sense of the term. The plasma focus proper can of course be defined arbitrarily as that part of the volume in which the plasma properties approach their maximum values (Fig. 15). Late in the process we have $T \approx 18$, $\rho \approx 3000$, $R \approx 0.004$, and $\Delta z = 0.06$ (§ 16). Then the dimensionless thermal energy is $E_{Tpf} \approx 1.5 \rho T \cdot 0.5R^2\Delta z \approx 4 \cdot 10^{-2}$. Even if we take account of the doubling of the focus energy due to the addition of E_{Kpf} and E_{Mpf}, we find a result which is a very small fraction of the total plasma energy: $E_{\Sigma pf} \approx 0.06E_\Sigma$. In version A we find approximately the same ratio, because of some redistribution of the energy and a decrease in E_Σ. We conclude that less than 3% of the bank energy is deposited in the plasma focus proper.

c. Neutron Yield of the Plasma Focus. The high neutron yield is one of the most important physical properties of the noncylindrical Z pinch. In gaseous deuterium the record reported neutron yield exceeds 10^{11} neutrons per discharge [5, 43].

An anisotropy in the neutron energy distribution has been reliably established experimentally [4, 43]. This anisotropy can be explained by assuming that thermonuclear reactions occur in the plasma, which is moving from the anode to the cathode with a high velocity (10^8 cm/sec). This interpretation of the effect was first advanced by Filippov and Fillippova [27] under the name of a "moving boiler." A detailed study of the neutron yield, on the other hand, leads to contradictions with the moving-boiler model.

Let us estimate the neutron yield on the basis of the calculations for the MHD model of the plasma. It is immediately clear that we cannot evaluate the neutron yield over the entire discharge, because the lifetime of the plasma focus is not determined (§ 16). We can assume that throughout most of its lifetime, which is measured in [44], the plasma focus is in a nonhydrodynamic state.[†] Since the measured lifetime (10^{-7} sec) is much longer than the transit time ($R_{pf}/v_T \lesssim 10^{-9}$ sec), we must assume the appearance of some nearly equilibrium nonhydrodynamic plasma configuration at the focus. Adopting this argument, we can estimate the lifetime of the plasma focus in a simple way because this lifetime can be equated to the current decay time in the equilibrium case [45]. There are grounds for assuming that a disconnected current loop forms in the plasma-focus stage, with a current J' described by Eq. (3.24), according to § 3. To estimate roughly the current decay time in (3.24) we consider Joule heating alone, and in the Joule-heating integral we only take account of the primary contribution of the axial region in the vicinity of the plasma focus. From (3.16) and (3.19) we find approximate values for the plasma inductance L'_p and the Joule heating W'_p:

$$\left.\begin{array}{l} L'_p \approx 2z_{pf} \ln \dfrac{R_{ex}}{R_{pf}}\,; \\[2mm] W'_p \approx \left(\dfrac{c}{4\pi}\right)^2 \dfrac{\pi H_\Phi^2}{\sigma_e\, d^2}\ (2R_{pf} - d)\, dz_{pf}, \end{array}\right\} \qquad (18.3)$$

[†] This point of view is consistent with the results of laser diagnostics of the plasma focus [7, 34]. The qualitative model for the process recently developed by Maisonnier et al. [46] is also based on a nonhydrodynamic model of the plasma focus. Here again, however, it is not possible to reconcile the picture with the uncertain rule played by the second compression of the plasma, which establishes the initial conditions for the plasma focus.

where R_{ex} is the external radius of the toroid with the magnetic field ($R_{ex} \gg R_{pf}$). According to the calculations (version A), we can assume $d \approx R_{pf}$ for the sheet thickness. From Eq. (3.24) we find an order-of-magnitude value for the lifetime of the plasma focus:

$$t_{pf} \approx \frac{4\pi\sigma_e}{c^2} R_{pf}^2 \ln \frac{R_{ex}}{R_{pf}}. \qquad (18.4)$$

Substituting the effective plasma conductivity σ_e into (18.4), so that $\lambda^{-1} = 4\pi\sigma_e c^{-2} = 5.9 \cdot 10^8 (kT_{pf})^{3/2}$ (§5), and substituting typical values of the parameters in the plasma focus, $kT_{pf} \approx 1.2$ keV and $R_{pf} \approx 0.65$ mm, we find $t_{pf} \approx 10^{-6}$ sec. This value is slightly longer than the experimental lifetime, but if the coefficient k_e in σ_e is neglected the value is higher by two orders of magnitude. The value found for t_{pf} can be reduced by assuming that the radius R_{pf} is less than the value used above. The time t_{pf} can be found as a function of R_{pf} from (18.4) by using the following relations:

(1) $\rho \sim R_{pf}^{-6/5}$ in the equilibrium configuration [9];
(2) $T \sim \rho^{2/3} \sim R_{pf}^{-4/5}$ in the adiabatic case;
(3) a coefficient $k_e \sim (\rho R_{pf}^2)^3 R_{pf} \sim R_{pf}^{7/5}$ (§1).

Then $t_{pf} \sim R^{11/5}$. For a radius of $R_{pf} \approx 0.2$ mm, which is consistent with the measurements, we have $t_{pf} \approx 10^{-7}$ sec. The current decay time is estimated under the assumption that no macroscopic turbulence develops in the plasma focus in this interval but that microscopic turbulence is generated to the extent that the actual plasma conductivity is lower than the classical value by a factor of k_e^{-1}.

Furthermore, in estimating the neutron yield we have assumed that a thermonuclear mechanism is responsible for the neutron production. Although this approach cannot lead to a complete explanation for the observed anisotropy, which is the crudest characteristic of the effect, we can estimate the total number of neutrons emitted during the discharge on the basis of the thermonuclear equations. We note in this connection that at a plasma temperature exceeding a few keV the effective energy of the deuterons involved in the thermonuclear reaction is comparable to the average thermal energy of these deuterons [1]. For this reason, the integrated neutron yield is relatively insensitive to the details of the neutron distribution, provided, of course, that this distribution has no exotic singularities. Using the dimensionless variables, we can write the following integral equation for the neutron yield [1]:

$$W_n = 10^{-50} \; \frac{1}{2\,(2\pi)^{5/2}} \cdot \frac{m_D^2}{k^4 c^7} \cdot \frac{J_0^7}{\rho_0^{3/2} R_0^3} \int\limits_0^\infty dt \int\limits_{(V_p)} T^4 \rho^2 \, rd\,rdz. \qquad (18.5)$$

The function $f\,(t) = \int\limits_{(V_p)} T^4 \rho^2 rdrdz$ is calculated theoretically [f(t) is only calculated in version B; we make an estimate for version A below]. There is a weak peak in f(t) near the first compression but this peak is much smaller than the values of f(t) during the second compression. In accordance with the discussion above, we take the value of f(t) at time t* = 0.829, at which point the plasma-focus parameters listed above obtain: f(t*) $\approx 3 \cdot 10^5$. Substituting in (18.5) all units from § 5 and $t_{pf} \approx 10^{-6}$ sec, we find $W_n = 6 \cdot 10^5 \cdot 7 \cdot 10^{-1} \cdot 3 \cdot 10^5 \approx 10^{11}$ neutrons/discharge. Since the density in version A is lower than that in version B by a factor of about three (§ 16), and since the estimated time is longer by an order of magnitude, we can reduce W_n by two orders of magnitude. We then find $W_n \approx 10^9$ neutrons/discharge, in complete agreement with experiment.

In conclusion it is appropriate to say a few words regarding the actual mechanism for the neutron production in the plasma focus. The target mechanism undoubtedly makes some contribution (although it is apparently small in absolute value); the moving target here is the front of the axial shock wave [9, 30]. This effect can explain the large axial dimension of the neutron source and the partial anisotropy of the neutron emission. To find a complete explanation for this anisotropy it is necessary to examine the kinetic properties of the plasma focus. The anisotropy of the deuteron distribution can be primarily responsible for the observed high velocities of the neutron source. The importance of this effect for an explanation of the neutron emission has been pointed out in [31]. The role played by direct acceleration of deuterons in the induced electric fields, which is responsible for the neutron emission in linear Z pinches, is apparently unimportant under the conditions that prevail in a noncylindrical Z pinch. We note that the difference between the electric potentials at the surface of the plasma focus in the axial direction has been estimated theoretically [47] to be 10^6 V; this potential difference is probably responsible for the formation of the very fast deuterons detected by nuclear emulsions. The number of these deuterons turns out to be small, so that their contribution to the neutron yield is unimportant.

CONCLUSION

The point of primary interest in this work is the spatial and temporal structure of the plasma focus, which is to be understood as an extremely dense, hot plasma formation which occurs during the second compression of the plasma near the axis. The calculations show that this plasma focus is of an irregular nature: Its parameters are governed by an instability of the plasma boundary with the magnetic field. In the nonlinear stage this instability is manifested as a competition between adjacent constrictions of the pinch. The maximum parameter values (density and temperature) are ultimately reached at one of these competing constrictions. The calculations show convincingly that at this "dominant" constriction there is an uninterrupted reduction in the plasma radius and that the conditions for the applicability of the MHD approximation are violated.

The main task of future theoretical work on the plasma focus (or, to state the problem more generally, on the plasma constriction in a pinch) is to develop and study a kinetic model for the process. We do not believe that the possible refinements of dissipative effects in the MHD model are very important. To a certain extent, this belief is supported by an auxiliary calculation, mentioned in [30], in which the electrode polarity is reversed.

For the system of equations and the corresponding initial and boundary conditions used here (Chap. 1), the electrode polarity obviously plays no role, since all the additional terms in the equations for the electron component from [30] (the Hall terms in Ohm's law, the correction to the average electron velocity, and the transverse terms in the thermal conductivity) make reverse contributions when the electrode polarity is reversed. This contribution in fact turns out to be insignificant within the error of the calculations of [30] (about 1.5%). It was pointed out previously (§ 1) that there is no point in incorporating the ion magnetization in the ion transport coefficients, although it is difficult to find quantitative estimates of the sensitivity of the calculated results to the various descriptions of the dissipative effects in the ion fluid.

Potter and Haines [31] have taken an important step toward the formulation of the kinetic problem. In particular, they point out that the conditions for the applicability of the MHD description are initially violated only in the ion fluid. At the plasma focus the results show that $l_i \gtrsim R_{pf}$ and $r_{\Lambda i} \sim d \lesssim R_{pf}$, where, as before l_i is the ion mean free path, $r_{\Lambda i}$ is the ion gyroradius, and d and R_{pf} are the sheet thickness and the pinch radius, respectively. At the same time, $l_e \sim l_i \gg r_{\Lambda e}$ for the electrons, which are strongly magnetized, and the condition for adiabatic motion, $r_{\Lambda e} \ll d$, is satisfied. Accordingly, the electrons behave as a continuous medium in the transverse direction (except in a very thin region near the axis). The electrons also carry a current which obeys Ohm's law; this law can, in general, incorporate an effective conductivity σ_e which is different from the classical value $\sigma_\perp$. Accordingly, the first step is to consider a mixed model for the plasma; i.e., the model should use a kinetic treatment for the ions and an MHD treatment for the electrons. It is of fundamental importance to incorporate the two-dimensional nature of the system in order to study the dynamics of the plasma focus. Actually, the kinetic problem should be a continuation of the MHD calculations, since the initial conditions for the kinetic problem are two-dimensional. Furthermore, the kinetic model should be capable of describing the plasma flow away from a constriction, which leads to the uninterrupted reduction in the plasma radius in the MHD stage of the process. The question of plasma neutrality is of importance in this mixed model. For plasma motion at nonrelativistic velocities, the neutrality condition is apparently satisfied at the plasma focus. It is not difficult to show that $\Delta n / n \sim v_e v_z / c^2$, where Δn is the excess of particles of one charge, v_e is the directed electron velocity, and v_z is the axial velocity of the plasma. It is quite possible that at some stage it will also be necessary to take account of the kinetic effects in the electron fluid. Finally, we note that a recent experiment [34] raises hope regarding the formulation of the kinetic problem.

However, no matter what happens in the future, the two-dimensional MHD model for a plasma has already led to important progress in the theoretical understanding of the complicated physical processes which occur in a noncylindrical Z pinch (Chap. 3). The satisfactory agreement of theory and experiment has made it possible to predict the optimum changes in the parameters of the experimental device for increasing the neutron yield [45]. This prediction of course rests on the scaling law based on the MHD model; it is supported in part by previous measurements of the neutron yield as a function of the energy of the capacitor bank [7]. (Unfortunately, the measurements of [7, 28] are only carried out over a narrow energy range.)

The comprehensive and careful experiments carried out by Filippov et al. and by our non-Soviet colleagues have opened up a broad field for the application of plasma theory. Both here in the Soviet Union and elsewhere, the plasma focus has become an important object of theoretical and mathematical analysis over the past decade. The interesting and difficult fundamental problems of the two-dimensional focus in a plasma have stimulated the development and implementation of new numerical methods for solving MHD problems as well as a careful study of various physical effects. It is reasonable to expect that experiment and theory will continue to stimulate each other in this field of plasma physics.

In conclusion the authors thank V. V. Paleichik for his assistance in the computer calculations.

REFERENCES

1. V. F. D'yachenko and V. S. Imshennik, Reviews of Plasma Physics, Vol. 5 (ed. M. A. Leotovich), Consultants Bureau, New York (1970), p. 447.
2. N. V. Filippov, T. I. Filippova, and V. P. Vinogradov, Nucl. Fusion Suppl., 2:571 (1962).
3. D. P. Petrov, N. V. Filippov, T. I. Filippova, and V. A. Khrabrov, in: Plasma Physics and the Problem of Controlled Thermonuclear Reactions [in Russian], Vol. IV, Izd. Akad. Nauk SSSR, Moscow (1958), p. 170.
4. V. I. Agafonov et al., Plasma Physics and Controlled Nuclear Fusion Research, Vol. 2, IAEA, Vienna (1969), p. 21.
5. C. Maisonnier, M. Samuelli, J. G. Linhart, and C. Gourlan, Plasma Physics and Controlled Nuclear Fusion Research, Vol. 2, IAEA, Vienna (1969), p. 77.
6. M. Kruskal and M. Schwarzschild, Proc. Roy. Soc. A223:348 (1954).
7. N. V. Filippov et al., Plasma Physics and Controlled Nuclear Fusion Research, Vol. 1, IAEA, Vienna (1971), p. 573.
8. V. S. Imshennik, Dokl. Akad. Nauk SSSR, 204:1335 (1972) [Sov. Phys. Dokl. 17: 576 (1972)].
9. V. F. D'yachenko and V. S. Imshennik, Zh. Eksp. Teor. Fiz., 56:1766 (1969) [Sov. Phys. JETP 29:947 (1969)].
10. J. W. Mather, Phys. Fluids, 8:366 (1965).

11. K. V. Roberts and D. E. Potter, in: Methods in Computational Physics. Vol. 9.
    Plasma Physics (ed. B. Alder, S. Fernbach, and M. Rotenberg), Academic Press,
    New York (1970), p. 339.
12. G. Basque, A. Jolas, and J. P. Watteau, Phys. Fluids, 11:1384 (1968); 12:1529
    (1969).
13. B. V. Robuoch and G. DiCola, "A one-dimensional model of a two-dimensional
    snow-plough Mirpa focus discharge," LGI, 70/5, E. Frascati, 1970.
14. L. Spitzer, Physics of Fully Ionized Gases, Interscience, New York (1956).
15. L. A. Artsimovich, Controlled Thermonuclear Reactions, Gordon and Breach,
    New York (1968).
16. L. V. Dubovoi, V. P. Fedyakov, and V. P. Fedyakova, Zh. Eksp. Teor. Fiz, 59:1475
    (1970) [Sov. Phys. JETP 32:805 (1971)].
17. L. V. Dubovoi, A. V. Komin, and V. P. Fedyakov, Zh. Eksp. Teor. Fiz. 62:1335
    (1972) [Sov. Phys. JETP 35:703 (1972)].
18. A. Kuckes, Phys. Fluids, 7:1511 (1964).
19. A. B. Mikhailovskii, Theory of Plasma Instabilities, Vol. 1, Consultants Bureau,
    New York (1973).
20. N. V. Chudin, Zh. Tekh. Fiz. 41:60 (1971) [Sov. Phys. Tech. Phys.].
21. S. I. Braginskii, Reviews of Plasma Physics, Vol. 1 (ed. M. A. Leontovich), Con-
    sultants Bureau, New York (1965), p. 205.
22. N. E. Kochin, Vector Calculus and Introduction to Tensor Calculus [in Russian],
    Izd. Akad. Nauk SSSR, Moscow (1951).
23. V. F. D'yachenko and V. S. Imshennik, Zh. Vychslit. Matem. i Matem. Fiz.,
    3:915 (1963).
24. L. D. Landau and E. M. Lifshits (Lifshitz), Fluid Mechanics, Addison-Wesley, Read-
    ing, Mass. (1959).
25. L. D. Landau and E. M. Lifshits (Lifshitz), Electrodynamics of Continuous Media,
    Addison-Wesley, Reading, Mass. (1960).
26. V. S. Imshennik, I. V. Otroshchenko, V. V. Palichik, and K. V. Khodataev,
    "Magnetohydrodynamic fluctuations in a plasma subjected to an intense rf elec-
    tromagnetic field," Preprint No. 35, Institute of Applied Mathematics, Academy
    of Sciences of the USSR, Moscow, 1970.
27. N. V. Filippov and T. I. Filippova, "Phenomena accompanying the formation of a
    dense plasma focus upon the collapse of a noncylindrical Z pinch," Preprint No.
    913, I. V. Kurchatov Institute of Atomic Energy, Moscow, 1965.
28. V. I. Agafonov et al., "Study of and improvements in the properties of the plasma
    in the noncylindrical Z pinch," Preprint No. 2017, I.V. Kurchatov Institute of
    Atomic Energy, Moscow, 1970.
29. V. F. D'yachenko, Zh. Vychslit. Matem. i Matem. Fiz. 5:680 (1965).
30. D. E. Potter, Phys. Fluids, 14:1911 (1971).
31. D. E. Potter and M. G. Haines, Paper CN-28/D-8, Fourth Conference on Plasma
    Physics and Controlled Nuclear Fusion Research, Madison, Wisconsin, 1971.
32. G. Guderley, Luftfahrtforschung, 19:302 (1942).
33. V. G. D'yachenko and V. S. Imshennik, Prikl. Matem. i Mekh. 29:993 (1965).
34. V. A. Gribkov et al., Pis'ma Zh. Eksp. Teor. Fiz. 15:329 (1972) [JETP Lett. 15:232
    (1972)].

35. T. H. Sitx, The Theory of Plasma Waves, McGraw-Hill, New York (1962).

36. A. B. Mikhailovskii, Theory of Plasma Instabilities, Vol. 2, Consultants Bureau, New York (1973).

37. S. Chandrasekhar, Hydrodynamic and Hydromagnetic Stability, Claredon Press, Oxford (1961), Ch. 10.

38. R. J. Tayler, Proc. Phys. Soc. B70:31 (1957).

39. V. D. Shafranov, in: Plasma Physics and the Problem of Controlled Thermonuclear Reactions [in Russian], Izd. Akad. Nauk SSSR, Moscow (1958), p. 130.

40. Yu. I. Galushkin, Multiply Charged Impurities in the Plasma Focus. Dissertation [in Russian], MFTI (1971).

41. Yu. I. Galushkin and V. I. Kogan, Nucl. Fusion, 11:597 (1971).

42. R. W. P. McWhirter and A. G. Hearn, Proc. Phys. Soc. B82:641 (1963). R. W. P. McWhirter, in: Plasma Diagnostic Techniques (ed. R. E. Huddlestone and S. L. Leonard), Academic Press, New York (1965).

43. R. J. Bottoms, et al., Plasma Physics and Controlled Nuclear Fusion Research, Vol. 2, IAEA, Vienna (1969), p. 67.

44. P. D. Morgan, H. J. Peacock, R. J. Speer, and P. D. Wilcock, Plasma Physics and Controlled Nuclear Fusion Research, Vol. 2, IAEA, Vienna (1969).

45. V. S. Imshennik, T. I. Filippova, and N. O. Filippov, Nucl. Fusion (in press).

46. C. Maisonnier, F. Pecorella, J. P. Rager, and M. Samuelli, "A model for the dense plasma focus," Fifth European Conference on Controlled Fusion and Plasma Physics, Grenoble, 1972.

47. V. S. Imshennik, S. M. Osovets, and I. V. Otroshchenko, Zh. Eksp. Teor. Fiz. 64: 2057 (1973) [Sov. Phys. JETP 37:1037 (1973)].

PLASMA OPTICS

A. I. Morozov and S. V. Lebedev

Chapter 1

INTRODUCTION

§1. Basic Concepts of Plasma Optics

The simplest plasma-dynamic system is evidently a stream of unmagnetized ions with a small velocity spread whose space charge is neutralized by electrons [1]. We refer to such a stream as a "neutralized" ion beam.

Here "unmagnetized" means that the ion gyroradius ρ_i is comparable to or larger than the scale dimensions of the system:

$$\rho_i \gtrsim \Lambda. \tag{1.1}$$

The assumption of a small velocity spread is equivalent to the following inequality in the laboratory coordinate system:

$$\langle |\mathbf{v}| \rangle = \frac{1}{n} \int |\mathbf{v}| f_i \, d\mathbf{v} \gg \frac{1}{n} \int |\mathbf{v} - \langle \mathbf{v} \rangle| f_i \, d\mathbf{v}. \tag{1.2}$$

Here f_i is the ion velocity distribution. Finally, "neutralization of the space charge" means that the stream can be assumed to be quasineutral:

$$n_i \equiv \int f_i \, d\mathbf{v} \approx n_e. \tag{1.3}$$

We assume everywhere below that the ions are singly charged. We refer to that branch of plasma physics which deals with the properties of neutralized ion beams as "plasma optics" [2-4] if (a) collisions of ions with other ions and with other particles can be

301

neglected in a first approximation and (b) the electron temperature T_e is much lower than the average ion kinetic energy,

$$kT_e \ll M \langle v^2 \rangle / 2. \tag{1.4}$$

Here, k is the Boltzmann constant. In this Review we only treat plasma-optical systems in which, to a first approximation, the magnetic field produced by the moving particles (ions and electrons) can be neglected.

The name "plasma optics" is applied to this class of phenomena because in the overwhelming majority of practical problems involving neutralized ion beams it is necessary to deal with the same problems as those which arise in the conventional ion optics of unneutralized, low-density streams. We refer to the classical ion optics of unneutralized streams as "vacuum" or "classical" optics.

As a rule, in classical particle optics the main problem is the "focusing" of the beam. Here "focusing" means a transformation of the stream such that the particles emerging from a single point P_0 of the source are collected at a single point, P_1, after they pass through the optical system (Fig. 1).

With a few exceptions, the focusing problem has been solved so far only in the case in which the spread of the particles with respect to a given parameter is modest. The pertinent parameters are obviously the total particle velocity, the angle made by the velocity with some fixed direction, and the mass-to-charge ratio of the particle.

Depending on the particular problem involved, we require focusing with respect to certain parameters and blurring (dispersion) of the image with respect to others. For example, in classical particle (electrons and ions) systems designed for producing an image (microscopes, kinescopes, etc.) the dispersion with respect

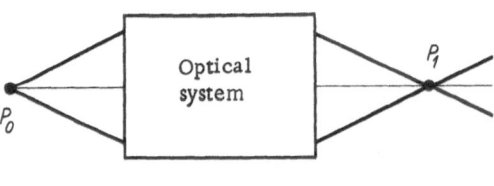

Fig. 1

to all parameters, with the possible exception of the mass, must be negligibly low. In contrast, in energy analyzers we want a large energy dispersion and a dispersion as small as possible in terms of the other parameters.

In plasma optics it is also necessary to focus and isolate an ion stream, but in contrast with classical optics there must be no deviation from neutrality. This feature requires a review of classical systems.

If the neutralized ion beam moves in a magnetic field, and the electron temperature is very low, such a beam obviously behaves in the same way as a stream of single ions, with the exception of possible instabilities. This result is demonstrated convincingly in electromagnetic isotope separators. Only an external magnetic field is used in these devices [1].

The situation is much more complicated when the neutralized ion beam is also subjected to an external electric field. Obviously, the electrostatic elements (lenses, energy analyzers, etc.) of vacuum ion optics, designed for operation with beams of particles with the same sign, will not work if the beam is neutral, since the electrons will tend to shield the electric field from the beam volume. It is necessary to develop new approaches and this is the problem of the present Review.†

Plasma optics is interesting from various points of view. Physically, plasma optical systems are attractive because of their simplicity and amenability to experimental study. Unfortunately, these possibilities have not yet been properly exploited,‡ although neutralized ion beams are easily produced with ordinary ion sources. Physically, plasma optics is closely related to plasma electronics, which is the study of processes which occur when an electron beam moves through an ion or plasma background.

In plasma optics we have the opposite situation — the ions move through an electron background. Because of the finite dimensions of a neutralized beam, the problems of electron and ion beams

---

†The principles of plasma optics discussed below were first formulated in their general form in [2]. Among the earlier papers we cited [21-23].

‡There are a few exceptions to this statement, e.g., [4-7].

are not the same. It is this distinction which makes plasma electronics and plasma optics different.

Much experimental and theoretical work has been reported on plasma electronics, and this work has played an important role in the development of plasma physics in general [8-10].

The development of plasma optics is related to the increasing need for high-current ion beams. Such beams produce and incorporate electrons, which neutralize the space charge. The production of electrons can result from ionization of the residual gas by fast ions and electrons or from the ejection of secondary electrons from the walls of the channel through which the beam moves. To get an idea of the tendency toward neutralization of the beam space charge, we estimate the electric field E_0 at the surface of the beam, and the distance L over which the radius of an unneutralized circular beam increases by a factor of two as a result of self-repulsion due to the space charge.

The field at the beam surface is

$$E_0 = \frac{2\pi R^2\, en}{R} = \frac{2}{R} \cdot \frac{\mathcal{J}}{v} = \frac{2}{R} \cdot \frac{\mathcal{J}}{(Mv^2/2e)^{1/2}} \left(\frac{M}{2e} \right)^{1/2}. \qquad (1.5)$$

It is evident that the distance in which the beam radius increases by a factor of two is

$$L = N2R \approx 2R\, \frac{(Mv^2/2e)^{3/4}}{\sqrt{\mathcal{J}}} \left(\frac{e}{2M} \right)^{1/4} = \frac{2v}{\omega_{0i}};$$
$$\omega_{0i}^2 = 4\pi e^2\, n/M. \qquad (1.6)$$

Here N is L divided by the beam diameter, $\mathcal{J}$ is the ion current in the beam, M is the ion mass, v is the ion velocity, and n is the ion density.

For an ion energy of 1 keV and a beam current of 1 mA, the beam radius is doubled at $N \approx 5$ for hydrogen ions or $N \approx 12$ for argon ions.

At the surface of a hydrogen beam we find $E_0 \approx 45$ V/cm; for an argon beam we find $E_0 \approx 290$ V/cm. At a current of only 1 mA the beam diverges rapidly and produces a strong electric field in its vicinity. On the other hand, currents of the order of amperes or tens or hundreds of amperes are required in many present-day scientific and engineering applications; these beams can exist only in a neutralized state.

Ion currents of the order of milliamperes or tens of milliamperes are easily produced with ordinary ion sources. In these devices the ions are produced by surface ionization or in a gas discharge [11]. The ions are extracted through a narrow slit by means of a potential difference which is applied between the ion source and an electrode provided for the purpose. The ions are not only extracted but are also accelerated by this potential difference. Since the ion space charge is not neutralized by electrons in the acceleration zone, the maximum current that can be obtained per unit area of emitter is limited by the Childs—Langmuir "three-halves" law:

$$j_i = \frac{1}{9\pi} \sqrt{\frac{2e}{M}} \cdot \frac{U^{3/2}}{d^2}. \tag{1.7}$$

Here U is the extracting voltage, and d is the distance between the emitter and the extracting electrode. The space charge of the ion beam taken from the source is neutralized outside the source. A few ion sources capable of producing currents of about 1 A have been produced, but only through the use of high accelerating voltages (of the order of several tens of kilovolts) and long extracting gaps.

The limitations on ion sources have stimulated work on plasma accelerators, in which ions are accelerated under conditons of neutrality. The system which best approximates a plasma-optical system is the Hall-current accelerator (or "Hall accelerator" or "accelerator with a closed drift circuit"), which is described in § 2 below. These accelerators can produce essentially unlimited currents (up to 1000 A at particle energies ranging upward from 100 eV) over a broad energy range. The development of plasma accelerators has raised the problem of control and transformation of these high currents. The study of plasma optics must include all these elements.

In the present Review we focus on the problem of the transformation of an existing neutralized ion beam. The beam source is assumed to be given.

For the convenience of the reader we now define certain important terms.

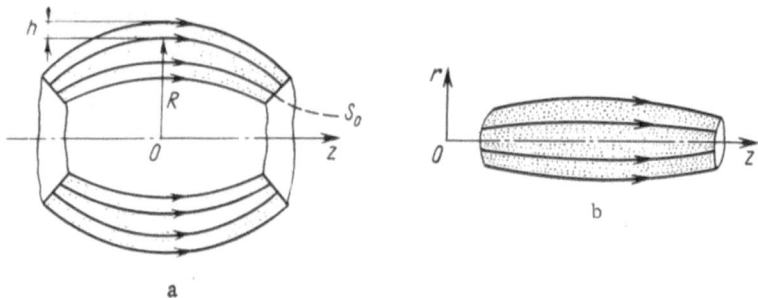

Fig. 2

1. We will treat point sources of infinitesimally narrow annular sources; we refer to the latter simply as "annular sources."

2. The beam produced by such an annular source is hollow (Fig. 2a). The beam produced by a point sources is "axial" (Fig. 2b). The corresponding plasma-optical systems are "annular" and "axial."

3. In the axisymmetric case the transverse dimensions of an axial beam are characterized unambiguously by the beam radius r. In a circular hollow beam two geometric characteristics are needed; the beam radius R and the beam half-width h (Fig. 2a). In an axial beam obviously $R = h$. The arrows in Fig. 2 show the ion trajectories.

4. The narrowest beam cross section is the "image" of the source. Broadening of the image of an infinitesimally narrow source is an "aberration."

5. We use the terms "single-particle" model or "skeleton" of the plasma-optical system to denote the set of fields and trajectories which constitute the solution of the problem that holds for $T_e = 0$.

6. A neutralized ion beam whose parameters can be written as a power series in the transverse coordinate in the single-particle approximation is "paraxial," regardless of whether the beam is axial or hollow. In this case a plasma configuration (model of a neutralized ion beam with a nonvanishing temperature T_e) that corresponds to a paraxial beam cannot, in general, be treated through a series expansion.

7. If the description of a paraxial beam is restricted to the first nonvanishing terms of the expansion in terms of the transverse coordinates, the beam is "narrow." Otherwise, the beam is "broad."

8. An ion trajectory selected on the basis of a given parameter in an axial system (single-particle model) is called the "principal" trajectory. In the axisymmetric case which is the simplest, this trajectory is the geometric axis of the system.

In annular systems we will frequently isolate a set of trajectories that form a "principal" surface. This surface can be complicated in shape, but still axisymmetric. Introducing a coordinate system in the channel in order to calculate the fields and trajectories, we would find it natural to use the principal surface as one of the coordinate surfaces; in an annular system this is frequently not the best choice, for two reasons: First, at the beginning of the calculation the dimensions and shape of the principal surface are not known; second, the shape of the main surface is generally too complicated to permit the use of this surface as a coordinate surface. Accordingly, we choose the "reference" coordinate surface to be a cylindrical surface of radius R_0, which is called the "initial" surface.

9. A calculation of the equilibrium configuration of a neutralized ion beam not only involves the nonvanishing value of T_e in the region occupied by the fast ions, but also a solution of the plasmadynamic equations in the volume between the core of the quasineutral ion beam (the "core" is the region occupied by the fast particles) and the channel wall.

If there is a plasma in this volume, the neutralized ion beam is said to have a "plasma sheath."

This plasma sheath can arise because of various factors, e.g., ionization of the residual gas and the ejection of the resulting ions from the core of the beam to the walls. A plasma sheath is also found in an ion source in which part of the beam is lost at the channel walls. We refer to these operating regimes of the plasmaoptical system as "coupled" regimes. Finally, it is possible to set up conditions such that the existence of ions between the beam core and the wall can be neglected. In this case we will speak in terms of an "electron sheath" of the neutralized ion beam, and refer to these regimes as "uncoupled."

§2. Electric Field in the Plasma for $T_e = 0$

1. Equipotential Nature of the Magnetic Lines of Force.

The basic factor that disrupts the electric field in the plasma is the high electron mobility. Accordingly, in order to determine the conditions under which an electrostatic field can exist in the neutralized ion beam it is necessary to examine the dynamics of the electron fluid. We describe the behavior of this fluid by means of the hydrodynamic equations:

$$\frac{m_e}{e} \cdot \frac{du}{dt} = -\left(\frac{\nabla p_e}{en} + \mathbf{E} + \frac{1}{c}[\mathbf{u}, \mathbf{H}] + \frac{\mathbf{j}}{\sigma} \right); \qquad (1.8a)$$

$$(\partial n/\partial t) + \operatorname{div} n\mathbf{u} = 0. \qquad (1.8b)$$

Here, m_e is the electron mass, $\mathbf{u}$ is the directed electron velocity, σ is the conductivity, and $p_e = kT_e n$ is the electron pressure.

In writing Eqs. (1.8) we have neglected the production and loss of electrons as well as collisions of electrons with neutrals. We have only taken account of collisions of electrons with ions, as described by the Coulomb conductivity σ.

It follows from Eq. (1.8a) that there are four factors that can produce an electric field in the plasma [12]: inertial effects $(m_e/e)(du/dt)$; the electron pressure, $-\nabla p_e/en$; the Lorentz force, $(1/c)[\mathbf{u}, \mathbf{H}]$; and the friction between electrons and ions, $\mathbf{j}/\sigma$. The inertial term and ohmic resistance term are extremely small in a neutralized ion beam. Specifically, for a characteristic electron velocity of $u \approx 10^7$ cm/sec and a scale length of 1 cm for the inhomogeneities, the equivalent electric field due to inertial effects is of the order of 0.1 V/cm.

If the electron temperature is of the order of 1 eV, and if the current density in the neutralized ion beam is of the order of 0.1 A/cm^2, the electric field due to the finite conductivity is again of the order of 0.1 V/cm. Accordingly, we discard these small terms and write the following dynamic equation for the electron component:

$$\mathbf{E} = -(1/c)[\mathbf{u}, \mathbf{H}] - (\nabla p_e/en). \qquad (1.9)$$

Thus, in a neutralized ion beam the electric field can only exist in the plasma as a result of the Lorentz force or a gradient of the electron pressure.

In defining a neutralized ion beam above we have noted that the electron temperature is assumed to be low. In many cases it is

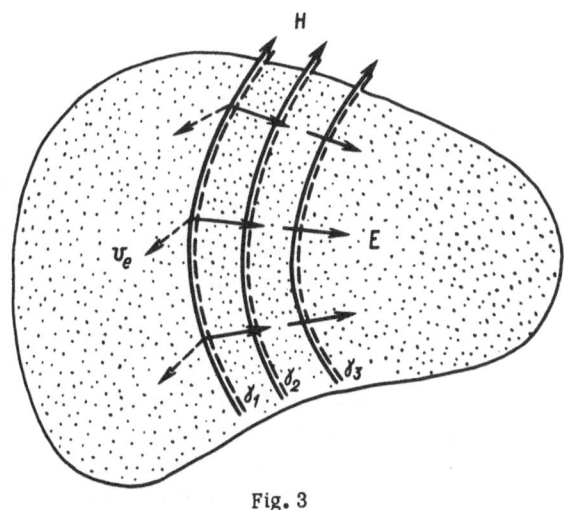

Fig. 3

then permissible to neglect the electron pressure in a first approximation. Then

$$\mathbf{E} = -(1/c)[\mathbf{u}, \ \mathbf{H}].$$ (1.10)

In order for an electric field to exist in a cold plasma there must be a transverse magnetic field with electron drift across **H**.

Equation (1.10) is the basic equation of the electric field when $T_e = 0$, the single-particle approximation, as follows from the definition above. Evidently, Eq. (1.10) does not include the particle density explicitly, so that the calculation of the ion dynamics can be carried out as if there were only a single particle.

The discussion below will be clearer if we assign each magnetic line of force an index γ. Assuming the electric field to be a potential field — clearly a legitimate assumption under equilibrium conditions — and using (1.10), we can write [2]

$$\Phi = \Phi \ (\gamma).$$ (1.11)

For the assumptions being used here this equation reflects an important feature of the electric field in the plasma: the magnetic lines of force are equipotentials† (Fig. 3).

---

†When ∇p_e is taken into account and it is assumed that $T_e = T_e (\gamma)$, of (1.11) is the following equation [2]:

$$\Phi - [k T_e \ (\gamma)/e] \ \ln \ (n/n_0) = \Phi^* \ (\gamma).$$

We call Φ^* the "thermalized potential" (see Chap. 5).

The role played by Eq. (1.11) in plasma optics with $T_e = 0$
is the same as that played by the Laplace equation $\Delta \Phi = 0$ in
vacuum ion optics. It might be said that plasma optics with $T_e = 0$
is the particle optics of systems with magnetic lines of force which
are equipotentials.

By choosing various configurations of the magnetic field and
various functions $\Phi(\gamma)$, we can produce electric fields with a wide
variety of configurations. The surfaces $\Phi = \text{const}$ are the surfaces
along which the electrons drift.

In principle, there are two classes of systems; these are dis-
tinguished by the nature of the electron drift or, equivalently, by
the shape of the surfaces $\Phi = \text{const}$. A system of the first kind is
called a system with an "open drift contour" while a system of the
second kind is a system with a "closed drift contour." In the first
case the electrons starting at the emitter (cathode) 2 move through

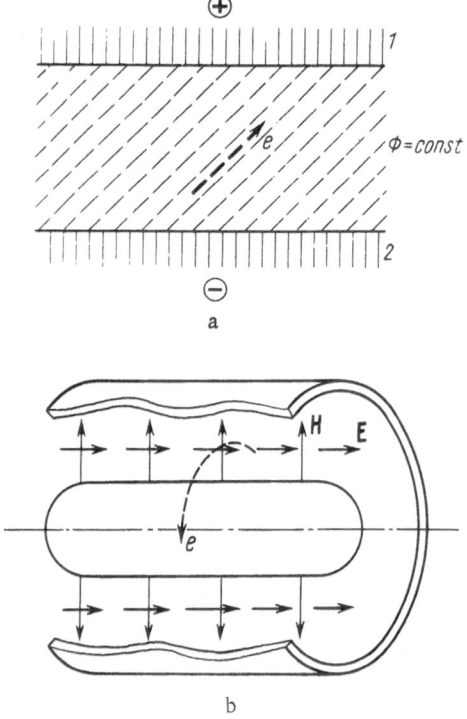

Fig. 4

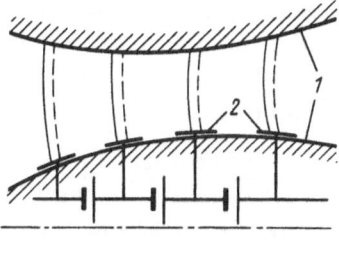

Fig. 5

the plasma and eventually arrive at a collector (the anode) 1 (Fig. 4a). In the second case, the electrons that enter the plasma volume ideally remain in this volume (Fig. 4b). It is this second case which is of primary interest for problems involving neutralized ion beams.† Both cases of electron drift occur naturally in axisymmetric systems.

The simplest example of an axisymmetric system with a closed drift contour is a device in which the magnetic lines of force lie in meridional planes. Under static conditions the electric field obviously has no azimuthal component, so that the electrons drift in the azimuthal direction. The electron gyroradius $R_{e\Lambda}$ must be smaller than the scale dimensions of the system, Λ,

$$R_{e\Lambda} \ll \Lambda, \tag{1.12}$$

while the ion gyroradius $R_{i\Lambda}$ must be comparable to or larger than Λ, by virtue of condition (1.1).

Up to this point we have assumed that the function $\Phi(\gamma)$ is given. Actually, this function is governed by the properties of the particular system and by the processes which occur in it. In principle, the simplest method for specifying $\Phi(\gamma)$ is to use a system of electrodes 2 that cover the channel walls 1 (Fig. 5). Within an error corresponding to the potential jump at the electrode in the case $T_e \rightarrow 0$ the electrode potential "propagates" along the entire line of force, intersecting it.

In general, experiment confirms the effectiveness of this method for obtaining a particular function $\Phi(\gamma)$ under actual conditions

---

†See [13] for the theory of the ideal system with an open drift circuit.

(Chap. 5, §1). If the channel walls are covered with a dielectric, this function is governed by the electron conductivity of the plasma-filled volume (Coulomb, wall, or anomalous conductivity). This question is discussed in detail in Chap. 5.

2.   Field Configuration in an Axial Lens.   To get a clear picture of the physical consequences of Eq. (1.11), we consider the axial lens [3] (Fig. 6). If an electric current J flows along a ring of radius R, which forms the lens, we have a "thin" magnetic lens with the lines of force shown in Fig. 6a. Below

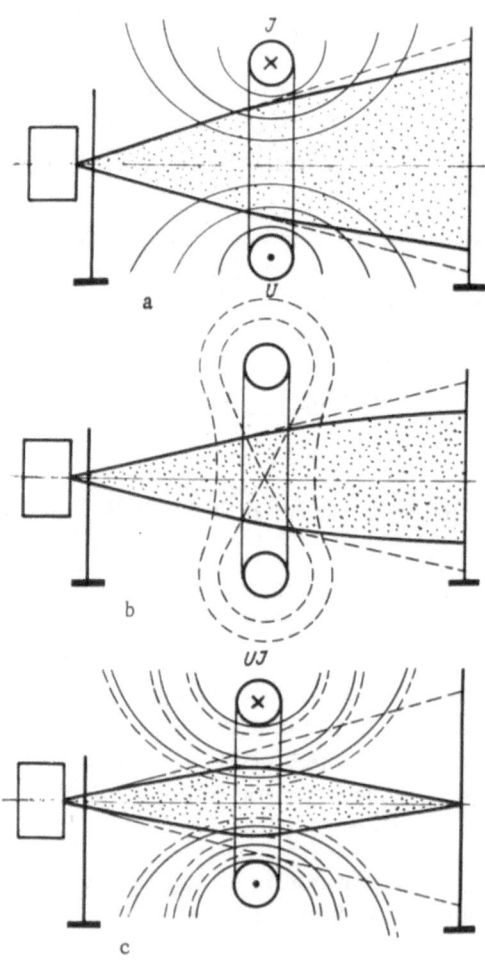

Fig. 6

(Chap. 3, § 1) we show that the focal length of this lens is given by the equation

$$\frac{1}{F_H} = \frac{3\pi^3}{16} \cdot \frac{e^2}{mc^2} \cdot \frac{1}{\mathscr{E}} \cdot \frac{\mathscr{J}^2}{c^2 R},$$ (1.13)

where $\mathscr{E} = mv^2/2$ is the ion energy.

This ring can also be converted into an electrostatic lens (a vacuum lens) if a potential U is applied to it. Figure 6b shows the resulting equipotential diagram. A characteristic feature of the particle motion in this field is that the particles are focused in certain regions (deflected toward the axis) and defocused in others.

As a result, the focal length of such a lens is quadratic in U, being given by

$$\frac{1}{F_E} = \frac{3e^2}{16\,(\mathscr{E})^2} \cdot \frac{\pi U^2}{8R}.$$ (1.14)

Since F is proportional to the square of U, the lens is a converging lens, regardless of the sign of the applied potential. If we now pass a current through the ring and simultaneously impose the same potential U on the ring, we find no qualitative change in the vacuum case, in the sense that the focusing effects of the magnetic and electric fields add, and the overall focal length becomes

$$F_0 = F_E F_H/(F_E + F_H).$$ (1.15)

The situation becomes qualitatively different if a dense ion beam, capable of neutralizing its own space charge, is passed through such a lens. In this case, because of the high electron mobility, the electric field in the beam volume is modified because the magnetic lines of force become equipotentials. The electric field configuration in this case is shown in Fig. 6c. In the beam volume the electric field is obviously directed either toward the axis or away from it, depending on whether the potential applied to the ring is positive or negative with respect to ground. As a result, the "plasma lens" can either focus or defocus the ion beam. Accordingly, the plasma lens has a qualitative advantage over its vacuum analog. Furthermore, its refractivity turns out to be far larger than that of the vacuum lens:

$$1/F_{E_p} = (2U/\mathscr{E}) \cdot (\theta/R).$$ (1.16)

The dimensionless parameter $\theta \approx 1$ and depends on the geometry of the system.

As an example, assume that 10-keV argon ions are to be focused. Choosing a magnetic field of 400 Oe and a potential of $U = 1$ keV, we find the focal lengths from Eqs. (1.13), (1.14), and (1.16), respectively, to be $F_0 = 10^4$ cm and $F_{Ep} = 50$ cm.

Experiments carried out by Zhukov [4] have confirmed both these arguments and Eq. (1.16). Zhukov has also shown that the function $\Phi(\gamma)$ can be controlled by wall electrodes (see Chap. 5).

3. Hall-Current Plasma Accelerators. With minor reservations, we can also classify Hall-current plasma accelerators (with a closed drift contour) as plasma-optical systems. Figure 7 shows the simplest "single-lens model" [14-16], of such an accelerator. In this accelerator, the magnetic circuit 1 and coils 3 produce a meridional magnetic field which is approximately radial at the main surface. The magnetic field at this surface has a maximum near the end of the channel. As a result, the magnetic lines of force in the channel have a lenslike geometry.

An electric field is imposed between anode 2, which lies within the channel formed by dielectric 4, and an electron emitter 5, which is outside the channel. The anode usually also serves as a gas distributor, supplying the working medium to the channel. Since the magnetic lines of force are equipotentials, the electric field in the channel is independent of the cathode position. Accordingly, the electrons in the channel drift in crossed **E** and **H** fields.

Neutral atoms coming from the anode enter the cloud of rotating electrons and are ionized. An electron resulting from this ioniza-

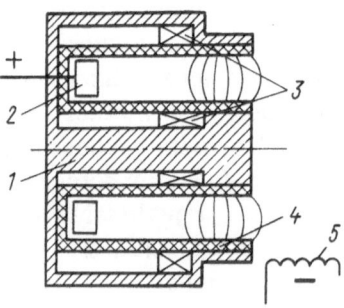

Fig. 7

tion process reaches the anode as the result of classical and anomalous conductivity, while an ion, accelerated by the electric field, leaves the channel. The accelerator parameters are such that the electrons are magnetized,[†] $R_{e\Lambda} \ll \Lambda$, while the ions are not, $R_{i\Lambda} \gg \Lambda$. Here R_Λ is the particle gyroradius and Λ is the length of the channel.

In addition to the single-lens accelerators, there is interest in two-lens accelerators [19, 40–42] and multiple-lens accelerators [20] (Chap. 3, § § 3 and 4; Chap. 4, § 1). It was in two-lens accelerators that ion currents of hundred of amperes at energies of several kilovolts were first achieved. We will examine the single-particle model of an accelerator with a very large number of lenses below (Chap. 3, § 3).

Some experimental data on the processes occurring in Hall-current accelerators are reported in Chap. 5.

4. C o n v e r t i b l e   a n d   N o n c o n v e r t i b l e   S y s t e m s .
The plasma-lens example discussed above gives an extremely clear picture of the particular features of plasma optics. It is clear from this example that in going from the vacuum regime to the plasma regime we generally find important changes in the properties of the system. Below we refer to these modified systems as "convertible" systems.

However, there is also great interest in optical systems whose optical properties do not change, in a first approximation, when the space charge of the ion beam is neutralized. We call these systems nonconvertible system. Let us analyze the condition under which a system is nonconvertible.

In the axisymmetric vacuum case the electric field is described by a potential $\Phi(r, z)$, which satisfies the Laplace equation

$$\Delta\Phi \equiv \frac{1}{r} \cdot \frac{\partial}{\partial r}\, r\, \frac{\partial\Phi}{\partial r} + \frac{\partial^2\Phi}{\partial z^2} = 0. \qquad (1.17)$$

[†]It should be noted that there is a particular type of Hall-current plasma accelerator in which the acceleration zone is $R_{e\Lambda}$ [17, 18, 42].

In the plasma case, the potential Φ satisfies Eq. (1.11), which can be written in the axisymmetric case as†

$$\Phi = \Phi(\psi). \tag{1.18}$$

Here ψ is the magnetic stream function.

Assuming that ψ satisfies the modified Laplace equation (2.9), and substituting (1.18) into (1.17), we find the general condition

$$\begin{cases} \Phi'' \left[\left(\dfrac{\partial \psi}{\partial r} \right)^2 + \left(\dfrac{\partial \psi}{\partial z} \right)^2 \right] + \Phi' \dfrac{2}{r} \cdot \dfrac{\partial \psi}{\partial r} = 0; & (1.19a) \\[3mm] \Delta^* \psi = 0; \quad \Delta^* = r \dfrac{\partial}{\partial r} \cdot \dfrac{1}{r} \cdot \dfrac{\partial}{\partial r} + \dfrac{\partial^2}{\partial z^2}. & (1.19b) \end{cases}$$

Here the prime denotes the derivative with respect to ψ. System (1.19) has one approximate but very general solution:

$$\begin{cases} \Phi = k\psi, \quad r \to \infty; \\ \partial^2 \psi / \partial r^2 + \partial^2 \psi / \partial z^2 = 0. \end{cases} \tag{1.20}$$

This solution corresponds to a two-dimensional field, that is to say, the channel width is much smaller than the distance from the channel to the axis of the system. The exact solutions of system (1.19), on the other hand, are limited to two particular cases [2]:

$$\psi = -(H_0 r^2/2); \quad \Phi = -E_0 a \ln(r/a); \tag{1.21a}$$

$$\psi = -H_0 z a; \quad \Phi = -E_0 z, \tag{1.21b}$$

where H_0, E_0, and a are arbitrary constants.

Figures 8a and 8b show the geometry of systems in which the fields in (1.21a) and (1.21b), respectively, occur.

We can show that solutions (1.21) exhaust the class of exact solutions of system (1.19). The first equation in system (1.19) is a nonlinear, first-order partial differential equation for the function ψ. Solving this equation by the Lagrange method, we find the

---

†See Chap. 2, §1, for further details.

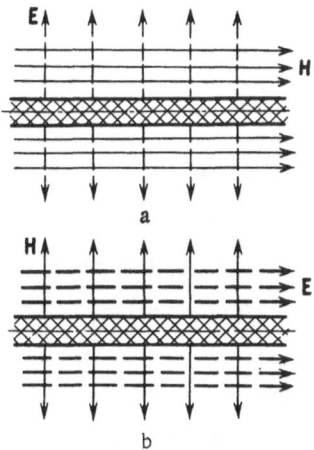

Fig. 8

characteristic system of equations:

$$\frac{dr}{2\,\dfrac{\partial z}{\partial x}-\dfrac{1}{r}\,f(\psi)}=\frac{dz}{2\,\dfrac{\partial z}{\partial y}}=\frac{dy}{2\left(\dfrac{\partial z}{\partial x}\right)^2+2\left(\dfrac{\partial z}{\partial y}\right)^2-\dfrac{1}{r}\cdot\dfrac{\partial z}{\partial x}\,f(\psi)}=$$

$$=\frac{d\,\dfrac{\partial z}{\partial x}}{\dfrac{\partial z}{\partial x}\cdot\dfrac{1}{r}\left(\dfrac{\partial z}{\partial x}\cdot\dfrac{\partial f}{\partial y}-\dfrac{f(\psi)}{r}\right)},\qquad (1.22)$$

where

$$f=-\frac{2}{(d/d\psi)\,\ln\,[\partial\Phi\,(\psi)/d\psi]}.\qquad (1.23)$$

Hence we can write the additional equation

$$\partial z/\partial y = Cf\,(\psi); \quad C = \text{const.}\qquad (1.24)$$

We note that solutions like that in (1.21) are degenerate in the sense of the Lagrange method because if the solution is independent of one of the coordinates the auxiliary equation is meaningless. Accordingly, solutions (1.21) are not included in the general solution of Eq. (1.19a) obtained by this method. Determining the quantity $\partial z/\partial x$ from Eq. (1.22) and substituting $\partial z/\partial x$ and $\partial z/\partial y$ into the

Pfaffian equation, we can reduce the solution to quadratures:

$$\int^{?} \frac{d\psi}{f(\psi)} = \frac{1}{2} \ln r \pm \frac{1}{4} \ln \frac{1 - \sqrt{1 - 4C^2 r^2}}{1 + \sqrt{1 - 4C^2 r^2}} \pm$$

$$\pm \frac{1}{2} \sqrt{1 - 4C^2 r^2} + Cz + \text{const.} \qquad (1.25)$$

Substituting (1.23) into (1.25) for $f(\psi)$, we find

$$d\Phi(\psi)/d\psi = F(r, z) = F(\psi), \qquad (1.26)$$

where

$$F(r, z) = \frac{\text{const}}{\sqrt[4]{r}} \left| \frac{1 + \sqrt{1 - 4C^2 r^2}}{1 - \sqrt{1 - 4C^2 r^2}} \right|^{\pm \frac{1}{2}} \times$$

$$\times \exp\left(\pm \frac{1}{4} \sqrt{1 - 4C^2 r^2}\right) \exp(-2Cz). \qquad (1.27)$$

To find the explicit function $\psi(r, z)$, we must know the function $\Phi(\psi)$. However, the function ψ determined by (1.26) and (1.27) must satisfy Eq. (1.19b). It follows from (1.26) that

$$\psi = \psi(F). \qquad (1.28)$$

Substituting this expression into (1.19b), we find the following equation for the function ψ:

$$d \ln (d\psi/dF)/dF = \Delta^* F/(\nabla F)^2, \qquad (1.29)$$

where

$$(\nabla F)^2 \equiv (\partial F/\partial r)^2 + (\partial F/\partial z)^2.$$

The right side of Eq. (1.29) must obviously also be a function of F. From Eqs. (1.26) and (1.29) we find

$$\Delta^* F/(\nabla F)^2 = 1/F. \qquad (1.30)$$

Equation (1.20) can be solved by separation of variables by writing the function F(r, z) in the form

$$F(r, z) = R(r)Z(z).$$

Substituting this product into Eq. (1.30), we find

$$r \frac{\partial}{\partial r} \frac{1}{r} \frac{\partial R}{\partial r} \frac{1}{R} - \frac{1}{R^2} \left(\frac{\partial R}{\partial r} \right)^2 = \frac{1}{Z^2} \left(\frac{\partial Z}{\partial z} \right)^2 - \frac{\partial^2 Z}{\partial z^2} \cdot \frac{1}{Z^2} = \mu = \text{const.}$$

(1.31)

Replacing $Z(z)$ by $Z = \exp(-2Cz)$ we have $\mu = 0$. Thus function $R(z)$ must satisfy

$$R \ (r) \ (d^2 R/dr^2) - (1/r)R \ (dR/dr) - (dR/dr)^2 = 0. \tag{1.32}$$

Direct substitution shows that the function $R(r)$ from (1.27) does not satisfy this equation. Consequently, when irrotational magnetic fields are used, the class of nonconvertible systems is exhausted by (1.21).

While the characteristics of nonconvertible systems do not change in going from a low-density, unneutralized ion beam to a dense, neutralized ion beam, the characteristics of convertible systems change only if the density of the neutralized ion beam exceeds a certain critical density n*. What is this density? In the convertible case the potential $\Phi(\psi)$ satisfies the Poisson equation

$$\Delta \Phi \ (\psi) = -4\pi e \nu. \tag{1.33}$$

Here $\nu \equiv n_i - n_e$ is the density of the unneutralized particles.

The critical density n* must obviously be at least an order of magnitude larger than ν, so that the following condition holds for the plasma-optical regime:

$$n > n^* \approx 10\nu. \tag{1.34}$$

Here n is the ion density in the neutralized ion beam.

Using Eqs. (1.17), (1.18), and (1.33), we find

$$\nu = \frac{1}{4\pi e} \left| \Phi'' \left[\left(\frac{\partial \psi}{\partial r} \right)^2 + \left(\frac{\partial \psi}{\partial z} \right)^2 \right] + \Phi' \frac{2}{r} \cdot \frac{\partial \psi}{\partial r} \right| \lesssim \frac{n}{10}. \tag{1.35}$$

If Φ is a linear function of ψ, the quantity ν is

$$\nu = -\frac{2}{r} \cdot \frac{\partial \Phi}{\partial r} \cdot \frac{1}{4\pi e} = \frac{2Er}{r} \cdot \frac{1}{4\pi e}. \tag{1.36}$$

Equation (1.36) is equivalent (in order of magnitude) to the estimate

$$v/n \approx (D/\Lambda)^2 ,$$ (1.37)

where D is the Debye length calculated from the ion velocity, and Λ is the scale dimension of the system. This estimate only holds at low values of T_e.

5. **Axisymmetric System with an Azimuthal Magnetic Field.** If the electric and magnetic fields have no azimuthal components, and if electrons do not flow along the magnetic lines of force, the only macroscopic motion of the electrons is a drift with velocity

$$u_{dr} = c \, [\mathbf{E,H}]/H^2,$$ (1.38)

which is strictly azimuthal.

It turns out that the closed-contour electron drift can also occur if the azimuthal magnetic field is nonvanishing. The physical meaning of this phenomenon can be illustrated by the following simple example: Assume an accelerator of the single-lens type, described above, and assume that the electric field is constant, being directed along the z axis, $\mathbf{E} = (0, 0, E_0)$. We also assume that the magnetic field has both radial and azimuthal components, $\mathbf{H} = [H_{01}(a/r), H_{02}(a/r), 0]$, with $H_{01} = \text{const}$, $H_{02} = \text{const}$. The channel has dielectric walls.

It would seem that the azimuthal magnetic field, along with the homogeneous electric field, would lead to radial drift, but this is

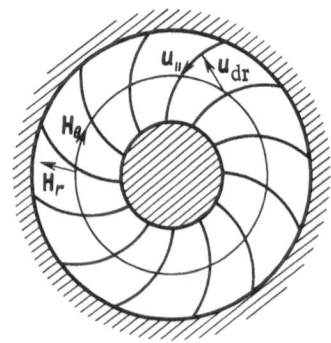

Fig. 9

not necessarily the case: The electrons drifting in the radial direction can also "slide" backward along the lines of force (Fig. 9). As a result, the resulting radial displacement can vanish. To prove this rigorously, we use the following system of equations for the electrons:

$$\begin{aligned} \operatorname{div} n\mathbf{u} &= (1/r)\,(\partial/\partial r)\,rnu_r = 0; \\ E_0 + (1/c)\,(u_r\,H_\theta - u_\theta\,H_r) &= 0. \end{aligned} \right\} \qquad (1.39)$$

It follows from the continuity equation that this regime occurs if $u_r = 0$. Substituting this value into the second equation in (1.39), we find

$$u_\theta = cE_0/H_r. \qquad (1.40)$$

We can now determine the velocity of the electrons along the line of force $u_{\|}$:

$$u_{\|} = \frac{cE_0}{H_r} \cdot \frac{H_{02}}{\sqrt{H_{01}^2 + H_{02}^2}}. \qquad (1.41)$$

From Eqs. (1.40) and (1.41) we see that as the azimuthal magnetic field tends toward zero ($H_{02} \to 0$) the velocity components behave in the following way: $u_{\|} \to 0$, $u_\theta \to (cE_0/H_r)$. If, on the other hand, the radial magnetic field tends toward zero ($H_{01} \to 0$), then $u_{\|} \to \infty$, $u_\theta \to \infty$. These qualitative results remain valid if the electron temperature and the z dependence of the radial field are taken into account.

Consider the system of equations

$$\operatorname{div} n\,\mathbf{u} = 0; \quad (\nabla p_e/en) + \mathbf{E} + (1/c)[\mathbf{u},\,\mathbf{H}] = 0; \quad p_e = p_e\,(n). \qquad (1.42)$$

and introduce the electron stream function χ and the stream function for the meridional magnetic field, ψ:

$$\begin{aligned} rnu_r &= -\partial\chi/\partial z; \quad rnu_z = \partial\chi/\partial r; \\ rH_r &= -\partial\psi/\partial z; \quad rH_z = \partial\psi/\partial r. \end{aligned} \right\} \qquad (1.43)$$

Using $\mathbf{E} = -\nabla\Phi$, we can rewrite (1.42) as

$$\begin{aligned} W_e - e\Phi &= U\,(\psi); \quad \chi = \chi\,(\psi); \\ u_\theta &= \left[\frac{H_\theta}{n}\,\chi'\,(\psi) - rU'\,(\psi)\,\frac{c}{e}\right]; \quad W_e = \int\frac{dp_e}{n}. \end{aligned} \right\} \qquad (1.44)$$

where p_e is the electron pressure.

The second equation in (1.44) shows that if the system is axisymmetric the electrons can only move along the magnetic surfaces ψ = const. If we also require that $u_r = 0$ in the general case (i.e., that the drift contour is closed), then $\chi = 0$.

Thus, $u_z = 0$ and

$$u_\theta = -(rc/e)U'(\psi). \tag{1.45}$$

It is evident that (1.40) follows as a particular case from (1.45): Setting $W_e = 0$, $\Phi = -Ez$, we find

$$U'(\psi) = \frac{dU}{d\psi} = \frac{1}{d\psi/dz} \cdot \frac{dU}{dz} = -\frac{eE}{rH_z}, \tag{1.46}$$

and substituting (1.45) into the third equation in (1.44), we find (1.40).

Systems with a closed drift contour and an azimuthal magnetic field are undoubtedly more complicated to construct. In particular, in an accelerator with $H_\theta \neq 0$ the contour for the longitudinal current can be closed by means of conductors located in the emerging plasma stream. However, these complications are offset by new opportunities. For example, without any fundamental changes,[†] existing mass-separator designs with a triad of mutually perpendicular vectors $\mathbf{v}$, $\mathbf{E}$, $\mathbf{H}$ can be modified to separate intense ion currents, at least when $T_e \to 0$.

§3. Analysis of Plasma-Optical Systems

In conclusion we turn to a method for systematic analysis of plasma-optical systems. This analysis should be carried out in four steps.

The first step involves the construction of the single-particle model (skeleton) for the plasma-optical system. In the case of classical ion optics, the calculation obviously consists of this step alone, except that the Laplace equation is used instead of the condition that the magnetic lines of force be equipotentials. The single-particle model treats three tasks:

1. We "guess" at a general scheme for the system which is capable of solving the problem at hand.

[†] It is sufficient to make them axisymmetric, as in the approach taken in [2] with a cylindrical capacitor.

2. We choose an optimum principal trajectory and optimum fields $\psi(r, z)$, $\Phi(\psi)$ on the basis of this guess.

3. We analyze the aberrations of the system and methods for correcting them.

The reason for taking this approach to the design of a particle optical system is that a general method is not now available for choosing a fundamental scheme capable of solving the particular problem. Since the general layout of the optical system must be guessed at, it is important to have a certain number of crude models. In Chapters 2-4 below we deal with the choice of these crude models for plasma-optical systems.

The second step in the calculation for a plasma-optical system involves the construction of the equilibrium plasma configuration. This means that we must (a) take account of the nonvanishing electron temperature T_e, (b) "tie" the neutralized ion beam to the channel walls, and (c) explicitly take account of those mechanisms which lead to the steady-state function $\Phi^*(\gamma)$.

As noted above [Eq. (1.11a)], the consequence of a nonvanishing T_e in the simplest case is that it is the thermalized potential

$$\Phi^*(\gamma) = \Phi - (kT_e/e) \ln(n/n_0)$$

rather than the electric potential $\Phi(\gamma)$ which remains constant on a magnetic line of force.

The density dependence of Φ^* clearly shows that with a nonvanishing T_e the properties of the system cannot be determined in the single-particle approximation. Accordingly, in the second step of the calculation the distribution of the particle density must be determined all the way out to the walls. The nonvanishing value of T_e is taken into account and the neutralized ion beam is "tied" to the walls in Chap. 5.

We emphasize that it is necessary to analyze the volume and wall processes which govern the function $\Phi^*(\gamma)$. As noted above, we only consider plasma-optical systems in which all the magnetic lines of force intersect the walls, so that it is feasible to control the volume of the neutralized ion beam by wall electrodes which absorb or emit electrons. In this way the desired field can be established in the beam volume; furthermore, it is possible to sta-

bilize the oscillations by means of passive or active systems such
as the feedback systems proposed by Morozov and Solov'ev [19].

Under these conditions it is important to calculate the potential drop at the wall; this drop is examined in Chap. 5.

Under actual conditions, both wall effects manifested in $\Phi^*(\gamma)$
and volume effects modify the potential distributions. If $\Lambda \gg \rho_{e\Lambda}$,
volume processes are only important if oscillations capable of
causing an anomalous conductivity are excited in the system. Experiment shows that these effects can dominate, e.g., in a Hall-
current accelerator with simple dielectric walls [15, 20]. If $\Lambda \approx \rho_{e\Lambda}$, on the other hand, two-body collisions become important [17].
In the present paper we restrict the discussion to the wall methods
for controlling the function $\Phi^*(\gamma)$.

The equilibrium configurations of a neutralized ion beam are
complicated, so that extensive experimental work is required. The
available experimental data only deal with particular systems and
do not given an overall picture.

The third step in the analysis of a plasma-optical system is
to study the stability of the neutralized ion beam and to develop
methods for maintaining stability.

Again, stability can only be analyzed by working closely with
experiment. This statement holds for most plasma systems, and
it seems to be particularly appropriate for plasma-optical systems.
The discussion above shows that the plasma configuration in a plasma-optical system is fundamentally two-dimensional and highly
sensitive to wall processes.

Analytic calculations for the stability of two-dimensional systems can only be carried out in isolated cases. Since wall processes
must be taken into account, the possibilities of analytic studies
seem to be extremely limited. For this reason, we will not treat
the behavior of perturbations in plasma-optical systems in the
present review.

Finally, the last step in a study of plasma-optical systems,
which we will not cover in this Review, should include a) the construction of the equilibrium pattern, incorporating turbulent processes which cannot be suppressed, and b) a calculation of the final
optical characteristics of the system.

In summary, in the present Review we examine in detail the construction of single-particle models for plasma-optical systems and analyze certain features of equilibrium configurations of neutralized ion beams.

Chapter 2

ANALYSIS OF A PARAXIAL BEAM IN THE SINGLE-PARTICLE APPROXIMATION

§ 1. Specification of the Magnetic Fields

All plasma-optical systems use a magnetic field. Let us assume that a neutralized ion beam moves in a channel and that all the magnetic lines of force intersect the walls of this channel. We assume the magnetic field to be axisymmetric (Figs. 2a and 2b). Since the magnetic fields produced by the currents flowing in the plasma-filled volume can be neglected, we describe the magnetic field by the scalar magnetic potential Ω,

$$H_p = \nabla\Omega, \tag{2.1}$$

which satisfies the Laplace equation

$$\Delta\Omega = \frac{1}{r} \cdot \frac{\partial}{\partial r} r \frac{\partial\Omega}{\partial r} + \frac{\partial^2\Omega}{\partial z^2} = 0. \tag{2.2}$$

The axisymmetric field is divided into two components:

$$H_p = H + H_\theta. \tag{2.3}$$

Here H_p is the total field in the channel; H is the meridional component of the field, which lies in the (r, z) planes, and is given by

$$H = H_r\,(r,\,z)r_0 + H_z\,(r,\,z)z_0; \tag{2.4}$$

where H_θ is the azimuthal component,

$$H_\theta = [H_0\,(R/r)]\theta_0; \quad H_0 = \text{const.} \tag{2.5}$$

In Eqs. (2.4) and (2.5), r_0, θ_0, and z_0 are the unit vectors in the direction of the corresponding coordinates. Writing Eqs. (2.1) and (2.2) for the meridional field component, we have

$$\Omega = \Omega\,(r, z). \tag{2.6}$$

In addition to using the scalar potential Ω, it is frequently convenient to use the magnetic stream function ψ, which is related to the magnetic field components by

$$H_r = -(1/r)(\partial\psi/\partial z); \quad H_z = (1/r)\,(\partial\psi/\partial r). \qquad (2.7)$$

It is evident that the lines of

$$\psi\,(r,\ z) = \text{const} \qquad (2.8)$$

are the equations of the projection of the magnetic lines of force in the meridional planes r, z. The function ψ satisfies the modified Laplace equation

$$\Delta^*\,\psi = r\,\frac{\partial}{\partial r}\cdot\frac{1}{r}\cdot\frac{\partial\psi}{\partial r} + \frac{\partial^2\psi}{\partial z^2} = 0. \qquad (2.9)$$

In axial systems, the external azimuthal field must vanish. In annular systems the external azimuthal field can be nonzero. In calculations for axial plasma-optical systems the magnetic field at the axis is frequently known:

$$H_r|_{r=0} = 0; \quad H_z|_{r=0} = H_0\,(z). \qquad (2.10)$$

Then Eqs. (2.2) and (2.9) can be used to express Ω and ψ at any distance from the axis as a power series in r:

$$\Omega = \Omega_0\,(z) - \frac{r^2}{4}\cdot\frac{\partial^2\Omega\,(z)}{\partial z^2} + \frac{r^4}{8}\cdot\frac{\partial^4\Omega\,(z)}{\partial z^4} + \ldots = \mathcal{J}_0\left(r\,\frac{\partial}{\partial z}\right)\Omega_0; \qquad (2.11)$$

$$\psi = \frac{r^2}{2}\,H_0\,(z) - \frac{3}{2}\cdot\frac{r^4}{4!}\cdot\frac{\partial^2 H_0\,(z)}{\partial z^2} + \frac{3\cdot 5}{2\cdot 4}\cdot\frac{r^6}{6!}\cdot\frac{\partial^4 H_0\,(z)}{\partial z^4} + \ldots =$$

$$= r\mathcal{J}_1\left(r\,\frac{\partial}{\partial z}\right)\Omega_0; \quad \Omega_0 \equiv \int H_0\,(z)\,dz. \qquad (2.12)$$

Here $\mathcal{J}_0\,(\alpha)$ and $\mathcal{J}_1\,(\alpha)$ are the Bessel functions of order zero and of order unity.

It is particularly convenient to specify the magnetic field by means of a scalar potential when it is necessary to determine the geometry of the pole-pieces of the iron core that produces the field. Since the magnetic field enters the unsaturated core essentially along the normal, to a first approximation the equation of the pole-pieces is

$$\Omega\,(r,\ z) = \text{const}. \qquad (2.13)$$

On the other hand, if the magnetic field is produced by a single-turn superconducting coil, the equation for the intersection of this coil with the θ = const plane is the same as the equation of the lines of force, (2.8). In analyzing annular plasma-optical systems we describe the magnetic field by the components H_{r0} and H_{z0} on the initial surface, which, as noted above, is taken to be a cylindrical surface of radius R. Knowing

$$H_{r0} = H_r\,(R,\ z);\ H_{z0} = H_z\,(R,\ z), \tag{2.14}$$

we can use Eqs. (2.2) and (2.9) to find the magnetic field within the channel as a power series in $y = r - R$:

$$\Omega = \Omega_0 + y\Omega_1 + \frac{y^2}{2}\left(-\Omega_0'' - \frac{1}{R}\Omega_1\right) + \frac{y^3}{3!}\left(-\Omega_1'' + \frac{1}{R}\Omega_0'' + \frac{1}{R^2}\Omega_1\right) +$$

$$+ \frac{y^4}{4!}\left(\Omega^{(\mathrm{IV})} + \frac{2}{R}\Omega_1'' + \frac{3}{R^2}\Omega_0'' - \frac{5}{R^3}\Omega_1\right) + \ldots; \tag{2.15a}$$

$$\psi = \psi_0\,(z) + \psi_1\,(z)\,y + \frac{1}{2}\,y^2\left(-\psi_0'' + \frac{\psi_1}{R}\right) + \frac{y^3}{3!}\left(-\psi_1'' - \frac{1}{R}\psi_0''\right) +$$

$$+ \frac{y^4}{4!}\left(\psi_0^{(\mathrm{IV})} + \frac{\psi_0''}{R^2} - 2\frac{\psi_1''}{R}\right) + \ldots\ . \tag{2.15b}$$

Here $\Omega_0 = \int H_{z0}\,dz;\ \ \Omega_1 = H_{r0};\ \ \psi_0 = -R\int H_{r0}\,dz;\ \ \psi_1 = RH_{z0}.$

The procedure given above is a basis for straightforward analysis of coils and cores in the synthesis of a desired magnetic field.

Let us consider another way to produce the desired field in an annular gap — two infinitesimally thin coaxial coils [24].

Assume for definiteness that on the initial surface

$$H_{z0} = 0;\ H_{r0} = H_0\,f\,(z/\Lambda). \tag{2.16}$$

Here $f(z/\Lambda)$ is a specified smooth function. We write the general solution of the Laplace equation $\Delta\Omega = 0$ as

$$\Omega = \int d\lambda\ [(b_1\,(\lambda)\cos\lambda z + b_2\,(\lambda)\sin\lambda z)I_0\,(\lambda r) +$$

$$+ (c_1\,(\lambda)\cos\lambda z + c_2\,(\lambda)\sin\lambda z)\,K_0\,(\lambda r)], \tag{2.17}$$

where I_0 and K_0 are modified Bessel functions. From (2.16) at the initial surface we find the coefficients $b(\lambda)$ and $c(\lambda)$:

$$\left.\begin{array}{ll} b_1(\lambda) = H_r^{(s)}(\lambda) K_0(\lambda R) \lambda R; & b_2(\lambda) = H_r^{(a)}(\lambda) K_0(\lambda R) \lambda R; \\ c_1(\lambda) = -H_r^{(s)}(\lambda) I_0(\lambda R) \lambda R; & c_2(\lambda) = -H_r^{(a)}(\lambda) I_0(\lambda R) \lambda R. \end{array}\right\} \quad (2.18)$$

Here $H_r^{(s)}(\lambda)$ and $H_r^{(a)}(\lambda)$ are respectively the symmetric and anti-symmetric parts of the Fourier components, and

$$H_0 f(z/\Lambda) = (2/\pi) \int\limits_0^\infty d\lambda \left[H_r^{(s)}(\lambda) \cos \lambda z + H_r^{(a)}(\lambda) \sin \lambda z \right].$$

We now assume that the magnetic field is produced by two infinitesimally thin coaxial coils of infinite length with radii a_1 and a_2 ($a_1 < R < a_2$). The current densities in these coils are

$$j_{\theta 1} = j_1(z)\delta(r - a_1); \quad j_{\theta 2} = j_2(z)\delta(r - a_2). \quad (2.19)$$

We are to determine the functions $j_1(z)$ and $j_2(z)$ from the assumption that the field between the coils is that described by (2.17) while the field outside the coils is regular: $\Omega \to 0$ as $r \to 0, \infty$. Then at $r < a_1$

$$\Omega = \int (B_1(\lambda) \cos \lambda z + B_2(\lambda) \sin \lambda z) I_0(\lambda r) d\lambda, \quad (2.20a)$$

and at $r > a_2$

$$\Omega = \int (C_1(\lambda) \cos \lambda z + C_2(\lambda) \sin \lambda z) K_0(\lambda r) d\lambda. \quad (2.20b)$$

When the current layers at $r = a_1$ and $r = a_2$ are crossed, the following conditions must hold:

$$[H_r] = 0; \quad [H_z] = (4\pi/c) j_\theta(z). \quad (2.21)$$

From the first of these conditions we can find relations between the coefficients $B(\lambda)$, $C(\lambda)$, and the coefficients $b(\lambda)$ and $c(\lambda)$:

$$\left.\begin{array}{l} B_1(\lambda) = b_1(\lambda) + c_1(\lambda) \dfrac{K_0'(\lambda a_1)}{I_0'(\lambda a_1)} ; \\[2mm] B_2(\lambda) = b_2(\lambda) + c_2(\lambda) \dfrac{K_0'(\lambda a_1)}{I_0'(\lambda a_1)} ; \\[2mm] C_1(\lambda) = b_1(\lambda) \dfrac{I_0'(\lambda a_2)}{K_0'(\lambda a_2)} + c_1(\lambda); \\[2mm] C_2(\lambda) = b_2(\lambda) \dfrac{I_0'(\lambda a_2)}{K_0'(\lambda a_2)} + c_2(\lambda). \end{array}\right\} \quad (2.22)$$

From the second condition in (2.21) we find the Fourier components of the current densities $j_{\theta 1}(\lambda)$ and $j_{\theta 2}(\lambda)$:

$$\left.\begin{aligned}
j_{\theta 1}^{(a)}(\lambda) &= \frac{c}{4\pi} H_r^{(s)}(\lambda) \frac{R}{a_1} \cdot \frac{I_0(\lambda R)}{K_1(\lambda a_1)}; \\
j_{\theta 1}^{(s)}(\lambda) &= \frac{c}{4\pi} H_r^{(a)}(\lambda) \frac{R}{a_1} \cdot \frac{I_0(\lambda R)}{K_1(\lambda a_1)}; \\
j_{\theta 2}^{(a)}(\lambda) &= \frac{-c}{4\pi} H_r^{(s)}(\lambda) \frac{R}{a_2} \cdot \frac{K_0(\lambda R)}{I_1(\lambda a_2)}; \\
j_{\theta 2}^{(s)}(\lambda) &= \frac{-c}{4\pi} H_r^{(a)}(\lambda) \frac{R}{a_2} \cdot \frac{K_0(\lambda R)}{I_1(\lambda a_2)}.
\end{aligned}\right\} \qquad (2.23)$$

The superscript "(s)" refers to the symmetric part of $j_\theta(\lambda)$, while "(a)" refers to the antisymmetric part.

From the known Fourier components of the currents we can calculate the current density in the coils:

$$j_i(z) = \frac{2}{\pi} \int_0^\infty (j_{\theta i}^{(a)} \sin \lambda z + j_{\theta i}^{(s)} \cos \lambda z)\, d\lambda, \qquad (2.24)$$
$$i = 1,\ 2.$$

As an example, we choose the function $f(z/\Lambda)$ to be

$$f(z/\Lambda) = \exp(-z^2/\Lambda^2), \qquad (2.25a)$$

so that

$$H^{(s)}/H_0 = \int_0^\infty f(k/\Lambda) \cos \lambda k\, dk = (\sqrt{\pi}/2) \exp(-\lambda^2 \Lambda/4), \quad H^{(a)} = 0. \qquad (2.25b)$$

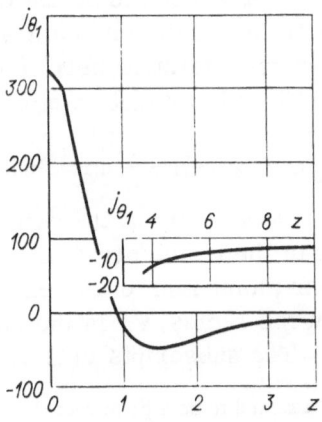

Fig. 10

Substituting (2.25b) into (2.23) and (2.24), we find equations for the current distributions in the coils. The function $j_{\theta 1}(z)$ found by computer calculations for the case $a_2 = 2a_1$, $R = 3/2$, $a_1 = 1$ is shown in Fig. 10. It is interesting to note that the sign of $j_{\theta 1}$ changes: The current densities fall off rapidly as $|z| \to \infty$, so that the coaxial coils can have finite dimensions without substantial field distortion.

§2.   Paraxial Approximation

The paraxial approximation is convenient for calculations in many optical systems. When this approximation is used some reference line (or surface) is chosen in an arbitrary way; in general, this line may not coincide with an actual trajectory (or trajectory surface formed by a family of trajectories). The focusing fields and the positions of the actual trajectories are found through an expansion in the transverse coordinate with respect to this reference surface. A reference surface which coincides with a trajectory is called the "principal" surface and if this surface does not coincide with a trajectory it is called the "initial" surface.

If $r_0(s)$ is the radius vector of the reference surface, then the radius vector of the trajectory of a particle which passes near the reference surface can be written $r(s) = r_0(s) + p n_0$. Here n_0 is the unit vector normal to the reference surface and p(s) is the deviation of the trajectory from this surface. All the beam quantities are sought as power series in p in the paraxial approximation.

We take the image of the source to be the cross section of the beam that corresponds to the minimum beam width,† $d = \min (p_{max} - p_{min})$. In the case under consideration here, focusing of axisymmetric beams, the quantity d is a function of eight arguments:

$$d = d(v_{min}, v_{max}, \alpha_{min}, \alpha_{max}, D_{min}, D_{max}, M_{min}, M_{max}). \qquad (2.26a)$$

Here v is the total particle velocity at the exit from the source $(z = z_0)$; α is the slope of the trajectory at $z = z_0$, $\tan \alpha = r/z|_{z=z_0}$, $D = MRv_\theta$ is the angular momentum of the particle at the exit from the source, R is the source radius, v_θ is the angular velocity, and M is the ion mass. The subscripts in Eq.(2.26a) represent

---

†We recall that the source is assumed to be a point source or an infinitesimally narrow source.

the limits of the range over which the corresponding parameters vary.

We assume that the parameters of a principal particle at the exit from the source are $v = v_0$, $\alpha = \alpha_0$, $D = D_0$, and $M = M_0$.

Assuming that the parameters of the other particles are not very different, we can write d as the series

$$d = \delta\alpha \frac{\partial d_0}{\partial \alpha} + \delta D \frac{\partial d_0}{\partial D} + \delta v \frac{\partial d_0}{\partial v} + \delta M \frac{\partial d_0}{\partial M} + \frac{(\delta D)^2}{2} \cdot \frac{\partial^2 d_0}{\partial D^2} + \frac{(\delta v)^2}{2} \cdot \frac{\partial^2 d_0}{\partial v^2} + \cdots.$$

(2.26b)

Here $\delta D \equiv |D_{max} - D_{min}|$; $\delta\alpha = |\alpha_{max} - \alpha_{min}|$, ...

If the system focuses the beam one of the first derivatives vanishes. It is usually assumed that

$$\partial d_0/\partial \alpha = 0. \qquad (2.26c)$$

The particular functional dependence of d on α, D, M, and v is governed by the particular features of the focusing system. For example, if a magnetic field is used to focus particles with an initial velocity $\mathbf{v}$,

$$d(M, v) = f\left(M|v|, \frac{\mathbf{v}}{|\mathbf{v}|} \right). \qquad (2.26d)$$

If the focusing is carried out by an electric field,

$$d(M, v) = g\left(Mv^2, \frac{\mathbf{v}}{|\mathbf{v}|} \right). \qquad (2.26e)$$

If one other term in series (2.26b) vanishes in addition to the coefficient $\partial d_0/\partial \alpha$, this corresponds to "two-parameter" focusing. Systems with triple, etc., focusing are defined similarly.

When a beam parameter changes, the image coordinate z and the image width can change. The image coordinate z_1 is obviously also a function of eight arguments:

$$z_1 = z_1(v_{max}, v_{min}, \alpha_{max}, \alpha_{min}, D_{max}, D_{min}, M_{max}, M_{min}). \qquad (2.27)$$

Thus, a system can exhibit "transverse" aberrations, associated with the quantity d, and "longitudinal" aberrations, which are a consequence of the dependence of z_1 on these arguments.

In a real system there will be certain specific aberrations which cause the most serious degradation of the solution of the problem at hand and which require special analysis if they are to be eliminated or minimized.

Clearly, the fields in the focusing system must be chosen in a special way in order to eliminate the aberration terms in (2.26b). Since d and z_1 are functionals of the configuration of the focusing fields, vanishing of the derivatives is equivalent to the imposition of certain additional conditions on the focusing fields. Since the fields are determined at the onset by an infinite number of parameters, it is possible, in principle, to design a system with arbitrarily weak aberrations. Several examples of systems in which certain aberrations are corrected are described in Chapters 3 and 4.

In analyzing the motion of a paraxial beam we must specify the configuration of the magnetic field on the initial surface or the principal surface, depending on the particular formulation of the problem. In §1 of this chapter we have described a method for calculating the spatial configurations of magnetic fields for a known field distribution on a cylindrical surface (the initial surface). Calculations of the configuration of a magnetic field on the basis of the field components specified on an axisymmetric surface of arbitrary shape are described in [27].

1. Equations of Trajectories near the Reference Surface. To calculate the ion trajectories in an optical system we must solve the equation of motion

$$\frac{M}{e} \cdot \frac{dv}{dt} = \mathbf{E} + \frac{1}{c} [\mathbf{v}, \, \mathbf{H}]. \qquad (2.28)$$

There are two possible approaches to the solution of optical problems. In the first the motion of the particle beam is studied in specified electromagnetic fields (which are actually guessed at or determined empirically). In the second we specify the shape of the focused beam, i.e., its geometric parameters (the "type of focusing"), and then we determine the necessary focusing electromagnetic fields.

In vacuum optics, both these problems have been studied in the most general case by Grinberg [25] for narrow beams and by Vandakurov [26] for broad beams. Grinberg's method can be ex-

tended to plasma-optical problems if the space charge is treated
as in (1.11).

Let us illustrate the situation for the case of paraxial axisymmetric systems. The equations of motion of an ion in an axisymmetric field reduce to the two equations [27]

$$M\ddot{r} = -(\partial U/\partial r) - (e/c)\,\dot{z}H_\theta; \bigg\}$$
$$M\ddot{z} = -(\partial U/\partial z) + (e/c)\,\dot{r}H_\theta. \bigg\}$$
(2.29a)

Here, U is the generalized potential

$$U(r, z) = \frac{[D - (e/c)\,\psi]^2}{2Mr^2} + e\Phi(\psi).$$
(2.29b)

Motion in the azimuthal direction is subject to conservation of
angular momentum,

$$Mr^2\dot{\theta} + (e/c)\,\psi = D = \text{const},$$
(2.29c)

where D is the generalized momentum. Equation (2.29) can also be
written

$$(M/2)(v_r^2 + v_z^2) + U = \mathcal{E};$$
(2.30a)

$$(M/\rho)(v_r^2 + v_z^2) = -(\partial U/\partial n) - (e/c)\,vH_\theta.$$
(2.30b)

Here $\mathcal{E}$ is the total particle energy and v is the total velocity in
the r, z plane.

Equation (2.30a) is an expression of energy conservation;
(2.30b) states that the centrifugal and centripetal forces are equal;
$\partial/\partial n$ is the derivative along the normal to the trajectory; and ρ
is the radius of curvature of the trajectory. The sign of ρ is chosen
in accordance with the Frenet equations:

$$\partial r/\partial s = \mathbf{t}; \quad d\mathbf{n}/ds = -\mathbf{t}/\rho; \quad d\mathbf{t}/ds = \mathbf{n}/\rho.$$
(2.31)

Here $\mathbf{t}$ and $\mathbf{n}$ are the unit tangent and normal vectors; we go from
$\mathbf{t}$ to $\mathbf{n}$ by a counterclockwise rotation. Also, ds is the arc element.
System (2.30) can be reduced to the single equation

$$2u/\rho = (\partial u/\partial n) - (e/c)\sqrt{-(2/M)u}\,H_\theta,$$
(2.32a)

where $u = U - \mathcal{E}$. Equation (2.32a) assumes a particularly simple
form if $H_\theta = 0$:

$$2u/\rho = \partial u/\partial n.$$
(2.32b)

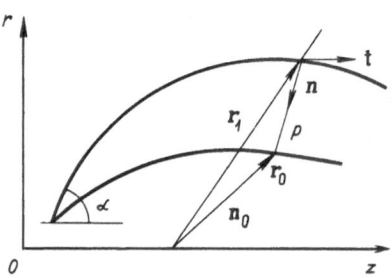

Fig. 11

Of the possible beam trajectories we consider a trajectory $r_0(s)$, which is called the principal trajectory (Fig. 11). All the other ("adjacent") trajectories are characterized by the value of p, the deviation from the main trajectory along the normal to the adjacent trajectory:

$$r_1 = r_0 - p n_0. \tag{2.33a}$$

Assuming quantities of p to be small, and neglecting terms of higher orders, we can relate the parameters of an adjacent trajectory to those of the principal trajectory. Using $ds_0^2 = (dr_0)^2$, $ds_1^2 = (dr_1)^2$, the Frenet equations, and (2.33a), we find

$$ds_1^2 = [1 + (p/\rho_0)]^2 \, ds_0^2 + dp^2 \approx [1 + (p/\rho_0)]^2 \, ds_0^2,$$

that is,

$$ds_1 = [1 + (p/\rho_0)] ds_0. \tag{2.33b}$$

To the same accuracy, we can relate t_0 and t_1:

$$t_1 = dr_1/ds_1 = d\,(r_0 - p n_0)/[ds_0\,(1 + p/\rho_0)] \approx t_0 - n_0\,(dp/ds_0). \tag{2.33c}$$

Similarly,

$$n_1 = n_0 + t_0\,(dp/ds_0); \tag{2.33d}$$

$$1/\rho_1 = 1/\rho_0\,[1 - (p/\rho_0) - \rho_0\,(d^2\,p/ds_0^2)]. \tag{2.33e}$$

We can now relate u and $\partial u_1/\partial n_1$ to u_0 and $\partial u_0/\partial n_0$.

Using (2.32) and (2.33) we find

$$u_1 = u_0\,(r_0 - p n_0) = u_0 - p\,(n_0\,\nabla)\,u_0 = u_0 - p\left[\frac{2u_0}{\rho_0} + \frac{e}{c}\,\sqrt{-\frac{2u_0}{M}}\,H_\theta\right] =$$

$$= u_0\left(1 - \frac{2p}{\rho_0}\right) - p\,\frac{eH_\theta}{c}\,\sqrt{-\frac{2u_0}{M}}. \tag{2.34a}$$

Similarly,

$$\frac{\partial u_1}{\partial n_1} \approx \left[\left(\mathbf{n}_0 + \mathbf{t}_0 \frac{dp}{ds_0}\right) \nabla\right] u_1 (\mathbf{r}_0 - p\mathbf{n}_0) =$$

$$= (\mathbf{n}_0 \nabla) u_0 + \frac{dp}{ds_0} (\mathbf{t}_0 \nabla) u_0 - p (\mathbf{n}_0 \nabla) u_0 =$$

$$= \frac{2u_0}{\rho_0} + \frac{du_0}{ds_0} + \frac{du_0}{ds_0} \cdot \frac{dp}{ds_0} - p \frac{\partial^2 u_0}{\partial n_0^2} + \frac{e}{c} \sqrt{-\frac{2u_0}{M}} H_\theta. \qquad (2.34b)$$

In certain cases it is convenient to express $\partial^2 u_0 / \partial n_0^2$ in terms of the two-dimensional Laplacian

$$\dot{\Delta} u = (\partial^2 u / \partial r^2) + (\partial^2 u / \partial z^2). \qquad (2.34c)$$

If the parameters s and p are treated as orthogonal coordinates, we can write

$$\dot{\Delta} u_0 = \frac{1}{1 + (p/\rho)} \cdot \frac{\partial}{\partial s} \cdot \frac{1}{1 + (p/\rho)} \frac{\partial u}{\partial s} + \frac{1}{1 + (p/\rho)} \cdot \frac{\partial}{\partial p} \left(1 + \frac{p}{\rho}\right) \frac{\partial u}{\partial p} =$$

$$= \frac{\partial^2 u_0}{\partial n_0^2} + \frac{\partial^2 u_0}{\partial s_0^2} - \frac{2u_0}{\rho_0^2} + 0(p^2). \qquad (2.34d)$$

Here, $\partial^2 u_0 / \partial p^2 \equiv \partial^2 u_0 / \partial n^2$ and $\partial u_0 / \partial n \equiv -\partial u_0 / \partial p$. Substituting Eqs. (2.34a), (2.34b), (2.33e), and (2.34d), for u_1, $\partial u_1 / \partial n_1$, p, and Δ into (2.32), and carrying out some simple manipulations, we find the desired Grinberg equation for this case:

$$(pu')' - \dot{\Delta} up + 2u\left(p'' - \frac{2p}{\rho^2}\right) + \frac{2p}{\rho} \cdot \frac{e}{c} \sqrt{-\frac{2}{M} u} H_\theta +$$

$$+ \frac{e^2}{Mc^2} H_\theta^2 p + \frac{e}{c} \sqrt{-\frac{2}{M} u} H_\theta \frac{\cos \alpha}{r} p = 0. \qquad (2.35)$$

All the quantities in (2.35) refer to the principal trajectory; the subscript "0" is omitted. In writing (2.35) we also use v = $(-2u/M)^{1/2}$.

Equation (2.35), together with the geometric equations

$$\partial \alpha / \partial s = 1/\rho; \quad \partial r / \partial s = \sin \alpha; \quad \partial r / \partial n = \cos \alpha \qquad (2.36)$$

and the equilibrium equation (2.32), written for the principal trajectory, determine the motion of a paraxial beam in axisymmetric electromagnetic fields. In Eq. (2.36), α is the angle between the axis and the tangent to the main trajectory.

If $H_\theta = 0$, Eqs. (2.35) and (2.36) are actually separable. In the case $H_\theta = 0$, Eq. (2.35) contains the four quantities $p(s)$, $u(s)$, $\dot{\Delta}u(s)$, and $\rho(s)$, three of which can be specified independently under plasma-optical conditions.

The reason that $\dot{\Delta}u$, like u, can be treated as an independent function of s is that u contains the two quantities $\psi(s)$ and $\Phi(\psi)$.

Accordingly, when $H_\theta = 0$ we can pose four classes of problems, depending on which of the quantities is treated as the unknown. Thus, the Grinberg equation can be used to find the beam parameters (p or ρ) for motion in specified fields (with known u and $\dot{\Delta}u$) or to determine the electromagnetic fields ($\dot{\Delta}u$ or u) for a given beam geometry (p, ρ).

We can eliminate the terms with p' in (2.35) by introducing the function ξ, which is defined by

$$p = \xi \, (u/u_0)^{-1/4}. \qquad (2.37)$$

Then (2.35) can be rewritten in the more symmetric form (for the case $H_\theta = 0$)

$$(\xi''/\xi) + (1/4)(u''/u) - (1/2)(\dot{\Delta}u/u) + (3/16)(u'^2/u^2) + (2/\rho^2) = 0. \qquad (2.38)$$

Systems described by Eq. (2.38) are obviously focusing systems if the boundary conditions

$$\xi(0) = \xi(L) = 0 \qquad (2.39a)$$

hold, where L is the distance from the source to the image.

No restrictions are imposed here other than the boundary conditions in (2.39a) and the requirement that the function $\xi(s)$ be smooth; in other words, the shape of the adjacent trajectories (the type of focusing) can be specified arbitrarily. In particular, if

$$\xi = \sin(\pi n/L)s, \qquad (2.39b)$$

where n is an integer, we refer to such systems as "harmonic." When $H_\theta \neq 0$, Eq. (2.35) contains r and α explicitly, so that (2.35) and (2.36) must be solved simultaneously.

2. Equations of the Trajectories near the Initial Surface. In the preceding subsection we have treated Grinberg equations that describe adjacent trajectories near a prin-

cipal trajectory. As noted in the Introduction, however, in annular systems the principal surface, formed by the set of main trajectories, can be extremely complicated, so that it is difficult to use this surface as a coordinate surface. Accordingly, we have not expanded the magnetic field in powers of the transverse coordinate near the principal surface, but near an auxiliary cylindrical surface (the initial surface).

It is also useful to treat the trajectory equations of the paraxial beam near the initial surface, rather than near the principal surface. Again, we work with (2.29). Since we are only interested in the geometric properties of the ion trajectories, it is convenient to transform from the variable t to the variable z in equations of motion (2.29). As a result we find

$$r'' + r'\left(\frac{\ddot{z}}{\dot{z}^2}\right) = -\frac{1}{M\dot{z}^2} \cdot \frac{\partial U}{\partial r} - \frac{e}{Mc} \cdot \frac{H_\theta}{\dot{z}}, \qquad (2.40a)$$

where $r' \equiv dr/dz$. An equation for $\dot{z}$ is found from energy conservation:

$$\frac{M}{2}\dot{z}^2\left(1 + r'^2\right) = -u. \qquad (2.40b)$$

Using (2.40b) to eliminate $\ddot{z}$ and $\dot{z}$, we can reduce system (2.40) to the single equation

$$r'' = -\frac{1 + r'^2}{(-2u)}\left(\frac{\partial U}{\partial r} + r'\frac{\partial U}{\partial z}\right) - \left(\frac{eH_\theta}{Mc}\right)^2 \frac{(1 + r'^2)^{3/2}}{(-2u/M)^{1/2}}. \qquad (2.41)$$

Assume that the expansion in (2.12) and (2.15) is known:

$$\psi = \psi_0(z) + y\,\psi_1(z) + (y^2/2)\psi_2(z) + \ldots, \qquad (2.42)$$

where $y = r$ in an axial system or $y = r - R$ in an annular system. Then we can also write expansions of $\Phi(\psi)$ in powers of y:

$$\Phi(\psi) = \Phi(\psi_0) + y\left(\frac{d\Phi}{d\psi}\bigg|_{\psi_0}\psi_1\right) + \ldots. \qquad (2.43)$$

To expand U in powers of y, we must analyze the axial and annular systems separately. In the axial case we must set $H_\theta = 0$ and assume $D = 0$. If $D \neq 0$, the calculations for the particle motion cannot be carried out on the basis of system (2.40) through expansions in powers of r. The reason is that the generalized potential increases without bound in the limit $r \to 0$, with $D \neq 0$:

$$U|_{r \to 0} \to \frac{D^2}{2Mr^2} \to \infty.$$

However, if $D = 0$, the expansion of u in powers of r is regular, since

$$\psi = (r^2/2)\psi_2\,(z) + (r^4/4!)\psi_4\,(z) + \dots, \tag{2.44}$$

and in all cases (vacuum and plasma) we have

$$\Phi = \Phi_0 + (r^2/2)\Phi_2 + (r^3/3!)\Phi_3 + (r^4/4!)\Phi_4 + \dots \tag{2.45}$$

In the vacuum case, in which Φ satisfies the Laplace equation $(\Delta\Phi = 0)$, the expansion coefficients in (2.45) are

$$\left. \begin{array}{c} \Phi_3(z) = 0; \quad \Phi_{2n+1}(z) = 0; \\[2mm] \Phi_2(z) = -\dfrac{1}{2^2}\,\Phi_0''(z); \quad \Phi_4(z) = \dfrac{1}{2^2 \cdot 4^2}\,\Phi_0^{IV}(z); \\[2mm] \Phi_{2n}(z) = -\dfrac{1}{(2n)^2}\,\Phi_{2n-2}''(z) \quad n = 1,\ 2,\ \dots\ . \end{array} \right\} \tag{2.46}$$

In the plasma case we have $\Phi = \Phi(\psi)$ so that all terms containing odd powers of r, except Φ_1, can be nonzero. If we assume that the function $\Phi(\psi)$ is analytic around $\psi = 0$, however, then the coefficients Φ_{2n+1} vanish even in the plasma case. Then $\Phi_0 = 0$ and

$$\Phi = k_1\psi + k_2\psi^2 + \dots = \frac{r^2}{2}\,(k_1\psi_2) + r^4\left(\frac{k_1\psi_4}{4!} + k_2\frac{1}{4}\psi_2^2\right) + \dots \tag{2.47}$$

Accordingly, when $\Phi_{2n+1} = 0$, the potential u near the z axis is

$$u = -\mathscr{E} + (r^2/2)U_2 + (r^4/4!)U_4 + \dots \tag{2.48}$$

The coefficients U_2 and U_4 in (2.48) are

$$U_2 = (e^2/Mc^2)(\psi_2^2/4) + e\Phi_2 = U_{2H} + U_{2E}; \tag{2.49a}$$

$$U_4 = (e^2/2Mc^2)\psi_2\psi_4 + e\Phi_4 = U_{4H} + U_{4E}. \tag{2.49b}$$

In annular systems the expansion of u in powers of r has no singularities. Unfortunately, the resulting equations are lengthy, so that we simply write the terms proportional to $y = r - R$. Using $y' = r'$ and setting $U = U_0 + yU_1 + (y^2/2)U_2 + \dots$, we can rewrite

(2.41) as

$$y'' + y' \left(\frac{1}{(-2U_0)} \cdot \frac{\partial U_0}{\partial z} + \frac{U_1 + yU_2}{(-2U_0 - 2U_1 y)} \right) +$$

$$+ \left(\frac{eH_{\theta 0}}{Mc} \right)^2 \frac{1 - 2\dfrac{y}{R}}{\left(-\dfrac{2U_0}{M} - y\dfrac{2U_1}{M} \right)^{1/2}} = P(y). \qquad (2.50)$$

Here, P(y) is the set of terms which are nonlinear in y, and $H_{\theta 0}$ = $H_\theta(R)$. Linearizing the left side of (2.50) we find

$$y'' + ay' + by + c = 0, \qquad (2.51)$$

where

$$a = -\frac{1}{M} \cdot \frac{U_0'}{\left(-\dfrac{2}{M} u_0 \right)}; \quad b = \frac{1}{M} \cdot \frac{1}{\left(-\dfrac{2}{M} u_0 \right)} \left(U_2 - \frac{U_1^2}{u_0} \right) +$$

$$+ \left(\frac{eH_{\theta 0}}{Mc} \right)^2 \frac{1}{\left(-\dfrac{2u_0}{M} \right)^{1/2}} \left(-\frac{2}{R} - \frac{1}{2} \cdot \frac{U_1}{u_0} \right); \qquad\qquad\left. \begin{array}{c} \\ \\ \\ \\ \\ \\ \\ \end{array} \right\} \quad (2.52)$$

$$c = \frac{1}{M} \cdot \frac{U_1}{\left(-\dfrac{2}{M} u_0 \right)} + \frac{\left(\dfrac{eH_{\theta 0}}{Mc} \right)^2}{\left(-\dfrac{2u_0}{M} \right)^{1/2}}.$$

In (2.52) we have introduced the notation $u_0 \equiv -\mathscr{E} + U_0$ and $U_0' \equiv$ dU_0/dz. The quantities U_0, U_1, U_2 here are obviously

$$U_0 = \frac{(D - (e/c)\,\psi_0)^2}{2MR^2} + e\Phi(\psi_0) = U_{0H} + U_{0E}; \qquad (2.53a)$$

$$U_1 = \frac{(D - (e/c)\,\psi_0)^2}{MR^3} - \frac{e}{Mc} \cdot \frac{\psi_1(D - (e/c)\,\psi_0)}{R^2} + e\frac{d\Phi}{d\psi}\Big|_{\psi_0} \psi_1 = U_{1H} + U_{1E}; \qquad (2.53b)$$

$$U_2 = 3\frac{(D - (e/c)\,\psi_0)^2}{MR^4} + \frac{1}{MR^2} \cdot \frac{e^2}{c^2}\psi_1^2 + \frac{e}{c} \cdot \frac{D - (e/c)\psi_0}{MR^2} \left(\frac{4\psi_1}{R} - \psi_2 \right) +$$

$$+ e\left(\frac{d\Phi}{d\psi}\Big|_{\psi_0} \psi_2 + \frac{d^2\Phi}{d\psi^2}\Big|_{\psi_0} \psi_1^2 \right) = U_{2H} + U_{2E} \qquad (2.53c)$$

For convenience in the analysis below, we have separated U_0, U_1, and U_2 into two parts, denoted by the subscripts "H" and "E", and these refer to terms associated with the magnetic and electric fields, respectively. Let us write the explicit expressions for the quadratic terms in $P(y)$, which we will need below. When $H_\theta = 0$, we have

$$P_{\text{quadr}} = \frac{y^2}{2} \left[-\frac{1}{M} \cdot \frac{1}{\left(-\frac{2}{M} u_0\right)} \left(U_3 - 3 \frac{U_1 U_2}{u_0} + 2 \frac{U_1^3}{u_0^2} \right) \right] +$$

$$+ y'y \left[\frac{1}{M} \cdot \frac{1}{\left(-\frac{2}{M} u_0\right)} \left(U_1' - \frac{U_0' U_1}{u_0} \right) \right] - y'^2 \frac{1}{M} \cdot \frac{1}{\left(-\frac{2}{M} u_0\right)} U_1. \qquad (2.54)$$

Here

$$U_3 = \frac{3}{MR^2} \cdot \frac{e^2}{c^2} \psi_1 \psi_2 - 12 \frac{(D - (e/c) \psi_0)^2}{MR^5} - \frac{e}{MR^3} \cdot \frac{e^2}{c^2} \psi_1^2 +$$

$$+ \frac{D - (e/c) \psi_0}{MR^2} \cdot \frac{e}{c} \left(6 \frac{\psi_2}{R} - \psi_3 - 18 \frac{\psi_1}{R^2} \right) + e \left(\frac{d\Phi}{d\psi} \Big|_{\psi_0} \psi_3 + \right.$$

$$\left. + 3 \frac{d^2 \Phi}{d\psi^2} \Big|_{\psi_0} \psi_1 \psi_2 + \frac{d^3 \Phi}{d\psi^3} \Big|_{\psi_0} \psi_1^3 \right) = U_{3H} + U_{3E}. \qquad (2.53d)$$

3. Gaussian Dioptrics.   Before solving Eqs. (2.49) and (2.51) we carry out a geometric analysis of the trajectories in the extremely interesting case in which the span of the electromagnetic fields is bounded along the beam trajectory, and the particle source and collector lie outside the field. Systems of this type include electromagnetic lenses. They can be thought of as analogs of "thick" optical lenses and described on the basis of Gaussian dioptrics, through the introduction of the cardinal elements — the focal lengths, nodal points, etc.

To be general, we discuss annular lenses; the transformation to axial lenses will be obvious.

To find the cardinal elements we neglect terms which are quadratic in y in (2.49) or (2.51). Then the function which serves as the solution of linearized equation (2.51) can be written

$$y(z) = A q_1(z) + B q_2(z) + y_{ps}, \qquad (2.55)$$

where y_{ps} is a particular solution of inhomogeneous equation (2.51),

q_1 and q_2 are linearly independent solutions of homogeneous equation (2.51) with $c = 0$, and A and B are arbitrary constants that depend on the initial conditions. The boundary conditions for the function y_{ps} are chosen as follows:

$$\frac{dy_{ps}}{dz}\bigg|_{-\infty} = \frac{dy_{ps}}{dz}\bigg|_{+\infty} = 0. \qquad (2.56a)$$

Then the surface defined by the line y_{ps} is the analog of the optic axis in conventional ion optics. The surface y_{ps} is obviously the principal surface.

The functions $q_1(z)$ and $q_2(z)$ are conveniently chosen as basis functions with respect to the principal surface by specifying the boundary conditions

$$q_1\bigg|_{-\infty} = 1; \quad \frac{dq_1}{dz}\bigg|_{-\infty} = 0; \quad q_2\bigg|_{+\infty} = 1; \quad \frac{dq_2}{dz}\bigg|_{+\infty} = 0. \qquad (2.56b)$$

The optical properties of a thick lens are governed completely by the asymptotic form of the functions q_1, q_2, and y_{ps}. The starting point and the focus are then conjugate points (in the Gaussian sense), as follows from the linearity of Eqs. (2.51). Introducing the focal lengths F_1 and F_2 (Fig. 12), we can write the asymptotic behavior of the functions $q_1(z)$ and $q_2(z)$ as

$$q_1 = \begin{cases} 1 & z \ll -\Lambda; \\ g\,[1-(z/F_1)] & z \gg \Lambda; \end{cases} \qquad (2.57a)$$

$$q_2 = \begin{cases} l\,[1+(z/F_2)] & z \ll -\Lambda; \\ 1 & z \gg \Lambda, \end{cases} \qquad (2.57b)$$

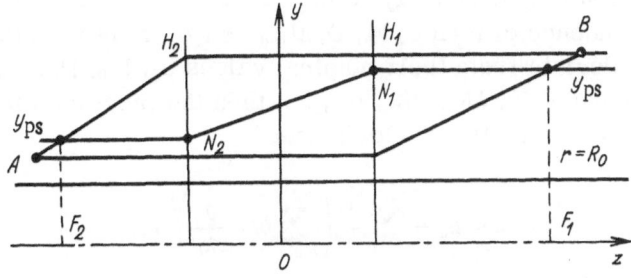

Fig. 12

where Λ is the scale range of the focusing fields along the initial surface, and F_1 and F_2 are reckoned from the point $z = 0$.

If g, l, and F_1, F_2 are known from the solution of Eq. (2.57), we can easily determine the other cardinal elements of the lens, e.g., the coordinates H_1 and H_2 of the main planes:

$$H_1(z) = (g - 1)(F_1/g); \qquad (2.58a)$$

$$H_2(z) = -(l - 1)(F_2/l). \qquad (2.58b)$$

The principal points N_1 and N_2 lie on the intersection of the main planes with the "lines" $y_{ps}(-\infty)$ and $y_{ps}(+\infty)$.

Figure 12 illustrates the construction of the image in a thick lens by means of cardinal elements. Frequently it is more convenient to find the asymptotic behavior of the functions $q_1(z)$ and $q_2(z)$, and $y_{ps}(z)$ rather than the quantities F_1, F_2, g, and l. Then the quantities g and l are found from the points at which the lines $q_1(-\infty)$, $q_1(+\infty)$ intersect $q_2(-\infty)$, $q_2(+\infty)$. The focal length is

$$1/F_1 = -q_1'(+\infty)/g; \quad 1/F_2 = -q_2'(-\infty)/l. \qquad (2.59)$$

§3. Incorporation of Aberrations in the Grinberg Scheme

To find the magnitudes of the aberrations in (2.35) we must take account of the higher-order terms in the expansion in p, i.e., the beam width. Furthermore, in the derivation of Eq. (2.35) we have only considered the dependence of u on the coordinates r and z; in other words, we have assumed M, D, and $\mathscr{E}$, to be the same for all particles. Because of the instability of the ion sources, the magnetic field, the presence of ions of various species, etc., in a real situation it is necessary to take account of the explicit functional dependence of u on e, M, D, $H_{\theta 0}$, $\mathscr{E}$, ψ, Φ, etc. If these quantities deviate from their nominal values e_0, M_0, D_0, $\mathscr{E}_0$, ψ_0, Φ_0... by amounts δe, δM, $\delta\mathscr{E}$, $\delta\psi$, ..., then the function u is conveniently written as the Taylor series

$$u = u_0 + \sum_{i=1}^{n} \frac{1}{i!} \left(\sum_{k=1}^{l} \delta\lambda_k \frac{\partial}{\partial\lambda_k} \right)^i u_0, \qquad (2.60)$$

where $\delta\lambda_1 = \delta v$, $\delta\lambda_2 = \delta D$, etc.

When $\delta M/M_0$ and $\delta D/D_0$ are small, we can restrict the analysis to the first terms in series (2.60).

Substituting the potential u as given in (2.60) into Eq. (2.32), and repeating the derivation procedure given above (to second order), we find the Grinberg equation with first and second order aberrations. After making a change of variables in (2.37), we can write this equation as†

$$\zeta'' + \omega^2\,\zeta = \frac{u^{-3/4}}{2}\left(\frac{2}{\rho} - \frac{\partial}{\partial n}\right)\frac{\partial u}{\partial \lambda_k}\,\delta\lambda_k + \frac{1}{2u}\left[\zeta f_1 - 2\zeta''\frac{u'}{u} - \right.$$

$$- \zeta'\left(\frac{d}{ds} - \frac{1}{4}\cdot\frac{u'}{u}\right)\frac{\partial u}{\partial \lambda_k}\,\delta\lambda_k\Bigg] + \frac{u^{-5/4}}{2}\Bigg[\zeta^2 f_2 + \frac{\delta u}{\rho}\,\zeta\zeta'' + $$

$$+ \zeta\zeta' f_3 + \zeta'^2\,\frac{2u}{\rho}\Bigg] + \frac{u^{-3/4}}{2}\Bigg[\left(\frac{\partial^2}{\partial\lambda^2_i} + \frac{\partial}{\partial\lambda_i}\cdot\frac{\partial}{\partial\lambda_k}\right)\left(\frac{2u}{\rho} - \frac{\partial u}{\partial n}\right)\Bigg]\delta\lambda_i\,\delta\lambda_k. \qquad (2.61)$$

Here

$$\zeta = \left(\frac{u}{u_0}\right)^{1/4}p; \quad \omega^2 = \frac{1}{4}\cdot\frac{u''}{u} - \frac{1}{2}\cdot\frac{\Delta u}{u} + \frac{3}{16}\cdot\frac{u'^2}{u^2} + \frac{2}{\rho^2}; \qquad (2.62a)$$

$$f_1 = \frac{\partial^2}{\partial n^2} + \frac{u''}{2u} - \frac{2}{\rho^2} - \frac{5}{8}\cdot\frac{u'^2}{u^2} - \frac{2}{\rho}\cdot\frac{\partial}{\partial n}; \qquad (2.62b)$$

$$f_2 = \frac{\Delta u}{\rho} - 3\,\frac{u''}{\rho} + \frac{8u}{\rho^3} - \frac{1}{2}\cdot\frac{\partial^3}{\partial n^3} + \frac{19}{8}\cdot\frac{u'^2}{u\rho} - $$

$$- \frac{1}{4}\cdot\frac{u'}{u}\cdot\frac{\partial^2}{\partial n\,\partial s} + \frac{1}{2}\cdot\frac{\rho'}{\rho}\,u'; \qquad (2.62c)$$

$$f_3 = \frac{\partial^2 u}{\partial n\,\partial s} - \frac{4u}{\rho} - \frac{2\rho'}{\rho}\,u. \qquad (2.62d)$$

In Eq. (2.61) a summation over identical indices is implied.

We seek a solution of Eq. (2.61) as a series in first and second order terms:

$$\zeta = \zeta_i\delta\lambda_i + \zeta_{ij}\delta\lambda_i\delta\lambda_j, \qquad (2.63)$$

where i, j = 0, 1, 2, ..., n; ζ_i and ζ_{ij} are zero-order quantities. The terms with i, j = 0 correspond to a spread in the radial com-

---

†For simplicity we assume $H_\theta = 0$.

ponent of the velocity, i.e., $\delta\lambda_0 \approx \delta v_{r0}$; the terms with i, j $\neq$ 0 correspond to the values $\delta\lambda_i \approx \delta m, \delta\mathcal{E}, \ldots; \delta\lambda_{ij} \approx \delta mp, \delta m\delta\mathcal{E}, \delta\mathcal{E}^2, \ldots$.

Substituting (2.63) into (2.61), and collecting terms of the same order, we find a chain of coupled equations

$$\zeta_i'' + \omega^2\zeta_i = F_i, \tag{2.64a}$$

$$\zeta_{ij}'' + \omega^2\zeta_{ij} = F_{ij}, \tag{2.64b}$$

where i, j = 0, 1, 2, 3, ..., n;

$$F_0 = 0; \quad F_i = \frac{u^{-3/4}}{2}\left(\frac{2}{\rho} - \frac{\partial}{\partial n}\right)\frac{\partial u}{\partial\lambda_k} \quad \text{for } i \neq 0; \tag{2.65a}$$

$$F_{ij} = \frac{1}{2u}\left[\zeta_i f_1 - 2\zeta_i'\frac{u'}{u} - \zeta_i'\left(\frac{d}{ds} - \frac{1}{4}\cdot\frac{u'}{u}\right)\frac{du}{d\lambda_i}\right] +$$
$$+ \frac{u^{-5/4}}{2}\left\{(2 - \Delta(i - j))\zeta_i\zeta_j f_2 + \left[\frac{8u}{\rho}(\zeta_i\zeta_j'' + \zeta_i''\zeta_j) + \right.\right.$$
$$+ (\zeta_i\zeta_j' + \zeta_i'\zeta_j)f_3\left]\frac{1}{1+\Delta(i-j)} + (2 - \Delta(i-j))\zeta_i'\zeta_j'\frac{2u}{\rho}\right\} +$$
$$+ \frac{u^{-3/4}}{2}\left[\left(\frac{\partial}{\partial\lambda_i}\right)^2 + \frac{\partial}{\partial\lambda_i}\cdot\frac{\partial}{\partial\lambda_j}\right]\left(\frac{2u}{\rho} - \frac{\partial u}{\partial n}\right). \tag{2.65b}$$

Let us examine system (2.64) in detail. The first equation in this system ($F_0 = 0$) determines the first-order geometric focusing. This is a homogeneous equation, and if its linearly independent solutions η_1 and η_2 are known we can write the solution of the other equations as

$$\zeta_i = -\eta_2\int_0^s ds\eta_1 F_i + \eta_1\int_0^s ds\eta_2 F_i; \tag{2.66a}$$

$$\zeta_{ij} = -\eta_2\int_0^s ds\eta_1 F_{ij} + \eta_1\int_0^s ds\eta_2 F_{ij}. \tag{2.66b}$$

Solution (2.66a) gives the first-order aberrations. The first-order aberrations are obviously determined completely once we know the solutions of the homogeneous equation and the function F_i. In calculating the second-order aberrations we must take account of the fact that the function F_{ij} can contain ζ_i for the first-order aberrations; i.e., the equations for the second-order aberrations are related to those for the first-order aberrations. The chain of equations in (2.64) is coupled in this way.

§4. Harmonic Systems with Equipotential Principal Trajectory

1. **Shape of the Principal Trajectories.** As an example of the use of the Grinberg equations in (2.35) we consider harmonic focusing systems, for which condition (2.39b) holds.

It follows from Eq. (2.35) that by specifying the quantities u(s) and $\dot{\Delta}$u(s) we can find the curvature of the main trajectory, ρ(s). Equation (2.36) can then be used to determine the explicit shape of the principal trajectory, i.e., the function r(z). In general, it is necessary to solve a differential equation in order to determine ρ(s) with specified fields, but there are two particular cases in which the differential equation involved in the determination of ρ(s) reduces to an algebraic equation. The first case, described in [28], refers to the problem of the motion of a particle beam in "two-dimensional" magnetic fields (**E** = 0). The second case arises in the motion of a particle beam in which the principal trajectory coincides with an equipotential u = const [29]. Let us examine this problem in detail.

Since the particle velocity does not change in motion along an equipotential, we can assume $u = -\mathscr{E}_0$, $\psi = 0$. without loss of generality. To find the quantity $\dot{\Delta}$u we assume

$$\Phi = (\varkappa/R)\psi + 0 \, (\psi^2),\tag{2.67}$$

where $\varkappa$ is a dimensionless constant and R is the scale distance from the symmetry axis to the main trajectory. Assuming $H_\theta = 0$ and using equilibrium equation (2.32), we find

$$\dot{\Delta}u = \left(\frac{\partial^2}{\partial r^2} + \frac{\partial^2}{\partial z^2}\right)\left[\frac{e^2}{2Mc^2}\cdot\frac{\psi^2}{r^2} + e\frac{\varkappa}{R}\psi\right] = \frac{4\mathscr{E}_0^2 R^2}{Mc^2 r^2 \rho^2 \varkappa^2} - \frac{\cos\alpha}{r\rho}\,2\mathscr{E}_0.\tag{2.68}$$

Assuming the system to be harmonic, i.e., $\zeta = \sin(\pi/L)s$, and taking L = 1, we find an algebraic relation among ρ, r, and α by substituting (2.68) into (2.33):

$$\pi^2 = \frac{2}{\rho^2} + \frac{v_0^2}{r^2\,(\varkappa^2/R^2)\,c^2\,\rho^2} - \frac{\cos\alpha}{r\rho}\,.\tag{2.69}$$

Using

$$eE = -\frac{c\varkappa}{R}\,rH = \frac{Mv_0^2}{\rho}\,,\tag{2.70}$$

where v_0 is the ion velocity along the principal trajectory, we can write (2.69) in the more compact form

$$\pi^2 = \frac{2}{\rho^2} + \frac{1}{R_{\Lambda i}^2} - \frac{\cos \alpha}{r\rho} \; ; \quad R_{\Lambda i} = \frac{v_0 Mc}{eH} . \qquad (2.71)$$

Together with the geometric equations (2.36), Eq. (2.71) determines the shape of the principal trajectory. The right side of (2.71) contains three terms of different physical origins. The first term is due to focusing in the external electric field (Hughes – Rojansky focusing), the second is due to focusing in the magnetic field (Busch focusing), and the third is due to focusing in the space-charge field. Equation (2.71) thus determines the shape of the principal trajectory in "mixed" focusing, for which the focusing modes listed above are limiting cases.

Let us follow the evolution of the shape of the principal trajectory due to changes in the ratios of the magnitudes of these three terms in Eq. (2.71). Solving Eq. (2.69) for ρ we find

$$\frac{1}{\rho_\pm} = \frac{\dfrac{\cos \alpha}{r} \pm \sqrt{\dfrac{\cos^2 \alpha}{r^2} + 4\pi^2 \left(2 + \dfrac{v_0^2 R^2}{r^2 \varkappa^2 c^2} \right)}}{2 \left(2 + \dfrac{v_0^2 R^2}{r^2 \varkappa^2 c^2} \right)} . \qquad (2.72)$$

The "$\pm$" in the subscript on ρ corresponds to the sign of the radical. For given values of α_0, v_0, $\varkappa$, and R, two curves ρ_+ and ρ_- pass through each point of the plane. If $\cos \alpha > 0$, these curves differ in the magnitude and sign of the radius of curvature. We note, however, that

$$\rho_+ (\alpha, \, r) = -\rho_- (\alpha + \pi, \, r), \qquad (2.73)$$

and it is concluded that the two curve families are identical, except that they differ in the sense in which they are traced out.

Turning to the calculation of the principal trajectories, we first consider the limiting case $H \to 0$ ($R_{\Lambda i} \to \infty$), assuming that the magnetic field is still strong enough to confine the space charge. As noted earlier, the limit $r \to \infty$ corresponds to Hughes–Rojansky focusing with $\rho_0 = \sqrt{2}/\pi$. For large but finite values of r, the ratio ρ_0 / R can be assumed to be a small quantity. Then

$$\rho = \rho_+ = \frac{\sqrt{2}}{\pi} \left(1 - \frac{\cos \alpha}{\pi 2 \sqrt{2} R} \right) . \qquad (2.74)$$

Substituting (2.74) into (2.36) and integrating, we find the explicit shape of the principal trajectory in parametric form:

$$r = R + \frac{V\overline{2}}{\pi} \cos \alpha + \frac{\cos 2\alpha}{8\pi^2 R} \; ; \\ z = -\frac{\alpha}{4\pi^2 R} + \frac{V\overline{2}}{\pi} \sin \alpha - \frac{\sin 2\alpha}{8\pi^2 R} \; . \quad \Bigg\} \tag{2.75}$$

Evidently the trajectory is trochoidal.

Curves calculated on a computer with arbitrary values of r are shown in Fig. 13a. Shown for comparison (Fig. 13b) are the principal trajectories in focusing of an unneutralized beam, in which the potential Φ satisfies the Laplace equation.

When the magnetic field is strong and cannot be neglected, it is still possible to study analytically the shape of the principal trajectory in the limiting case. For example, we take the zeroth approximation to be the value ρ_0 defined by ($R \to \infty$)

$$\rho_0^2 == \frac{1}{\pi^2} \left(2 + \frac{v_0^2}{c^2} \cdot \frac{1}{\varkappa^2} \right), \tag{2.76}$$

and assume

$$r = R - \rho_0 \cos \alpha . \tag{2.77}$$

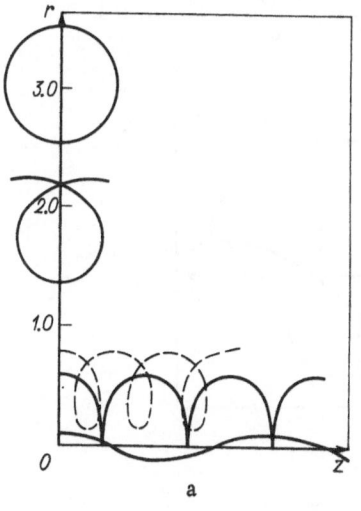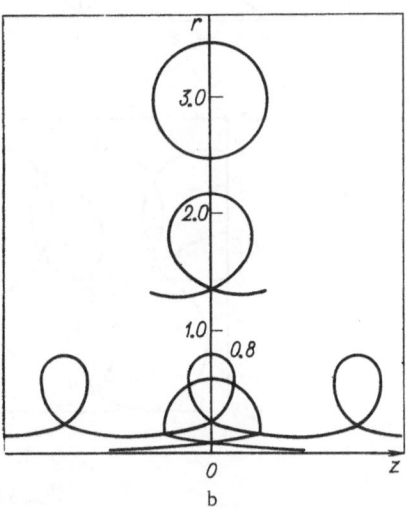

Fig. 13

Then

$$\rho = \frac{\cos\alpha}{2\pi^2 R}\left[2\,\frac{v_0^2}{c^2}\cdot\frac{1}{\varkappa^2}-1\right]+\rho_0 = \frac{\cos\alpha}{2\pi^2 R}\left[2\,\frac{v_0^2}{v_{dr}^2}-1\right]+\rho_0, \qquad (2.78)$$

where v_{dr} is the drift velocity, which satisfies $|v_{dr}| = c|\varkappa|$.

Accordingly, large values of r yield the three cases shown in Fig. 14a; these cases correspond to the ratios

$$v_0/v_{dr} = 1/\sqrt{2}; \quad v_0/v_{dr} < 1/\sqrt{2}; \quad v_0/v_{dr} > 1\sqrt{2}\,.$$

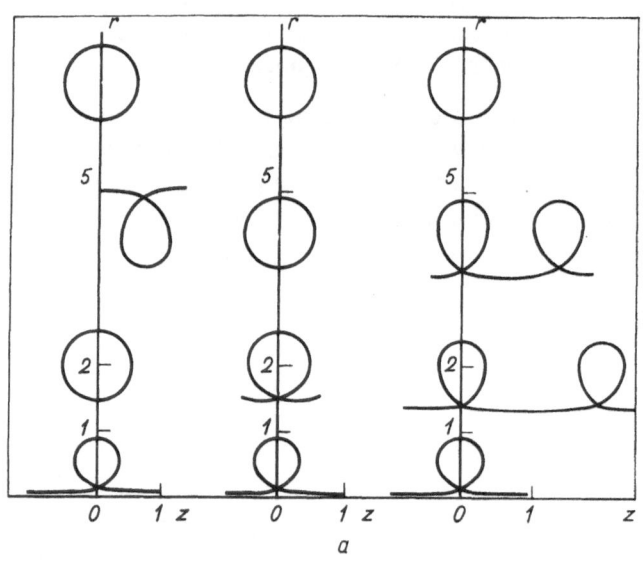

a

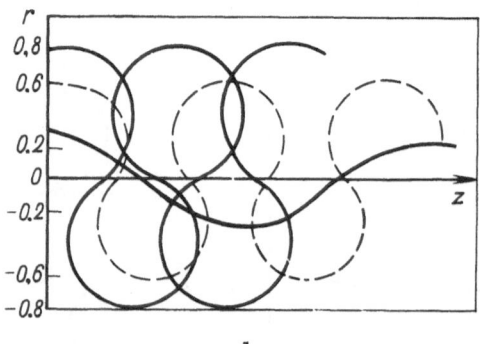

b

Fig. 14

As before, curves for smaller values of r have been calculated on a computer (Fig. 14b).

2. Aberrations in Harmonic Systems. Let us find the first-order aberrations of a harmonic system with an equipotential principal trajectory. Since $u = -\mathscr{E}_0$, the quantity ζ is equal to p [see Eq. (2.37)]. In this case the first equation in system (2.64) is

$$p_0'' + (\pi^2/L^2)p_0 = 0, \qquad (2.79a)$$

and its solution (for L = 1) is

$$\eta_1 = \sin \pi s; \; \eta_2 = \cos \pi s. \qquad (2.79b)$$

Then the first-order aberrations at the focus are, from (2.66),

$$\delta p_k = -\frac{1}{2\pi} \cdot \frac{1}{\mathscr{E}_0} \int_0^1 \left(\frac{2}{\rho} - \frac{\partial}{\partial n} \right) \frac{\partial u}{\partial \lambda_k} \delta\lambda_k \sin \pi s ds. \qquad (2.80)$$

We turn first to the momentum aberration associated with the spread in the values of the momentum D. Setting $D = D_0 + \delta D$ and assuming that D_0 can be set equal to zero without loss of generality, to terms linear in δD we have

$$\frac{\partial u}{\partial D} = -\frac{e}{c} \cdot \frac{\psi}{Mr^2}\Big|_{\psi=0} = 0; \; \frac{\partial}{\partial n} \cdot \frac{\partial u}{\partial D} = \frac{2e}{c} \cdot \frac{R}{Mr^2} \cdot \frac{\mathscr{E}_0}{e\rho\varkappa} . \qquad (2.81)$$

Substituting (2.81) into (2.80), we find

$$\delta p_D = \frac{R\delta D}{\pi M e \varkappa} \int_0^1 \frac{\sin \pi s}{r^2 \rho} ds. \qquad (2.82)$$

It is difficult to evaluate this integral in the general case, since we do not have explicit expressions for r(s) and ρ(s). Therefore we write this integral for a circle (r → ∞) and for a quasitrochoidal trajectory. Using $\delta D = Mr_0 v_{\theta 0}$ and $v_{dr} = -c\varkappa$, and assuming an initial radius $r_0 = R$, we find

$$\delta p_D = -\frac{v_{\theta 0}}{v_{dr}} \cdot \frac{1}{\rho_0} \cdot \frac{2}{\pi^2} \left[1 + \left(2\frac{\rho_0}{R} - a \right) \pi^2 \frac{1+\cos\frac{1}{\rho_0}}{\pi^2 - \frac{1}{\rho_0^2}} \right], \qquad (2.83a)$$

where $a = \dfrac{1}{2\pi^2 R\rho_0}\left[2\,\dfrac{v_0^2}{v_{dr}^2} - 1\right]$, and ρ_0 is given by Eq. (2.76). If, on the other hand, $R \to \infty$,

$$\delta p_D = -\frac{v_{\theta 0}}{v_{dr}}\cdot\frac{1}{\rho_0}\cdot\frac{2}{\pi^2}\,. \tag{2.83b}$$

We turn now to a calculation of the chromatic aberration, due to the spread in the longitudinal velocity component. In this case

$$\partial u/\partial\mathcal{E}_0 = -1 = \text{const}; \quad (\partial/\partial n)(\partial u/\partial\mathcal{E}_0) = 0. \tag{2.84}$$

Substituting these expressions into (2.8), we find

$$\delta p_{\mathcal{E}} = \frac{\delta\mathcal{E}_0}{\mathcal{E}_0}\cdot\frac{1}{\pi}\int_0^1\frac{\sin\pi s}{\rho}\,ds. \tag{2.85}$$

For the particular cases treated above, we find

$$\delta p_{\mathcal{E}} = \frac{\delta\mathcal{E}_0}{\mathcal{E}_0}\cdot\frac{2}{\pi^2}\cdot\frac{1}{\rho_0}\left[1 - \frac{\pi^2}{2}\,a\,\frac{1+\cos(1/\rho_0)}{\pi^2-(1/\rho_0^2)}\right]; \tag{2.86a}$$

$$\delta p_{\mathcal{E}} = \frac{\delta\mathcal{E}_0}{\mathcal{E}_0}\cdot\frac{2}{\pi^2}\cdot\frac{1}{\rho_0}, \quad R \to \infty. \tag{2.86b}$$

If the particle beam contains ions with different mass but whose energies are the same and whose azimuthal velocity is zero, the principal trajectory does not change if the mass differences are small. This conclusion is obvious from

$$\frac{\partial u}{\partial M} = -\frac{1}{M}\cdot\frac{\psi^2}{2Mr^2}\bigg|_{\psi=0} = 0; \quad \frac{\partial}{\partial n}\cdot\frac{\partial u}{\partial M}\bigg|_{\psi=0} = 0. \tag{2.87}$$

Accordingly, the transverse mass aberration is negligible. On the other hand, the distance from the source to the image changes, i.e., the longitudinal aberration is nonzero.

To calculate correctly the change in the distance we must take account of second-order aberrations as well as first-order aberrations. This requirement can be clarified by the simple example of a cylindrical capacitor (Hughes—Rojansky focusing), which, as noted above, is a particular case of a harmonic system. Figure 15 shows the trajectories (2) of ions with different energy emerging from source 4 and focused at collectors 3. The electric field is produced by the potential difference applied to capacitor plates 1.

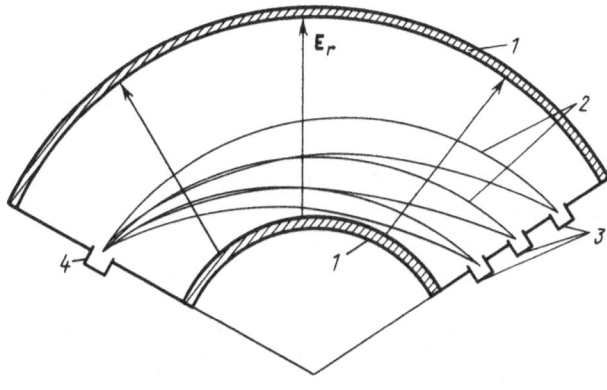

Fig. 15

The equations of motion of the particles in the cylindrical capacitor are

$$\ddot{r} = (v_\theta^2/r) + (e/M) E_0 (R_0/r);$$ (2.88a)

$$Mrv_\theta = MR_0 v_{\theta 0} = \text{const},$$ (2.88b)

where R_0 is the radius of the principal trajectory, and E_0 is the electric field at $r = R_0$.

The principal trajectory is obviously a circular trajectory of radius R if the equilibrium condition

$$(v_{\theta 0}^2/R_0) + (e/M)E_0 = 0$$ (2.89)

is satisfied. Here $v_{\theta 0}$ is the particle velocity along the equilibrium trajectory.

We assume a spread in initial velocities proportional to δv_θ and seek the spherical and chromatic aberrations of first and second order. For a paraxial beam we write r as

$$r = R_0 + y,$$ (2.90)

where y is a first-order quantity.

Replacing the time derivative in (2.88) by the derivative with respect to the polar angle θ and expanding quantities in this equation which are functions of r and v_θ in powers of y and δv_θ, we retain second-order terms. The result is an equation of motion with the second-order aberrations:

$$\frac{d^2 y}{d\theta^2} + 2y = 2\left(\frac{dy}{d\theta}\right)^2 \frac{1}{R_0} - 3\frac{y^2}{R_0} + 2\frac{\delta\lambda_1}{v_{\theta 0}} R_0 + 6y\frac{\delta\lambda_1}{v_{\theta 0}} - 3\frac{R_0}{v_{\theta 0}^2}\delta\lambda_1^2 \qquad (2.91)$$

Here $\delta\lambda_1$ represents the quantity δv_θ (see Chap. 2, § 3). As in § 3, we seek a series solution:

$$y = y_0\delta\lambda_0 + y_1\delta\lambda_1 + y_{00}\delta\lambda_0^2 + y_{01}\delta\lambda_0\delta\lambda_1 + y_{11}\delta\lambda_1^2, \qquad (2.92)$$

where $\delta\lambda_0$ is a first-order quantity (proportional to y or p). Substituting (2.92) into (2.91), and collecting terms of the same order, we find the chain of equations:

$$y_0'' + 2y_0 = 0; \qquad (2.93a)$$

$$y_1'' + 2y_1 = 2R_0; \qquad (2.93b)$$

$$y_{00}'' + 2y_{00} = 2(y_0')^2(1/R_0) - 3y_0^2 \, (1/R_0); \qquad (2.93c)$$

$$y_{01}'' + 2y_{01} = 4y_0'y_1'(1/R_0) - 6y_0y_1(1/R_0) + 6y_0; \qquad (2.93d)$$

$$y_{11}'' + 2y_{11} = 2(y_1')^2(1/R_0) - 3y_1^2 \, (1/R_0) + 6y_1 - 3R_0. \qquad (2.93e)$$

These equations have the same structure as in (2.64) (§ 3).

By solving system (2.93) using (2.66), we can find the quantity y to second order:

$$y = R_0 \, \delta\lambda_0 \sin\sqrt{2}\theta + R_0 \, v_{\theta 0} \, \delta\lambda_1 (1 - \cos\sqrt{2}\theta) +$$

$$+ R_0 \, v_{\theta 0}^2 \, \delta\lambda_1^2 \left(\frac{1}{4} + \frac{7}{2} \cos 2\sqrt{2}\theta\right) - \frac{7}{6} R_0 \, v_{\theta 0} \, \delta\lambda_0 \, \delta\lambda_1 \sin\sqrt{2}\theta +$$

$$+ R_0 \, \delta\lambda_0^2 \frac{8}{3\sqrt{2}} (\cos\sqrt{2}\theta - 1 - \frac{1}{16} \sin\sqrt{2}\theta \sin 2\sqrt{2}\theta). \qquad (2.94)$$

It follows from (2.94) that a change in the initial velocity causes transverse aberrations but does not affect the distance from the source to the image, to first order in $\delta\lambda_1$. We note that in deriving Eq. (2.93d) it is necessary to take account of terms of order $\delta\lambda_0^2$ in Eq. (2.91); at first glance, it would appear that these terms do not affect the chromatic aberration; however, if the equation of motion (2.91) is written without the terms that are quadratic in $\delta\lambda_0$, we find

$$\frac{d^2 y}{d\theta^2} - \left(1 + 3\frac{e}{M} \cdot \frac{E_0 R_0}{v_{\theta 0}^2}\right) y = 0.$$

In this equation we have not yet used the equilibrium conditions and we have not expanded in powers of $\delta\lambda_1$. It follows that changing v_θ changes source—image distance, in contradiction with the result above.

Thus, terms of order $\delta\lambda_0^2$ must be taken into account in calculations of the second-order aberrations and the source—image distance.

§5. The "Weak-Field Approximation" for Lenses

1. General Equations. Guassian dioptrics describes the properties of lenses with fields which have a bounded span along z, if we know the cardinal elements of the lens. In the present case, it is sufficient to know the asymptotic behavior of the functions $q_1(z)$, $q_2(z)$, and y_{ps}, but since these functions are governed by the boundary conditions, we must solve Eq. (2.57) for all z in order to find the asymptotic behavior of these functions. Unfortunately, it is generally not possible to find an exact solution of Eq. (2.57). Accordingly, in analyzing focusing properties we must resort to an approximation method. In particular, we shall use the "weak-field" approximation. In this method the ratio of the increment in the particle velocity in the lens, δv, to the total particle velocity v_0 is taken as a small parameter:

$$\mu = (\delta v/v_0) \ll 1. \tag{2.95}$$

We can therefore seek a solution of Eq. (2.41) or of its simplified versions in (2.50) and (2.51) as a series in inverse powers of v_0:

$$r = r_0 + (1/v_0)r_1 + (1/v_0^2)r_2 + \ldots \tag{2.96}$$

Writing Eq. (2.41) in the integrodifferential form

$$r(z) = r(z_0) + r'(z_0)(z - z_0) -$$

$$-\int_{z_0}^{z} dz_1 \int_{z_0}^{z_1} \left[\frac{1+r'^2}{(-2u)} \left(\frac{\partial u}{\partial r} + r' \frac{\partial u}{\partial z} \right) - \left(\frac{eH_\theta}{Mc} \right)^2 \frac{(1+r'^2)^{3/2}}{\left(-\frac{2u}{M} \right)^{1/2}} \right] dz_3 \tag{2.97a}$$

and using

$$\frac{1}{\left(-\frac{2u}{M} \right)} = \frac{1}{v_0^2 - \frac{2}{M}U} = \frac{1}{v_0^2} \left(1 + \frac{2U}{Mv_0^2} + \ldots \right), \tag{2.97b}$$

we see that the correction to the rectilinear motion due to the interaction of the lens field is small when (2.95) holds. It is not difficult to see that the correction is proportional to $1/v_0^2$ if $H_\theta = 0$ or to $1/v_0$ if $H_\theta \neq 0$. Below we consider the case $H_\theta = 0$.

Equation (2.97a) takes the form

$$y(z) = y(z_0) + y'(z_0)(z-z_0) + \varepsilon \int_{z_0}^{z} dz_1 \int_{z_0}^{z_1} dz_2 f(y, z_2), \qquad (2.98)$$

where ε is a small parameter. Substituting in $y(z)$ in series form,

$$y(z) = y_1(z) + \varepsilon y_2(z) + \varepsilon^2 y_3(z) + ..., \qquad (2.99a)$$

where $y_1(z) = y(z_0) + y'(z_0)(z - z_0)$, we find equations for $y_n(z_0)$:

$$\left.\begin{array}{l} y_2(z) = \int_{z_0}^{z} dz_1 \int_{z_0}^{z_1} dz_2 f(y_1, z_2); \\[3mm] y_3(z) = \int_{z_0}^{z} dz_1 \int_{z_0}^{z_1} dz_2 \dfrac{\partial f(y_2, z_2)}{\partial y_2} y_2(z_2); \\[3mm] \cdot \quad \cdot \quad \cdot \quad \cdot \quad \cdot \quad \cdot \quad \cdot \quad \cdot \quad \cdot \quad \cdot \quad \cdot \quad \cdot \end{array}\right\} \qquad (2.99b)$$

Accordingly, the solution of Eq. (2.97a) can be reduced to quadratures in the weak-field approximation.

Let us use the weak-field approximation to determine the functions q_1, q_2, y_{ps} in (2.55), working from linearized equation† (2.51) with $H_\theta = 0$.

In this case the coefficients a, b, and c [see Eq. (2.52)] can be written with the help of (2.97b):

$$a = \frac{a_1}{v_0^2} + \frac{a_2}{v_0^4} + ... = \frac{1}{v_0^2}\left(-\frac{1}{M}\frac{dU_0}{dz}\right) + \frac{1}{v_0^4}\left(-\frac{2U_0}{M^2}\cdot\frac{dU_0}{dz}\right) + ...; \qquad (2.100a)$$

$$b = \frac{b_1}{v_0^2} + \frac{b_2}{v_0^4} + ... = \frac{1}{v_0^2}\left(\frac{U_2}{M}\right) + \frac{1}{v_0^4}\left(\frac{2}{M^2}(U_1^2 + U_0 U_2)\right) + ...; \qquad (2.100b)$$

$$c = \frac{c_1}{v_0^2} + \frac{c_2}{v_0^4} + ... = \frac{1}{v_0^2}\left(\frac{1}{M}U_1\right) + \frac{1}{v_0^4}\left(\frac{2}{M^2}U_1 U_0\right) + ... \qquad (2.100c)$$

---

†The equation for axial lenses, (2.49), has a form corresponding to (2.51) in the particular case a = 0.

The functions q_1 and q_2 are the solution of the homogeneous equation (2.51),

$$q'' + aq' + bq = 0, \tag{2.101}$$

which satisfies boundary conditions (2.56b).

In order to use Eq. (2.99), we write Eq. (2.101) as

$$q_1 = 1 - \int_{-\infty}^{z} dz_1 \int_{-\infty}^{z_1} dz_2 \left[\left(\frac{a_1}{v_0^2} + \frac{a_2}{v_0^4} + \dots \right) q_1' + \left(\frac{b_1}{v_0^2} + \frac{b_2}{v_0^4} + \dots \right) q_1 \right]. \tag{2.102a}$$

This equation is written for q_1; the equation for q_2 is similar but with different integration limits:

$$q_2 = 1 - \int_{\infty}^{z} dz_1 \int_{\infty}^{z_1} dz_2 \left[\left(\frac{a_1}{v_0^2} + \frac{a_2}{v_0^4} + \dots \right) q_2' + \left(\frac{b_1}{v_0^2} + \frac{b_2}{v_0^4} + \dots \right) q_2 \right]. \tag{2.102b}$$

We can write the solution of (2.102a) as

$$q_1 = q_{10} + q_{11} + q_{12} + \dots, \tag{2.103}$$

where

$$q_{10} = 1; \tag{2.104a}$$

$$q_{11} = -\frac{1}{v_0^2} \int_{-\infty}^{z} dz_1 \int_{-\infty}^{z_1} dz_2 \, b_1; \tag{2.104b}$$

$$q_{12} = -\frac{1}{v_0^4} \left[\int_{-\infty}^{z} dz_1 \int_{-\infty}^{z_1} dz_2 \, b_2 - \int_{-\infty}^{z} dz_1 \int_{-\infty}^{z_1} dz_2 \, a_1 \int_{-\infty}^{z_2} dz_3 \, b_1 - \right.$$
$$\left. - \int_{-\infty}^{z} dz_1 \int_{-\infty}^{z_1} dz_2 \, b_1 \int_{-\infty}^{z_2} dz_3 \int_{-\infty}^{z_3} dz_4 \, b_1 \right]. \tag{2.104c}$$

A similar equation can be written for q_2.

We now consider the function y_{ps} in contrast with q_1 and q_2. This function is governed by boundary conditions (2.56a), so that it cannot be calculated by the Cauchy method, since (2.99) is not applicable. However, this function can be written in series form:

$$y_{ps} = y_{ps\,0} + \varepsilon y_{ps\,1} + \dots; \varepsilon \equiv 1/v_0^2, \tag{2.105}$$

where the higher-order terms fall off rapidly with increasing v_0 and the first term y_{ps0} can be assumed to be a constant.

We again write Eq. (2.51) in integral form, but now with the free term c(z). When (2.56a) holds we find

$$y_{ps} = y_{ps}(-\infty) - \int\limits_{-\infty}^{z} dz_1 \int\limits_{-\infty}^{z_1} dz_2 \, (ay'_{ps} + by_{ps} + c); \qquad (2.106a)$$

$$y'_{ps}(z) = - \int\limits_{-\infty}^{z} dz_1 \, (ay'_{ps} + by_{ps} + c). \qquad (2.106b)$$

We have written an equation for y'_{ps} here since the second boundary condition on this function is specified at z = +∞.

Substituting expansion (2.105) into (2.106), and noting that † $y_{ps0} \equiv$ const, we find the following equation:

$$0 = \int\limits_{-\infty}^{+\infty} (b_1 y_{ps\,0} + c_1) \, dz. \qquad (2.107a)$$

Hence we find [see Eq. (2.100)]

$$y_{ps\,0} = -\frac{\int\limits_{-\infty}^{+\infty} c_1 dz}{\int\limits_{-\infty}^{+\infty} b_1 \, dz} = -\frac{\int\limits_{-\infty}^{+\infty} U_1 dz}{\int\limits_{-\infty}^{+\infty} U_2 \, dz}. \qquad (2.107b)$$

It follows from (2.107b) that y_{ps0} is independent of the particle mass and velocity when $H_\theta = 0$; thus, the cylindrical surface defined by $y_{ps\,0}$ for a "thin" lens is the exact analog of the optical axis of an axial lens. Substituting the known value of y_{ps0} into (2.106a) we find the refined value $y_{ps\,1}$:

$$y_{ps\,1} = y_{ps\,1}(-\infty) - \frac{1}{v_0^2} \int\limits_{-\infty}^{z} dz_1 \int\limits_{-\infty}^{z_1} dz_2 \, (b_1 y_{ps\,0} + c_1) \equiv$$

$$\equiv y_{ps\,1}(-\infty) + \tilde{y}_{ps\,1}. \qquad (2.108a)$$

Here $y_{ps1}(-\infty)$ can be determined from the condition

$$y'_2(+\infty) = 0. \qquad (2.109)$$

---

†Obviously $y_{ps} = y_{ps\,0}$ at $\varepsilon = 0$, but in this case $v_0 \to \infty$, and the trajectory is not perturbed.

We write $y_{ps1}(-\infty)$ in the form

$$y_{ps\,1}(-\infty) = y_{ps\,0} + \delta y_1. \tag{2.108b}$$

Substituting (2.108b) into (2.106b), and using (2.109), we find an equation for δy_1:

$$\delta y_1 = -\frac{\int\limits_{-\infty}^{+\infty} dz(a_1 \tilde{y}'_{ps\,1} + b_1 \tilde{y}_{ps\,1} + c_1)}{\int\limits_{-\infty}^{+\infty} b_1\,dz}. \tag{2.110}$$

These equations can be simplified for an axial lens: According to (2.53) in the limit $R \to \infty$, the functions $U_0(z)$ and $U_1(z)$ vanish identically. Thus, in this case

$$q_1 = 1 - \frac{1}{Mv_0^2} \int\limits_{-\infty}^{z} dz_1 \int\limits_{-\infty}^{z_1} dz_2\, U_2 -$$

$$- \frac{1}{(Mv_0^2)^2} \int\limits_{-\infty}^{z} dz_1 \int\limits_{-\infty}^{z_1} dz_2\, U_2 \int\limits_{-\infty}^{z_2} dz_3 \int\limits_{-\infty}^{z_3} dz_4\, U_2; \tag{2.111}$$

$$y_{ps} = 0. \tag{2.112}$$

2. Focal Lengths. Knowing the functions q_1 and q_2, we can now derive equations for the focal lengths of the lenses, working from (2.59). Explicit expressions are needed for $q_1'(+\infty)$, $q_2'(-\infty)$, g, and l.

To be definite we consider the "right focus," finding q_1' and g. For generality we write homogeneous equation (2.101) in the form

$$q_1'' = f(q_1, q_1', z) \equiv f_1(q_1, z). \tag{2.113}$$

The exact solution of (2.113) is obviously

$$q_1(z) = 1 + \int\limits_{-\infty}^{z} dz_1 \int\limits_{-\infty}^{z_1} dz_2\, f_1(q_1, z_2), \tag{2.114}$$

so that

$$q_1'(+\infty) = \int\limits_{-\infty}^{+\infty} dz f_1(q_1, z). \tag{2.115}$$

Equation (2.57a) then provides an equation for g:

$$g = \lim_{z \to \infty} [q_1 - z(dq_1/dz)]. \qquad (2.116)$$

Substituting (2.114) and (2.115) into (2.116), and carrying out some simple manipulations, we find

$$g = 1 + \int_{-\infty}^{0} dz_1 \int_{-\infty}^{z_1} dz_2 f(z_2, q_1) - \int_{0}^{\infty} dz_1 \int_{z_1}^{\infty} dz_2 f(z_2, q_1). \qquad (2.117)$$

Accordingly, the required equation for F_1 is

$$\frac{1}{F_1} = -\frac{q_1'(+\infty)}{g} =$$

$$= -\frac{\displaystyle\int_{-\infty}^{+\infty} dz_1 f_1(z_1, q_1)}{1 + \displaystyle\int_{-\infty}^{0} dz_1 \int_{-\infty}^{z} dz_2 f(z_2, q_1) - \int_{0}^{\infty} dz_1 \int_{z_1}^{\infty} dz_2 f(z_2, q_1)}. \qquad (2.118a)$$

Similarly, the equation for F_2 is

$$\frac{1}{F_2} = -\frac{q_2'(-\infty)}{l} =$$

$$= -\frac{\displaystyle\int_{-\infty}^{+\infty} dz_1 f(z_1, q_2)}{1 + \displaystyle\int_{\infty}^{0} dz_1 \int_{\infty}^{z_1} dz_2 f(z_2, q_2) - \int_{0}^{\infty} dz_1 \int_{-\infty}^{z_1} dz_2 f(z_2, q_2)}. \qquad (2.118b)$$

Now, substituting (2.104) and the function $f(z, q)$, written as power series in v_0^2, we find the focal length, again as a power series in v_0^2. For F_1 this series is

$$\frac{1}{F_1} = \frac{1}{Mv_0^2} W\Big|_{-\infty}^{+\infty} + \frac{1}{(Mv_0^2)^2} A + 0\left(\frac{1}{(Mv_0^2)^3}\right), \qquad (2.119)$$

where

$$W = \int_{-\infty}^{z} U_2 \, dz = \int_{-\infty}^{+\infty} (U_{2H} + U_{2E}) \, dz = W_H + W_E; \qquad (2.120a)$$

$$A = 2 \int\limits_{-\infty}^{+\infty} U_1^2\, dz + 2\, [U_0\, W]_{-\infty}^{+\infty} - W \left.\right|_{-\infty}^{+\infty} \int\limits_{-\infty}^{+\infty} W\, dz + \int\limits_{-\infty}^{+\infty} W^2\, dz -$$

$$- \int\limits_{-\infty}^{+\infty} W U_0'\, dz - W \left.\right|_{-\infty}^{+\infty} \left[\int\limits_{-\infty}^{0} W\, dz + \int\limits_{0}^{\infty} dz_1 \int\limits_{z_1}^{\infty} U_2\, dz \right]. \qquad (2.120b)$$

The equation for F_2 is analogous, with

$$W \to W \equiv \int\limits_{z_1}^{\infty} U_2\, dz. \qquad (2.120c)$$

If the focused beam is of finite width, the nonlinear terms in Eq. (2.50) must be taken into account. These terms are taken into account in § 3 of this chapter in the calculation of the aberrations in the Grinberg scheme. We can now use the results of that section. The solution of paraxial equation (2.51) has the form of (2.55). Adopting this solution as a zeroth approximation, and writing the solution of nonlinear equation (2.50) in form (2.63) we find a solution of Eq. (2.50) with nonlinear terms in form (2.66). In calculating the focal length we assume that a parallel ion beam coming from infinity along the negative z axis is incident on the lens. Then the initial conditions are

$$y_{max}\, (-\infty) = y_0;\ y_0'\, (-\infty) = 0, \qquad (2.121)$$

where $y_{max}(-\infty)$ is the maximum beam width at $z = -\infty$. Restricting the discussion to lenses with $F \approx v_0^2$ and using (2.118a), we find the focal length F_1:

$$\frac{1}{F_1} = \frac{1}{Mv_0^2}\, W_1 \left.\right|_{-\infty}^{+\infty}, \qquad (2.122a)$$

where

$$W_1 = \int\limits_{-\infty}^{z} \left(U_2 - \frac{y_0}{2}\, U_3 \right) dz. \qquad (2.122b)$$

Equation (2.122) shows that when the spherical aberration is taken into account the focal length depends on the beam geometry as well as the lens fields. Equations for the focal lengths of axial lenses are easily found from (2.119) by noting that $U_0 = U_1 = 0$ in an axial lens. Specifically, for an axial lens

$$\frac{1}{F_1} = \frac{1}{Mv_0^2} \left. W \right|_{-\infty}^{+\infty} + \frac{1}{(Mv_0^2)^3} A + 0 \left(\frac{1}{Mv_0^2} \right)^3, \qquad (2.123a)$$

where

$$A = \int_{-\infty}^{+\infty} W^2 \, dz + W \left.\right|_{-\infty}^{+\infty} \left(\int_{-\infty}^{+\infty} W dz - \int_{-\infty}^{0} W dz + \int_{0}^{\infty} dz_1 \int_{z_1}^{\infty} U_2 \, dz \right). \qquad (2.123b)$$

When the finite beam width (the spherical aberration) is taken into account for an axial lens, we must also take account of the fact that the coefficients of the cubic terms in (2.48) are zero, that is, the corrections for the deviation from a paraxial beam are cubic. Carrying out calculations for lenses with $F \approx v_0^2$ in an analogous manner we find the focal length for a beam of finite width in form (2.122a), where

$$W_1 = \int_{-\infty}^{z} \left(U_2 - \frac{y_0^2}{12} U_4 \right) dz. \qquad (2.124)$$

Everywhere below we will refer to a lens as "thin" if its focal length is given accurately by the first term in (2.119). If, on the other hand, the first term in (2.119) vanishes, or if the subsequent terms in the expansion have important affects on F, the lens is "thick."

Finally, if the focusing effect of the lens is governed by both electric and magnetic fields, we say that the lens is "electromagnetic." An interesting question in connection with electromagnetic lenses is that of the coupling between the **E** and **H** focusing. The focusing is uncoupled if

$$1/F = (1/F_E) + (1/F_H). \qquad (2.125)$$

Here F_E is the focal length of the lens in the limit **H** $\rightarrow$ 0, and F_H is the focal length when **E** = 0. If (2.125) does not hold, the focusing effects are coupled.

Let us examine in detail the independence condition for **E** and **H** focusing in annular lenses. Separating those terms in the expansion coefficients of generalized potential u which are associated with the electric and magnetic fields, (2.120a), we can write the focal length of an annular lens as

$$\frac{1}{F_1} = \left\{ \frac{1}{Mv_0^2} W_H \Big|_{-\infty}^{+\infty} + \frac{1}{(Mv_0^2)^2} \left[\int_{-\infty}^{+\infty} (W_H^2 - W_H U_{0H}' + 2U_{1H}^2)\, dz + \right. \right.$$

$$+ 2\,[U_{0H} W_H]_{-\infty}^{+\infty} - W_H \Big|_{-\infty}^{+\infty} \int_{-\infty}^{+\infty} W_H\, dz -$$

$$\left. \left. - W_H \Big|_{-\infty}^{+\infty} \left(\int_{-\infty}^{0} W_H\, dz + \int_{0}^{\infty} dz_1 \int_{z_1}^{\infty} U_{2H}\, dz_2 \right) \right] \right\} +$$

$$+ \left\{ \frac{1}{Mv_0^2} W_E \Big|_{-\infty}^{+\infty} + \frac{1}{(Mv_0^2)^2} \left[\int_{-\infty}^{+\infty} (W_E^2 + W_E U_{0E}' + 2U_{1E}^2)\, dz + \right. \right.$$

$$+ 2\,[U_{0E} W_E]_{-\infty}^{+\infty} - W_E \Big|_{-\infty}^{+\infty} \int_{-\infty}^{+\infty} W_E\, dz -$$

$$\left. \left. - W_E \Big|_{-\infty}^{+\infty} \left(\int_{-\infty}^{0} W_E\, dz + \int_{0}^{\infty} dz_1 \int_{z_1}^{\infty} U_{2E}\, dz_2 \right) \right] \right\} +$$

$$+ \frac{1}{(Mv_0^2)^2} \left[4 \int_{-\infty}^{+\infty} U_{1E} U_{1H}\, dz - W_E \Big|_{-\infty}^{+\infty} \int_{-\infty}^{+\infty} W_H\, dz - \right.$$

$$\left. - W_H \Big|_{-\infty}^{+\infty} \int_{-\infty}^{+\infty} W_E dz + 2 \int_{-\infty}^{+\infty} W_E W_H\, dz - \int_{-\infty}^{+\infty} (W_E U_{0H}' + W_H U_{0E}')\, dz \right] + \ldots .$$

$$(2.126)$$

Here we have separated the terms associated with the field **E** alone, those associated with the field **H** alone, and the cross terms associated with **E** and **H**. In general the focal length in (2.126) cannot be written in form (2.125). If, however, the critical velocity is high and if terms of order v_0^{-4} can be neglected (the lens is thin), then the electric and magnetic focusing are independent [see condition (2.125)]. If the terms with v_0^{-2} vanish $(W|_{-\infty}^{+\infty} = 0)$, then the focal length of the lens is $F \approx v_0$. If the **E** and **H** focusing are to be uncoupled, the fields must be chosen so that the cross terms are zero,

$$\int_{-\infty}^{+\infty} W_E W_H\, dz = 0; \qquad \int_{-\infty}^{+\infty} W_E \frac{dU_{0H}}{dz}\, dz = 0;$$

$$\int_{-\infty}^{+\infty} W_H \frac{dU_{0E}}{dz}\, dz = 0; \qquad \int_{-\infty}^{+\infty} U_{1E} U_{1H}\, dz = 0. \qquad (2.127)$$

If (2.127) holds the focal length is

$$\frac{1}{F} = \frac{1}{F_E} + \frac{1}{F_H} = \frac{1}{(Mv_0^2)^2} \left[\int_{-\infty}^{+\infty} (W_E^2 - W_E U_{0E}' + 2U_{1E}^2) \, dz + \right.$$

$$\left. + \int_{-\infty}^{+\infty} (W_H^2 - W_H U_{0H}' + 2U_{1H}^2) \, dz \right]. \qquad (2.128)$$

Introducing the functions $f_1(z)$, $f_2(z)$, and $f(z)$ by means of

$$\left. \begin{array}{c} W_E W_H = f(z); \quad \dfrac{f(z)}{W_H} U_{0H}' = f_1(z); \\[3mm] \dfrac{f(z)}{W_E} U_{0E}' = f_2(z), \end{array} \right\} \qquad (2.129)$$

we can rewrite (2.127) as

$$\int_{-\infty}^{+\infty} f(z) \, dz = \int_{-\infty}^{+\infty} f_1(z) \, dz = \int_{-\infty}^{+\infty} f_2(z) \, dz = 0. \qquad (2.130)$$

Fields that provide uncoupled focusing are governed by the following equations:

$$W_H = \frac{f(z)}{f_1(z)} U_{0H}'; \quad W_E = \frac{f(z)}{f_2(z)} U_{0E}'; \quad \int_{-\infty}^{+\infty} U_{1E} U_{1H} \, dz = 0. \qquad (2.131)$$

Since U_{0E}, U_{0H}, U_{1E}, U_{1H}, W_E, and W_H in (2.131) are related to the nature of the fields $\mathbf{E}$ and $\mathbf{H}$ on the initial surface [see Eqs. (2.53) and (2.120)] it follows that Eq. (2.131) contains differential equations for ψ and $\Phi(\psi)$. The only restrictions on the functions $f_1(z)$ and $f_2(z)$ are (2.130) and the natural requirement of smoothness, so that these functions can be chosen in the most convenient manner.

In the particular case $\psi_1 = 0$, we can take a symmetric function W_E, an antisymmetric W_H, and a symmetric $\Phi_0(\psi_0)$ to satisfy conditions (2.127). In this case the focusing is independent in terms of the fields $\mathbf{E}$ and $\mathbf{H}$ [see (2.125)].

§6. Variational Formulation of Particle-Optical Problems

1. Figure of Merit. We have discussed several methods for the design of particle-optical systems. However, these

methods suffer from two shortcomings: First, they require an educated guess or the a priori choice of the general arrangement of the focusing device. Second, the classification of effects as focusing proper and aberrations is intimately related to the paraxial approximation, while this paraxial approximation is generally adopted simply on the basis of the calculation scheme, rather than physical considerations. From the calculation standpoint, however, the paraxial approximation is valid only when higher-order aberrations are inconsequential.

We have seen that the procedure required for analyzing aberrations is laborious even for second-order aberrations, so that it is natural to seek a method for avoiding the classical methods.

Physically it is obvious that a systematic formulation of a particle-optical problem must be variational. In other words, it must reduce to seeking the extremum of some functional which is a measure of the quality of the optics system. This extremum must be sought, under certain restrictions (e.g., strength of the fields, etc.). Let us examine this question in more detail for the example of the spherical aberration of a cylindrical lens. We assume that the particle source is an infinitesimally thin filament at the origin, $x = y = 0$; all the ions leaving this source have the same velocity v_0. The ions are to be collected at the point $x = L$, $y = 0$.

In general, the resulting beam image obviously has some finite width d. The magnitude of d depends, in particular, on the width of the limiter diaphragm defining the beam, that is, the limits $-\alpha_0$, $+\alpha_0$ of the angular range within which the particles emerge from the source.

If we specify L, the distance from the source to the image, and the angular spread α_0, the problem is to choose fields such that d = min.

The functional for the image width can be determined in various ways; one way is

$$d^{2n} = \frac{1}{A} \int_0^L \int_{-\alpha_0}^{\alpha_0} a(\alpha) \, d\alpha \, dx \, (2ny^{2n-1} y') =$$

$$= \int_0^L \int_{-\alpha_0}^{\alpha_0} Q(\alpha, y, y') \, d\alpha \, dx, \quad y' \equiv \frac{dy}{dx}, \qquad (2.132)$$

where y is the distance of the particle from the x axis, $a(\alpha)$ is the "weight function," and A is a normalization factor.

The choice of n, a (α), and A is governed by the requirements imposed on a particular system. The quantity y which appears in (2.132) is a function of the variables x and α. An equation for y follows from Newton's law; this equation can be written

$$y'' = G\,(y',\ y,\ x,\ [f]). \tag{2.133}$$

Here $[f]$ is the set of functions that govern the electromagnetic fields in the channel. In the two-dimensional case under consideration here there are three such functions; these are specified on the initial surface:

$$\psi\,(x)\,|_{y=0};\quad \frac{\partial\psi}{\partial y}\bigg|_{y=0};\quad \Phi\,(\psi). \tag{2.134}$$

Equation (2.133) is second order, so that it has two integration constants:

$$y\,|_{x=0}\ \text{and}\ \frac{\partial y}{\partial x}\bigg|_{x=0}. \tag{2.135}$$

Since the function $y(\alpha,\ x)$ depends on $[f]$, we write

$$y = y\,(x,\ \alpha,\ [f]),$$

so that d is a functional of f:

$$d^{2n}[f] = \int_{0}^{L} \int_{-a_0}^{+a_0} Q\,d\alpha\,dx. \tag{2.136}$$

The problem thus reduces to a minimization of the functional (2.136) under the additional constraint (2.133). In the case under consideration here, the field in the beam volume can be expressed with the help of these three functions in the form of series, as in (2.42) and (2.43). This system of equations should be supplemented with integral equations that relate Ψ and Φ with f.

We see that a "frontal" attack on the particle-optical problem, in which an attempt is made to reduce it to a variational problem of standard form, results in a very complicated system, even in the simplest case.

Clearly, the problem as formulated has the solution d = 0 if fields of unlimited strength are allowed. To find a regular solution it may be necessary to satisfy some condition stating that the fields and their gradients are "reasonable," e.g.,

$$\int b\,(\alpha)\,(E^2 + \alpha^2\,H^2 + \beta^2\,(\nabla E)^2 + \gamma^2\,(\nabla H)^2)\,dx\,d\alpha = \text{const}.\qquad (2.137)$$

At the present time no method is available for solving this variational problem; in fact we do not even have a serious formulation of this problem. Accordingly, in addition to pursuing the analysis and improving the general analytic formulation of the variational problems, it is important to develop appropriate variational formulations. We turn now to a complete numerical solution of a special variational problem.

We have already noted that a prediction of the initial arrangement of the focusing device is presently of considerable importance. The problem of constructing the image of a filament without spherical aberration, noted above, was solved a long time ago by L. A. Artsimovich. He showed that if a homogeneous magnetic field H_0 is applied along the filament, and if an electric field E_0 satisfying the conditions

$$\frac{2\pi}{\omega_H}\,c\,\frac{E}{H} = L, \qquad \omega_H = \frac{eH}{Mc}\qquad (2.138)$$

is applied along the y axis (perpendicular to the magnetic field), then all the particles emerging from the filament meet at the point x = L, y = 0, regardless of the angle α and the particle velocity v_0.

2. **Shapiro Method.** The Shapiro method makes it possible to optimize the paraxial system if equations are available for the aberrations. This method has been adapted for computer calculation.

The method can be summarized as follows: We know that the Grinberg equation for a narrow beam contains several quantities which characterize the geometry of the principal trajectory, e.g., p(s), the deviation of the particles from the mainly trajectory; r(s); and ψ(s), Φ(s),...., which are characteristic of the magnetic and electric fields.

We denote all these quantities by ζ_k(s). Then the Grinberg equation can be written

$$\hat{\Gamma}\,[\zeta_k\,(s)] = 0, \quad k = 1,\,2,\,\ldots,\,n.\qquad (2.139)$$

We assume that the position of the particle source corresponds to s = 0 and that the image position corresponds to s = L.

The various types of aberrations $\delta\zeta_i$ are functionals of ζ_k [see Eq. (2.66)]:

$$\delta\zeta_i = A_i[\zeta_k], \qquad i = 1, 2, ..., l. \qquad (2.140)$$

The values of ζ_k at s = 0 and s = L must satisfy certain boundary conditions

$$\hat{\gamma}_j [\zeta_k]_{s=0} = 0; \quad 1 \leqslant j \leqslant n_1 \leqslant n; \qquad (2.141a)$$

$$\hat{\gamma}_j [\zeta_k]_{s=|L} = 0; \quad 1 \leqslant j \leqslant n_2 \leqslant n, \qquad (2.141b)$$

including

$$p(0) = 0; \; p(L) = 0. \qquad (2.141c)$$

The method uses the following approach: We approximate the quantities $\zeta_k(s)$ by finite sums $\widetilde{\zeta}_k$ of some known functions $\psi_{ki}(s)$:

$$\widetilde{\zeta}_k(s) = \sum_{i=1}^{\nu_k} \alpha_{ik} \psi_{ki}(s). \qquad (2.142)$$

Here α_{ik} is a constant which is to be determined. Substituting (2.142) into (2.141), we find $n_1 + n_2$ algebraic equations for the $n_1 + n_2$ coefficients α_{ki}.

To find the other coefficients we must require, first, that the value of ζ_k governed by Eq. (2.142) satisfy the Grinberg equation (2.139) as accurately as possible, and second, that the aberration in (2.140) be minimized. These requirements can be reduced to the requirement that the following functional be minimized:

$$F(\alpha_{ki}) = \int_0^L \hat{\Gamma}^2 \left[\widetilde{\zeta}_k \right] ds + \sum_i^l q_i A_i^2 \left[\widetilde{\zeta}_k \right]. \qquad (2.143)$$

Here q_i are coefficients which are specified a priori.

Since the functional (2.143) is an algebraic equation in terms of α_{ki}, it can, in principle, be minimized by the standard methods of differential calculus, through the use of additional conditions (2.141).

However, the resulting expressions for $F(\alpha)$ turn out to be so cumbersome that the calculation can be carried out on a computer using a search method like the "ravine" method.

Shapiro [30] has reported an illustrative calculation of the magnetic field of a cathode lens which produces, in combination with a homogeneous electric field, an image with corrected anisotropic distortion, rotated through a given angle and lying in a given plane. The magnification is assumed to be unity. The search process is continued until the value of F reaches 10^{-4}. This requires about ten "ravine" steps; in the first step, F is of the order of 10^4, and in the third it is of the order of unity.

The magnetic field found in this way is substituted into the equation of motion, and then the particle trajectories are calculated on a computer. This calculation shows that both the distance to the Gaussian plane and the magnification at the center agree with the given values; the relative error is less than 0.02.

Chapter 3

PLASMA LENSES

§1. Axial Plasma Lens

The plasma lens is the simplest plasma-optical system and merits special attention for this reason. Plasma lenses are the subject of the present chapter.

1. Focusing Properties of an Axial Lens. From the example of the axial plasma lens (Chap. 2, §2) we see that the focusing properties of the plasma lens is very different from those of the vacuum. The reason is that plasma lenses are convertible systems in the sense defined earlier so that they can perform focusing with space charge. The focal length of the axial lens is given in the "weak-field approximation" in Eq. (2.119).

Separating the terms in (2.119) associated with the magnetic and electric fields, we can rewrite the equation for the focal length as

$$
\frac{1}{F_1} = \frac{1}{Mv_0^2} (W_H + W_E) \Big|_{-\infty}^{+\infty} + \frac{1}{(Mv_0^2)^2} \Bigg\{ \int_{-\infty}^{+\infty} (W_H + W_E)^2 \, dz -
$$

$$
- (W_H + W_E) \Big|_{-\infty}^{+\infty} \int_{-\infty}^{+\infty} (W_H + W_E) \, dz - (W_H + W_E) \Big|_{-\infty}^{+\infty} \times
$$

$$
\times \Bigg[\int_{-\infty}^{0} (W_H + W_E) \, dz + \int_{0}^{\infty} dz_1 \int_{z_1}^{\infty} (U_{2H} + U_{2E}) \, dz \Bigg] \Bigg\} + 0 \left(\frac{1}{(Mv_0^2)^3} \right), \qquad (3.1a)
$$

where

$$W_H = \int_{-\infty}^{z} U_{2H}\, dz; \quad W_E = \int_{-\infty}^{z} U_{2E}\, dz. \tag{3.1b}$$

For the axial lens U_{2H} is given by Eq. (2.49a).

Since the magnetic field distribution H(z) does not change (if the single-particle approximation is valid), the focal length of the magnetic lens is the same in the vacuum and plasma cases.

In the plasma case the function U_{2E} is [see Eq. (2.49a)]

$$U_{2E} = \frac{e}{2} \cdot \frac{d\Phi}{d\psi}\, H(z), \tag{3.2a}$$

in the vacuum case it is

$$U_{2E} = \frac{e}{2} \cdot \frac{d^2\Phi}{dz^2}. \tag{3.2b}$$

It follows from (3.2b) that $W_E|_{+\infty}^{-\infty} = 0$, in an electrostatic vacuum lens; this is the case because with the limited range of the fields as $|z| \to \infty$, we find $(d\Phi/dz) = -E_z \to 0$.

Consequently, a "thin" electrostatic vacuum lens (a lens with $F \approx v_0^2$) is not possible. On the other hand, with a plasma lens it is possible to achieve $F \approx v_0^2$ through an appropriate choice of focusing fields, e.g., with $\Phi = k\psi$, k = const, H > 0. As in the case of an axial lens, **E** focusing and the **H** focusing are not coupled in a thin axial lens with $F \approx v_0^2$.

According to Eq. (3.1), the thin magnetic axial lens must be a converging lens. However, a thin axial electrostatic plasma lens cen be a converging lens or a diverging lens, depending on the sign of $\Phi(\psi)$ [Eq. (3.2a)].

Let us examine an axial lens in which

$$W_E\Big|_{-\infty}^{+\infty} = W_H\Big|_{-\infty}^{+\infty} = 0. \tag{3.3a}$$

In this case

$$\frac{1}{F} = \frac{1}{(Mv_0^2)^2} \int_{-\infty}^{+\infty} (W_H + W_E)^2\, dz + \frac{1}{(Mv_0^2)^3} \times$$

$$\times \left[\int_{-\infty}^{+\infty} W^2 dz_1 \int_{-\infty}^{z_1} W\, dz_2 - \int_{-\infty}^{+\infty} dz_1\, W \int_{-\infty}^{z_1} dz_2\, W^2 \right] + \ldots, \tag{3.3b}$$

where $W = W_H + W_E$.

A lens with $F \approx v_0^4$ can be only converging, even in the plasma case, since the integrand in (3.3b) is nonnegative when $F \approx v_0^4$.

We see from (3.3b) that a pure magnetic or pure electrostatic lens with focal length $F \approx v_0^6$ is not possible since this would imply

$$\int_{-\infty}^{+\infty} W_H^2 \, dz = 0 \qquad (3.4a)$$

or

$$\int_{-\infty}^{+\infty} W_E^2 \, dz = 0 \qquad (3.4b)$$

which is not possible with $H \neq 0$ and $E \neq 0$.

If the lens is electromagnetic, a weaker condition is sufficient to ensure the vanishing of the term proportional to v_0^4:

$$\int_{-\infty}^{+\infty} (W_H + W_E)^2 \, dz = 0. \qquad (3.4c)$$

It can be shown that when (3.4c) holds a lens with $F \approx v_0^6$ cannot exist. Accordingly no lens is possible with $(v_0^2)^n$, where $n > 3$.

2. **Example of an Axial Lens.** Let us consider in detail the simplest axial plasma lens (Chap. 1, § 2). Specifically, this lens is a ring that carries a current. A voltage u is applied between the ring and the ion source. The magnetic field at the axis of the ring in the vacuum case is known to be

$$H_0 = \frac{2\pi R^2 I}{c(z^2 + R^2)^{3/2}} . \qquad (3.5a)$$

Here I is the current in the ring and R is its radius.

If the ion current density is low, the electric field near the ring satisfies the Laplace equation (vacuum regime). In this case

$$\Phi(z) = \Phi_0 = \frac{u_0 R}{(z^2 + R^2)^{1/2}} \qquad (3.5b)$$

at the axis of the system, where u_0 is the potential on the axis at $z = 0$.

Substituting (3.2) into (3.1) in the vacuum regime we find the focal length for this lens to be†

$$\frac{1}{F_H} = \frac{3\pi^3}{16} \cdot \frac{e}{Mc^2} \cdot \frac{1}{\dfrac{Mv_0^2}{2e}} \cdot \frac{l^2}{c^2 R};$$ (3.6a)

$$\frac{1}{F_E} = \frac{3}{16} \cdot \frac{1}{\left(\dfrac{Mv_0^2}{2e}\right)^2} \cdot \frac{\pi u_0^2}{8R}.$$ (3.6b)

In the plasma case, F_H obviously does not change. On the other hand, the value of F_E does change, because the magnetic lines of force become equipotentials. Specifically, taking the simplest function $\Phi(\psi)$,

$$\Phi = k\psi; \quad k = \text{const},$$ (3.7)

we find

$$1/F_E = 2u_0 \Big/ \left(\frac{Mv_0^2}{2e}\right).$$ (3.8)

Clearly, a plasma lens with $F_H \gg F_E$ can be either converging or diverging, depending on the sign of u_0.

3. Chromatic Aberration of an Axial Lens.

It follows from (3.6) and (3.8), that chromatic aberration cannot be corrected in an axial lens in either the vacuum case or the plasma case: $(\partial F / \partial v_0) \neq 0$ if $F \neq \infty$. This result is based on the weak-field approximation, but it can also be derived exactly for the plasma case. A distinguishing feature of this latter case is the fact that the electric potential is constant along the axis if $T_e \to 0$.

We can show that chromatic aberration is inherent in any axial system if the velocity of a particle remains constant along the axis. We begin with the equation for small radial displacements, (2.49):

$$(d^2r/dz^2) + \lambda^2 \varkappa(z)r = 0.$$ (3.9)

Here $\lambda = 1/v$, where v is the particle velocity, and $\varkappa(z)$ is a func-

---

†Since the E focusing and the H focusing are not coupled, the quantities F_H and F_E are related to the focal length by (2.122), and can be calculated independently.

tion which depends on the electromagnetic fields. We assume that λ_0 is the reciprocal of the velocity for some principal particle and that $\lambda = \lambda_0 + \delta\lambda$ is the same for a "secondary" particle. Then the equation of motion for the latter can be written

$$\left.\begin{array}{l} r'' + \lambda_0^2 \varkappa(z)\, r = -2\lambda_0\, \delta\lambda\varkappa(z)\, r; \\ r'' \equiv d^2 r/dz^2. \end{array}\right\} \tag{3.10}$$

Assuming that $\delta\lambda$ is a first-order quantity, we solve Eq. (3.10) by successive approximations. We write r as

$$r = r_0 + r_1 + \dots \tag{3.11}$$

and only retain the first terms in the expansion. Using (3.9) we obtain the following equation for r_1:

$$r_1'' + \lambda_0^2 \varkappa(z)r_1 = 2\lambda_0\delta\lambda r_0'', \tag{3.12}$$

where r_0 is the solution of the equation

$$r_0'' + \lambda_0^2 \varkappa(z)r_0 = 0. \tag{3.13}$$

If we know linearly independent solutions η_{01}, η_{02} of Eq. (3.13) which satisfy the boundary conditions

$$\eta_{01}(0) = \eta_{01}(L) = 0; \tag{3.14a}$$

$$\eta_{02}'(0) = \eta_{02}'(L) = 0, \tag{3.14b}$$

where L is the distance from the source to the image, we can require, without any loss of generality, that r_0 satisfy boundary conditions (3.14a). Then

$$r_0 = \text{const } \eta_{01}(z), \tag{3.15}$$

and the particular solution of Eq. (3.12) can be written

$$r_1 = 2\text{const}\,\lambda_0\,\delta\lambda\left[-\eta_{01}\int_0^z \eta_{01}''\,\eta_{02}\,dz + \eta_{02}\int_0^z \eta_{01}''\,\eta_{01}\,dz\right]. \tag{3.16a}$$

Alternatively, after integrating the second term in (3.16a) by parts, we can write r_1 as

$$r_1 = 2\text{ const }\lambda_0\,\delta\lambda\left[-\eta_{01}\int_0^z \eta_{01}''\,\eta_{02}\,dz + \eta_{02}\,[\eta_{01}'\,\eta_{01}]_0^z - \eta_{02}\int_0^z (\eta_{01}')^2\,dz\right].$$

$$\tag{3.16b}$$

Using boundary conditions (3.14), we find the following at the focal point z = L:

$$r_1 = 2 \, \text{const} \, \lambda_0 \, \delta\lambda\eta_{02}(L) \int_0^L (\eta_{01}')^2 \, dz. \qquad (3.16c)$$

The function $\eta_{02}(z)$ does not vanish at the point z = L by virtue of boundary conditions (3.14) [the Wronskian is $W(\eta_{01}, \eta_{02}) = 1$]. Thus $r_1 \neq 0$ at the focal point of $\delta\lambda \neq 0$, so that chromatic aberration cannot be corrected in an axial plasma system with $T_e = 0$.

4. **Spherical Aberration of an Axial Plasma Lens.** It is known [31] that spherical aberration cannot be corrected in an axial vacuum lens. However, as noted by Scherzer [32] a long time ago, space charge can be used to modify the electric field in the lens and to correct various aberrations which cannot be corrected in a vacuum axial lens. In a plasma lens it is possible to produce a space charge with an arbitrary density distribution, in accordance with condition (1.11). Let us see how this feature affects the spherical aberration of axial plasma lenses.

In contrast with the case of chromatic aberration, the spherical aberration of an axial plasma lens can be corrected.

The focal length of a thin axial lens with chromatic aberration taken into account is given by Eq. (2.122a), where W_1 is given by (2.124). Substituting the explicit expressions for U_2 and U_4 into these equations, we find the focal length to be

$$\frac{1}{F} = \frac{1}{v_0^2} \int_{-\infty}^{+\infty} \left[\frac{e^2}{M^2 c^2} \cdot \frac{H^2}{4} + \frac{e}{M} \cdot \frac{d\Phi}{d\psi} H \right] dz - \frac{r_0^2}{v_0^2} \times$$

$$\times \int_{-\infty}^{+\infty} \left[-\frac{e^2}{M^2 c^2} \cdot \frac{1}{12} HH'' \frac{e}{M} \frac{d\Phi}{d\psi} \frac{H''}{6} + \frac{e}{M} \frac{d^2\Phi}{d\psi^2} \cdot \frac{H^2}{2} \right] dz. \qquad (3.17)$$

We can show that the second integral in (3.17), which describes spherical aberration, can be made to vanish. For this purpose we choose the function $\Phi(\psi)$ in the form

$$\Phi(\psi) = k_1\psi + k_2\psi^2, \quad k_1, \ k_2 - \text{const}. \qquad (3.18)$$

Substituting (3.18) into (3.17) we find

$$\frac{1}{F} = \frac{1}{F_0} - \frac{r_0^2}{v_0^2} \int_{-\infty}^{+\infty} \left(-\frac{e^2}{M^2 c^2} \frac{1}{12} HH'' - \frac{e}{M} k_1 \frac{H''}{6} + \frac{e}{M} k_2 H^2 \right) dz, \qquad (3.19)$$

where F_0 is the focal length of the lens when spherical aberration is neglected.

Equating the integral in (3.19) to zero, and noting that $H'|_{-\infty}^{+\infty}=0$, we find

$$k_2 = -\frac{1}{12} \cdot \frac{e}{Mc^2} \frac{\int\limits_{-\infty}^{+\infty} H'^2\, dz}{\int\limits_{-\infty}^{+\infty} H^2\, dz}. \tag{3.20}$$

Accordingly, by choosing k_0 from condition (3.20) with the function $\Phi(\psi)$ from (3.18), we can correct the spherical aberration in a thin axial lens. The focal length of the lens is not changed. In the axial lens treated above the quantity k_2 is

$$k_2 = -3 \cdot 10^{-2} \frac{1}{R^2} \cdot \frac{e}{Mc^2}. \tag{3.21}$$

§2. Annular Magnetic Lens

Let us examine an annular magnetic plasma lens in connection with the focusing of hollow neutralized ion beams.

1. General Properties of Magnetic Lenses.
The focusing properties of the magnetic lenses, which we assume to be "approximately thin," can be analyzed on the basis of Eq. (2.119) and (2.123), under the assumption that the field is $E = 0$, that is to say, U_{0E}, U_{1E}, and U_{2E} are zero. In this case the focal length can be written

$$\frac{1}{F_H} = \frac{1}{Mv_0^2} W_H \Big|_{-\infty}^{+\infty} + \frac{1}{(Mv_0^2)^2} \left\{ \int\limits_{-\infty}^{+\infty} W_H^2\, dz - W_H \Big|_{-\infty}^{+\infty} \int\limits_{-\infty}^{+\infty} W_H\, dz - \right.$$

$$- W_H \Big|_{-\infty}^{+\infty} \left[\int\limits_{-\infty}^{0} W_H\, dz + \int\limits_{0}^{\infty} dz_1 \int\limits_{z_1}^{\infty} U_{2H}\, dz_2 \right] +$$

$$\left. + 2 \int\limits_{-\infty}^{+\infty} U_{1H}^2\, dz + 2\, [U_{0H} W_H]_{-\infty}^{+\infty} \right\} + \cdots, \tag{3.22}$$

where, for $D = 0$ [see Eq. (2.53)],

$$W_H = \int\limits_{-\infty}^{z} U_{2H}\, dz = \frac{e^2}{Mc^2} \int\limits_{-\infty}^{z} \left(\frac{3\psi_0^2}{R^2} + \psi_1^2 - 3\frac{\psi_0\,\psi_1}{R} - \psi_0\,\psi_0'' \right) \frac{1}{R^2}\, dz. \tag{3.23}$$

We first assume that the particle velocity is high and that terms proportional to v_0^4 can be neglected. In this case the focal length of the magnetic lens is $F \approx v_0^2$ and Eqs. (3.22) and (3.23) show that

$$\frac{1}{F} = \frac{e^2}{(Mv_0)^2 c^2} \int_{-\infty}^{+\infty} \left(\psi_1^2 - 3\frac{1}{R}\psi_0\psi_1 + \psi_1^2 + 3\frac{\psi_0^2}{R^2} \right) \frac{1}{R^2}\, dz. \quad (3.24)$$

We now show that a magnetic lens with $F \approx v_0^2$ cannot be a diverging lens. First note that the integrand in (3.24) is a quadratic trinomial in $1/R$. Calculating its discriminant, we find the condition under which this discriminant is negative:

$$4\,\overline{\psi_0^2}\,(\overline{\psi_1^2} + \overline{\psi_0'^2}) > 3\,(\overline{\psi_0\psi_1})^2, \quad (3.25)$$

where $\bar{a} \equiv \int_{-\infty}^{+\infty} a\,dz$. Inequality (3.25) clearly holds,[†] i.e., F is positive everywhere. Inequality (3.25) becomes an equality only if $\psi_0 = 0$, $\psi_1 = 0$. Accordingly, it is not possible to make an approximately thin lenses for which $F \approx v_0^4$, i.e., $W|_{-\infty}^{+\infty} = 0$.

2. E x a m p l e   o f   a   S p l i t   M a g n e t i c   L e n s . As an example we consider the focusing properties of the split magnetic lens, shown schematically in Fig. 16a. For this lens $\psi_1 = 0$ and

$$\psi_0 = H_0 R\Lambda \sqrt{\frac{e_0}{2}} \exp(-z^2/\Lambda^2), \quad (3.26)$$

where e_0 is the base of the natural logarithm system, H_0 is the peak value of the magnetic field, and Λ is the span of the field along the initial surface.

The current distribution in the coils that produce the magnetic field corresponding to (3.26) is calculated in Chap. 2, §1. The configuration of the lines of force[‡] is shown in Fig. 16b. Taking account of the curvature of the lines of force and the direction of the vector **H** in the split lens, we conclude that the central zone of the split lens, $|z| \le \Lambda/2$, is converging, while the regions $|z| > \Lambda/2$ are diverging. A particle can traverse the split lens if its energy is $(Mv_0^2/2) > U = (e^2/Mc^2)(\psi_0^2/2R^2)$. Otherwise, the split lens reflects the particle.

[†]Inequality (3.25) is actually a strengthened form of the Schwarz inequality $\bar{a}^2\bar{b}^2 > (\overline{ab})^2$.
[‡]The magnetic field can be produced by magnet poles as well as by current-carrying coils. The shape of the pole tips is shown in Fig. 16b.

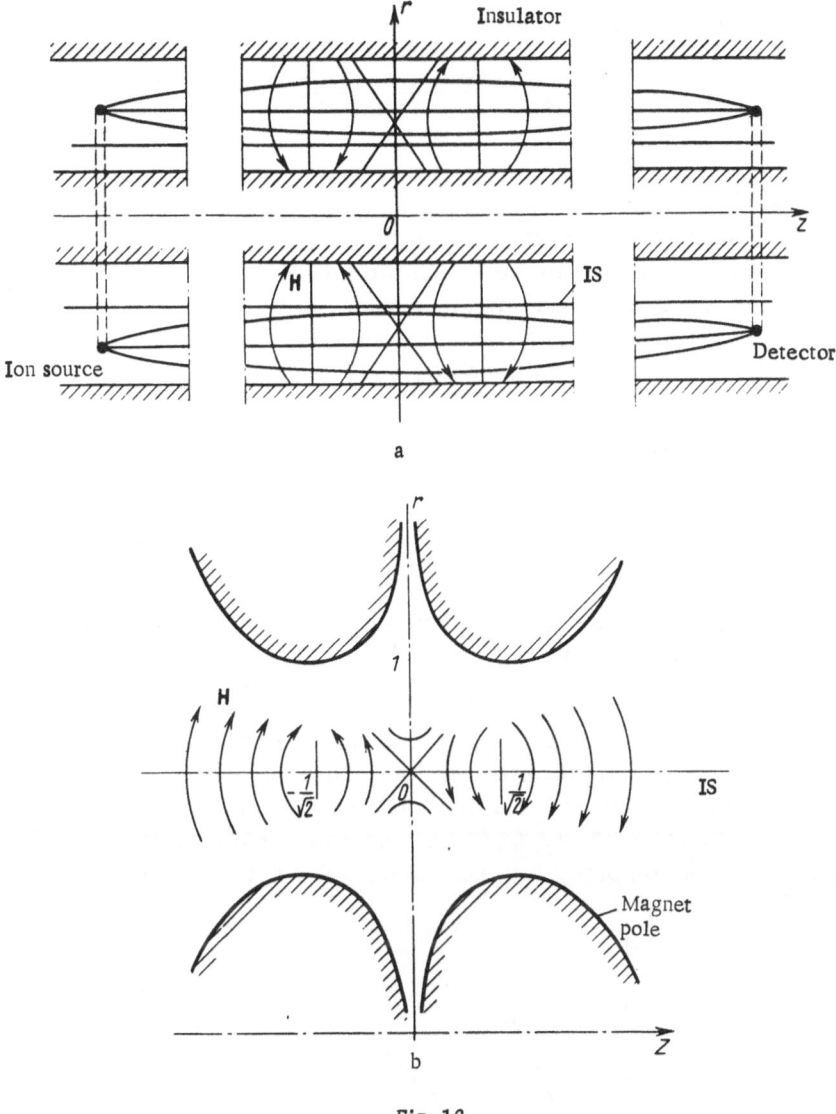

Fig. 16

For those particles which do traverse the lens, the focal length is found from (3.24) in the thin-lens approximation:

$$\frac{1}{F} = \frac{H_0^2 \, e^2 \, \Lambda^3}{c^2 \, (Mv_0)^2 \, R^3} \cdot \frac{e_0}{2} \sqrt{\frac{\pi}{2}} \left(1 + 3\, \frac{\Lambda^2}{R^2}\right). \qquad (3.27)$$

The position of the principal surface is governed by Eqs. (2.107b):

$$y_{ps\,0} = \frac{\Lambda^2}{R} \cdot \frac{1}{1 + 3\Lambda^2/R^2} \,. \tag{3.28}$$

Since we are dealing with paraxial beams that pass near the initial surface, obviously $y_{ps0} \ll R$; the value of R must roughly satisfy the condition $(\Lambda/R)^2 \ll 1$. As larger R, we find $y_{ps} \to 0$, and the principal surface coincides with the initial surface. This statement holds for a lens with any distribution of the field H_r on the initial surface (if $\psi_1 = 0$), as can be seen from the general equation for y_{ps} (2.107b), which now becomes

$$y_{ps\,0} = \frac{\Lambda^2}{R} \frac{\displaystyle\int_{-\infty}^{+\infty} \psi_0^2 \, dz}{\displaystyle\int_{-\infty}^{+\infty} \left(\psi_0'^2 + 3\frac{\Lambda^2}{R^2} \psi_0^2 \right) dz} \,. \tag{3.29}$$

We note that in such lenses the principal surface is always above the initial surface ($y_{ps0} > 0$).

As v_0 decreases, the focal length also decreases, becoming comparable to the scale range of the focusing fields. In this case the thin-lens approximation is not applicable and equation of motion (2.51) must be solved exactly. Figure 17 shows F (curves 1 and 2) as a function of $v_0^2/\omega_H^2\Lambda^2$, the lens power of the split lens, calculated on a computer for the value $\Lambda/R = 0.4$. Shown for com-

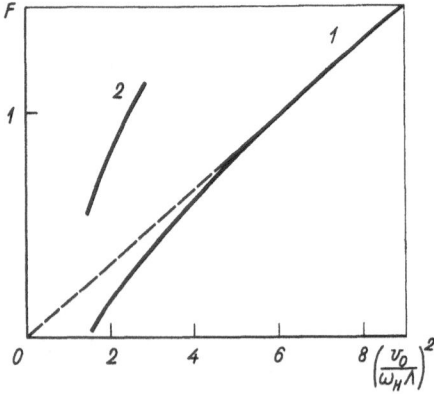

Fig. 17

parison (dashed line) are the values of F found in the thin-lens approximation, (3.27). At large values of $v_0/\omega_H\Lambda$ this approximation gives fairly accurate values of the focal length. Also shown in this figure are the multiple foci 2, which appear as $v_0/\omega_H\Lambda$ decreases. If, on the other hand, we have[†] $v_0 \to \lim v_0$, then the number of multiple foci increases without bound. Finally, if the ratio $v_0/\omega_H\Lambda$ becomes smaller than $\min v_0$, the particle is reflected from the split lens.

§3. Electrostatic Plasma Lens

1. General Properties of the Electrostatic Plasma Lens.

As noted in §1, the fundamental difference between the plasma electrostatic lens and its vacuum analog is the convertible nature of the plasma lenses. When $T_e \to 0$, condition (1.11) holds and the magnetic lines of force determine the shape of the equipotentials. We assume that the electric field distribution is established by an external system of electrodes to which a potential difference is applied. In this way we can choose the electric field configuration most suitable for a given problem. Let us examine in more detail the properties of the annular electrostatic plasma lens.

According to (2.123), the focal length of a plasma electrostatic lens is[‡]

$$\frac{1}{F} = \frac{1}{Mv_0^2} W_E \Big|_{-\infty}^{+\infty} + \frac{1}{(Mv_0^2)^2} \left\{ \int_{-\infty}^{+\infty} W_E^2 \, dz - W_E \Big|_{-\infty}^{+\infty} \int_{-\infty}^{+\infty} W_E \, dz + \right.$$

$$+ 2 \int_{-\infty}^{+\infty} U_{1E}^2 \, dz + 2 [U_{0E} W_E]_{-\infty}^{+\infty} - \int_{-\infty}^{+\infty} W_E U_{0E}' \, dz -$$

$$\left. - W_E \Big|_{-\infty}^{+\infty} \left[\int_{-\infty}^{0} W_E \, dz + \int_{0}^{\infty} dz_1 \int_{z_1}^{\infty} U_{2E} \, dz_2 \right] \right\} + \dots, \qquad (3.30)$$

[†] We can define v_0 as the minimum velocity for which the particle can still traverse the lens:

$$\min v_0 \approx \omega_H \Lambda \sqrt{\frac{\Lambda}{2R}}.$$

[‡] As before, we need consider only the "right focus" F_1, because the properties of F_1 and F_2 are analogous.

where W_E can be written explicitly as [see Eqs. (2.120b) and (2.53c)]

$$W_E = \int_{-\infty}^{z} U_{2E}\, dz = e \int_{-\infty}^{z} \left(\frac{d\Phi}{d\psi} \cdot \frac{\psi_1}{R} + \frac{d^2\Phi}{d\psi^2}\, \psi_1^2 - \frac{d\Phi}{d\psi}\, \psi_0'' \right) dz, \qquad (3.31)$$

and U_{0E} and U_{1E} are, according to Eqs. (2.53a) and (2.53b),

$$U_{0E} = \Phi_0(\psi_0); \quad U_{1E} = \left.\frac{d\Phi}{d\psi}\right|_{\psi_0} \psi_1. \qquad (3.32)$$

If the lens are thin ($F \approx v_0^2$), it is possible to make it a converging lens or a diverging lens through the appropriate choice of **E**. In a split lens, it is sufficient to remove the electric field from the focusing regions or to remove the electric field from the defocusing regions (by shorting the electrodes which establish the distribution of the electric potential).

When $W_E|_{-\infty}^{+\infty} = 0$ and the focal length $F \approx v_0^4$, the possibility of making diverging electrostatic plasma lenses is less obvious. We write the focal length as

$$\frac{1}{F} = \frac{1}{(Mv_0^2)^2} \int_{-\infty}^{+\infty} \left(W_E^2 - W_E \frac{dU_{0E}}{dz} + 2U_{1E}^2 \right) dz. \qquad (3.33)$$

For simplicity we treat the case $\psi_1 = 0$. Then

$$\frac{1}{F} = \frac{1}{(Mv_0^2)^2} \int_{-\infty}^{+\infty} (W_E^2 + W_E E_z)\, dz, \qquad (3.34)$$

since $dU_{0E}/dz = d\Phi_0/dz = -E_z$ and $U_{1E} = 0$. In a diverging lens $F < 0$, so that the lens fields must be chosen to satisfy

$$\int_{-\infty}^{+\infty} (W_E^2 + W_E E_z)\, dz < 0 \qquad (3.35)$$

and simultaneously to satisfy the condition $W_E|_{-\infty}^{+\infty} = 0$, in order to make a diverging lens with $F \approx v_0^4$. We introduce the auxiliary function

$$f(z) = W_E^2 + W_E E_z, \qquad (3.36)$$

which satisfies the conditions

$$\int_{-\infty}^{+\infty} f(z)\,dz < 0; \quad f(z)\Big|_{-\infty}^{+\infty} = 0. \tag{3.37}$$

Then assuming $E_z(z)$ and $f(z)$ to be given, we find the function W_E from (3.36):

$$W_E(z) = E_z/2 \pm \sqrt{\frac{E_z^2}{4} + f(z)}. \tag{3.38}$$

Knowing that

$$dW_E/E_z = dH_r/H_r, \tag{3.39}$$

and substituting (3.38) into (3.39), we find that magnetic field distribution required for devising a diverging lens with $F \approx v_0^4$

$$H_r = H_{r0} \exp\left\{\int_{z_0}^{z} \frac{dW_E}{dz} \cdot \frac{1}{E_z}\,dz\right\}, \tag{3.40}$$

where H_{r0} is the peak value of the magnetic field. This completes the solution of the problem of finding the focusing fields for a diverging electrostatic annular lens.

As an example we consider a lens with $E_z = \text{const} = k$, specifying the function $f(z)$ to be

$$f(z) = k^2 \frac{z^2}{z_0^2}\left(1 + \frac{z^2}{z_0^2}\right); \quad -z_0 < z < z_0. \tag{3.41}$$

Substituting (3.41) and E_z into (3.40), we find

$$H_r = H_{r0} \exp(-z^2/z_0^2); \quad |z| < z_0. \tag{3.42}$$

2. Examples of a Split Electrostatic Lens.

Let us examine the split lens with the magnetic field in (3.26), assuming that the electric field in the electrostatic lens is governed by the magnetic field configuration. The function $\Phi(\psi)$ in a split lens is governed by the system of electrodes and is arbitrary, in principle. We first assume that the electric field in the split lens is an accelerating field everywhere and that $E \approx H^2$. Then the fields E_z and H_r on the initial surface are

$$H_r = H_{r0}\sqrt{2e_0}\,\frac{z}{\Lambda}\exp(-z^2/\Lambda^2); \quad E_z = 2E_{z0}\frac{z^2}{\Lambda^2}\exp(1-2z^2/\Lambda^2), \qquad (3.43)$$

where H_{r0} and E_{z0} are the peak values of the fields **H** and **E**.

In this case the function W_E is

$$W_E = e \int_{-\infty}^{z} E_z \frac{1}{H_r} \cdot \frac{dH_r}{dz}\, dz = \frac{eE_z}{2}. \qquad (3.44)$$

Since $E_z|_{-\infty}^{+\infty} = 0$, (3.30) shows that the focal length of the split lens, F_E, can be proportional to v_0^4. Let us assume $F_H \gg F_E$, i.e., that the split lens is electrostatic. Substituting (3.44) into (3.33), we find the focal length to be

$$\frac{1}{F_E} = \frac{3\sqrt{\pi}}{512} \cdot \frac{1}{(Mv_0^2/2e)^2} \cdot \frac{e_0^2\,\Phi_0^2}{\Lambda}, \qquad (3.45)$$

where $\Phi_0 = E_{z0}\Lambda$.

It can be shown that the focal length of the electrostatic split lens is strongly affected by the electric field configuration. Specifically, assuming that $E \approx H$ [$E_z = \sqrt{2e_0}E_{z0}z/\Lambda \exp(-z^2/\Lambda^2)$] in the split lens, in the same way we find the focal length to be

$$\frac{1}{F_E} = \frac{1}{8}\frac{\sqrt{\pi}\sqrt{e_0}\,\Phi_0^2}{\Lambda\left(\dfrac{Mv_0^2}{2e}\right)^2}. \qquad (3.46)$$

Comparing (3.45) and (3.46), we see that in a lens in which energy is recovered, (3.46), the focal length is roughly an order of magnitude shorter than in a lens with acceleration (3.45), for given peak values of the focusing fields.

In the examples discussed above, the electric field is zero everywhere and the particle passes through the split lens, being focused and defocused. As a result, the focal length of the split lens is inversely proportional to the square of Φ_0. If, on the other hand, the electric field is zero only in the focusing regions (or only in the defocusing regions) as noted above, the focal length is inversely proportional to Φ_0. As an example we consider a split focusing lens with energy recovery, with the function $\Phi(\psi)$ determined as follows:

$$E_z = \sqrt{2e_0}\, E_{z0} \frac{z}{\Lambda} \exp\left(-z^2/\Lambda^2\right); \qquad -\frac{\Lambda}{\sqrt{2}} > z; \quad z > \frac{\Lambda}{\sqrt{2}}; \qquad (3.47a)$$

$$E_z = 0; \qquad\qquad\qquad -\frac{\Lambda}{\sqrt{2}} < z < \frac{\Lambda}{\sqrt{2}}. \qquad (3.47b)$$

Then

$$\frac{1}{F_E} = \frac{\sqrt{2}}{\dfrac{Mv_0^2}{2e}} \cdot \frac{\Phi_0}{\Lambda}. \qquad (3.48)$$

It follows immediately from (3.48) that an electric lens with $F_E \approx v_0^2$ can be either converging or diverging depending on the sign of Φ_0.

If the particle velocity v_0 is low, the weak-field approximation does not apply. In this case the configuration of the focusing and defocusing fields in the lens as well as the integrated effect of the electric fields on the particles are all important. Figure 18 shows the particle trajectories in such a split electric lens found by computer calculations for $(v_0/\omega_H\Lambda) = 2$, $\omega_H = (eH_{z0}/Mc)$. We note that at low velocities each part of the split lens focuses independently, and multiple foci are produced.

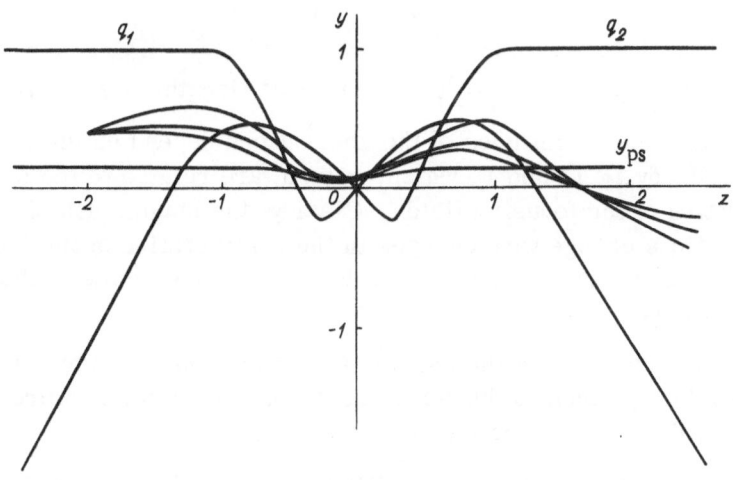

Fig. 18

§4. First-Order Aberrations of an Annular Plasma Lens

1. General Equations. The blurring of the focal points (Chap. 2, § 2) is due to the nonvanishing terms in series (2.26b) (for transverse aberrations) or series (2.27) (for longitudinal aberrations). In the present section we only take account of the linear terms in series (2.26b) and (2.27), the first-order aberrations. The explicit focusing functions d and z_1 can easily be found in the Gaussian-dioptrics approximation (Chap. 2, § 2), which we have been using throughout. Using the obvious relations

$$\frac{1}{z_B} = \frac{1}{F} - \frac{1}{z_A} \ ; \quad y_E = y_A - \frac{z_A}{z_A - F}(y_A - y_{ps\,0}), \qquad (3.49)$$

where $\{y_A, z_A\}$ are the coordinates of the source with respect to the initial surface, and $\{y_B, z_B\}$ are the image coordinates, and varying (3.49) with respect to the parameters F and y_{ps0} (which depend on the configuration of the focusing field), we find the following equations for the longitudinal and transverse aberrations, respectively:

$$z_1 = L\delta F \frac{1}{z_A - F} \ ; \qquad (3.50a)$$

$$d = -\delta y_{ps\,0}\frac{z_A}{z_A - F} - \frac{y_A - y_{ps\,0}}{(z_A - F)^2}z_A\frac{\delta F}{F} \ , \qquad (3.50b)$$

where $L = z_A + z_B = \dfrac{z_A^2}{z_A - F}$ is the focal length. We see from (3.50) that the magnitudes of the aberrations depend on the variables δF, δy_{ps0}, L, and F, which, in turn, are related to the configuration of the focusing fields. Clearly, the magnitudes of the aberrations change with changes in the configuration of the focusing field, and we can eliminate, or at least minimize, some of these aberrations.

The correction of the chromatic and momentum aberrations of annular systems is discussed in Chap. 4; here we examine the spherical aberration of annular lenses.

To find the spherical aberration we must take account of higher-order terms in the expansion in powers of y in the paraxial-approximation equation; in Eq. (2.50) we must take account of the explicit form of P. Solving Eq. (2.50) in the weak–field approximation, we then find the focal length to be [see Eq. (2.122)]

$$1/F_1 = 1/F_0 + 1/F_{sp}. \tag{3.51}$$

Here F_0 is the focal length for a narrow beam, (2.119) (we neglect P), and F_{sp} is the correction to F_0 when spherical aberration is taken into account. We see from (2.122a) that the consequence of incorporating the terms P_{quad} for thin lenses is that the quantity $W = \int_{-\infty}^{z} U_2 \, dz$, (2.120b), in Eq. (2.119) for the focal length is replaced by

$$W_1 = \int_{-\infty}^{z} \left(U_2 + \frac{y_0}{2} U_3 \right) dz, \tag{3.52}$$

where y_0 is the scale value of the beam half-width. Accordingly, the value of F_{sp} for a thin lens is

$$\frac{1}{F_{sp}} = \frac{1}{Mv_0^2} \cdot \frac{y_0}{2} \int_{-\infty}^{+\infty} U_3 \, dz. \tag{3.53}$$

The quantity F_{sp} depends on the configuration of the focusing fields, the lens power, $v_0/\omega_H \Lambda$, and the beam geometry. It is con-convenient to introduce the new quantity γ:

$$\gamma = F_0 \Lambda / F_{sp} \ h, \tag{3.54}$$

which is a universal dimensionless parameter, being a function of the configuration of the focusing fields alone. Here, h is the scale value of the beam half-width. Vanishing of γ corresponds to the correction of spherical aberration in the lens.

2. Spherical Aberration of a Thin Annular Magnetic Lens. It follows immediately from Eqs. (2.119) and (2.122) that

$$\gamma_H = \frac{\Lambda \int_{-\infty}^{+\infty} \left(-3 \frac{\psi_1^2}{R} + 3\psi_1 \psi_0'' + 5 \frac{\psi_0 \psi_0''}{R} - \psi_0 \psi_1'' + 12 \frac{\psi_0 \psi_1}{R^2} - 12 \frac{\psi_0^2}{R^3} \right) dz}{2 \int_{-\infty}^{+\infty} \left(3 \frac{\psi_0^2}{R^2} + \psi_1^2 - 3 \frac{\psi_0 \psi_1}{R} + \psi_0'^2 \right) dz}. \tag{3.55}$$

As before, the fields of an annular magnetic lenses are specified by the components H_r and H_z, that is, ψ_0 and ψ_1 on the initial surface.

We first consider a lens with "transverse" fields, for which $H_z = 0$ ($\psi_1 = 0$) on the initial surface. Then (3.55) yields

$$\gamma_H = \frac{\Lambda \int\limits_{-\infty}^{+\infty} \left(5\psi_0'^2 + 12 \frac{\psi_0^2}{R^2} \right) dz}{2R \int\limits_{-\infty}^{+\infty} \left(3 \frac{\psi_0^2}{R^2} + \psi_0'^2 \right) dz}. \tag{3.56a}$$

Since the integrands do not change sign, $\gamma_H \neq 0$ in such lenses, and the spherical aberration cannot be corrected. Furthermore, the quantity γ_H in (3.56a) is a weak function of the field configuration, since the terms in the integrands have different constant coefficients. For a split lens like that in (3.26) we have

$$\gamma_H = \frac{\Lambda}{R} \frac{5 + 12 \frac{\Lambda^2}{R^2}}{1 + 3 \frac{\Lambda^2}{R^2}}. \tag{3.56b}$$

If $H_r = 0$ ($\psi_0 = 0$; a lens with "longitudinal" fields) but $H_z \neq 0$ on the initial surface, then

$$\gamma_H = -3\Lambda/R \tag{3.57}$$

and this quantity is independent of the field configuration in the lens. It is obviously impossible to correct the spherical aberration.

We turn now to the general case in which $H_r \neq 0$, $H_z \neq 0$ on the initial surface. For certain relations between H_r and H_z the spherical aberration can be corrected. Let us assume that ψ_0 and ψ_1 are related by

$$\psi_1 = k\psi_0; \quad k = \text{const.} \tag{3.58}$$

A relation of this type holds in a lens in which

$$H_r \approx \frac{z}{\Lambda} \exp\left(-z^2/\Lambda^2\right); \quad H_z \approx \exp\left(-z^2/\Lambda^2\right). \tag{3.59}$$

Then

$$\gamma_H = \frac{\Lambda \int\limits_{-\infty}^{+\infty} (p\psi_0^2 - q\psi_0'^2)\, dz}{2R \int\limits_{-\infty}^{+\infty} \left[\psi_0^2 \left(3\frac{\Lambda^2}{R^2} + k^2 - 3\frac{k\Lambda}{R}\right) + \psi_0'^2\right] dz}, \qquad (3.60)$$

where

$$p = 12k\Lambda^2/R^2 - 3k^2\Lambda/R - 12\Lambda^3/R^3; \qquad (3.61a)$$

$$q = 5\Lambda/R - k\Lambda^2/R^2 - 3k. \qquad (3.61b)$$

We choose ψ_0 in the form in (3.26). Substituting (3.26) into (3.60), and integrating, we find that the spherical aberration is corrected if the parameters k and Λ/R satisfy

$$p - q = 0. \qquad (3.62)$$

Solving (3.62) for k, we find

$$k = \frac{6\frac{\Lambda^2}{R^2} + 4 \pm \sqrt{9\frac{\Lambda^2}{R^2} + 4}}{3\frac{\Lambda}{R}}. \qquad (3.63)$$

3. Correction of Spherical Aberration by an Electric Field.

Since the choice of $\Phi(\psi)$ is arbitrary, we can choose the electric field in the electromagnetic lens so that spherical aberration is corrected. We write γ as follows for the particular case $\psi_1 = 0$ for a lens with E and H fields [see Eq. (2.122)]:

$$\gamma = \frac{\dfrac{e^2}{Mc^2} \int\limits_{-\infty}^{+\infty} \left(12\frac{\Lambda^2}{R^2}\psi_0^2 - 5\psi_0\psi_0''\right) dz + e \int\limits_{-\infty}^{+\infty} \left.\frac{d\Phi}{d\psi}\right|_{\psi_0} \psi_0'' \, dz}{\dfrac{e^2}{Mc^2} \int\limits_{-\infty}^{+\infty} \left(\psi_0'^2 + 3\frac{\Lambda^2}{R^2}\psi_0^2\right) dz + e \int\limits_{-\infty}^{+\infty} \left.\frac{d\Phi}{d\psi}\right|_{\psi_0} \psi_0'' \, dz}. \qquad (3.64)$$

The integrand in the numerator of (3.64) is set equal to zero. As a result we find an equation from which $\Phi(\psi)$ can be determined:

$$12\Lambda^2/R^2\psi_0^2 - 5\psi_0\psi_0'' + (d\Phi/d\psi)\psi_0'' = 0; \qquad (3.65)$$

thus

$$\Phi = \int_{-\infty}^{z} \frac{1}{\psi_0''} \left(5\psi_0 \psi_0'' - 12 \frac{\Lambda^2}{R^2} \psi_0^2 \right) \psi_0' \, dz. \qquad (3.66)$$

If $(\Lambda/R) \to 0$, (3.66) is simplified and we have

$$\Phi(\psi) \approx \text{const } \psi_0^2. \qquad (3.67)$$

It is thus possible to correct spherical aberration in thin magnetic and electrostatic plasma lenses (other aberrations are generally not corrected; in fact, they may be aggravated).

§ 5. Multiple-Lens, Multiply Connected Charged-Particle Accelerators

In existing charged-particle accelerators the cross section of the accelerating channel is a singly connected figure — circular, rectangular, or elliptic. It is of interest to examine an accelerators in which the channel cross section is doubly connected, e.g., a ring. These accelerators, proposed by Morozov and Shchepkin [33], are called multiply connected here.

Figure 19 shows linear and cyclic (toroidal) versions of such an accelerator. We call the object in the interior of a channel the inner core 1 or core. In a toroidal accelerator this core must be

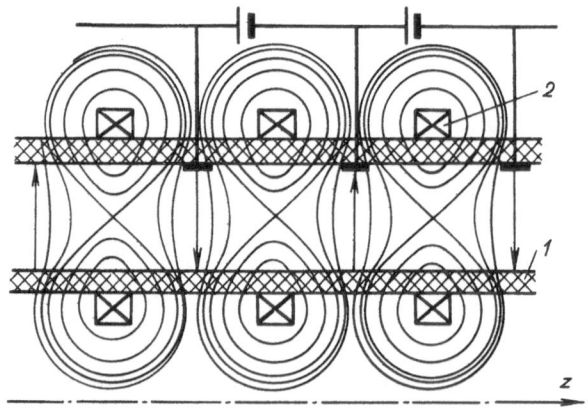

Fig. 19

suspended by a magnetic field or the accelerator must operate in
short pulses, so that the core does not have time to fall. The construction of a toroidal fusion system with a suspended core (levi-
tron) has already received much study. Although multiply connected
systems are structurally more complicated, they do permit the acceleration of neutralized ion beams.

Let us consider a multiply connected system in whose accelerating gap a meridional sign-changing magnetic field is produced by an inner core 1 and an outer magnetic circuit 2 (Fig. 19).
The lines of force of this field obviously form a lenslike configuration (Fig. 20), so that such an accelerator might be called a "multiple-lens, multiply connected accelerator." Under the influence
of the longitudinal electric field the ions are accelerated in the
channel and if the ion gyroradius is larger than the length of the lens,

$$R_{\Lambda i} > \Lambda, \tag{3.68}$$

the ion acceleration proceeds continuously.

A linear multiple-lens, multiply connected accelerator operates
in a direct-acceleration configuration similar to a Van de Graaf
generator [34], so that the particle energy which can be achieved
is low (1-5 MeV), although the ion currents can be large, in principle.

If much higher energies are required it is necessary to resort
to a cyclic system.

In discussing the multiply connected accelerator below we will
be thinking only of linear systems. The acceleration of neutral

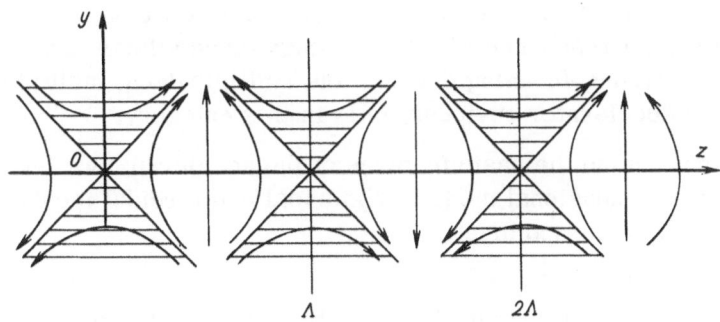

Fig. 20

beams in a cyclic multiple-lens, multiply connected accelerator is probably realized best with the betatron mechanism.

The acceleration of ions to high energies opens up new possibilities for the cyclic multiple-lens, multiply connected accelerator. The space charge plays a particularly important role at low particle energies, in which case the stabilizing magnetic field is weak.[†] As the ion energy is increased, however, it is possible to use stronger focusing magnetic fields, which are capable of confining unneutralized ion beams. Noting that $\omega_f \approx \omega_H \approx H$ in such an accelerator, we see that the density of a nonneutral beam is

$$n \approx H^2. \tag{3.69}$$

In other words, it is possible to arrange things so that the initial acceleration occurs in the quasineutral regime; subsequently the electrons are removed from the beam. The remaining unneutralized ion beam is then accelerated further, e.g., by an rf field.

It is also possible, in principle, to extract the beam from a cyclic accelerator. When the accelerating electric field is removed (i.e., when electron drift is "switched off"), it is necessary to constrict the beam to a narrow filament. After the sheath has been removed, the beam can be extracted by the usual methods.

1. Particle Dynamics in a Two-Dimensional Multiple-Lens, Multiply Connected Accelerator.

We consider the single-particle model, neglecting all plasma effects except the fact that the magnetic lines of force become equipotentials. We assume the magnetic and electric fields to be ideal and neglect all resonance effects, although these effects are important in real strong-focusing devices; the multiple-lens, multiply connected accelerator obviously falls in this category.

To determine the basic features of the accelerator we introduce a two-dimensional model; cylindrical and toroidal models will also be discussed briefly.

[†]An approximate upper limit on the density of a beam which can be confined by the focusing field is set by $\omega_0 < \omega_f$, where ω_0 is the plasma frequency and ω_f is the frequency of transverse oscillations.

Specification of the Fields in the Two-Dimensional Model.

The initial surface is the plane y = 0. We assume that on this surface

$$H_x = 0; \quad H_y = H_0 \sin \alpha z; \quad H_z = 0; \quad \alpha = \pi/\Lambda. \qquad (3.70)$$

In this case the Laplace equation can be integrated exactly:

$$\Omega = (H_0/\alpha)\sinh \alpha y \sin \alpha z; \qquad (3.71)$$

$$A_x \equiv \psi = (H_0/\alpha)\cosh \alpha y \cos \alpha z. \qquad (3.72)$$

To determine the function $\Phi = \Phi(\gamma)$ we specify the electric field on the initial surface as

$$E_x = 0; \quad E_y = 0; \quad E_z = E_0 (H/H_0)^2. \qquad (3.73)$$

This functional dependence of E_z on H is only a model dependence but it does reflect two important physical points: the fact that the sign of the longitudinal component of the electric field is fixed and the fact that the electric field vanishes at the point at which H = 0.

Using (3.71) and (3.72) we can write the condition for equipotential magnetic lines of force as

$$\Phi = E_0 \left(-\frac{z_0(f)}{2} + \frac{1}{4\alpha} \sin 2\alpha z_0(f) \right), \qquad (3.74)$$

where

$$z_0(f) \equiv \frac{1}{\alpha} \arccos \frac{\psi_\alpha}{H_0}; \quad \psi < \frac{H_0}{\alpha}. \qquad (3.75)$$

With the function Φ specified as in (3.74), the value of Φ on lines of force which do not intersect the initial surface (y = 0) remains undetermined. The region corresponding to these lines of force is shown cross-hatched in Fig. 19. This region corresponds to $\psi > (H_0/\alpha)$. To complete the determination of $\Phi(\gamma)$ we can assume these cross-hatched regions to be equipotential regions. In the region of the neutralized ion beam we write

$$\Phi = E_0 \left(-\frac{z_0}{2} + \frac{1}{4\alpha} \sin 2\alpha z_0 \right) + \dots \qquad (3.76)$$

Particle Dynamics in the Two-Dimensional
Model.   The ion motion is described by

$$\dot{x} + \frac{e}{Mc}\psi = D = \text{const};$$ (3.77a)

$$M\ddot{y} + \frac{\partial U}{\partial y} = 0;$$ (3.77b)

$$M\ddot{z} + \frac{\partial U}{\partial z} = 0;$$ (3.77c)

$$U = M/2\left(D - \frac{e}{Mc}\psi\right)^2 + e\Phi(\psi).$$ (3.77d)

Since a general solution for (3.77) is not available we consider certain particular cases from which an overall picture can be obtained. We begin with a calculation of the principal trajectories, by which we mean the two-dimensional trajectories in the $y = 0$ plane. For these trajectories, (3.77) can be reduced to quadratures:

$$\dot{x} + \frac{e}{Mc} \cdot \frac{H_0}{\alpha}\cos\alpha z = D;$$

$$\dot{z}^2 + \left(D - \frac{eH}{Mc\alpha}\cos\alpha z\right)^2 + e\Phi = v_0^2.$$ (3.78)

These equations can only be integrated in terms of elementary functions if $D = 0$ and $\Phi = 0$. In this case the trajectory equation $x = x(z)$ is

$$x = -\frac{1}{\alpha}\ln\left[k\sin\alpha z + \sqrt{1 - k^2\cos^2\alpha z}\,\right] + \text{const};$$
$$k = \omega_H/\alpha v_0.$$ (3.79)

With these assumptions, the average value of x remains constant.

We now assume $D \neq 0$ but $\Phi = 0$. Then the average particle displacement along the x axis is no longer zero. In this case the function $x = x(t)$ and $z = z(t)$ can be expressed in terms of elliptic functions, but the resulting equations are too cumbersome to deal with here. We therefore restrict the discussion to expansions of the functions $x(t)$ and $z(t)$ at small values of k and for $\beta = D/v_0$. These expansions are

$$x = Dt - \frac{k}{\alpha} \sin \alpha v_0 t + 0 (k^{3-n} \beta^n); \tag{3.80a}$$

$$z = v_0 t - \frac{v_0}{2} \left[\left(\beta^2 + \frac{k^2}{2} \right) t - \frac{2k\beta}{v_0 \alpha} \sin \alpha v_0 t + \right.$$

$$\left. + \frac{k^2}{4\alpha v_0} \sin 2\alpha v_0 t + ... \right]. \tag{3.80b}$$

When $D \neq 0$ and $\Phi \neq 0$, we can again find equations for $x(t)$ and $z(t)$ in a straightforward manner by noting that in a system with a large number of lenses the increment in the particle energy in a single lens is comparatively small:

$$\gamma \equiv \frac{eE_0}{2} \Lambda \bigg/ \frac{Mv_0^2}{2} \ll 1. \tag{3.81}$$

Accordingly, within a few lenses we can write expansions like (3.80) in powers of k, β, and γ. We omit the resulting equations because they are lengthy and elementary in content.

In general, an analytic calculation of the particle dynamics along the initial surface is complicated. For example, even with $D = 0$ and $\Phi = 0$ the linearized equation for oscillations along y is

$$y'' + k^2 \left(\frac{cn(t, k)}{dn(t,k)} \right)^2 y = 0. \tag{3.82}$$

Here, cn (t, k) and dn (t, k) are the Jacobi elliptic functions. We know that the exact solution of an equation like (3.82) is interesting only in connection with resonance effects, so that we will not analyze these effects here; we restrict the discussion to a calculation of the particle motion by the method of averaging, which gives correct results when there are no resonances.

In the case under consideration here the averaging procedure is carried out over the high-frequency oscillations associated with the finite dimensions of the lens; it is carried out over an oscillation period of the order of the transit time through a single lens.

In general, if it is assumed that the magnetic field does not have a constant component, we can write an averaged system of equations of motion for the particle in the following form (first approximation in t) [35]:

$$\frac{M}{e} \cdot \frac{dv}{dt} = -\nabla \Phi_0 + \frac{v_{z\sim}}{v_z} \nabla \Phi_\sim - \frac{1}{c}\left(\frac{v_{z\sim}}{v_z}\right)[v, H_\sim] + \left.\phantom{\frac{M}{e}}\right\}$$

$$\left. + \frac{1}{c}[v_\sim, H_\sim]; \right\} \tag{3.83}$$

$$d\mathbf{r}/dt = \mathbf{v}.$$

Here the tilde denotes the value of a quantity averaged over the spatial period, while the subscript "0" denotes the constant component of the potential.

Equation (3.83) contains the components of the rf velocity modulation,

$$\mathbf{v}_\sim = \frac{e}{Mv_z}\left(-\nabla\hat{\Phi} + \frac{1}{c}[v, H]\right). \tag{3.84}$$

Here $\hat{\Phi}$ is the variable part of the integral over z of the variable part of the corresponding quantity.

Assuming $\Phi = \Phi(\gamma)$, we find the following equation for the averaged quantities from (3.74) when (3.81) applies:

$$\frac{M}{e} \cdot \frac{dv}{dt} = -\nabla \Phi_0 + F_H - \frac{e}{2Mv_z^2} \nabla \Phi_\sim^2, \tag{3.85}$$

where Φ_0 is the constant component of the electric potential and $\Phi_\sim$ is the component which oscillates along z. In deriving (3.85) from (3.83) we note that the variable part of $\Phi_\sim$ is at twice the frequency of the magnetic field.

The force F_H has the components

$$F_{Hx} = -\frac{M}{e}\omega_H^2 \frac{v_y v_x}{v_z^2}\beta; \quad F_{Hy} = \frac{M}{e}\omega_H^2 \frac{v_x^2 + v_z^2}{v_z^2}\beta; \left.\right\}$$

$$F_{Hz} = -\frac{M}{e}\omega_H^2 \frac{v_y}{v_z}\beta; \quad \beta = \frac{1}{4\alpha}\sinh 2\alpha y. \left.\right\} \tag{3.86}$$

Substituting $\Phi_\sim$ from (3.74) into (3.85), and linearizing with respect to y, we find the following system of equations

$$dv_x/dt = 0; \tag{3.87a}$$

$$v_z^2 + (2e\Phi_0/M) = v_0^2; \tag{3.87b}$$

$$\frac{dv_y}{dt} + \Omega^2 y = 0; \quad \Omega^2 = \left(\frac{v_x^2}{v_z^2} + 1\right)\frac{\omega_H^2}{2} + \frac{1}{16} \cdot \frac{e^2 E_0^2}{M^2 v_z^2}. \tag{3.87c}$$

Evidently the displacements along the y axis are stabilized in this system. This stabilization is caused by the magnetic field and the electric field. The relative importance of the two focusing effects is governed by the particle velocity; specifically, if

$$v > v_* = \frac{1}{2\sqrt{2}} c \frac{E_0}{H_0}, \tag{3.88}$$

the magnetic focusing dominates; otherwise the electrostatic focusing dominates. We note that with $E_0 = 1$ kV/cm and $H = 3 \cdot 10^3$ Oe the critical velocity is $v_* \approx 10^7$ cm/sec. This value of v_* means that magnetic focusing should dominate in the most interesting cases. This is a very important result, since it means that the perturbations of the electric field which accompany various plasma instabilities will not affect the particle dynamics to a first approximation.

2. Particle Dynamics in a Linear Cylindrical Multiple-Lens, Multiply Connected Accelerator. As in the previous discussion of the two-dimensional model, we restrict the present discussion to the "central" part of the accelerator, neglecting end effects and the changes in the parameters along the axis of the system. A qualitatively new feature in the cylindrical case is the appearance of centrifugal effects. As a result, radial particle oscillations appear; these are not about the initial surface,[†] but about a principal surface which does not coincide with the initial surface.

In principle, centrifugal effects can be suppressed — exactly or as averaged over z — by means of "constant" magnetic fields: a homogeneous longitudinal field $H_{0z} = H_\parallel$ and an azimuthal field

$$H_\theta \sim 1/r. \tag{3.89}$$

In the first case the centrifugal force is balanced by the Lorentz

---

[†]We assume that the tangential component of the magnetic field vanishes at the initial surface.

force caused by the regular azimuthal motion; in the second case the centrifugal force is balanced by the Lorentz force caused by the longitudinal motion. For simplicity we restrict this analysis to centrifugal effects with no electric field.

The equations of motion for the particles in an azimuthally symmetric magnetic field

$$M\ddot{r} = -\frac{\partial U}{\partial r} - \frac{e}{c}\dot{z}H_\theta, \quad M\ddot{z} = -\frac{\partial U}{\partial z} + \frac{e}{c}\dot{r}H_\theta, \qquad (3.90)$$

show that an exact principal surface exists if, at some $r = r_0$, the radial velocity vanishes for all z, i.e., if

$$\partial U/\partial r + (e/c)\dot{z}H_\theta = 0. \qquad (3.91)$$

It is not difficult to see that for a magnetic field specified by the conditions $H_z = 0$ and $H_r = H_0 \sin \alpha z$ on the initial surface an exact magnetic surface exists if

$$H_\theta = 0; \quad q/r_0 = \partial q/\partial r_0; \quad D = 0. \qquad (3.92)$$

Here, q(r) is related to ψ by

$$\psi = \frac{H_0}{\iota\alpha}\cos(\alpha z)q(r) = \frac{H_0}{\alpha}\cos(\alpha z)\left(1 + \frac{\alpha^2 y^2}{2} + \ldots\right). \qquad (3.93)$$

If the radius of the main surface, r_0, only differs slightly from the radius of the initial surface, R, we can use (3.93) to write condition (3.92) as

$$r - R = 1/R\alpha^2 = \Lambda^2/\pi^2 R.$$

In the general case in which there is an electric field, and in which conditions (3.91) do not hold, an average principal surface exists; it is defined by

$$(\partial\bar{U}/\partial r) + (e/c)\dot{z}H_\theta = 0.$$

§6. Toroidal Multiply Connected, Multiple-Lens Accelerator

1. Specification of the Magnetic Field. We follow the general scheme for specifying the magnetic field with the help of the initial surface used above [see (2.11) and (2.12)]. We choose the initial surface to be a torus of circular cross

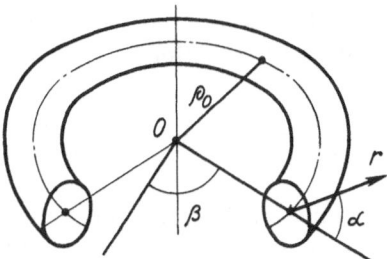

Fig. 21

section, with minor radius a and major radius b. We construct a toroidal coordinate system r, β, α in this torus (Fig. 21). The arc element in these coordinates is given by

$$ds^2 = dr^2 + r^2 d\alpha + (\rho_0 + r \cos \alpha)^2 d\beta^2. \qquad (3.94)$$

Noting that the external magnetic field in such an accelerator is inherently a three-component field, we describe this field by means of the magnetic scalar potential Ω, which satisfies the Laplace equation

$$\Delta\Omega = 0; \quad \mathbf{H} = \nabla\Omega. \qquad (3.95)$$

In terms of these toroidal coordinates, the magnetic field components are related to Ω by

$$H_r = \frac{\partial\Omega}{\partial r}; \quad H_\beta = \frac{1}{(\rho_0 + r \cos \alpha)} \cdot \frac{\partial\Omega}{\partial\beta};$$
$$H_\alpha = \frac{1}{r} \cdot \frac{\partial\Omega}{\partial\alpha}, \qquad (3.96)$$

and the Laplace equation is

$$\frac{1}{rR} \cdot \frac{\partial}{\partial r} rR \frac{\partial\Omega}{\partial r} + \frac{1}{r^2 R} \cdot \frac{\partial}{\partial\alpha} R \frac{\partial\Omega}{\partial\alpha} + \frac{1}{R^2} \cdot \frac{\partial^2\Omega}{\partial\beta^2} = 0. \qquad (3.97)$$

We specify the periodic magnetic field on the initial surface in the form

$$H_\alpha|_{\text{IS}} = 0; \quad H_\beta|_{\text{IS}} = 0; \quad H_r|_{\text{IS}} = h_0(\alpha) \sin N\beta. \qquad (3.98)$$

Here, 2N is the number of lenses in the torus.

Using (3.98), we can write the following equation for the magnetic potential of the periodic field:

$$\Omega = yh_0(\alpha)\sin N\beta - \frac{y^2}{2} \cdot \frac{h_0(\alpha)\sin N\beta}{R_1} \cdot \frac{\rho_0 + 2R_1\cos\alpha}{\rho_0 + R_1\cos\alpha} + \ldots \qquad (3.99)$$

In addition to the periodic field it is necessary to produce magnetic fields independent of β. We call these fields "constant fields."

A constant magnetic field with components $H_{\alpha 0}$, H_{r0}, which we call a "transverse field," causes the particles to move along the major azimuthal angle, β. This field is analogous to the transverse magnetic field in conventional cyclic accelerators (betatrons, cyclotrons, etc.).

In general, the fields $H_{\alpha 0}$, H_{r0} do not vanish on the initial surface. If we denote their values on the initial surface by $h_\alpha(\alpha)$ and $h_r(\alpha)$, we can write the following equations for the potential of the transverse field, $\Omega_\perp$, near the initial surface:

$$\Omega_\perp = R_1\int h_\alpha\, d\alpha + yh_r(\alpha) - \frac{y^2}{2}\left(\frac{h_r(\rho_0 + 2R_1\cos\alpha)}{R(\rho_0 + R_1\cos\alpha)} + \right.$$
$$\left. + \frac{1}{R}\cdot\frac{\partial h_\alpha}{\partial\alpha} - h_\alpha\frac{\sin\alpha}{\rho_0 + R_1\cos\alpha}\right) + \ldots \qquad (3.100)$$

In addition to the transverse magnetic field we are interested in a longitudinal constant magnetic field with the single component $H_{\beta 0}$ in a linear multiply connected, multiple-lens accelerator. This field can also result in a suppression of the centrifugal force. Evidently

$$H_{\beta 0} \approx \frac{\text{const}}{\rho_0 + R_1\cos\alpha}. \qquad (3.101)$$

2. Particle Dynamics in a Toroidal Multiply Connected, Multiple-Lens.

The particle motion is described in toroidal coordinates by the following system of equations:

$$M\frac{dv_r}{dt} = \frac{e}{c}\left[\frac{v_\beta}{R}H_\alpha - \frac{v_\alpha}{r}H_\beta\right] + \frac{Mv_\alpha^2}{r^3} + M\frac{v_\beta^2\cos\alpha}{R^3}; \qquad (3.102a)$$

$$M\frac{dv_\beta}{dt} = R\frac{e}{c}\left[\frac{v_\alpha}{r}H_r - v_r H_\alpha\right]; \qquad (3.102b)$$

$$M \frac{dv_\alpha}{dt} = r \frac{e}{c} \left[v_r H_\beta - \frac{v_\beta}{R} H_r \right] - \frac{Mr \, v_\beta^2 \sin \alpha}{R^3} ; \quad (3.102c)$$

$$\frac{dr}{dt} = v_r; \quad \frac{d\beta}{dt} = \frac{v_\beta}{R^2} ; \quad \frac{d\alpha}{dt} = \frac{v_\alpha}{r^2} . \quad (3.102d)$$

Breaking up the magnetic field into constant and periodic parts, and treating the transit time through the lens as a small parameter, we find the following equations for the rapidly varying modulation of v:

$$\tilde{v}_r = \frac{e}{Mc} \left[R \hat{H}_\alpha - \frac{v_\alpha}{v_\beta} \cdot \frac{R^2}{r} \hat{H}_\beta \right] ; \quad (3.103a)$$

$$\tilde{v}_\beta = \frac{e}{Mc} R^3 \frac{1}{v_\beta} \left[\frac{v_\alpha}{r} \hat{H}_r - v_r \hat{H}_\alpha \right] ; \quad (3.103b)$$

$$\tilde{v}_r = \frac{e}{Mc} r R^2 \left[\frac{v_r}{v_\beta} \hat{H}_\beta - \frac{\hat{H}_r}{R} \right] . \quad (3.103c)$$

The averaged equations are

$$M \frac{dv_r}{dt} = \frac{e}{c} \left(\frac{v_\beta}{R} H_{\alpha 0} - \frac{v_\alpha}{r} H_{\beta 0} \right) + \frac{Mv_\alpha^2}{r^3} + \frac{M v_\beta^2 \cos \alpha}{R^3} +$$
$$+ \frac{e^2}{Mc^2} R \left(1 + \frac{R^2}{r^2} \frac{v_\alpha^2}{v_\beta^2} \right) \overline{\tilde{H}_\beta \hat{H}_r} ; \quad (3.104a)$$

$$M \frac{dv_\beta}{dt} = R \frac{e}{c} \left(\frac{v_\alpha}{r} H_{r0} - v_r H_{\alpha 0} \right) +$$
$$+ \frac{e^2}{Mc^2} R^3 \left(\frac{v_r}{H_\beta} \overline{\hat{H}_\beta \tilde{H}_r} + \frac{v_\alpha}{v_\beta} \cdot \frac{1}{r} \overline{\hat{H}_\beta \tilde{H}_\alpha} \right) ; \quad (3.104b)$$

$$M \frac{dv_\alpha}{dt} = r \frac{e}{c} \left(v_r H_{\beta 0} - \frac{v_\beta}{R} H_{r0} \right) -$$
$$- \frac{Mr \, v_\beta^2 \sin \alpha}{R^3} + rR \frac{e^2}{Mc^2} \overline{\hat{H}_\alpha \tilde{H}_\beta} \left(1 + \frac{v_r^2}{v_\beta^2} R^2 \right) -$$
$$- \frac{e}{Mc^2} \cdot \frac{v_r v_\alpha}{v_\beta^2} \overline{\hat{H}_r \tilde{H}_\beta} . \quad (3.104c)$$

Restricting the analysis to the linear approximation in y, we can omit the terms that contain the explicit averaging in the last two equations.

3. Principal Trajectories. As in the cylindrical case, there are averaged principal trajectories. We consideer accelerators in which these trajectories lie on the initial surface, $r = R_1$.

Setting $r = R_1 = \text{const}$, $\alpha = \alpha_0 = \text{const}$ in (3.104), we find conditions on the fields $H_{\beta 0}$, $H_{\alpha 0}$, H_{r0} which ensure that the main and initial surfaces coincide:

$$\frac{e}{c}\left[\frac{v_\beta}{R_0}H_{\alpha 0} - \frac{v_\alpha}{R_1}\right] + \frac{Mv_\alpha^2}{R_1^3} + M\frac{v_\beta^2\cos\alpha}{R_0^3} = 0; \qquad (3.105a)$$

$$M\frac{dv_\beta}{dt} = R_0\frac{e}{c}H_{r0}\frac{v_\alpha}{R_1}; \qquad (3.105b)$$

$$0 = -\frac{e}{c}R_1\frac{v_\beta}{R_0}H_{r0} - M\frac{R_1\,v_\beta^2\sin\alpha}{R_0^3}; \qquad (3.105c)$$

$$R_0 \equiv \rho_0 + R_1\cos\alpha. \qquad (3.105d)$$

It is evident that the next-to-last equation is equivalent to the obvious condition

$$\omega_\beta \equiv v_\beta/R_0 = v_0 = \text{const}. \qquad (3.106)$$

With given values of v_0 and v_α, conditions (3.105) determine the field components $H_{\beta 0}$, H_{r0}, $H_{\alpha 0}$. One of these components turns out to be arbitrary. For example, if we require that the current along the core be zero,

$$\oint H_{\alpha 0}d\alpha = 0, \qquad (3.107)$$

then the component $H_{\beta 0}$ must be nonzero if $v_\alpha = 0$. In the particular case $v_\alpha = 0$ the principal trajectories are circles on the initial surface. In this case Eq. (3.105b) means that

$$v_\beta = \text{const}, \qquad (3.108)$$

so that the azimuthal velocities w_β on different trajectories are not related.

The two other equations determine the components of the transverse magnetic field on the principal surface:

$$H_{\alpha 0} \equiv h_\alpha = -(Mc/e)\dot{\beta}(\alpha)\cos\alpha; \qquad (3.109a)$$

$$H_{r0} = h_r = -(Mc/e)\dot{\beta}(\alpha)\sin\alpha. \qquad (3.109b)$$

If $\beta(\alpha)$ = const, the transverse field is homogeneous and directed along the z axis. If $\beta(\alpha) \neq 0$, the field is again directed along the z axis on the main surface. Near the main surface, the field can be determined from (3.100):

$$\Omega_0 = - \frac{R_1 Mc}{e} \int \dot{\beta}(\alpha) \cos \alpha d\alpha - y \frac{Mc}{e} \dot{\beta}(\alpha) \sin \alpha +$$

$$+ \frac{y^2}{2} \cdot \frac{Mc}{R_1 e} \cdot \frac{\partial \beta(\alpha)}{\partial \alpha} \cos \alpha + \dots . \tag{3.110}$$

4. Stability of Principal Trajectories when $v_\alpha = 0$.

A general analysis of the behavior of the solutions of system (3.102) near the principal trajectories, (3.105), is complicated. However, it is physically obvious that the motion of particles with $v_\alpha \neq 0$ is stable over broad ranges of the parameters.

The case with $v_\alpha = 0$ should be less stable, so that it is important to study this case; furthermore, the analysis is comparatively simple.

Linearization of system (3.102) under condition (3.105) leads to the following equations:

$$\left[\frac{d^2}{dt^2} + 2 \cos^2 \alpha_0 \beta_0^2 + \omega_0^2 - \left(\frac{\partial \omega_{a0}}{\partial r} \right) \frac{v_{\beta 0}}{R_0} \right] y_1 -$$

$$- \left[R_0 \beta \left(\frac{\partial \omega_{a0}}{\partial \alpha} - \dot{\beta} \sin \alpha_0 + \frac{R_1}{R_0} \dot{\beta} \sin 2\alpha_0 \right) - \right.$$

$$\left. - R_1 \frac{e}{c} H_{\beta 0} \frac{d}{dt} \right] \alpha_1 - \dot{\beta} \cos \alpha_0 \frac{v_{\beta 1}}{R_0} = 0; \tag{3.111a}$$

$$v_{\beta 1} = R_0 R_1 \omega_{r0} \alpha_1 - R_0 \omega_{a0} y_1; \tag{3.111b}$$

$$\left[\frac{d^2}{dt^2} + \frac{v_{\beta 0}}{R_1 R_0} \left(\frac{\partial \omega_{r0}}{\partial \alpha} + 2 \frac{R_1}{R} \sin^2 \alpha_0 \dot{\beta} + \dot{\beta}_0 \cos \alpha \right) \right] \alpha_1 +$$

$$+ \left[\frac{v_{\beta 0}}{R_1 R_0} \left(\frac{\partial \omega_{r0}}{\partial r} - \frac{\sin 2\alpha_0}{R_0} \dot{\beta}_0 \right) - \frac{1}{R_1} \cdot \frac{e H_{\beta 0}}{Mc} \cdot \frac{d}{dt} \right] y +$$

$$+ \frac{\sin \alpha_0}{R_0^2} v_{\beta 0} \frac{v_{\beta 1}}{R_0 R_1} = 0. \tag{3.111c}$$

Here the subscript "0" denotes the value of a quantity on the principal trajectory; we have also introduced the following notation:

$$y\omega_0^2 = \frac{e^2}{M^2 c^2} R_0 \tilde{H}_\beta \hat{H}_r; \ \omega_{r0} = \frac{eH_{r0}}{Mc}; \ \omega_{a0} = \frac{eH_{a0}}{Mc}. \tag{3.112}$$

Restricting the analysis to the case β = const, we find the following characteristic equation (y_1, α_1, $v_{\beta 1} \approx \exp i\omega t$):

$$\omega^4 - (\omega_0^2 + \dot{\beta}_0^2 + \omega_{\beta 0}^2)\,\omega^2 + \omega_0^2\,\dot{\beta}^2 \sin^2\alpha = 0. \tag{3.113}$$

As expected, the oscillations are stable everywhere. The only exception is represented by oscillations along the transverse field with $\sin \alpha = 0$, in which case the oscillations are marginally stable. This result also makes it more probable that the particle motion will be stable over broad ranges of the parameters in the case $v_\alpha \neq 0$.

Chapter 4

PLASMA-OPTICAL SYSTEMS WITH TWO-PARAMETER FOCUSING

§1. Systems with Stabilized Focus

The term "two-parameter" or "double" focusing is usually understood as meaning focusing in terms of any two parameters of the beam.[†] In this chapter we examine systems which focus a diverging ion beam and which are free of chromatic aberration or momentum aberration. Such systems have the advantage over systems with single-parameter focusing that the requirements on the stability of the ion source are relaxed.

1. Achromatic Lens System. By "stabilization" of the focus we mean the elimination of the longitudinal chromatic aberration from the lenses while geometric focusing is retained. This situation corresponds to the vanishing of the term $\partial z_1 / \partial \alpha$ in (2.27). In ray optics achromatic lenses are fabricated by cementing two lenses together, one converging and the other di-

---

[†]Examining Eq. (2.26b), we see that two-parameter focusing corresponds to the vanishing of any two terms in the series in (2.26b).

verging, with a relatively large dispersion (an Abbe apochromat).
As is shown in Chap. 3, § 1, such a system cannot be fabricated
with vacuum or plasma axial systems. In an annular plasma lens,
the situation is different. It is possible to construct a converging
lens with $F \approx v_0^2$ and a diverging lens with $F \approx v_0^4$, i.e., it is possible to develop a ray stabilization scheme. Let us examine the
situation in more detail. We assume that the focal lengths of the
two lenses are

$$F_1 = F_{10}(v/v_0)^{n_1}; \quad F_2 = F_{20}(v/v_0)^{n_2}. \tag{4.1}$$

Here v_0 is the typical particle velocity, n_1 and n_2 are nonnegative numbers, and F_{10} and F_{20} are the focal lengths of the
lenses for the velocity v_0. If the scale distance between the lens
is L, and if the source lies a distance a_1 from the first lens (Fig.
22), the focal point can be determined from the familiar equations

$$1/a_1 + 1/b_1 = 1/F_1; \; 1/a_2 + 1/b_2 = 1/F_2; \; a_2 = L - b_1. \tag{4.2}$$

Introducing the auxiliary variables

$$\alpha = 1/a_1; \; \beta = 1/b_2; \; \omega_1 = 1/F_1; \; \omega_2 = 1/F_2, \tag{4.3}$$

we can rewrite (4.2) together with the stabilization condition
$\partial b_2/\partial v = 0$ in the form

$$\frac{1}{\omega_1 - \alpha} + \frac{1}{\omega_2 - \beta} = L; \quad \left(\frac{\omega_1 - \alpha}{\omega_2 - \beta}\right)^2 = -\frac{n_1}{n_2} \cdot \frac{\omega_1}{\omega_2}. \tag{4.4}$$

Equation (4.4) shows that if $n_1 > 0$, $n_2 > 0$ it is necessary to
use one converging lens and one diverging lens (e.g., with $\omega_1 > 0$,
$\omega_2 < 0$) in order to stabilize the focus. System (4.4) can be simplified by introducing the notation $1/(\omega_1 - \alpha) = \xi$, $1/(\omega_2 - \beta) = \eta$,
$n_1/n_2 = \theta$, and by setting L = 1. Then (4.4) becomes

$$\xi + \eta = 1. \quad \frac{\eta}{\xi} = -\theta \frac{1 + \xi\alpha}{1 + \eta\beta}. \tag{4.5}$$

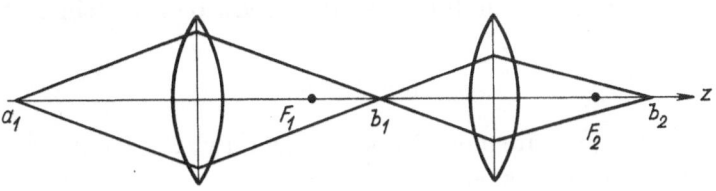

Fig. 22

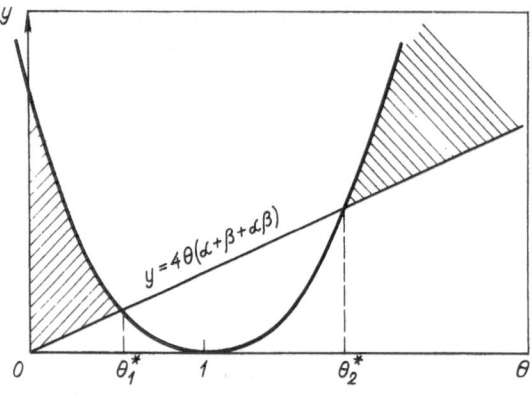

Fig. 23

Hence

$$\eta = \frac{-1 + \theta(1 + 2\alpha) \pm \sqrt{(1-\theta)^2 - 4\theta(\alpha + \beta + \alpha\beta)}}{2(\beta + \alpha\theta)}. \qquad (4.6)$$

From the condition that the discriminant be nonnegative,

$$D = (1 - \theta)^2 - 4\theta(\alpha + \beta + \alpha\beta) > 0, \qquad (4.7)$$

we find the ratio $n_1/n_2 = \theta$ at which stabilization is possible.

The regions $D \geq 0$ are shown in Fig. 23. The slope of the line $y = 4\theta(\alpha + \beta + \alpha\beta)$ depends on the lens power. Evidently stabilization is possible if $\theta < \theta_1^*$, $\theta > \theta_2^*$. The regions of θ for which stabilization is possible correspond to the ray scheme. If $\theta < \theta_1^*$, the first lens should be converging, $\omega_1 > 0$, and the second should be diverging, $\omega_2 < 0$. If a real image is required we impose the further condition $\omega_1 > |\omega_2|$. In order to stabilize the focus in the region $\theta > \theta_2^*$ we must use lenses with $\omega_1 < 0$, $\omega_2 > 0$, and $|\omega_1| < \omega_2$. The converging and diverging lenses can be superimposed; i.e., $L = 0$ is allowed. In this case the condition for stabilization of the focus is

$$\omega_1 = -\theta\omega_2, \qquad (4.8)$$

and a real image is formed if $\theta < 1$. Actually, the focal length of such a superimposed system is

$$\beta = \omega_2(1 - \theta); \quad \alpha = 0. \qquad (4.9)$$

Systems with a stabilized focus made from thin lenses have a long focal length. This focal length can be reduced by using lenses with fields that extend over a large range or by using lenses which are capable of modifying the particle velocity in the gap between the lenses.

2. **Three-Element System with Stabilized Focus.** As an example of a short-focus device we consider a system that consists of three elements (Fig. 24): a thick magnetic lens with a longitudinal field, 1, a retarding element 2, which reduces the velocity of the particles without affecting their direction, and a thin diverging lens 3.

If the source is located at the origin, z = 0, and if the thin lens is located at a distance L from the source, the focal point is determined from

$$1/z_2 = 1/(z_1 - L) - 1/F_2, \qquad (4.10)$$

where z_2 is the focal point reckoned from the center of the thin lens, z_1 is the z coordinate of the focal point of the thick lens, and F_2 is the focal length of the thin lens.

For simplicity we restrict the discussion to the two-dimensional case, which corresponds to $R \to \infty$. It is not difficult to show that particles emerging from the thick lens with a longitudinal homogeneous field are focused at the point

$$z_1 = z_0 - \frac{1}{2} R_{\Lambda i} \tan \frac{2z_0}{R_{\Lambda i}}, \qquad (4.11)$$

where z_0 is the span of the field of the thick lens, $R_{\Lambda i} = v_0/\omega_{Hi}$ is the ion gyroradius, $\omega_{Hi} = eH_0/Mc$, and H_0 is the longitudinal magnetic field.

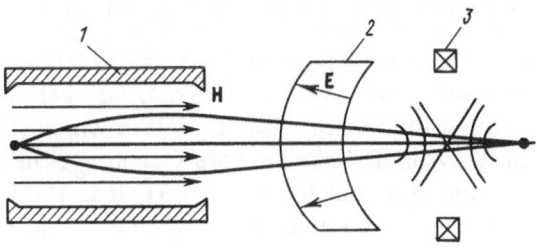

Fig. 24

If the thin diverging lens is replaced by a split electrostatic lens the focal length is [see Eq. (3.45)]

$$F_2 = F_0 \, (v_2/v_0)^4, \tag{4.12a}$$

where v_2 is the velocity of the ions incident on the split lens. The focus stabilization condition is obviously

$$\partial z_2/\partial v_0 = 0. \tag{4.13a}$$

Using (4.10), we can rewrite (4.13a) as

$$\partial z_1/\partial F_2 = (z_1 - L)^2/F_2^2. \tag{4.13b}$$

Here the variation in z_1 and F_2 is carried out with respect to the velocities. Substituting (4.11) and (4.12a) into (4.13b), we find a relationship between the focal length z_1 and F_2:

$$F_2 = \frac{4\,(z_1 - L)^2}{z_1 + k} \cdot \frac{v_1}{v_2}; \quad k = z_0 \, \tan^2 \frac{2z_0}{R_{\Lambda i}}. \tag{4.12b}$$

Using (4.13b), (4.12a), and (4.10), we find

$$\frac{1}{z_2} = \frac{1}{z_1 - L} \left(1 - \frac{z_1 + k}{4\,(z_1 + L)} \cdot \frac{v_2}{v_1} \right). \tag{4.14}$$

We require that z_2 be positive (a converging system) and find a condition on the ion retardation in the retarding system:

$$\frac{v_1}{v_2} > \frac{z_1 + k}{4\,(z_1 - L)}. \tag{4.15}$$

The retarding system must be thin along the initial surface and its lines of force must coincide with the particle trajectories.

§2. Focusing in Terms of the Radial and Azimuthal Velocities (Whirler System)

1. **Structure of the Whirler System.** In this section we consider an axisymmetric plasma-optical system capable of focusing a neutralized ion beam with $T_e = 0$ in which the particles have spreads in the velocity components v_r and v_θ. In contrast with stabilized-focus systems (Chap. 4, §1), it is the transverse momentum aberration rather than the longitudinal chromatic aberration which is corrected. The field configuration in the system

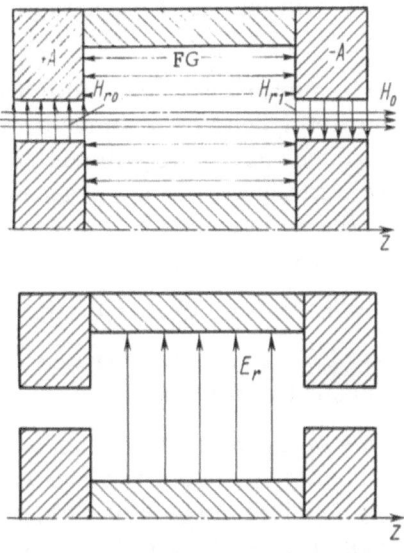

Fig. 25

is as shown in Fig. 25. We should distinguish three elements in this system: the Whirler ("azimuth adder") (+A), the focusing gap (FG), and the dewhirler ("azimuth subtractor") (−A). Since elements (+A) and (−A) govern the properties of the system and its operating principle, we call the overall system a "whirler system" (or "A system"). Let us examine the operation of each element separately.

In the whirler the magnetic field is approximately radial, $H_r \sim 1/r$. Traversing this whirler, the particles intersect a magnetic flux ψ_A and acquire an angular velocity

$$v_{\theta A} = -\frac{e}{Mc} \psi_A, \qquad (4.16)$$

where ψ_A is the magnetic flux in the whirler.

The field configuration in the dewhirler is the same as that in the whirler, except that the magnetic field vector is in the opposite direction. Since a particle does not acquire an additional angular velocity δv_θ in the focusing gap, the angular velocity acquired in the whirler is cancelled in the dewhirler if the magnetic fluxes in the whirler and the dewhirler are equal in magnitude. Thus, the particle emerges from the system with its original angular velocity, $v_{\theta 0}$.

The focusing gap is a cylindrical capacitor with a radial electric field $E_r \approx 1/r$ superimposed on a homogeneous longitudinal magnetic field $H_z = H_0 =$ const. The equilibrium trajectory in the focusing gap is a circle in the $r\theta$ plane. A necessary condition is that the centrifugal force be balanced by the Lorentz force and by the force exerted by the electric field:

$$Mv_\theta^2/R = eE_0 + (e/c)v_\theta H_0; \quad v_r = 0, \tag{4.17}$$

where R is the radius of the equilibrium trajectory, and E_0 and H_0 are the electric and magnetic fields at $r = R$.

If the magnetic flux of the whirler is such that the particles leaving it acquire a velocity v_θ, in accordance with (4.17), the particles move along a cylindrical surface of radius $r = R$ in the focusing gap. In this case this surface is the principal surface. The focusing gap has the same focusing capability as an ordinary cylindrical capacitor with a homogeneous magnetic field parallel to the symmetry axis. As shown in [2], a diverging particle beam with an azimuthal velocity v_θ which satisfies condition (4.17) in a first approximation is focused as the particles are rotated through an angle

$$\theta = \frac{\pi}{\sqrt{2 + \omega_H^2/\dot\theta^2}}. \tag{4.18}$$

Here $\omega_H = eH_0/Mc$ and $\dot\theta = v_\theta/R$. The distance from the source to the image along the initial surface is determined from the simple equation

$$L = v_z T = v_z \frac{R\theta}{v_\theta}, \tag{4.19}$$

where T is the time required for the rotation of the particle through an angle θ.

This whirler system is nearly nonconnective; i.e., its properties remain essentially the same in the transition from the vacuum case to the plasma case. In both cases the electric field in the focusing gap is $E \sim 1/r$, i.e., satisfies the Laplace equation. Accordingly, this part of the whirler system is nonconnectible.

Strictly speaking, the configuration of the electric and magnetic fields in the whirler and dewhirler is convertible, but if the gap between the poles is sufficiently small, the field in the gap can be assumed to be two-dimensional, so that it can be made nonconvertible (Chap. 2, § 2).

The focusing properties of the system depend on the ion mass. If ions with mass $M_0 + \delta M$ emerge from the source, they acquire a velocity v_θ in the whirler which is independent of the initial ion velocity, but which depends on the mass, (4.16). Equilibrium condition (4.17) for the given fields **E** and **H** only holds for a single value of $v_{\theta 0}$, which corresponds to a certain mass M_0; it does not hold for ions of other mass, so that ions with mass different from M are displaced from the equilibrium surface r = R and focused at other radii.

2. Fields in the Whirler System. Since the motion of the particle beam in the whirler system is examined near the initial surface, we use the equations of motion near the initial surface in the form in (2.51). It can be shown that the influence of the fringing fields between the various parts of the system can be neglected. Then we can assume that the boundaries between these regions are sharp, find the nature of the trajectories in each of the three regions, and match the solutions at the boundaries.

In the calculations below we adopt the following simple field configuration (Fig. 25). The magnetic field has a constant z component H_0 through the whirler system. In the whirler and dewhirler there is a radial magnetic field

$$H_r = H_{r0}R/r, \qquad (4.20)$$

whose magnitude is independent of z.

There is also an electric field $E_r = E_{r0}/r$ in the focusing gap. Since the whirler and dewhirler do not have focusing capabilities, we assume (+A) and (−A) to be infinitesimally short to simplify the analysis, and assume the fields themselves to be δ-shaped. Then the displacement of the particle trajectory in the whirler and dewhirler can be neglected. Since the particle motion in the focusing gap occurs near the initial surface, the magnetic stream function ψ can be written as the series (2.15b). In the split of the paraxial approximation, we only retain the first terms of the series. For the magnetic field in (4.20) we find

$$\psi_0 = 0; \quad \psi_1 = H_0 R; \quad \psi_2 = H_0. \tag{4.21}$$

The electric scalar potential in the focusing gap can also be written in series form (2.43), with the coefficients

$$\Phi_0 = 0; \quad \Phi_1 = -E_0; \quad \Phi_2 = E_0/R. \tag{4.22}$$

3. Focusing in Terms of the Radial Velocity.

We first consider the focusing of a beam of particles emerging from an annular source with coordinates $y = 0$, $z = 0$. The velocity of the particles emerging from the source is $\mathbf{v}_0 = \{v_{r0}, 0, v_{z0}\}$.

We restrict the discussion to the paraxial approximation using beam equation (2.51). Also using (4.21) and (4.22), we find the coefficients a, b, and c, assuming that the particles leaving the δ-field of the whirler have acquired an azimuthal velocity $v_{\theta 0}$:

$$a_0 = 0; \tag{4.23a}$$

$$b_0 = \frac{1}{Mv_z^2} \left(\frac{3Mv_{\theta 0}^2}{R^2} + M\omega_H^2 + \frac{3M\omega_H v_{\theta 0}}{R} + \frac{eE_0}{R} \right); \tag{4.23b}$$

$$c_0 = \frac{1}{Mv_z^2} \left(\frac{Mv_{\theta 0}^2}{R} + M\omega_H v_{\theta 0} + eE_0 \right). \tag{4.23c}$$

Here $\omega_H = eH_0/Mc$ and $v_z^2 = v_0^2 - v_{\theta 0}^2$; the subscript "0" represents the zero-order quantities.

It follows from Eq. (2.51) that to satisfy the equilibrium condition we must set c_0 equal to zero:

$$Mv_{\theta 0}^2/R + M\omega_H v_{\theta 0} + eE_0 = 0. \tag{4.24}$$

As will be shown below, this condition is precisely the same as condition (4.17). The solution of homogeneous equation (2.51) is

$$y = A \sin \sqrt{b_0}\, z + B \cos \sqrt{b_0}\, z, \tag{4.25}$$

where, for the initial conditions adopted above, we have

$$A = y'_{00}/\sqrt{b_0}; \quad B = 0; \quad y'_{00} = v_{r0}/v_z. \tag{4.26}$$

Clearly, the whirler system performs focusing in terms of the radial velocity; the focal length is

$$L = \pi/\sqrt{b_0}. \tag{4.27}$$

Using equilibrium condition (4.24), we can write b_0 as

$$b_0 = \frac{1}{Mv_z^2}\left(\frac{2Mv_{\theta 0}^2}{R^2} + M\omega_H^2 + \frac{2M\omega_H v_{\theta 0}}{R}\right), \qquad (4.28)$$

thereby eliminating the explicit dependence of the focal length on the strength of the electric field.

4. Focusing in Terms of the Azimuthal Velocity. We assume that the particles entering the system have an azimuthal-velocity spread δv_θ; thus, it is assumed that $\mathbf{v} = \{v_{r0}, \delta v_\theta, v_{z0}\}$. For generality we assume that the source has finite dimensions y_{00}. In the paraxial approximation the coefficient c in Eq. (2.51) is

$$c = c_0 + c_1, \qquad (4.29)$$

where c_1 is a quantity of first order with respect to D_0 and y_{00}:

$$c_1 = -\frac{1}{Mv_z^2}\left(2\,\frac{v_{\theta 0}}{R} + \omega_0\right)\left(\frac{D_0}{R} + y_{00}\,M\omega_0\right). \qquad (4.30)$$

To find the coefficient c_1 we assume that the particles leaving the δ-field whirler have an azimuthal velocity

$$v_{\theta 0} = \frac{e}{Mc}\cdot\frac{R}{(y_{00}+R)}\int_0^\lambda H_{r0}\,dz + \delta v_\theta, \qquad (4.31)$$

where $\lambda \to 0$ is the length of the whirler.

To first order the generalized momentum in the focusing gap is

$$D = D_0\left(1 + \frac{\delta v_\theta}{v_{\theta 0}}\right) + M\omega_H R y_{00} = \text{const},$$

where $D_0 = MRv_{\theta 0}$. The general solution of Eq. (2.51) is

$$y = \frac{y_{00}'}{\sqrt{b_0}}\sin\sqrt{b_0}\,z + y_{00}\cos\sqrt{b_0}\,z + \frac{c_1}{b_0}\left(\cos\sqrt{b_0}\,z - 1\right). \qquad (4.32)$$

It follows from (4.32) that focusing in terms of the azimuthal velocity is possible only if $z = 2\pi/\sqrt{b_0}$ (two-half-period focusing) or if $c_1 = 0$. This latter condition leads to the following relation between $v_{\theta 0}$ and ω_H:

$$2\,\frac{v_{\theta 0}}{R} + \omega_H = 0. \qquad (4.33)$$

This condition is equivalent to the condition

$$D_0 = 0. \tag{4.34}$$

Condition (4.34) means that the magnetic fluxes in the whirler and the focusing gap are equal. The source–image distance under condition (4.33) is

$$L = \frac{\pi \sqrt{2}}{\omega_H} v_z. \tag{4.35}$$

Using solution (4.32) we can find the transverse dimension of the source image. If the half-width of the source is y_{00}, the half-width of the image at the focus is

$$y|_{z=L} = -y_{00} \left(1 + 2 c_1/b_0\right). \tag{4.36}$$

Substituting the explicit expressions for b_0 and c_1 into (4.36), we find

$$y|_{z=L} = \left[-1 + 2 \frac{\dfrac{2v_{\theta 0}\,\omega_H}{R} + \omega_H^2}{2\dfrac{v_{\theta 0}^2}{R} + \omega_H^2 + 2\dfrac{\omega_H\,v_{\theta 0}}{R}} \right] y_{00}. \tag{4.37}$$

When (4.33) holds the magnification of the system is $y|_{z=L}/y_{00} = -1$, so that the whirler system is similar to a thin lens if the object is at twice the focal length.

§3. Energy Recovery

An energy recovery system extracts energy from the particles by means of an electric field, thereby retarding the particles. The simplest retarding element is the inverted single-lens accelerator described above.

As a rule, a neutralized ion beam that experiences energy recovery exhibits a broad energy spectrum, so that it is advantageous to energy-analyze the beam before the recovery process. In order to improve the recovery efficiency, we take account of the finite beam width and allow for possible correction of spherical aberration. In an axisymmetric system, energy resolution of the beam can be hindered by the spread in the azimuthal velocity (momentum

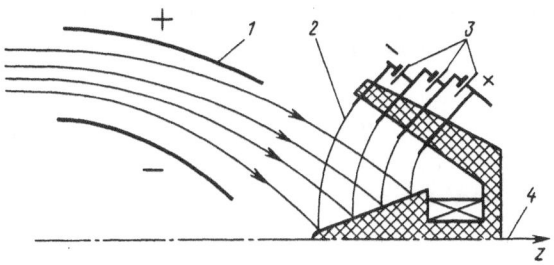

Fig. 26

aberration). This aberration must also be corrected. One of the simplest recovery schemes that meets these conditions is shown in Fig. 26. It consists of a toroidal capacitor 1 and a device which serves as a collector and as a retarding device, 4, with separate power supplies 3. The equipotentials 2 coincide with magnetic lines of force in the case $T_e = 0$. The magnetic lines of force of the capacitor are directed along the major and minor circumferences of the torus.

Let us examine the operation of this system in the limit $r \to \infty$. Choosing a cylindrical coordinate system in which the z axis passes through the center of the minor circumference of the torus, and the r axis coincides with the direction of the electric field $\mathbf{E}$, we can write the Lagrangian in the form

$$L = \frac{M}{2}\left(\dot{r}^2 + (r\dot{\theta})^2 + \dot{z}^2\right) + \frac{e}{c}(\mathbf{Av}) - e\Phi(rA_\theta), \qquad (4.38)$$

where $\mathbf{A} = \{0, A_\theta, A_z\}$. Since the coordinates z and θ are cyclic, we find the Routh function

$$R = M\dot{r}^2/2 - U(r), \qquad (4.39)$$

where

$$\left.\begin{array}{c} U(r) = \dfrac{M}{2}\left(p - \dfrac{e}{Mc}A_z\right)^2 + \dfrac{\left(D - \dfrac{e}{c}rA_\theta\right)^2}{2Mr^2} + e\Phi(rA_\theta); \\[3mm] p = \dot{z} + \dfrac{e}{Mc}A_z; \quad D = Mr^2\dot{\theta} + \dfrac{e}{c}rA_\theta. \end{array}\right\} \qquad (4.40)$$

Thus we have one–dimensional motion in the potential field U(r). We write the equation of motion as

$$M\ddot{r} = -\frac{\partial U(r)}{\partial r}, \qquad (4.41)$$

and the focusing condition as

$$\frac{\partial U}{\partial r}\bigg|_{r=R} = 0.$$

We assume that the beam motion occurs near the circular trajectory r = R (the equilibrium condition for this motion is given below) and expand the potential U(r) in powers of y = r − R. Terms of order y^3 are retained; these are associated with spherical aberration:

$$U(y) = U(R) + \frac{\partial U}{\partial r}\bigg|_{r=R} y + \frac{1}{2}\frac{\partial^2 U}{\partial r^2}\bigg|_{r=R} y^2. \qquad (4.42)$$

Writing the coefficients A_θ and A_z in the form

$$A_\theta = -rH_0/2; \quad A_z = -H_\theta R \ln r/R \qquad (4.43)$$

and writing the function $\Phi(rA_\theta)$ as

$$\Phi(rA_\theta) = k\psi + q\psi^2/2R^2, \quad \psi = rA_\theta, \qquad (4.44)$$

we find the following equation of motion near the circular principal trajectory:

$$\ddot{y} + \Omega^2 y = Q + sy^2 + \delta Q, \qquad (4.45)$$

where

$$\delta Q = -\frac{\omega_\theta\,\delta p + 2D\,\delta D}{M^2 R^3} ; \quad \omega_0 = \frac{eH_0}{Mc} ; \quad \omega_\theta = \frac{eH_\theta}{Mc} ; \qquad (4.46a)$$

$$\Omega^2 = -\frac{\omega_0 p}{R} + \omega_\theta^2 + \frac{3D^2}{M^2 R^4} + \frac{\omega_0^2}{4} - \frac{e}{M}\cdot\frac{k}{R}H_0 + \frac{e}{M}\frac{3}{2}H_0^2 q; \qquad (4.46b)$$

$$Q = \omega_\theta\, p - \frac{D^2}{M^2 R^3} + \frac{\omega_0^2 R}{4}\cdot\frac{e}{M}kH_0 + \frac{e}{M}\frac{1}{2}qRH_0^2; \qquad (4.46c)$$

$$s = \frac{\omega_\theta\, p}{R^2} - \frac{3}{2}\cdot\frac{\omega_0^2}{R} - 6\frac{D^2}{M^2 R^5} + \frac{e}{M}\frac{3}{2}q\frac{H_0^2}{R}. \qquad (4.46d)$$

Here δp and δD are the spreads of values of p and D, respectively.

Equation (4.45) contains the four free parameters ω_0, ω_θ, q, and k in addition to the parameter p, which can, in principle, be modified by means of a whirler (Chap. 4, § 2).

The equilibrium condition for motion along the main trajectory takes the form Q = 0. Finding the value of q and requiring that the spherical aberration correspond to s = 0, we find the following solution for Eq. (4.45):

$$y = B \sin \Omega t + \frac{-\omega_\theta \, \delta p + 2D \delta D \dfrac{1}{M^2 R^3}}{\Omega^2} (1 - \cos \Omega t), \qquad (4.47)$$

where B is a constant which depends on the initial conditions.

Since k and ω_0 are arbitrary, they can be used to modify the source–image distance or to eliminate other aberrations. It follows from (4.47) that the momentum aberration associated with δp is not eliminated. Accordingly, if we wish to achieve energy resolution of the beam we must meet the additional condition

$$(\delta p / \delta D) \ll (2D/MR^3 \omega_\theta) \qquad (4.48)$$

or

$$(\delta v_z / \delta v_\theta) \ll 2 \frac{R_{\Lambda\theta}}{R}, \qquad (4.49)$$

where $R_{\Lambda\theta} = v_\theta / \omega_\theta$. When $\delta v_z / \delta v_\theta \approx 1$ there are restrictions on $R_{\Lambda\theta}$:

$$2R_{\Lambda\theta} \gg R. \qquad (4.50)$$

Eliminating p and q from Ω^2 by means of the conditions Q = 0, s = 0, when $\omega_\theta \approx 0$ we have

$$\Omega^2 = \frac{12D^2}{M^2 R^4} + \omega_0^2 - 4 \frac{e}{M} \cdot \frac{k}{R} H_0. \qquad (4.51)$$

In deriving (4.51) we have assumed that the values of p can be varied with the help of a whirler. If there is no whirler, and p = 0, the focusing and the energy resolution are retained, but

$$\Omega^2 = 8 \frac{D^2}{M^2 R^4}. \qquad (4.52)$$

In this case the source–image distance is

$$L = \pi R / 2\sqrt{2}. \qquad (4.53)$$

Thus this distance is independent of the initial particle velocity and the values of k and q are fixed[†]:

$$k = \frac{1}{\omega_0 c} \left(\frac{\omega_0^2 R}{4} + \frac{1}{2} \omega_0^2 R + \frac{D^2}{M^2 R^3} \right); \qquad (4.54)$$

$$q = \frac{2e}{\omega_0^2 c^2 M} \left(\frac{1}{2} \omega_0^2 R - 2 \frac{D^2}{M^2 R^3} \right). \qquad (4.55)$$

Chapter 5

EQUILIBRIUM NEUTRALIZED ION BEAMS

§1. Thermalized Potential

As noted above, the term equilibrium neutralized ion beam means a model of a neutralized ion beam which takes account of the nonzero electron temperature and the boundary conditions at the channel walls (including the ion source and the ion collector).

Except in § 2, we will assume that T_e is known because this quantity depends on many factors which have not yet been identified.

If $T_e \neq 0$ and the system is collisionless, Ohm's law can be written

$$\frac{\nabla p_e}{en} - \nabla \Phi + \frac{1}{c} [\mathbf{u}, \mathbf{H}] = 0. \qquad (5.1)$$

Assuming that T_e = const or $T_e = T_e(\gamma)$ along a line of force, (5.1) shows that when $T_e \neq 0$ it is not the electric potential Φ, but the "thermalized" potential

$$\Phi^* = \Phi - (kT_e(\gamma)/e)\ln n/n_0 \qquad (5.2)$$

that is conserved along a magnetic line of force. Here n_0 is a constant.

[†]Here ω_0 is a free parameter, and ω_Θ should be as small as possible in order to correct the momentum aberration.

The difference between the thermalized potential and the electric potential lies in the fact that those parts of a line of force at which the density is relatively high must be at a higher electric potential in order to ensure neutrality. Thermalization of the potential obviously leads to an additional expansion of the ion beam. This effect is nothing more than the expansion of the beam caused by the electron pressure. It is particularly strong near the beam boundaries, where ∇n is large. For example, with $T_e \sim 1$ eV and $(n_{min}/n_{max}) \sim 10^{-3}$, the potential drop is about 7 eV. If the beam is not surrounded by even a low-density "coat," the boundary potential drop calculated from (5.2) is extremely large for any value of T_e; in fact, Eq. (5.2) no longer applies (cf. § 4). The difference between Φ and Φ^* becomes obvious if

$$\left| \nabla_{\|} \left(\frac{kT_e}{e} \ln \frac{n}{n_0} \right) \right| \gtrsim | \nabla_{\perp} \Phi |. \qquad (5.3)$$

Here the subscripts "$\|$" and "$\perp$" indicate that the derivatives are evaluated along and across the magnetic field respectively.

Condition (5.3) is obviously equivalent to

$$\left| \frac{1}{\delta_{\|}} \frac{kT_e}{e} \right| \gtrsim | E_{\perp} |, \qquad (5.4)$$

where $\delta_{\|}$ is the scale length of the inhomogeneity in the density distribution along the magnetic field.

Experiments confirm these results clearly. Figure 27 shows the distribution of the electric potential (a) in the channel of a "single-lends" Hall-current accelerator, diagrams of the magnetic lines of force (b), and the current density and the T_e distribution (c). Evidently the equipotential are very different from the magnetic lines of force. However, if the known values of Φ, n, T_e are used to calculate Φ^*, within the experimental error, the lines of $\Phi^* = $ const coincide with the lines of $\psi = $ const, or the magnetic lines of force [42].

We find an analogous situation in an axial plasma lens. Here again the curves of $\Phi = $ const and $\psi = $ const are quite different (Fig. 28). In this figure, 2 is the coil that produces the magnetic

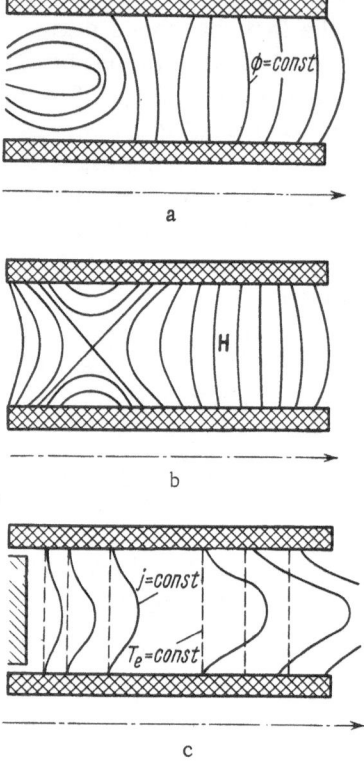

Fig. 27

field, 3 is the "comb" of electrodes that establish the potential of
the magnetic lines of force, 1 is the potential distribution over the
electrodes of the comb, and 4 are the equipotentials. Lacking data
on n_0 and T_e we cannot calculate Φ^* accurately. Nevertheless, experiments with this lens have shown clearly that by varying the
potential distribution on the electrode system around the beam it
is possible to modify the potential distribution in the beam itself,
thereby modifying the focusing properties of the system. In particular, the potential distribution shown in Fig. 28a suffers from
a clearly defined spherical aberration; however, the system in Fig.
28b shows essentially no spherical aberration [4].

These data on the lens properties pertain to a strong magnetic
field. If the magnetic field is reduced the focusing properties

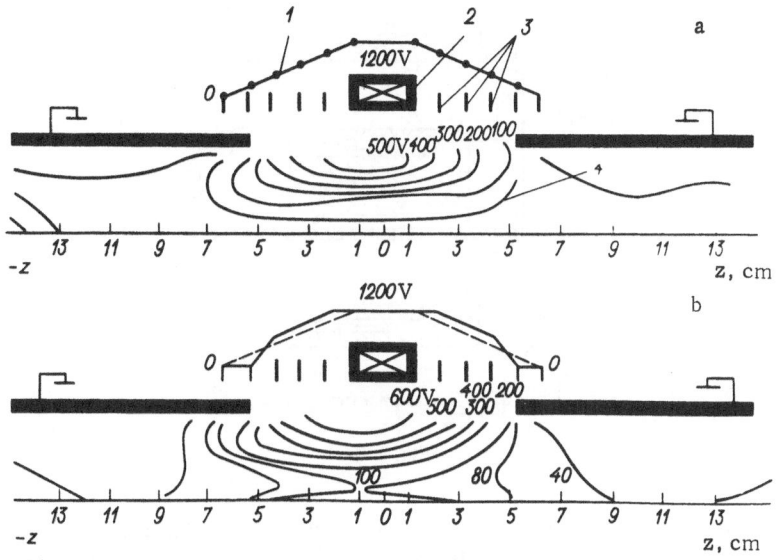

Fig. 28

vanish abruptly at some H_{min}. It has been found experimentally that the focusing no longer occurs when the electron gyroradius becomes comparable to the diameter of the neutralized ion beam.

The plasma-optical regime in the lens is also violated when the beam density is reduced so that the Debye length becomes comparable to the dimensions of the system.

We should also note the two-lens Hall-current accelerator (Fig. 29). This accelerator contains an anode 1 and a cathode 2, and the working medium is supplied through the anode. The ions are accelerated in dielectric channel 3, which is penetrated by the magnetic field produced by the core and by coils 4 [20, 40, 41]. The inequality opposite to (5.4) holds for the accelerated beam, since $E_\perp \approx 100\text{-}200 \text{ V/cm}$, $T_e \approx 10 \text{ eV}$, and $\delta \sim 3 \text{ cm}$. It is then not surprising that the distribution of the electric potential is approximately the same as that of the magnetic lines of force.

When the working medium is hydrogen magnetic focusing is more important than electrostatic focusing (Chap. 3). The drift velocity in this system (with $E = 200 \text{ V/cm}$ and $H = 800 \text{ Oe}$) $u = cE/H = 2.5 \cdot 10^7 \text{ cm/sec}$, while the directed ion velocity is

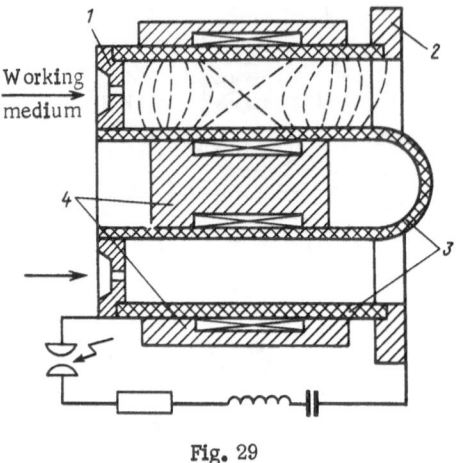

Fig. 29

about 10^8 cm/sec. Accordingly, the focusing of the neutralized
ion beam in a two-lens accelerator is governed by the magnetic
field. Figure 30 shows the radial profile of the beam density in
various cross sections of the system as obtained experimentally.
Figure 31 shows computer calculations of the trajectories of H^+
ions found under the assumption that the particles are produced at
the same point. In the calculation of **E** and **H** the values of the
fields are taken from experiment. Good qualitative agreement ob-
tains between the theoretical and experimental results.

§2. Distributions of the Thermalized Potential and the Electron Temperature in a Steady-State Plasma System

1. **Autonomous† Idealized Electron Fluid.** Let
us examine the properties of the distribution Φ^* for arbitrary
systems, including asymmetric systems. We can show that the
surfaces $\Phi^* = $ const are related to the surfaces $T_e = $ const and
to the density distribution in the system.

---

† "Autonomous" means that there is no exchange of electrons between the neutralized
ion beam and the walls.

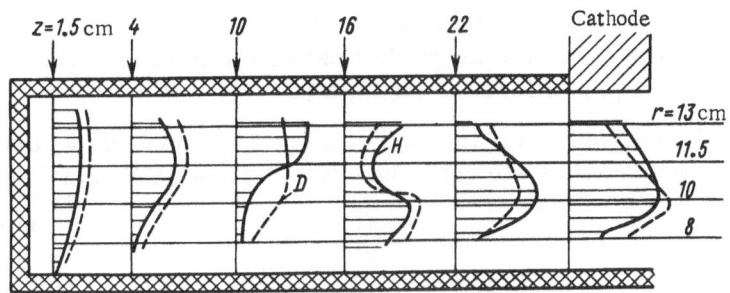

Fig. 30

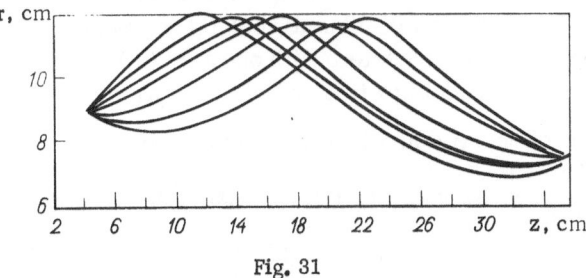

Fig. 31

Neglecting the conductivity of the plasma volume, we write the following system of equations for the electrons:

$$\operatorname{div} n\, \mathbf{u} = 0, \qquad (5.5a)$$

$$0 = \frac{\nabla p_e}{en} + \mathbf{E} + \frac{1}{c}\, [\mathbf{u},\, \mathbf{H}]; \quad \mathbf{E} = -\nabla\Phi, \qquad (5.5b)$$

$$c_V\, [n\, (\mathbf{u}\, \nabla)\, T_e - T_e\, (\gamma_a - 1)\, (\mathbf{u}\, \nabla n)] = \operatorname{div} \overleftrightarrow{\varkappa}\, \nabla T_e. \qquad (5.5c)$$

Here $\mathbf{u}$ is the directed electron velocity, $\overleftrightarrow{\varkappa}$ is the thermal conductivity tensor, c_V is the specific heat (per electron), and γ_a is the ratio of specific heats.

We assume that n, $\mathbf{H}$, and $\overleftrightarrow{\varkappa}$ in Eq. (5.5) are given; the unknowns are Φ, $\mathbf{u}$, and T_e.

Since the ohmic resistance is neglected, the discussion is limited to the idealized heat-conduction model, so that

$$\varkappa_{\|} \approx \infty; \quad \varkappa_{\perp} = 0. \qquad (5.6a)$$

In other words, it is assumed that the thermal conductivity is arbitrarily large along the magnetic lines of force while that across the magnetic lines of force is negligible.

It follows from (5.6a) that the electron temperature and the thermalized potential are constant along a magnetic line of force:

$$T_e = T_e(\gamma); \tag{5.6b}$$

$$\Phi^*(\gamma) = \Phi - \frac{kT_e(\gamma)}{e} \ln \frac{n}{n_0}. \tag{5.6c}$$

Our problem is to derive from (5.5) equations that relate the functions $T(\gamma)$ and $\Phi^*(\gamma)$ to the magnetic field and density distribution. For this purpose we assume that the magnetic field allows an orthogonal coordinate system with one axis (x_3) that coincides with the lines of force. The other two coordinates are x_1 and x_2 (Fig. 32).

In terms of these coordinates we can write

$$T_e = T_e(\gamma) = T_e(x_1, x_2); \tag{5.7a}$$

$$\Phi^*(x_1, x_2) = \Phi - \frac{kT(x_1, x_2)}{e} \ln \frac{n}{n_0}. \tag{5.7b}$$

The metric, gradient, and divergence in an arbitrary orthogonal coordinate system can be written

$$\left. \begin{aligned} ds^2 &= h_1^2 \, dx_1^2 + h_2^2 \, dx_2^2 + h_3^2 \, dx_3^2; \\ \nabla &= \left(\frac{1}{h_1} \cdot \frac{\partial}{\partial x_1}, \ \frac{1}{h_2} \cdot \frac{\partial}{\partial x_2}, \ \frac{1}{h_3} \cdot \frac{\partial}{\partial x_3} \right); \\ \operatorname{div} \mathbf{a} &= \frac{1}{h_1 h_2 h_3} \left[\frac{\partial}{\partial x_1} h_2 h_3 a_1 + \frac{\partial}{\partial x_2} h_1 h_3 a_2 + \frac{\partial}{\partial x_3} h_2 h_1 a_3 \right], \end{aligned} \right\} \tag{5.8}$$

where h_1, h_2, h_3 are the Lamé parameters.

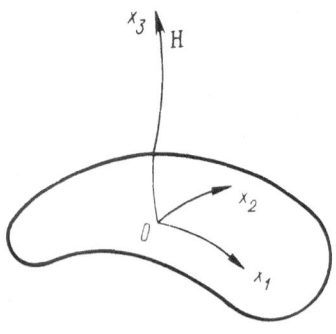

Fig. 32

Using (5.5b), (5.7), and (5.8), we can derive the following equations for the components of the electron velocity across the magnetic field:

$$u_1 = -\frac{c}{Hh_2}\left(\frac{\partial \Phi^*}{\partial x_2} + \frac{k}{e}\left(\ln\frac{n}{n_0 e_0}\right)\frac{\partial T_e}{\partial x_2}\right);$$ (5.9a)

$$u_2 = \frac{c}{Hh_1}\left(\frac{\partial \Phi^*}{\partial x_1} + \frac{k}{e}\left(\ln\frac{n}{n_0 e_0}\right)\frac{\partial T_e}{\partial x_1}\right).$$ (5.9b)

Here e_0 is the base of the natural logarithm system.

To find the third component of $\mathbf{u}$, we substitute (5.9) into continuity equation (5.5a). After some manipulations, we find

$$u_3 = \frac{1}{h_1 h_2 n}\left[\frac{D(\omega, \Phi^*)}{D(x_1, x_2)} + \frac{k}{e}\cdot\frac{D(\lambda, T_e)}{D(x_1, x_2)}\right],$$ (5.10)

where

$$\omega(x_3) = \int_{x_3(0)}^{x_3} dx_3 \frac{h_3 n}{H}; \quad \lambda(x_3) = \int_{x_3(0)}^{x_3}\frac{n h_3 \, dx_3 \, L}{H};$$

$$L = \ln\frac{n}{n_0 e_0},$$ (5.11)

and $x_3(0)$ is the coordinate of the "origin" of the line of force, at which $u_3 = 0$. Here $D(f, g)/D(\alpha, \beta)$ represents the Jacobian

$$\frac{D(f, g)}{D(\alpha, \beta)} = \begin{vmatrix} \dfrac{\partial f}{\partial \alpha} & \dfrac{\partial f}{\partial \beta} \\ \dfrac{\partial g}{\partial \alpha} & \dfrac{\partial g}{\partial \beta} \end{vmatrix}.$$

Assuming that the velocity u_3 vanishes at the "origin" and end of the line of force, from (5.10) we find the first of the required equations, which relates Φ^* and T_e:

$$\frac{D(\Phi^*, W)}{D(x_1, x_2)} + \frac{k}{e}\cdot\frac{D(T_e, \Lambda)}{D(x_1, x_2)} = 0;$$ (5.12)

$$W \equiv \int_{x_3(0)}^{x_3(\infty)}\frac{n\,dl}{H}; \quad \Lambda \equiv \int_{x_3(0)}^{x_3(\infty)}\frac{nL\,dl}{H}.$$ (5.13)

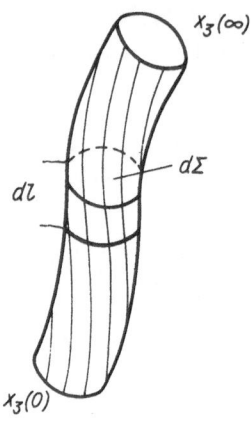

Fig. 33

Here $x_3(0)$ is the coordinate of the "origin" of the line of force, and $x_3(\infty)$ is the coordinate of the end. The differential $dl = h_3 dx_3$ in (5.13) is the arc element along a magnetic line of force.

We now consider the energy equation. We multiply this equation by the volume element of an infinitesimally thin magnetic tube (Fig. 33),

$$dV = dl d\Sigma, \tag{5.14}$$

and use the conservation of magnetic flux

$$H d\Sigma = \text{const}; \tag{5.15}$$

using (5.6a) and the fact that there is no heat flux out of the regions in which the magnetic line of force begins and ends, we find

$$\int \frac{dl}{H} [n (\mathbf{u}\ \nabla) T_e] = (\gamma_a - 1) T_e \int \frac{dl}{H} [\mathbf{u}\ \nabla) n]. \tag{5.16}$$

Substituting the equations for the velocity components (5.9) and (5.10) into (5.16), and noting that (5.15) is equivalent to the condition

$$h_2 h_3 H = \text{const}, \tag{5.17}$$

we find a second relation between T_e and Φ^*:

$$W \frac{D(T_e, e\Phi^*)}{D(x_1, x_2)} = T_e (\gamma_a - 1) \left[\frac{D(\Lambda, \Phi^* e)}{D(x_1, x_2)} + \frac{D(Y^* - W, T_e k)}{D(x_1, x_2)} \right], \tag{5.18}$$

where [see Eqs. (5.11)]

$$\Lambda \equiv -\int \frac{dl}{H} \cdot \frac{\partial n}{\partial l} \, \omega \, (l); \quad Y^* \equiv \int \frac{dl}{H} \cdot \frac{\partial n}{\partial l} \, \lambda \, (l); \quad dl = h_3 \, dx_3. \qquad (5.19)$$

System (5.12), (5.18) provides the desired relation between $T_e(\gamma)$ and $\Phi^*(\gamma)$ and the distributions of H and n for an idealized autonomous electron component. It should be noted that x_1 and x_2 in these equations can now be understood to be any coordinates.

The complete system of equations in (5.12) and (5.18) shows that if the plasma is at a negligibly low temperature, i.e., if $T \to 0$, or if the plasma is isothermal ($\gamma = 1$), then the potential is constant on the surfaces W = const; thus,

$$\Phi = \Phi \, (W); \quad W = W \, (\gamma). \qquad (5.20)$$

Here Φ^* has been replaced by Φ, since $\Phi^* \to \Phi$ when $T_e \to 0$.

Equations (5.20) refine the conditions derived above for equipotential magnetic lines of force.

If the system is axisymmetric its parameters are independent of the coordinate $x_2 = 0$, and Eqs. (5.12) and (5.18) are satisfied identically by

$$T_e = T_e \, (x_1); \quad \Phi^* = \Phi^* \, (x_1). \qquad (5.21a)$$

The orthogonal coordinate x_1, which "aligns" the coordinate system with the magnetic field, is a function of the magnetic stream function ψ. Accordingly, (5.21a) is equivalent to

$$T_e = T_e \, (\psi); \quad \Phi^* = \Phi^* \, (\psi). \qquad (5.21b)$$

Examining Eqs. (5.13) and (5.20) for W, we note two features. First, the equation for W can be treated as a generalization of the "freezing" of the magnetic tube, which is used in the theory of equilibrium plasma configurations:

$$n/H = \text{const}, \qquad (5.22)$$

and an extension of the equation for the specific volume of the magnetic tube, which is used in theory of equilibrium plasma configurations:

$$U = \int \frac{dl}{H} . \qquad (5.23)$$

The hydrodynamic equilibrium condition

$$\nabla p = \frac{1}{c} [\mathbf{j}, \mathbf{H}],\tag{5.24}$$

implies that the pressure is constant along a magnetic line of force. Accordingly, under the reasonable condition $T_e = T_e(\gamma)$, we can take n outside the integral for W in (5.13) to find Eq. (5.23).

Second, Eq. (5.22) holds along a streamline of the medium in idealized single-fluid† magnetohydrodynamics. However, the direction of the velocity of the medium is related to the electric field by

$$\mathbf{E} + \frac{1}{c} [\mathbf{v}, \mathbf{H}] = 0.\tag{5.25}$$

Evidently the medium flows along equipotentials, in agreement with Eqs. (5.20).

To pursue the general analysis of systems (5.12) and (5.18) we assume that the functions $W(x_1, x_2)$ and $\Lambda(x_1, x_2)$ are independent of each other, so that the Jacobian satisfies‡

$$\frac{D(W, \Lambda)}{D(x_1, x_2)} \neq 0.\tag{5.26}$$

Then transforming from the variables x_1, x_2 to W, Λ, we can replace (5.12) and (5.18) by

$$\frac{D(W, e\Phi^*)}{D(W, \Lambda)} = \frac{D(T_e, \Lambda)}{D(W, \Lambda)} ;\tag{5.27a}$$

$$W \frac{D(T_e, e\Phi^*)}{D(W, \Lambda)} = T_e(\gamma_a - 1)\left[\frac{D(\Lambda, e\Phi^*)}{D(W, \Lambda)} + \frac{D(Y^* - W, T_e)}{D(W, \Lambda)}\right].\tag{5.27b}$$

These equations are convenient since they do not contain the coordinates. The first of these equations is obviously equivalent to

$$\partial(e\Phi^*)/\partial\Lambda = \partial T/\partial W.\tag{5.28}$$

---

† In single-fluid magnetohydrodynamics it is assumed that $|\mathbf{v} - \mathbf{u}| \to 0$.
‡ In particular, we are thus assuming that the system is asymmetric.

The existence of a relation between Φ^* and T means that we can reduce (5.27) to a single equation for a function χ, which we call the "electric potential of the plasma." This potential is related to Φ^* and T_e by

$$e\Phi^* = \partial\chi/\partial W; \quad T_e = \partial\chi/\partial\Lambda. \tag{5.29}$$

Substituting (5.29) into (5.27b) we find a first-order equation for χ:

$$W \frac{D\left(\frac{\partial\chi}{\partial\Lambda}, \frac{\partial\chi}{\partial W}\right)}{D(W, \Lambda)} = \frac{\partial\chi}{\partial\Lambda}(\gamma_a - 1) \times \left[\frac{D\left(\Lambda, \frac{\partial\chi}{\partial W}\right)}{D(W, \Lambda)} + \frac{D\left(Y^* - W, \frac{\partial\chi}{\partial\Lambda}\right)}{D(W, \Lambda)}\right]. \tag{5.30}$$

If T_e is sufficiently low or if $\gamma_a \to 1$, we can find a general solution of Eq. (5.30) as a power series in the small parameter $\varepsilon \sim T_e(\gamma_a - 1)$:

$$\chi = \chi_0(W) + \varepsilon\chi_1(\Lambda, W) + \varepsilon^2\chi_2(\Lambda, W) + \dots \tag{5.31}$$

Substituting (5.31) into (5.28) we have

$$W \frac{\partial^2 \chi_1^2}{\partial\Lambda^2} = \frac{\partial\chi_1}{\partial\Lambda}\frac{\partial\Lambda^*}{\partial\Lambda}(\gamma_a - 1). \tag{5.32}$$

Integration of this equation yields

$$\ln\frac{\partial\chi_1}{\partial\Lambda} = (\gamma_a - 1)\frac{\Lambda^*}{W} + g_1(W). \tag{5.33}$$

Here $g_1(W)$ is an arbitrary function of its argument. When χ_1 is known we can determine χ_2, etc.

The next approximations can be found by integrating the equation

$$W \frac{\partial^2 \chi_n}{\partial\Lambda^2} - \frac{\partial\chi_n}{\partial\Lambda}(\gamma_a - 1)\frac{\partial\Lambda^*}{\partial\Lambda} = F_n(\Lambda, W, \chi_1, \dots, \chi_{n-1}). \tag{5.34}$$

Accordingly, χ can be calculated for known $\mathbf{H}(\mathbf{r})$ and $n(\mathbf{r})$. We assume that the magnetic field is known at all times and find n from the ion dynamic equations or from experiment. In general, we are confronted with an extremely complicated system of equations, but if $kT_e \ll Mv^2/2$ or if there is some symmetry the problem is simplified and can frequently be solved.

It should be kept in mind that the relation found here between $T_e(\gamma)$, $\Phi^*(\gamma)$, W, Λ, and Y improves our picture of the processes that occur in plasma-optical systems.

2. Nonautonomous Idealized Electron Fluid. We have assumed above that the electron fluid is autonomous so that energy or particles are not exchanged with the wall. As a rule, however, such an exchange occurs so that it is useful to generalize the equations of the preceding subsection to this case.

We again assume the plasma to be ideal so that it can be described by system (5.5) with condition (5.6a).

In this case the functions $T_e = T_e(\gamma)$ and $\Phi^*(\gamma)$ are valid, and Eq. (5.9) can again be used for the transverse components of the electron velocity. However, Eq. (5.10) assumes a different form. If the wall coordinates are $x_3(0)$ and $x_3(\infty)$, and if the density and velocity u_3 at $x_3(0)$ are $n^{(0)}$ and $u_3^{(0)}$, respectively, then

$$u_3 n h_1 h_2 = u_3^{(0)} h_1^{(0)} h_2^{(0)} n^{(0)} + \left[\frac{D(W, \Phi^*)}{D(x_1, x_2)} + \frac{k}{e} \cdot \frac{D(\lambda, T_e)}{D(x_1, x_2)} \right]. \qquad (5.35)$$

Here $\omega(x_3)$ and $\lambda(x_3)$ are given by Eq. (5.11).

If the electron flux density through the upper wall, $u_3^{(\infty)}$, $n^{(\infty)}$ is known, (5.9) can be replaced by

$$q(x_1, x_2) = \frac{D(\Phi^*, W)}{D(x_1, x_2)} + \frac{k}{e} \cdot \frac{D(T_e, \Lambda)}{D(x_1, x_2)}, \qquad (5.36)$$

where

$$q \, dx_1 \, dx_2 = u_3^{(\infty)} n^{(\infty)} ds^{(\infty)} - u_3^{(0)} n^{(0)} ds^{(0)}; \qquad (5.37)$$

$$ds^{(\infty)} \equiv (h_1 h_2)^{(\infty)} dx_1 dx_2; \quad ds^{(0)} = (h_1 h_2)^{(0)} dx_1 dx_2. \qquad (5.38)$$

The quantity $q dx_1 dx_2$ is obviously equal to the number of electrons which leave the magnetic tube per unit time.

The fact that a magnetic tube continuously transfers a certain number of electrons to the wall does not contradict the condition for a steady state, since other magnetic tubes of the system are continuously acquiring electrons from the wall and these electrons are transferred to the tube under consideration as a result of drift.

Similarly, we can generalize energy equations (5.18):

$$W \frac{D(T_e, e\,\Phi^*)}{D(x, x_2)} = T_e(\gamma_a - 1)\left[\frac{D(\Lambda^*, e\,\Phi^*)}{D(x_1, x_2)} + \frac{D(Y^* - W, kT_e)}{D(x_1, x_2)}\right] + \frac{1}{c_V}\,Q(\gamma). \qquad (5.39)$$

Here Qdx_1dx_2 is the heat flux carried away from the given tube by electrons.

System (5.37), (5.39) shows that it is possible, in principle, to specify arbitrary functions $\Phi(\gamma)$ and $T_e(\gamma)$, but then the electron fluid must exchange particles and energy with the wall.

§ 3. Ion Dynamics

1. Calculation Methods. Since collisions are neglected the Vlasov equation must be used to describe the ion dynamics:

$$\frac{\partial f_i}{\partial t} + (\mathbf{v}\,\nabla)f_i + \frac{e}{M}\left(\mathbf{E} + \frac{1}{c}\,[\mathbf{v}, \mathbf{H}]\right)\frac{\partial f_i}{\partial \mathbf{v}} = 0. \qquad (5.40)$$

In this equation the electric field is governed by the thermalized potential $\Phi^*(\gamma)$, which is assumed to be fixed, e.g., by electrodes at the channel walls:

$$\mathbf{E} = -\nabla\Phi = -\nabla\left(\Phi^*(\gamma) + \frac{kT_e}{e}\ln\frac{n}{n_0}\right); \quad n = \int f\,d\mathbf{v}. \qquad (5.41)$$

If the thermalization effect is negligible, the electric field can be assumed to be known (Φ^* is fixed). The magnetic field is also known, and the solution of the Vlasov equation reduces to a calculation of the motion of a single particle in the specified fields, since the characteristic equations are

$$\frac{d\mathbf{r}}{dt} = \mathbf{v}; \quad M\frac{d\mathbf{v}}{dt} = e\left(\mathbf{E} + \frac{1}{c}\,[\mathbf{v}, \mathbf{H}]\right). \qquad (5.42)$$

In the nonstationary case, (5.42) has six integrals,

$$g_k(\zeta_1, \ldots, \zeta_6, t) = \zeta_{k0} \quad \text{for} \quad k = 1, \ldots, 6; \atop \zeta_{1,2,3} = z, x, y;\ \zeta_{4,5,6} = v_z, v_x, v_y. \qquad (5.43)$$

If $f_0(\zeta_k)$ is the value of the distribution function f_i at t = 0, at any other time

$$f_i(\zeta,\ t) = f_0\left(g\left(\zeta,\ t\right)\right);\quad \left.\right\}$$
$$\zeta = (\zeta_1,\ ...,\ \zeta_6);\ g = (g_1,\ ..,\ g_6).\quad \left.\right\} \tag{5.44}$$

In plasma optics there is particular interest in stationary problems, in which the beam properties are independent of t. In this case it is useful to take the independent variable to be one of the coordinates, say, $\zeta_1 \equiv z$. Then, using the first equation from (5.43) to eliminate t, we find five integrals

$$G_k(\zeta_j,\ ...,\ z) = \zeta_{k0};\ k,\ j = 2,\ ...,\ 6, \tag{5.45a}$$

so that if the distribution function for ζ_{10} is f_0, we have

$$f_i(\zeta,\ z) = f_0(G(\zeta,\ z)) \tag{5.45b}$$

in any other cross section.

The situation is more complicated when the thermalization effect must be taken into account. When (5.41) is used in (5.40), we obtain a complicated nonlinear integrodifferential equation for f_i.

As noted earlier, the temperature T_e is usually low in plasma-optical systems, so that the thermalization can be taken into account by successive approximations, with the single-particle model being used as a first approximation. The subsequent approximations can be calculated by various methods; here we consider two such methods. In the first, we seek the function f in series form,

$$f = f_0 + f_1 + f_2 + ... , \tag{5.46}$$

where $f_1 \sim T_e$, $f_2 \sim T_e^2$, etc. Substituting (5.46) into (5.40) and (5.41), we find

$$
\begin{aligned}
(\mathbf{v}_0\,\nabla)f_1 + \frac{e}{M}\left(-\nabla\Phi^*(\gamma) + \frac{1}{c}\,[\mathbf{v}_0,\,\mathbf{H}]\right)\frac{\partial f_1}{\partial \mathbf{v}} = \\
= \frac{e}{M}\,\nabla\left(\frac{kT_e(\gamma)}{e}\,\ln\frac{\int f_0\,d\mathbf{v}}{n_0}\right)\frac{\partial f_0}{\partial \mathbf{v}}\,; \\
(\mathbf{v}_0\,\nabla)f_2 + \frac{e}{M}\left(-\nabla\Phi^*(\gamma) + \frac{1}{c}\,[\mathbf{v}_0,\,\mathbf{H}]\right)\frac{\partial f_2}{\partial \mathbf{v}} = \\
= \frac{e}{M}\,\nabla\left(\frac{kT_e}{e}\cdot\frac{\int f_1\,d\mathbf{v}}{\int f_0\,d\mathbf{v}}\right)\frac{\partial f_0}{\partial \mathbf{v}} + \frac{e}{M}\,\nabla\left(\frac{kT_e}{e}\,\ln\frac{\int f_0\,d\mathbf{v}}{n_0}\right)\frac{\partial f_1}{\partial \mathbf{v}}\,; \\
\cdots\cdots\cdots\cdots\cdots\cdots\cdots
\end{aligned}
\quad \right\} \tag{5.47}
$$

Here f_0 is the first approximation, which is calculated on the basis of the known values of Φ and $\mathbf{H}$ with thermalization neglected.

Inhomogeneous equations like (5.47) are easily integrated by a transformation of variables. Specifically, if we replace ζ_i by the variables ζ_1, ζ_{j0}, where j = 2, ..., 6, then the inhomogeneous equation

$$\frac{Df}{Dz} = \frac{\partial f}{\partial z} + \frac{v_x}{v_z}\frac{\partial f}{\partial x} + \frac{v_y}{v_z}\frac{\partial f}{\partial y} + \frac{e}{M}\frac{E_*}{v_z}\frac{\partial f}{\partial v} = Q, \qquad (5.48a)$$

where $\mathbf{E}_* = -\nabla\Phi(\gamma) + \frac{1}{c}[\mathbf{v}, \mathbf{H}]$, becomes

$$\partial f/\partial z = Q \qquad (5.48b)$$

and can be integrated immediately:

$$f = F_0(\zeta_{0i}) + \int_0^z Q(\zeta_{0i}, z)\,dz. \qquad (5.48c)$$

This method of successive approximations presupposes that the functions $f_0(\mathbf{v})$ have good analytic properties, since the calculation of f_n requires the existence of the n-th derivative of f_0 with respect to v.

The second method is physically more transparent and less restrictive; this second method involves seeking corrections to the characteristic equations rather than to the distribution functions.

Specifically, we construct a sequence of approximations

$$f_{(0)}, f_{(1)}, f_{(2)}, \cdots ,$$

where $f_{(n)}$ satisfies

$$(\mathbf{v}\,\nabla)f_{(n)} + \frac{e}{M}\left\{\left[-\nabla\left(\Phi^*(\gamma) + \frac{kT_e}{e}\ln\frac{\int f_{(n-1)}\,dv}{n_0}\right)\right] + \frac{1}{c}[\mathbf{v}, \mathbf{H}]\right\}\frac{\partial f_{(n)}}{\partial v} = 0. \qquad (5.49a)$$

The solution of this equation reduces to a calculation of the motion in the known field of the forces

$$\mathbf{H}, \ \Phi_n = \Phi^*(\gamma) + \frac{kT_e}{e}\ln\frac{\int f_{(n-1)}\,dv}{n_0}, \qquad (5.49b)$$

as can be done by the weak-field method.

Let us illustrate this method with a simple example. We assume that a neutralized ion beam is incident on the free half-space $z > 0$; at $z = 0$, the beam ions have the distribution function

$$f_{(0)\,0}\,|_{z=0} = n_{00} \exp\left(-y^2/b^2\right) \delta\left(v_z - v_0\right) \delta\left(v_y\right). \tag{5.50}$$

The electron temperature T_e is assumed to be constant. In this case the kinetic equation for the ions is

$$(\mathbf{v}\,\nabla)f - \frac{kT_e}{M} \cdot \frac{1}{n}\left(\nabla n\,\frac{\partial f}{\partial v}\right) = 0; \quad n = \int f\,d\mathbf{v}. \tag{5.51}$$

The zeroth approximation of the distribution function satisfies

$$(\mathbf{v}\,\nabla)f_{(0)} = 0, \tag{5.52}$$

and the characteristics of this equation describe uniform rectilinear motion:

$$v_{y0} = v_y;\; v_{z0} = v_z;\; y_0 = y - (v_y z/v_z). \tag{5.53}$$

Accordingly, for an arbitrary value of x we have

$$f_0 = n_{00} \exp\left(-y^2/b^2\right)\delta\left(v_z - v_0\right)\delta\left(v_y\right). \tag{5.54}$$

Hence, in the zeroth approximation the density distribution of the beam is

$$n_0 = n_{00} \exp\left(-y^2/b^2\right), \tag{5.55}$$

and the first-approximation equation is

$$v_z\,\frac{\partial f_{(1)}}{\partial z} + v_y\,\frac{\partial f_{(1)}}{\partial y} - \frac{kT_e}{M} \cdot \frac{2y}{b^2} \cdot \frac{\partial f_0}{\partial v_y} = 0. \tag{5.56}$$

Its characteristic system,

$$\frac{dz}{v_z} = \frac{dy}{v_y} = \frac{dv_z}{0} = \frac{dv_y}{\dfrac{kT_e}{M}\cdot\dfrac{2y}{b^2}}, \tag{5.57a}$$

has the solutions

$$v_{z0} = v_z;\quad v_{y0} = v_y\,\cosh\frac{qz}{v_z} - qy\,\sinh\frac{qz}{v_z};$$

$$y_0 = y\,\cosh\frac{qz}{v_z} - \frac{v_y}{q}\,\sinh\frac{qz}{v_z};\quad q^2 = \frac{2kT_e}{Mb^2}. \tag{5.57b}$$

Substituting (5.57b) into (5.50), we find

$$f_1 = n_{00} \exp\left[(-y^2/b^2)\left(y \, \cosh\frac{qz}{v_z} - \frac{v_y}{q}\sinh\frac{qz}{v_z}\right)^2\right] \times$$

$$\times \delta(v_{z0} - v_0)\delta\left(v_y \cosh\frac{qz}{v_z} - qy \sinh\frac{qz}{v_z}\right). \qquad (5.57c)$$

Using

$$\delta(f(\alpha)) = \delta(\alpha - \alpha_0)/|f'(\alpha_0)|, \quad f(\alpha_0) = 0, \qquad (5.58)$$

we can simplify (5.57c):

$$f_1 = n_{00} \frac{\exp\left(-y^2/b^2 \, \cosh^2\frac{qz}{v_z}\right)}{\cosh(qz/v_z)} \delta(v_{z0} - v_0)\delta\left(v_y - qy \, \tanh\frac{qz}{v_z}\right). \qquad (5.59)$$

Evidently the electron pressure causes the neutralized ion beam to expand.

The beam density corresponding to the first approximation, (5.59), is

$$n_{(1)} = n_{00} \frac{\exp\left(-y^2/b^2 \, \cosh^2\frac{qz}{v_z}\right)}{\cosh\frac{qz}{v_z}}. \qquad (5.60)$$

Knowing $n_{(1)}$, we can find $f_{(2)}$, etc. The structure of solution (5.59) shows that the dimensionless small parameter is

$$\frac{qz}{v_z} = \frac{c_T}{v_z} \cdot \frac{z}{b}; \quad c_T = \sqrt{\frac{2kT}{M}}. \qquad (5.61)$$

If the method based on (5.46)-(5.48) is used in this example, the following correction to the functions in (5.54) is obtained:

$$f_1 = \frac{kT}{M} N(y)\delta(v_{z0} - v_0)\frac{\delta_0'(v_y)}{v_y}.$$

2. Focusing of an Ion Beam with a Velocity Spread by an Axial Lens.

How does the ion velocity spread affect the structure of the beam focused by an axial lens?

We assume that the electron temperature is zero, so that the motion for each ion can be computed in the single-particle approximation. Knowing the integrals of the equations of motion and the distribution function at z = 0, we then use Eq. (5.45) to find the particle distribution for any z and, thus, any desired property of the ion beam. In particular, the density distributions $n_{(0)}$ found in this way can be used with the known value of T_e to calculate the distribution functions $f_{(1)}$ so that the effect of T_e on the dynamics of the ion beam can be evaluated.

Restricting the discussion to the linearized model of the axial system, we use the equation of motion in the following form (Chap. 2):

$$v_0^2 r'' + \omega_0^2 (z) r = 0. \tag{5.62a}$$

To simplify the equations we replace Eq. (5.62) by the equivalent equation

$$r_1'' + \Omega^2 (z) r_1 = 0, \tag{5.62b}$$

where r_1, Ω_1, z_1 are the dimensionless quantities corresponding to r, ω_0/v_0, z. Below we omit the subscript "1," which indicates a dimensionless quantity. The fundamental system of solutions $R_1(z)$, $R_2(z)$ of Eq. (5.62b) is chosen to satisfy the boundary conditions

$$\left. \begin{array}{l} R_1(0) = 1; \quad R_1'(0) = 0; \\ R_2(0) = 0; \quad R_2'(0) = 1. \end{array} \right\} \tag{5.63a}$$

Then the Wronskian is

$$W = \begin{vmatrix} R_1 & R_2 \\ R_1' & R_2' \end{vmatrix} = 1. \tag{5.63b}$$

If the particle parameters at z = 0 are

$$r_0, \ r'_0 = v_{r0}/v_0 \equiv dr_0/dz, \tag{5.63c}$$

the equations can be written

$$\left. \begin{array}{l} r = r_0 R_1(z) + r_0' R_2(z); \\ r' = r_0 R_1'(z) + r_0' R_2'(z). \end{array} \right\} \tag{5.64a}$$

Solving system (5.64) for r_0 and r_0', we find

$$\left. \begin{array}{l} r_0 = r R_2'(z) - r' R_2(z); \\ r_0' = -r R_1'(z) + r' R_1(z) \end{array} \right\} \tag{5.64b}$$

or

$$r_0 = rR_2'(z) - \frac{v_r}{v_0} R_2(z); \left.\begin{array}{c} \\ \end{array}\right\} \tag{5.64c}$$
$$v_{r0} = -rv_0 R_1'(z) + v_r R_1(z).$$

The first equation in (5.64a) shows that the image of the object is determined from the condition

$$R_2(z_*) = 0. \tag{5.65a}$$

In this case, regardless of the spread in radial velocity r_0' all the particles emerging from the points of a circle are collected on a circle with radius

$$r_* = r_0 R_1(z_*). \tag{5.65b}$$

Let us use these solutions to construct the distribution function. We must know the distribution function f_0 at z = 0, i.e., the properties of the ion sources. Since we are interested primarily in the qualitative behavior, we choose functions f_0 for which the calculations are particularly simple. For example, it is convenient to assume that the beam is emitted by the entire z = 0 plane but that the particle density falls off radially as a Gaussian:

$$n|_{z=0} = n_0 \exp(-z^2/q^2). \tag{5.66}$$

The ion velocity distribution at z = 0 is that of a simple model of a "parallel" beam with a Maxwellian velocity spread along all coordinates; although simple, this model is physically interesting. The most general distribution function is

$$f_0 = \frac{n_0 |r_0| \Pi(r_0)}{b_0 (\sqrt{\pi})^3 C_1 C_2 C_3} \exp\left\{ -\frac{r_0^2}{q^2} - \frac{(v_{z0} - v_0)^2}{C_1^2} - \frac{(r_0 v_{\theta 0})^2}{b_0^2 C_3^2} - \frac{v_{r0}^2}{C_2^2} \right\}. \tag{5.67}$$

Here $C_1(r)$, $C_2(r)$, $C_3(r)$ are the scale values of the velocity spread along the corresponding directions; $b_0(r_0)$ is a function with the dimensions of length; and $\Pi(\alpha)$ is the unit step function which is equal to unity at $\alpha > 0$ and zero at $\alpha < 0$. The normalization factor in (5.67) is chosen to satisfy (5.66).

If C_1, C_2, $C_3 \to 0$ and b = const, distribution (5.67) obviously becomes

$$f_0 = n_0 |r_0| e^{-r_0^2/q^2} \delta(v_{z0} - v_0) \delta(r_0 v_{\theta 0}) \delta(v_{r0}) \Pi(r_0), \tag{5.68}$$

which represents a monochromatic ion beam that moves along the z axis.

We first consider the case of a cold parallel beam. Without complicating the calculations we can replace the initial function in (5.68) by the more general function

$$f_0 = n_0 r_0 Q(r_0) \delta(v_{r0}) \delta(v_{\theta 0} r_0) \delta(v_z - v_0) \Pi(r_0), \qquad (5.69)$$

where $Q(r_0)$ is an arbitrary function of r_0.

To find the function f for any value of z we substitute (5.64b) and make use of the conservation of the angular momentum and the longitudinal velocity component:

$$r v_\theta = r_0 v_{\theta 0}, \quad v_z = v_{z0}. \qquad (5.70)$$

As a result we find

$$f = n_0 \left(r R_2' - \frac{v_r}{v_0} R_2 \right) Q \left(r R_2' - \frac{v_r}{v_0} R_2 \right) \delta(r v_\theta) \delta(v_z - v_0) \delta(-r v_0 R_1' +$$

$$+ v_r R_1) \Pi \left(r R_2' - \frac{v_r}{v_0} R_2 \right). \qquad (5.71)$$

The last δ-function in Eq. (5.71) is nonvanishing at $v_r = r v_0 R_1'/R_1$, so that Eq. (5.71) can be rewritten as follows, where the unit value of the Wronskian is taken into account:

$$f = \frac{n_0 r}{R_1 |R_1|} Q \left(\frac{r_1}{R_1} \right) \delta(v_z - v_0) \delta(r v_\theta) \delta \left(v_r - \frac{r v_0 R_1'}{R_1} \right) \Pi \left(\frac{r}{R_1} \right). \qquad (5.72)$$

By integrating the resulting expression for f over v_z, v_θ, v_r, we find the density distribution:

$$n = n_0 \frac{1}{R_1^2} Q \left(\frac{r}{R_1} \right); \quad R_1 > 0. \qquad (5.73a)$$

In particular, in the case in (5.68) we have

$$n = \frac{n_0}{R_1^2} \exp \left(-\frac{r^2}{q^2 R_1^2} \right); \quad R_1 > 0. \qquad (5.73b)$$

Evidently the beam is focused to a point at $z = z_1$, where

$$R_1(z_1) = 0. \qquad (5.74)$$

It is interesting to note that the coordinate of the image, (5.56), differs from that of the focus, (5.74). The reason is that the minimum beam cross section, whose position is given by condition (5.74), does not necessarily coincide with the image of a source which is extended along r (in the z = 0 plane). If the source is a point source, however, these positions coincide.

These equations have another noteworthy feature: Eq. (5.73) shows that the function f vanishes at r > 0 and R_1 < 0, since the factor $\Pi(r/R_1)$ vanishes. Obviously, the reason is that at $z = z_1$, the particles cross the axis r = 0; (5.62b) implies that r is then negative at $z > z_1$. In a cylindrical coordinate system, however, we always require r > 0. This formal contradiction is a consequence of the special nature of this case with D = 0. To avoid these formal complications, we write the function f as

$$f = F\,(v_z,\ v_\theta\, r,\ v_r,\ r,\ z)\ \Pi(r/R_1) + F\,(v_z,\ rv_\theta,\ -\, v_r,\ -\, r,\ z)\ \Pi\,(-r/R_1).$$

$$(5.75a)$$

Here F represents the coefficient of Π in (5.75), and r is assumed to be the ordinary (positive) cylindrical coordinate. We note that when the axis r = 0 is crossed there are changes in the sign of v_r and r as well as v_θ. Thus the sign of the product rv_θ remains unchanged.

Making use of the parity of the function f with respect to sign changes of r and v_r found with the help of (5.68), we can write (5.75a) as

$$f = F\,(\Pi\,(r/R_1) + \Pi(-r/R_1)) = F,\qquad (5.75b)$$

since

$$\Pi\,(\alpha) + \Pi\,(-\alpha) = 1.\qquad (5.75c)$$

Let us now generalize this example, assuming that the particles have a spread in radial velocity ($C_2 \neq 0$), but that the longitudinal and azimuthal components of the velocities are the same ($C_1 = C_3 = 0$).

In this case Eq. (5.67), which gives the ion distribution function at z = 0, becomes ($b_0 \equiv$ const)

$$f_0 = \frac{n_0\, r_0}{\sqrt{\pi}\, C_2}\, \exp\!\left(-\frac{r_0^2}{q^2} - \frac{v_{r0}^2}{C_2^2}\right) \delta\,(v_{z0} - v_0)\, \delta\,(r v_{\theta 0})\, \Pi\,(r_0).\qquad (5.76)$$

Substituting Eq. (5.64b) and the conservation relations (5.70) into this equation, and taking account of the comments regarding (5.75a), we find the following equation for the ion distribution function at any z:

$$f = \frac{n_0 e^{-\frac{r^2}{C_2^2 s q^2}}}{r \sqrt{\pi} C_2} \left\{ \left(rR_2' - \frac{v_r}{v_0} R_2 \right) e^{-s\left(v_r - r\frac{v_0}{2}\frac{s'}{s}\right)^2} \Pi \left(rR_2' - \frac{v_r}{v_0} R_2 \right) + \right.$$

$$\left. + \left(-rR_2' + \frac{v_r}{v_0} R_2 \right) e^{-s\left(v_r - r\frac{v_0}{2}\frac{s'}{s}\right)} \Pi \left(-rR_2' + \frac{v_r}{v_0} R_2 \right) \right\} \times$$

$$\times \delta (v_z - v_0) \delta (rv_\theta). \tag{5.77}$$

The argument of the exponential function transforms as follows:

$$\frac{1}{q^2} \left(rR_2' - \frac{v_r}{v_0} R_2 \right)^2 + \frac{1}{C_2^2} (v_r R_1 - rv_0 R_1')^2 = s\left(v_r - r\frac{v_0}{2}\frac{s'}{s} \right)^2 + \frac{r^2}{C_2^2 q^2 s},$$

$$\tag{5.78a}$$

where

$$s(z) \equiv \left(\frac{R_2^2}{v_0^2 q^2} + \frac{R_1^2}{C_2^2} \right); \quad s' = \frac{ds}{dz}. \tag{5.78b}$$

Integrating the function f over velocity, we find the particle density distribution in the beam to be

$$n = \frac{n_0 R_2}{\sqrt{\pi} C_2 v_0 r} \exp \left[-\frac{r^2}{C_2^2 s q^2} \left\{ \int_{-\infty}^{r\alpha} (r\alpha - x) \exp \left(-s(x - r\beta)^2 \right) dx + \right. \right.$$

$$\left. \left. + \int_{r\alpha}^{\infty} (-r\alpha + x) \exp \left(-s(x - r\beta)^2 \right) dx \right\} \right]. \tag{5.79}$$

In (5.79) we have introduced

$$\alpha \equiv v_0 R_2'/R_2; \quad \beta \equiv (v_0/2)(s'/s); \quad x \equiv v_r,$$

and it is evident that

$$\alpha - \beta = v_0 R_1/s R_2 C_2^2.$$

After some simple manipulations, Eq. (5.79) can be rewritten as

$$n = \frac{n_0 R_2}{\sqrt{\pi} C_2 v_0} e^{-\frac{r^2}{C_2^2 s q^2}} \left[(\alpha - \beta) \int_{-r(\alpha-\beta)}^{r(\alpha-\beta)} e^{-s^2 y^2} dy + \frac{e^{-s^2 r^2(\alpha-\beta)^2}}{sr} \right]. \tag{5.80a}$$

Multiplying (5.80a) by $2\pi r\, dr$, and integrating from zero to infinity, we find a quantity which is independent of z, as expected.

In contrast with the case of a beam without a velocity spread [(5.69)], for which the density becomes infinite only at a single point [see Eqs. (5.73b) and (5.74)], Eq. (5.80a) shows that the density in this case becomes infinite at essentially all $z \neq 0$.

Except at $z = 0$, the density singularity disappears when $R_2(z*) = 0$ In this cross section, $z = z*$, the density is

$$n = \frac{n_0}{R_1^2} \exp\left(-\frac{r^2}{R_1^2 q^2}\right); \tag{5.80b}$$

and this represents an image of the "source" at $z = 0$ which is magnified by a factor $R_1(z*)$.

This result is in complete agreement with Eq. (5.65) and the related discussion. The obvious reason for the infinite beam density at $r = 0$ is that we have neglected the angular-momentum spread of the particles.

We now consider a "parallel" beam with spreads in both the radial and azimuthal velocity components. The function f_0 is written

$$f_0 = \frac{n_0}{\pi c_T^2}\, e^{-\frac{1}{q^2}(x_0^2 + y_0^2) - \frac{v_{x0}^2 + v_{y0}^2}{c_T^2}}\, \delta(v_{z0} - v_0). \tag{5.81a}$$

Cartesian coordinates are used because the equations of motion (5.62) remain linear when $v_\theta \neq 0$

$$x'' + \Omega^2 x = 0; \quad y'' + \Omega^2 y = 0. \tag{5.81b}$$

Substituting equations for x_0 and y_0 analogous to (5.64b) into (5.81a), we find

$$f_0 = \frac{n_0}{\pi c_T^2}\, \delta(v_{z0} - v_0)\, e^{-s(v_x - xa)^2 - s(v_y - ya)^2 - \frac{x^2 + y^2}{q^2 c_T^2 s}}, \tag{5.81c}$$

where s is given by Eq. (5.78b) with $c_T \equiv C_2$. Integrating (5.81c) over v_x and v_y we find the density distribution

$$n = \frac{n_0}{\left(R_1^2 + \frac{c_T^2}{v_0^2 q^2} R_2^2\right)} \exp\left[-\frac{r^2}{q^2\left(R_1^2 + \frac{c_T^2}{v_0^2 q^2} R_2^2\right)}\right]. \tag{5.81d}$$

In the image plane at $z = z^*$, where $R_2(z^*) = 0$, the density distribution is independent of c_T and v_0.

3. **Influence of T_e on the Focusing of a Neutralized Ion Beam in a Multiply Connected Multiple Lens Accelerator.** Taking the density distributions of the neutralized ion beam found in the preceding subsection as an example we now use the methods developed above to evaluate thermal effects in focusing. We examine the effect of T_e on the multiply connected, multiple lens accelerator.

The theory for ion focusing in the two-dimensional model with $T_e = 0$ reduces to Eq. (3.87c):

$$\left.\begin{aligned}
\ddot{y} + \Omega^2\, y = 0; \\
\Omega^2 = \frac{1}{2}\,\frac{e^2\, H_0^2}{M^2\, c^2} + \frac{1}{16}\,\frac{e^2\, E^2}{M^2\, v^2}\,.
\end{aligned}\right\} \tag{3.82}$$

In the derivation of (5.82) it has been assumed that $\varphi = \varphi(\gamma)$. If $T_e \neq 0$, it is necessary to take account of the electric field component, which repels the beam.

Using the solutions of the equations of motion found in Chap. 3 we can construct the particle distribution function in the single-particle approximation, f_0, and use the method described above to calculate the corrections to this distribution by means of successive approximations.

However, we will not use this approach; instead we introduce an a priori specification of the particle density distribution in the beam:

$$n\,(y,\, z) = n_0\,(z)\,\exp\left(-\frac{y^2}{\delta^2(z)}\right), \tag{5.83}$$

where the z dependence is assumed to be weak.

Substituting (5.83) into the equation for the thermalized potential, we find

$$\varphi = \varphi^*\,(\gamma) - \frac{kT_e}{e}\,\frac{y^2}{\delta} + \frac{kT_e}{e}\,\ln\frac{n_0\,(z)}{n_{00}}\,. \tag{5.84}$$

Using the function $\varphi^*(\gamma)$ introduced earlier and substituting (5.84) into the equations for the averaging method, (3.83), we find that (5.82) is replaced by the following equation for the transverse oscillations:

$$\ddot{y} + \left(\Omega^2 - \frac{2kT_e}{M} \frac{1}{\delta^2} \right) y = 0. \tag{5.85}$$

Thus, at high electron temperatures the beam becomes unstable and no focusing occurs. A similar result can also be found, through the use of (5.81), for an axial lens.

§4. Electron Sheath of a Neutralized Ion Beam Detached from the Walls

1. General Discussion. The most attractive application of plasma optics is in the production of a neutralized ion beam which is detached from the walls. It seems likely that such beams can be achieved, although it may be difficult to control the beam and stabilize it by means of electrodes at the channel walls when there are only electrons between the beam and the walls. Without going into the stability question here, we will construct models for such detached equilibrium configurations which are axisymmetric. We must first investigate the electron sheath of the neutralized ion beam.

We consider an axial plasma lens and assume that the vacuum in the chamber is so high that ionization in the beam can be neglected. It follows from Eq. (5.1) for the thermalized potential that in the beam

$$n = n_0 \exp\left[-\frac{e}{kT_e} \left(\varphi^*(\psi) - \varphi \right) \right] = \nu(\psi) \exp \frac{e\varphi}{kT_e}. \tag{5.86}$$

However, Eq. (5.86) is not valid for the particle density in the electron shell of the neutralized ion beam because this equation only holds for electrons which are in thermodynamic equilibrium with each other inside a narrow magnetic tube.

Since a wide variety of potential differences can be established between the beam and the wall, the electron sheath is generally not in equilibrium.

In order to analyze the electron sheath (i.e., the Φ and n_e distributions) it is necessary to have an extremely accurate knowledge of the electron distribution function in the beam volume. Here "ex-

tremely accurate" means that it is important to know the behavior
of the tails of the distribution, since these tails determine the sheath
structure. Generally speaking, an accurate calculation of f_e under
these conditions is extremely complicated since the inevitable oscillations can affect the result. We will therefore simply assume
that the electron distribution is known, say from experiment.

Under typical conditions, the motion of an electron in the beam
and in its transit to the walls can be assumed to be collisionless.
Furthermore, since we assume a strong magnetic field in the system, the electron motion is a drift motion.

In this case two quantities are conserved: the energy and the
transverse adiabatic invariant:

$$\left.\begin{aligned}
\frac{m}{2}\,(v_\parallel^2 + v_\perp^2) - e\varphi = \mathcal{E} = \text{const;} \\
v_\perp^2 / H = I = \text{const.}
\end{aligned}\right\}\qquad (5.87)$$

Here $v_\parallel$ and $v_\perp$ are the components of the electron velocity respectively along and across the field.

Using the subscript "0" to denote the value of a quantity at the
surface of the neutralized ion beam, we find from (5.87) that the
longitudinal velocity at any point on a line of force is

$$v_\parallel^2 = v_{0\parallel}^2 + \left(1 - \frac{H}{H_0}\right) v_{0\perp}^2 - \frac{2e}{m}\,(\varphi_0 - \varphi). \qquad (5.88)$$

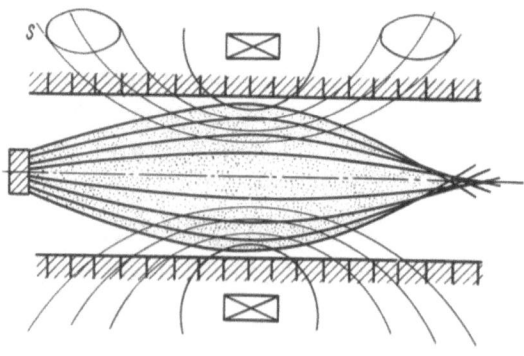

Fig. 34

Knowing the variation of $v_\parallel$ along the coordinates, we can calculate the density of those electrons between the beam and the wall which have escaped from the beam. For this purpose we isolate a narrow magnetic tube around the line of force under consideration (Fig. 34) and examine the electrons whose initial velocities lie in the intervals $(v_{0\parallel}, v_{0\parallel} + dv_{0\parallel})$, $(v_{0\perp}, v_{0\perp} + dv_{0\perp})$.

We first assume that the electrons of this group reach the wall, so that $v_\parallel > 0$ everywhere. By virtue of the continuity equation we then write

$$dn_{0k} S_0 v_{0\parallel} = dn_k S v_\parallel. \qquad (5.89a)$$

Here S is the cross-sectional area of the magnetic tube, and dn_k is the density of electrons of the given group which has emerged from the beam (subscript "k"). Introducing the distribution function $F_0(v_{0\parallel}, v_{0\perp})$ for the electrons near the beam surface,

$$dn_{0k}(v_{0\parallel}, v_{0\perp}) = F_0(v_{0\parallel}, v_{0\perp}) dv_{0\parallel} dv_{0\perp}, \qquad (5.89b)$$

and using the conservation of magnetic flux,

$$S_0 H_0 = SH,$$

we find

$$dn_k = \frac{H}{H_0} \frac{(F_0 dv_{0\parallel} dv_{0\perp}) v_{0\parallel}}{\sqrt{v_{0\parallel}^2 + \left(1 - \frac{H}{H_0}\right) v_{0\perp}^2 - \frac{2e}{m}(\varphi_0 - \varphi)}}. \qquad (5.90a)$$

If, on the other hand, the denominator becomes imaginary at some point along the magnetic line of force, this group of particles is reflected from an effective potential barrier (the wall potential is $\varphi = 0$):

$$\left.\begin{aligned}
\chi &= -\left[v_{0\perp}^2\left(1 - \frac{H}{H_0}\right) - \frac{2e}{M}(\varphi_0 - \varphi)\right]; \\
0 &< \chi < \chi_{max} \equiv \frac{2e\varphi_0}{M} - v_{0\perp}^2\left(1 - \frac{H_{wa}}{H_0}\right).
\end{aligned}\right\} \qquad (5.90b)$$

As a result

$$dn_k = \begin{cases} 0, & \text{if } v_{0\parallel}^2 - \chi < 0; \\ \dfrac{2H}{H_0} \dfrac{v_{0\parallel} dn_0}{\sqrt{v_{0\parallel}^2 - \chi}}, & \text{if } \chi_{max} > v_{0\parallel}^2 > \chi. \end{cases} \qquad (5.91)$$

We can thus write the following equation for the electron density within the magnetic tube under consideration:

$$n_e = 2 \frac{H}{H_0} \int_{\Sigma_1} \int_{-\Sigma} \frac{F_0(v_{0\parallel}, v_{0\perp})\, v_{0\parallel}\, dv_{0\parallel}\, dv_{0\perp}}{\sqrt{v_{0\parallel}^2 + \left(1 - \frac{H}{H_0}\right) v_{0\perp}^2 - \frac{2e}{m}(\varphi_0 - \varphi)}} +$$

$$+ \frac{H}{H_0} \int_{\Sigma_2} \int \frac{F_0(v_{0\parallel}, v_{0\perp})\, v_{0\parallel}\, dv_{0\parallel}\, dv_{0\perp}}{\sqrt{v_{0\parallel}^2 + \left(1 - \frac{H}{H_0}\right) v_{0\perp}^2 - \frac{2e}{m}(\varphi_0 - \varphi)}} + n_{\text{wa}}. \qquad (5.92)$$

The first integral determines the density of electrons which have been reflected between the given point on the line of force (the region of parameters Σ) and the wall (Σ_1), while the second integral determines the density of electrons which are not reflected and which strike the wall. Here n_{wa} is the density of electrons emitted by the wall.

The range of integration in (5.92) can be refined by examining the following curves in the $(v_{0\perp}, v_{0\parallel})$ plane:

$$Y(v_{0\perp}, v_{0\parallel}) = v_{0\parallel}^2 + \left(1 - \frac{H}{H_0}\right) v_{0\perp}^2 - \frac{2e}{m}(\varphi_0 - \varphi) = 0. \qquad (5.93)$$

Depending on the sign of the quantity $(1 - H/H_0)$, these curves will be either ellipses ($H/H_0 < 1$) or hyperbolas ($H/H_0 > 1$). In the first case, an electron moving away from the beam toward the wall along a magnetic line of force enters a region in which the magnetic field becomes progressively weaker; in the second case this electron moves into an increasing magnetic field. The first case arises in an axial lens, while the second arises in a multiply connected, multiple lens accelerator and in a split annular lens.

Knowing n_e, we can write equations for the potential distribution in the electron sheath,

$$\Delta\varphi = 4\pi e n_e, \qquad (5.94)$$

and in the beam volume,

$$\Delta\varphi = 4\pi e (n_e - n_i). \qquad (5.95)$$

Here n_e can be written in Boltzmann form and n_i can be taken from the single-particle model.

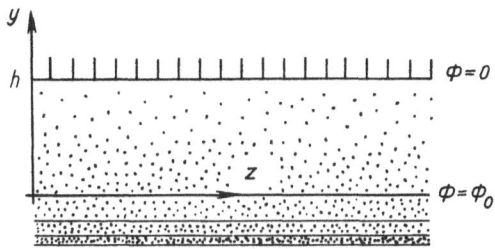

Fig. 35

The equations found here that contain the function $\varphi(r, z)$ must be supplemented with boundary conditions at the wall.

We have assumed that the ionization of neutrals in the beam volume is negligible.[†] If this is the case the electrons will gradually escape from the beam because of their finite temperature, leaving the beam with a net positive charge. The optical properties of the system will change as a result.

To prevent the loss of electrons from the neutralized ion beam we must use an emitting wall. To carry out calculations for the electron sheath in the case of wall emission we must know: 1) the potential distribution at the wall, 2) the emission law at the wall (e.g., E = 0 for thermionic emission), and 3) the condition under which the electron flux from the beam to the wall is equal to the flux in the opposite direction.

2. **Plane Model for the Electron Shell.** Even if the electron distribution in the beam, $F_0(v_{0\parallel}, v_{0\perp})$, is known the equations for the electron sheath around the beam are extremely complicated. We therefore use a simple one-dimensional model.

It is assumed that the beam surface is two-dimensional, that the magnetic field is homogeneous and perpendicular to the plane of the beam, and that the emitting wall is parallel to the beam and separated from it by a distance h (Fig. 35).

[†] This assumption actually means that we consider n_e but neglect n_i in the gap between the beam and the wall.

In this case the electron density in the shell is [see Eq. (5.91)]

$$n_e \equiv n_k + n_{wa} = n_{wa} + 2 \int\limits_{\frac{2e}{m}(\varphi_0 - \varphi) < v_0^2 < \frac{2e\varphi}{m}} \frac{F_0(v_0)\,v_0\,dv_0}{\sqrt{v_0^2 - \frac{2e}{m}(\varphi_0 - \varphi)}} +$$

$$+ \int\limits_{v_0^2 > \frac{2e\varphi_0}{m}} \frac{F_0(v_0)\,v_0\,dv_0}{\sqrt{v_0^2 - \frac{2e}{m}(\varphi_0 - \varphi)}}; \qquad (5.96a)$$

$$\varphi(0) = \varphi_0; \quad \varphi(h) = 0, \qquad (5.96b)$$

where n_{wa} is the density of electrons emitted by the wall. Here $F(v_0)$ is the electron distribution with respect to velocity component $v_{0\parallel}$, since the transverse component, $v_{0\perp}$, does not affect the particle density distribution, by virtue of our assumption of a homogeneous magnetic field. In writing (5.96) we have assumed the beam potential to be φ_0 and the wall potential to be zero.

If the function $F_0(v_0)$ is strictly a Maxwellian at y = 0, i.e.,

$$F_0(v_0) = \frac{n_0}{\sqrt{\pi}}\left(\frac{m}{2kT_e}\right)^{1/2} \exp\left(-\frac{mv^2}{2kT_e}\right), \quad v_0 > 0, \qquad (5.97)$$

the electron density n_k emitted by the beam in (5.96) becomes

$$n_k = \frac{n_0}{2}\exp\left(-\frac{e(\varphi_0 - \varphi)}{kT_e}\right)\left[1 + \frac{2}{\sqrt{\pi}} \int\limits_0^{\sqrt{\frac{e\varphi}{kT_e}}} \exp(-\alpha^2\,d\alpha)\right]. \qquad (5.98)$$

In the gap 0 < y < h, where

$$e\varphi/kT_e \gg 1, \qquad (5.99a)$$

the n_k distribution is described more accurately as an equilibrium (Boltzmann) distribution,

$$n_k \approx n_0 \exp\left(-\frac{\varphi_0 - \varphi}{kT_e}\right). \qquad (5.99b)$$

Near the wall, however (the limit $\varphi \to 0$), the electron density n_k is half that of an equilibrium distribution. This is understandable physically because in the limit y → 0 electrons move away from the beam and toward the beam, whereas when y → h the electrons move toward the wall only.

Up to this point we have been considering n_k, the electrons emitted by the beam. However, the electron shell also contains electrons emitted by the wall (n_{wa}).

The Langmuir model is used to describe the emission, so that the wall has an unlimited emissivity; hence the following condition holds at $y = h$:

$$\frac{d\varphi}{dy}\bigg|_{y=h} = 0. \tag{5.100}$$

Furthermore, we assume the electrons emitted by the wall to be cold, so that

$$j_{wa} = en_{wa}\sqrt{\frac{2e\varphi}{m}}. \tag{5.101a}$$

Making use of the conservation of j_{wa} from (5.101a) we find an equation for the density n_{wa} in the electron sheath:

$$n_{wa} = \frac{j_{wa}}{e\sqrt{\frac{2e\varphi}{m}}}. \tag{5.101b}$$

Using (5.98) and (5.101b), we find an equation for

$$\frac{d^2\varphi}{dy^2} = 4\pi e(n_k + n_{wa}) = 4\pi\left\{\frac{j_{wa}}{\sqrt{\frac{2e\varphi}{m}}} + e\frac{n_0}{2}\times\right.$$

$$\times \exp\left(-\frac{e(\varphi_0 - \varphi)}{kT_e}\right)\left[1 + \frac{2}{\sqrt{\pi}}\int_0^{\sqrt{\frac{e\varphi}{kT_e}}} \exp(-\alpha^2)d\alpha\right]\bigg\}. \tag{5.102}$$

This equation contains the two unknown constants j_{wa} and φ_0 as well as two constants which are governed by the beam and which are assumed to be known: n_0 and T_e. To obtain a unique solution of Eq. (5.102) we require four boundary conditions; two are related to the fact that the equations contain a second derivative, and two are related to the uncertainty regarding j_{wa} and φ_0. At this point we have the three conditions in (5.96) and (5.100). The fourth condition is the condition that the electron flux from the beam to the wall be equal to that in the reverse direction:

$$j_{wa} = e \int_{\sqrt{\frac{2e\varphi_0}{m}}}^{\infty} v_0 F(v_0) \, dv_0 = \frac{en_0}{2\sqrt{\pi}} \left(\frac{2kT_e}{m} \right)^{1/2} \exp\left(-\frac{e\varphi_0}{kT_e} \right). \qquad (5.103)$$

Substituting (5.103) into (5.102) and introducing the dimensionless quantities

$$\eta^2 = y^2 \frac{2\sqrt{\pi}e^2 n_0}{kT_e} \exp(-\chi_0); \quad \chi_0 \equiv \frac{e\varphi_0}{kT_e}; \quad \chi = \frac{e\varphi}{kT_e}, \qquad (5.104)$$

we can convert (5.102) to the more compact form

$$\frac{d^2\chi}{d\eta^2} = \frac{1}{\sqrt{\chi}} + \exp(\chi)\left(\sqrt{\pi} + 2 \int_0^{\sqrt{\chi}} \exp(-\alpha^2) \, d\alpha \right). \qquad (5.105)$$

Multiplying both sides by $d\chi/d\eta$, integrating, and using boundary conditions (5.96b) and (5.100), we find

$$\frac{1}{2} \left(\frac{d\chi}{d\eta} \right)^2 = \sqrt{\pi} \exp(\chi) + 2 \exp(\chi) \int_0^{\sqrt{\chi}} \exp(-\alpha^2) \, d\alpha - \sqrt{\pi}. \qquad (5.106)$$

Hence we have χ as a function of η:

$$\frac{1}{\pi^{1/4}} \int_0^{\chi} \frac{d\chi \exp(-\chi/2)}{\sqrt{\left[1 + \frac{2}{\sqrt{\pi}} \int_0^{\sqrt{\chi}} \exp(-\alpha^2) \, d\alpha \right] - e^{-\chi}}} = \sqrt{2}\eta. \qquad (5.107)$$

Setting $y = h$, we can then relate χ_0 or, equivalently, φ_0, to the parameters of the system:

$$G(\chi_0) = \exp\left(\frac{\chi_0}{2} \right) \int_0^{\chi_0} \frac{d\chi \exp\left(-\frac{\chi}{2} \right)}{\sqrt{\left[1 + \frac{2}{\sqrt{\pi}} \int_0^{\sqrt{\chi}} \exp(-\alpha^2) \, d\alpha \right] - e^{-\chi}}} =$$

$$= h \sqrt{\frac{4\pi e^2 n_0}{kT_e}}. \qquad (5.108)$$

The calculations show that this integral behaves in the following way in the limits $\chi \to 0$ and $\chi \to \infty$:

$$Z(\chi) \equiv \int_0^\chi \frac{d\chi \exp(-\chi/2)}{\sqrt{1 + \dfrac{2}{\sqrt{\pi}} \displaystyle\int_0^{\sqrt{\chi}} \exp(-\alpha^2)\, d\alpha - \exp(-\chi)}} =$$

$$= \begin{cases} Z(\infty) - \dfrac{1}{\sqrt{2}}\left\{ 2\exp(-\chi/2) + \dfrac{1}{6}\exp(-3/2\chi) + \ldots \right\}; & \chi \to \infty; \\[2ex] \sqrt[4]{\dfrac{\pi}{4}}\left(\dfrac{4}{3}\chi^{3/4} - \dfrac{\sqrt{\pi}}{5}\chi^{5/4} + \ldots \right); & \chi \to 0. \end{cases} \qquad (5.109)$$

Knowing the function $Z(\chi)$, we can now determine the potential drop for the known values of n_0, h, and T_e.

Turning to the most interesting case, in which h is large in comparison with the Debye length,

$$D = \left(\sqrt{\frac{4\pi e^2 n_0}{kT_e}} \right)^{-1}, \qquad (5.110)$$

we can write $Z(\chi) \approx Z(\infty) \approx 2$ by virtue of (5.109). Substituting this value of $Z(\chi)$ into (5.108), we find the required expression for the potential difference between the beam and the wall:

$$\varphi_0 \simeq \frac{kT_e}{e} \ln \frac{h}{D\sqrt{2}}. \qquad (5.111)$$

The logarithmic dependence of φ_0 on the ratio h/D is obviously a consequence of our assumption that F_0 is a Maxwellian. Determining the potential drop over the electron sheath, we can use the known potential distribution at the wall to find the potential distribution at the beam surface and, thus, to calculate the field in the beam volume.†

---

†Actually, we must solve Eq. (5.95) in the beam volume in order to match the non-neutral region near the beam surface to the neutral region in the beam volume. However, the potential drop in this region is small-of order kT_e/e (see §5, subsection 2).

§5. Grazing a Neutralized Ion Beam in an Axial
Lens

1. Wall Conductivity. By "wall conductivity" here
we mean the conductivity due to the scattering of electrons from
the walls. The basic equation of the wall conductivity in a strong
magnetic field can be written [39]

$$i_s = \lambda E \equiv \left(\frac{\varkappa n_t}{H^2} \right) E. \qquad (5.112)$$

Here, i_s is the current near the wall (in a layer with thickness of
the order of the electron gyroradius), expressed per centimeter in
the direction perpendicular to E, n_t is the plasma density near the
wall, H is the magnetic field, and $\varkappa$ is a coefficient which is governed
by the wall roughness, the electron temperature, and the angle between the magnetic lines of force and the normal to the wall.

We consider the equilibrium configuration of a neutralized
ion beam in a "primitive" axial lens: this is a dielectric tube of
radius a with a rough surface (Fig. 36). At the surface of the tube
at $z = z_0$ and $z = z_1$ there are two narrow rings, between which a
potential difference is applied. Let ψ_0 be the magnetic surface
that passes through ring z_0 and assume that $\varphi^* = 0$ for $\psi < \psi^*$.
We assume that all the ions in the system have come from the
source and that the ion motion is collisionless; once they strike
the wall, the ions are lost.

Neutrality is assumed throughout the tube for $z < z_1$; for $z > z_1$,
we assume that the beam is surrounded by an electron shell. In

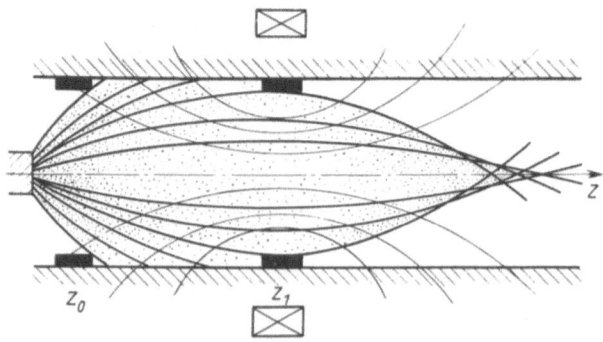

Fig. 36

the region $z > z_1$ we can carry out calculations by the method described in the preceding section, noting that the distribution $\varphi^*(\psi)$ is governed by the wall conductivity; in the steady state, the wall must be charged in such a way that there is no electron flux from the beam to the wall. This problem only has a rigorous solution if the electron energy in the neutralized ion beam is cut off at some high value.

We are then interested in the calculation for the region $z_0 < z < z_1$. To avoid complicating the problem we assume that T_e is low, so that thermalization in the beam volume can be neglected; thus, the ion motion is treated in the single-particle approximation. The equilibrium is then governed by the following factors: (1) the magnetic field configuration, $\psi(r, z)$; (2) the angular distribution of particles emerging from the source,

$$dN = \nu(\alpha)d\alpha; \tag{5.113}$$

(3) the wall-conductivity parameter $\varkappa(z)$; (4) the applied potential difference U_p; and (5) the geometric dimensions, a, z_0, and z_1.

If ψ, $\nu(\alpha)$, and $\varkappa(z)$ are given, the calculation reduces to a determination of the function $\varphi(\psi)$. To find the appropriate equation, we first write the charge conservation law,

$$j_{ir} - j_{er} = \partial i_s/\partial z; \quad i_s = \varkappa n_i E/H^2. \tag{5.114}$$

Here j_{ir} is the ion current density to the wall and j_{er} is the modulus of the electron current density to the wall.

Since the potential drop near the wall is proportional to T_e and is thus small, the change in the radial component of the ion velocity in this potential drop can be neglected. Then if the total velocity v_0 of the ions incident on the wall and the angle $(\pi/2 - \beta)$ of incidence are known (we neglect the ion velocity spread), the particle density near the wall can be written

$$n_t = j_{ir}/v_0 e \sin \beta. \tag{5.115}$$

Under these conditions, we have [see Eqs. (5.114) and (5.115)]

$$j_{ir} - j_{er} = \frac{\partial}{\partial z} \left(\frac{\varkappa(z)}{v_0 e} \frac{E j_{ir}}{H^2 \sin \beta} \right). \tag{5.116}$$

We assume that the wall conductivity of the channel is negligible. Since the number of electrons within a given magnetic tube is fixed, $j_{er} = 0$. Then, Eq. (5.116) becomes

$$j_{ir} = \frac{\partial}{\partial z} \left(\frac{\varkappa}{v_0 \sin \beta} \frac{E j_{ir}}{H^2 e} \right).$$

(5.117)

Let us relate $j_{ir}(z)$ and $\beta(z)$ to the angular distribution of particles $\nu(\alpha)$ from the source [see Eq. (5.113)]. Restricting the discussion to the simplest case, a narrow beam moving through an electrostatic plasma lens, we can write

$$\varphi = k\psi \approx r^2 \varphi_1(z)/2; \quad \varphi_1(z) \equiv kH_0(z).$$

(5.118)

The solution of the equation of motion

$$\frac{d^2 r}{dz^2} + \frac{e}{Mv_0^2} \varphi_1(z) r = 0$$

(5.119)

with the initial conditions $r\big|_{z=0} = 0$, $\dfrac{dr}{dz}\big|_{z=0} = \alpha$ is†

$$r = \alpha R_2(z).$$

(5.120)

Hence we can relate the angle α at which the particle emerges from the source to the coordinate z of the point at which the particle collides with the tube:

$$\alpha(z) = a / R_2(z).$$

(5.121)

If the system is capable of focusing, the function $R_2(z)$ has a maximum at some $z = z_1$. Accordingly, there exists a minimum α_{min} for which the particles can reach the wall; if $\alpha < \alpha_{min}$, they traverse the entire channel. Knowing $\alpha(z)$ and the angular distribution of particle emerging from the source, $\nu(\alpha)$, we can find the flux density of ions to the wall. Obviously

$$j_{ir} = e \frac{v_0(\alpha)\, d\alpha}{2\pi a dz} = e \frac{v_0(\alpha)}{2\pi} \frac{d}{dz} \frac{1}{R_2(z)}.$$

(5.122)

The angle at which the ions strike the wall is governed by

$$\tan \beta = \frac{dr}{dz}\bigg|_{r=a}.$$

---

†In contrast with the earlier discussion [see Eqs. (5.63) and (5.64)], we assume that the function $R_2(z)$ is dimensional.

Substituting (5.121) into this equation, we find

$$\tan \beta = \alpha \frac{dR_2}{dz} = \frac{a}{R_2} \frac{dR_2}{dz}.$$ (5.123)

If $\beta \to 0$, then $\tan \beta \approx \sin \beta$. In this case, as follows from Eqs. (5.115), (5.122), and (5.123), the ion density near the wall is

$$n_r = \frac{j_{ir}}{e \, v_0 \sin \beta} = -\frac{v_0(\alpha)}{2\pi v_0 a} \frac{1}{R_2(z)}.$$ (5.124a)

These equations have a singularity at $z = 0$ because it has been assumed that β is small in the derivation of (5.124). In the opposite case we find

$$n_r = -\frac{v_0(\alpha)}{2\pi v_0 R} \frac{\sqrt{R_2^2 + a^2 \left(\frac{dR_2}{dz} \right)^2}}{R_2^2 + a^2}.$$ (5.124b)

This expression is regular at $z = 0$. Substituting (5.124) and (5.122) into (5.117) we find the basic equation for the lens in the wall–conductivity regime:

$$\frac{e \, v}{R_2^2} \frac{dR_2}{dz} = -\frac{\partial}{\partial z} \varkappa \left(\frac{\partial \varphi}{\partial z} \right) \frac{1}{H^2} \frac{v}{v_0 a} \frac{1}{R_2}.$$ (5.125)

This equation must be supplemented with Eqs. (5.118) and (5.119), which relate H and φ to R_2, and which can be written

$$\left. \left(\frac{\partial \varphi}{\partial z} \right) \right|_{r=a} = -\frac{a^2}{2e} M v_0^2 \frac{\partial}{\partial z} \frac{R_2''}{R_2};$$
$$H = -\frac{1}{k} \frac{M v_0^2}{e} \frac{R_2''}{R}.$$ (5.126)

Equations (5.125), (5.126) show that we have an explicit equation for $\varkappa(z)$, a linear differential equation for $v(\alpha)$, and a fourth–order differential equation for $R_2(z)$, with the other quantities known. In the simplest case, in which $v(\alpha)$ and $R_2(z)$ are given, we can determine completely the equilibrium configuration of the beam, finding $\varkappa$ from (5.125) and (5.126).

2. Emitting Wall Case. We now consider a different model for the axial lens. It is assumed that the beam moves within a multisection tube, whose inner surface has an unlimited emis-

sivitiy. This sectioning makes it possible to specify an arbitrary potential distribution $\varphi_t(z)$ on the tube surface. However, because of the potential drop which unavoidably occurs near the wall, the potential distribution in the beam volume differs from that which follows formally from the condition $\varphi^* = \varphi^*(\psi)$.

The potential drop near the wall appears because the wall and the plasma volume are not in thermodynamic equilibrium, so that the electron inertia plays an important role (because of the macroscopic electron flux).

Assume that at $z < z_1$ the ion beam grazes the wall; treating the wall as smooth we neglect the wall conductivity† as well as the volume conductivity. Then the electron-balance equation in the neutralized ion beam is

$$-j_{es} + j_{er} = 0. \tag{5.127}$$

Here j_{es} is the flux of electrons emitted by the wall that enters the beam volume and j_{er} is the electron flux from the beam to the wall.

As before, we assume that the ion beam has no velocity spread and that the change in the ion radial velocity $v_\perp$ within the potential drop can be neglected. Then the potential distribution near the wall satisfies

$$d^2\varphi/dy^2 = -4\pi e\,(n_i - n_{er} - n_{es}). \tag{5.128}$$

Here, the ion density is $n_i \equiv j_{ir}/v_\perp = \text{const}$ and $y = a - r$ is the distance reckoned from the wall into the beam. The left side of (5.128) is the one-dimensional Laplacian, since this wall layer is thin.

For convenience in the calculations we assume that the wall potential vanishes at a selected point. Following the Langmuir scheme, (5.101), we write

$$n_{es} = \frac{|j_{es}|}{e\sqrt{\dfrac{2e\varphi}{m}}}; \quad |j_{es}| \equiv j_0. \tag{5.129}$$

---

† This assumption is reasonable since the bulk of the electrons incident on the wall are reflected within a Debye layer, which can be very "smooth."

Assuming that the electrons in the beam volume are in equilibrium, we can approximate density distribution (5.98) by the simple function

$$n = n_0 \exp\left(-e\,\frac{\varphi_0 - \varphi}{kT}\right). \tag{5.130a}$$

Here φ_0 is the potential of the neutralized ion beam near the wall (up to the wall boundary layer). Then [see Eqs. (5.103) and (5.127)]

$$j_{er} = \frac{e n_0 \mathrm{v}}{2\sqrt{\pi}} \exp\left(-\frac{e\varphi_0}{kT_e}\right) = j_0; \quad \mathrm{v}^2 = \frac{2kT_e}{m}. \tag{5.130b}$$

Now Eq. (5.128) becomes

$$\frac{d^2\chi}{d\zeta^2} = -\left(1 - \frac{b}{\sqrt{\chi}} - 2\sqrt{\pi}be^\chi\right), \tag{5.131a}$$

where

$$\chi \equiv \frac{e\varphi}{kT_e}; \quad b \equiv \frac{j_0}{en_{0i}\mathrm{v}}; \quad \zeta^2 = \frac{4\pi e^2 n_{0i}}{kT_e}\,y^2. \tag{5.131b}$$

We have thus derived an equation which depends on a single parameter. Integrating (5.131) and using the Langmuir boundary conditions at $y = 0$,

$$\chi = 0; \quad d\chi/d\zeta = 0, \tag{5.132}$$

we find

$$\frac{1}{2}\left(\frac{d\chi}{d\zeta}\right)^2 = 2b\sqrt{\chi} + 2\sqrt{\pi}b\,(e^\chi - 1) - \chi \equiv Y(\chi). \tag{5.133}$$

The value of b can be determined as follows: For sufficiently small values of b the function $Y(\chi)$ is not monotonic (Fig. 37). If we require that the potential increase monotonically, tending toward a finite value in the limit $\zeta \to \infty$, we must choose the parameter b to be that value b* at which the curve $Y(\chi)$ touches the χ axis at the point χ^*. We emphasize, however, that although the choice b = b* is natural, we must bear in mind the fact that bounded solutions are also possible when b < b*. However, these other solutions correspond to oscillating potential distributions which go to infinity. Ac-

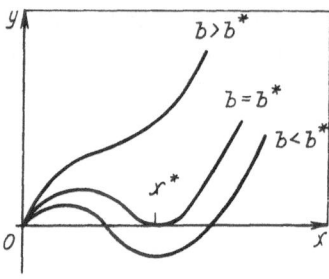

Fig. 37

cordingly, without other information we conclude that this static model can correspond to a broad class of models that satisfy the single indisputable assumption that φ remain bounded as $\chi \to \infty$. If b is to be single-valued, we must analyze the stability of the solution, but this analysis goes beyond the scope of the present paper.

We now consider the case $b = b^*$. In addition to the boundary conditions at $\zeta = 0$, in this case the following conditions must be satisfied, since χ tends toward χ^* in the limit $\zeta \to \infty$:

$$\frac{d\chi}{d\zeta}\bigg|_{\chi \to \chi} \to 0; \quad \frac{d^2\chi}{d\zeta^2}\bigg|_{\chi \to \chi^*} \to 0. \tag{5.134}$$

Hence we can determine both b^* and the quantity χ^*, the potential at infinity. From (5.131), (5.133), and (5.134) we obtain the equations

$$\left.\begin{array}{l} b^*\left[2\sqrt{\chi^*} + 2\sqrt{\pi}\left(e^{\chi^*} - 1\right)\right] = \chi^*; \\[2mm] b^*\left[\dfrac{1}{\sqrt{\chi^*}} + 2\sqrt{\pi}e^{\chi^*}\right] = 1. \end{array}\right\} \tag{5.135a}$$

Using these equations and the definition of b, we see that the potential drop φ_0 introduced in (5.130a) is related to χ^* by $e\varphi_0/kT_e = \chi^*$. Eliminating b^* from system (5.135a), we find an equation for χ^*:

$$\sqrt{\chi^*} + 2\sqrt{\pi}\left(e^{\chi^*} - 1\right) = 2\sqrt{\pi}e^{\chi^*}\chi^*. \tag{5.135b}$$

This equation has the two roots $\chi^* \approx 0$ and $\chi^* \approx 0.5$; for the present purposes, the second root is the interesting one. Accordingly, under these assumptions the potential drop is

$$\varphi_0 \approx 0.5kT_e/e. \tag{5.136}$$

Knowing $\chi^* \approx 0.5$, we can find b^* [see Eq. (5.135a)]:

$$b^* = \left(\frac{1}{\sqrt{\chi^*}} + 2\sqrt{\pi} \exp(\chi^*) \right)^{-1} \approx 0.14 \ .$$

Also, using (5.130b) and (5.135b) we find the emission current from the wall to be

$$j_{es} = e n_0 \mathbf{v} \cdot 0.14 = 0.14 \, e n_0 \, v_\perp \frac{\mathbf{v}}{v_\perp} \ ; \quad j_{es} = j_{i\perp} \left(0.14 \frac{\mathbf{v}}{v_\perp} \right). \qquad (5.137)$$

With $v \sim 10^8$ cm/sec and $v_\perp \sim 10^6$ cm/sec, the electron current from the emitter is nearly 14 times the ion current to the emitter.

Knowing the potential drop φ_0 near the wall, we can finally calculate the plasma configuration.

Specifically, if $\varphi_t(z)$ is the potential distribution near the wall, then

$$\varphi^*(\psi) = \varphi_t[z(\psi)] + \varphi_0[T_e(\psi)]. \qquad (5.138)$$

It follows in particular that if T_e = const then $\varphi_0^*(\psi)$ only differs from $\varphi_t[z(\psi)]$ by an inconsequential constant.

There may be an objection to the replacement of the density distribution in (5.98) by the simple function in (5.130a) in this calculation of the potential drop, especially since χ^* turns out to be 0.5 (so that it is clearly incorrect to treat the electron exchange between the beam and the wall as weak). Under these conditions, the assumption that the electrons in the beam volume are in equilibrium is not valid.

However, if we assume the electrons in the beam volume to be in equilibrium,[†] we can calculate the potential distribution in the layer, using (5.98). The calculations show that the values of χ^* and b^* found in this way are essentially the same as those given above.

[†] In this case the potential distribution is governed by the "bulk" part of the distribution, rather than by the tail.

§6. Magnetic Fields Produced by the Currents Flowing in a Multiply Connected, Multiple-Lens Accelerator

In carrying out calculations for the plasma-optical system up to this point we have neglected the magnetic fields produced by the currents that flow in the volume of the neutralized ion beam. As a rule, this simplification is valid if the beam momentum is much smaller than the magnetic pressure, i.e., if

$$\rho v^2 \ll (H^2/8\pi). \tag{5.139}$$

There are plasma-optical systems, however, in which the magnetic fields produced by the currents are the only magnetic fields. A situation of this type arises in a pulsed coaxial accelerator in which a stationary plasma focus exists [42]. We have not treated devices of this type in the present paper, so that the present discussion is limited to an analysis of the magnetic fields produced by the currents in plasma-optical systems with an external magnetic field.

As an example we consider the magnetic field produced by the current that flows in the volume of the neutralized ion beam under the conditions that prevail in a multiply connected, multiple-lens accelerator. In this accelerator, as in other axisymmetric devices, the magnetic field produced by the current has three components which are conveniently resolved into azimuthal and meridional fields. The azimuthal magnetic field is produced by the longitudinal current of ions and electrons. Let us assume $I \approx 10^3$ A and a beam diameter of 30 cm; then $H_\theta \approx 7$ Oe. This field is obviously very small in comparison with the magnetic field in the lenses, H_0, which is of the order of kiloersteds.

The azimuthal drift currents produce much stronger magnetic fields H_{cur}. Let us examine these fields in more detail in the two-dimensional model, assuming that the fields produced by the azimuthal currents can be calculated with the boundary condition

$$|H_{cur}| \to 0 \quad \text{as} \quad y \to \pm \infty. \tag{5.140}$$

Using $E = E_0(H/H_0)^2$ and $H = H_0 \sin \alpha z$, and neglecting the ion component of the current, we find

$$j_x = enc\, E/H = enc\, (E_0/H_0) \sin \alpha z. \tag{5.141}$$

Assuming $n = n_0(y)$, we can write an equation for the magnetic stream functions of the magnetic field produced by the neutralized ion beam:

$$\Delta \psi_{cur} = -\frac{4\pi}{c} j_x = -4\pi e n_0(y) \frac{E_0}{H_0} \sin \alpha z.$$

The solution of this equation, which is symmetric in y_1, is

$$\psi_{cur} = A \cosh \alpha y \sin \alpha z - \frac{4\pi e E_0}{H_0} \sin \alpha z \frac{1}{\alpha} \int_0^y n_0(\eta) \sinh \alpha (\eta - y) \, d\eta,$$

where A is a constant chosen on the basis of (5.140b).

In particular, if

$$n_0(y) = n_0 |_{|y| < y_0}; \quad n(y) = 0 |_{|y| > y_0},$$

then

$$\psi_{cur} = \frac{4\pi e E_0 n_0}{H_0 \alpha^2} \sin \alpha z \, [\exp(\alpha y_0) \cosh \alpha y - \cosh \alpha (y_0 - y)]_{|y| > y_0};$$

$$\psi_{cur} = \left[-\frac{4\pi e E_0 n_0}{H_0} \sin \alpha z \frac{1}{\alpha^2} (1 - \exp(-\alpha y_0)) \cosh \alpha y - \right.$$

$$\left. - \frac{4\pi e E_0}{H_0} n_0 \frac{\sin \alpha z}{\alpha^2} (1 - \cosh \alpha y) \right]_{|y| < y_0}.$$

Evidently the magnetic field produced by the current is $\pi/2$ out of phase with the main field.

The ratio of the peak value of the magnetic field produced by the current to the peak external field, $\varkappa$, is

$$\varkappa = \frac{|\psi_J|_{ampl}}{|\psi_{ex}|_{ampl}} \approx \frac{4\pi e E_0 n_0}{H_0^2 \alpha} = \frac{4 e E_0 \Lambda_0 n_0 M}{H_0^2 M} \approx \frac{\delta(v^2)}{c_{A0}^2}. \qquad (5.142)$$

Here $\delta(v^2)$ is the change in the square of the particle velocity whic occurs as the particle passes through one lens, and $c_{A0}^2 = H_0^2 4\pi M n_1$ is the square of the Alfvén velocity. Estimate (5.142) obviously has the same structure as that in (5.139).

For the example considered earlier, a proton accelerator with a current of 10^3 A, $Mv^2/2 = 10$ keV, and $H_0 = 10^4$ Oe, the parameter $\varkappa$ is of the order of 10^{-4}. This result means that if the distortion of the magnetic field is the only factor degrading the accelerator operation it is possible to increase the accelerated current to nearly 10^6 A in this system holding all other parameters constant.

REFERENCES

1. L. A. Artsimovich et al., "Electromagnetic separation of isotopes of heavy elements," At. Energ., 3:483-491 (1957).

2. A. I. Morozov, "Focusing of cold quasineutral beams in electromagnetic fields," Dokl. Akad. Nauk SSSR 163:1363 (1965) [Sov. Phys. Dokl. 10:775 (1966)].

3. S. V. Lebedev and A. I. Morozov, "Focusing of an ion beam in the field of a charged current-carrying ring," Zh. Tekh. Fiz. 36:960 (1966) [Sov. Phys. Tech. Phys. 11:707 (1966)].

4. V. V. Zhukov, A. I. Morozov, and G. Ya. Shchepkin, "Experimental study of plasma focusing of ion beams," Pis'ma Zh. Eksp. Teor. Fiz. 9:24 (1969) [JETP Lett. 9:14 (1969)].

5. M. V. Nezlin, "Plasma instabilities and the compensation of space charge on an ion beam," Plasma Phys., 10:337-358 (1965).

6. V. I. Raiko, "Scaling laws for ion sources," Zh. Tekh. Fiz. 33:244 (1963) [Sov. Phys. Tech. Phys. 8:175 (1963)].

7. M. D. Gabovich and G. S. Kirichenko, "Two-stream instability in a system of mutually penetrating beams," Zh. Eksp. Teor. Fiz., 50:1183 (1966) [Sov. Phys. JETP 23:785 (1966)].

8. A. I. Akhiezer and Ya. B. Fainberg, "High-frequency oscillations of an electron plasma," Dokl. Akad Nauk SSSR 64:555 (1949).

9. E. K. Zavoiskii, "Collective interactions and the problem of producing a high-temperature plasma," At. Energ., 14:57 (1963).

10. Ya. B. Fainberg and V. D. Shapiro, "Interaction of a modulated beam with a plasma," At. Energ., 19:336 (1965).

11. M. D. Gabovich, Plasma Sources of Ions [in Russian], Naukova Dumka, Kiev (1964).

12. A. I. Morozov, E. V. Artyushkov, L. S. Solov'ev, and A. P. Shubin, "Certain properties of the flow of a conducting gas in a magnetic field," in: Low-Temperature Plasmas [in Russian], Mir, Moscow (1967).

13. A. I. Morozov and L. S. Solov'ev, "Steady-state plasma flow in a magnetic field," this volume, p. 1.

14. E. C. Lary, "Ion acceleration in a space charge neutral plasma," UAC Res. Lab. Report UAR-A125, June 1962; G. R. Seikel and E. Reshotko, "Hall current ion accelerator," Bull. Amer. Phys. Soc., Ser. II 7:414 (1962).

15. G. S. Janes and J. Dotson, "Experimental studies of oscillations and accompanying anomalous electron diffusion," Proceedings of the Fifth Symposium on Engineering Aspects of MHD, M II, 1964, April 1-2, pp. 135-148.

16. A. I. Morozov, Yu. V. Esipchuk, et al., "Experimental study of a Hall-current plasma accelerator with an extended acceleration zone," Zh. Tekh. Fiz. 42:54 (1972) [Sov. Phys. Tech. Phys. 17:38 (1972)].

17. A. V. Zharinov and Yu. S. Popov, "Plasma acceleration by a closed Hall current," Zh. Tekh. Fiz. 37:294 (1967) [Sov. Phys. Tech. Phys. 12:208 (1967)].

18. N. A. Kervalishvili, "Influence of anode orientation on the properties of a low-pressure discharge in a transverse magnetic field," Zh. Tekh. Fiz. 38:637 (1968) [Sov. Phys. Tech. Phys. 13:476 (1968)].

19. A. I. Morozov and L. S. Solov'ev, "Cybernetic stabilization of plasma instabilities," Zh. Tekh. Fiz. 34:1566 (1964) [Sov. Phys. Tech. Phys. 9:1214 (1965)].

20. I. P. Zubkov, A. Ya. Kislov, S. V. Lebedev, and A. I. Morozov, "Ion motion in a two-lens Hall-current accelerator," Zh. Tekh. Fiz. 41:526 (1971) [SOv. Phys. Tech. Phys. 16:409 (1971)].

21. B. Borries and E. Ruska, Z. Phys. 76:649 (1932).

22. D. Gabor, Proc. Roy. Soc. A 183:992-935 (1945).

23. V. I. Ivanov, Dissertation [in Russian], KhFTI (1956).

24. S. V. Lebedev, "Current configurations producing rapidly decaying axisymmetric magnetic fields," Pribory i Tekh. Eksperim. No. 1:32 (1971).

25. G. A. Grinberg, Mathematical Theory of Electric and Magnetic Phenomena [in Russian], Izd. Akad. Nauk SSSR, Moscow (1947).

26. Yu. V. Vandakurov, "Equations of electron optics for broad beams with chromatic aberration," Zh. Tekh. Fiz. 25:1412 (1955).

27. A. I. Morozov and L. S. Solov'ev, "Motion of charged particles in electromagnetic fields," Reviews of Plasma Physics, Vol. 2 (ed. M. A. Leontovich), Consultants Bureau, New York (1966), p. 201.

28. S. V. Lebedev, "Focusing of charged particles by magnetic fields," Zh. Tekh. Fiz. 41:545 (1971) [Sov. Phys. Tech. Phys. 16:423 (1971)].

29. A. I. Morozov and S. V. Lebedev, "Theory of the focusing of quasineutral beams by axisymmetric electromagnetic fields," Zh. Tekh. Fiz. 37:633 (1967) [Sov. Phys. Tech. Phys. 12:455 (1967)].

30. Yu. A. Shapiro, "A direct method for calculating electromagnetic lenses," Zh. Tekh. Fiz. 34:1747-1751 (1964) [Sov. Phys. Tech. Phys. 9:1352 (1965)].

31. V. M. Kel'man and S. Ya. Yavor, Electron Optics [in Russian], Izd. Akad. Nauk SSSR, Moscow (1963).

32. O. Scherzer, Beiträge zur Electronenoptik., V. H. Busch, E. Brüche, Berlin (1937).

33. A. I. Morozov and G. Ya. Shchepkin, USSR Patent No. 115/435; Byull. Izobret., No. 4:64 (1969).

34. A. A. Kolomenskii and A. I. Lebedev, Theory of Cyclic Accelerators [in Russian], Fizmatgiz, Moscow (1962).

35. T. F. Volkov and A. I. Morozov, "Magnet system of tubular multilens accelerator," Zh. Tekh. Fiz. 41:1247 (1971) [Sov. Phys. Tech. Phys. 16:981 (1971)].

36. T. F. Volkov, "Theory of the motion of charged particles in a tubular multilens accelerator," Zh. Tekh. Fiz. 41:1257 (1971) [Sov. Phys. Tech. Phys. 16:988 (1971)].

37. B. B. Kadomtsev, "Hydromagnetic stability of a plasma," Reviews of Plasma Physics, Vol. 2 (ed., M. A. Leontovich), Consultants Bureau, New York (1966), p. 153.

38. L. N. Dobretsov and M. V. Gomoyukova, Emission Electronics [in Russian], Nauka, Moscow (1966).

39. A. I. Morozov, "Wall conductivity of well-magnetized plasma," Priklad. Matem. i Teor. Fiz., No. 3:19 (1968).

40. A. I. Morozov, A. Ya. Kislov, and I. P. Zubkov, "High-current Hall-current plasma accelerator," Pis'ma Zh. Eksp. Teor. Fiz. 7:224 (1968) [JETP Lett. 7:172 (1968)].

41. M. P. Zubkov, A. Ya. Kislov, and A. I. Morozov, "Experimental study of a two-lens accelerator," Zh. Tekh. Fiz. 40:2301 (1970) [Sov. Phys. Tech. Phys. 15:1796 (1971)].

42. L. A. Artsimovich (editor), Plasma Accelerators [in Russian], Mashinostroenie, Moscow (1973).